D1243997

Forest Succession

Concepts and Application

Edited by
Darrell C. West · Herman H. Shugart
Daniel B. Botkin

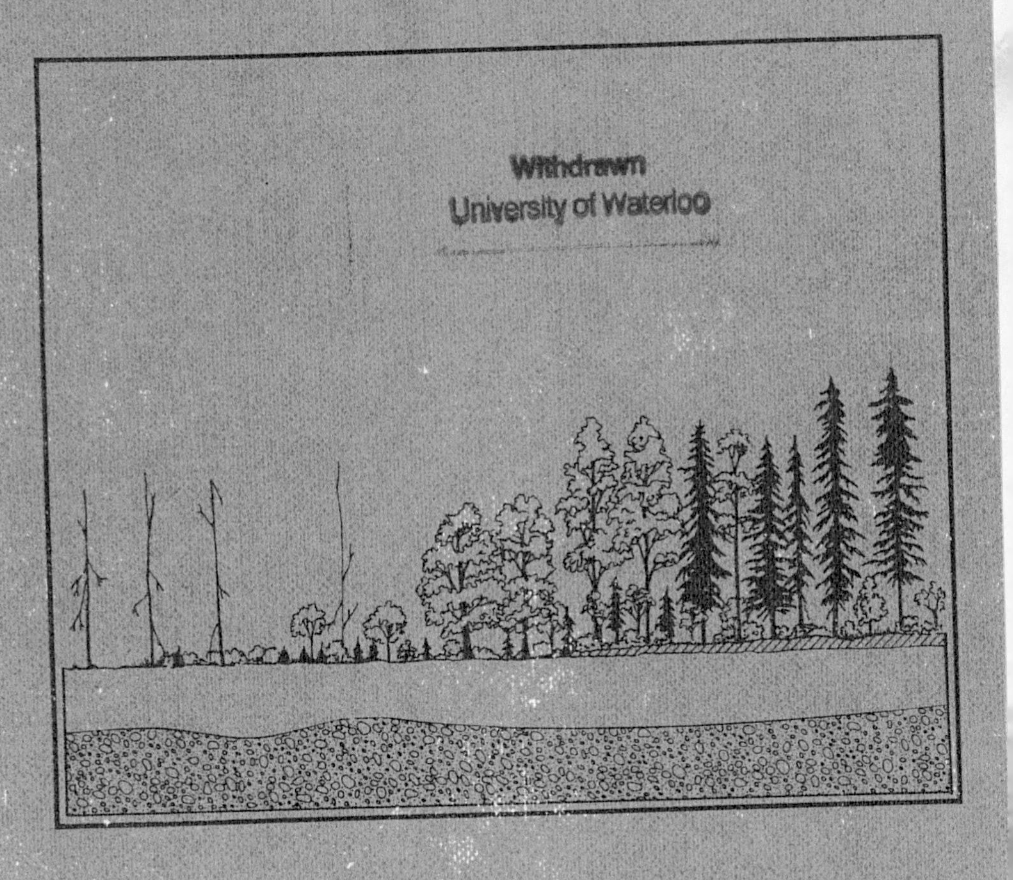

Springer-Verlag
New York Heidelberg Berlin

Springer Advanced Texts in Life Sciences

David E. Reichle, Editor

Forest Succession

Concepts and Application

Edited by

Darrell C. West
Herman H. Shugart
Daniel B. Botkin

With 112 Figures

Springer-Verlag
New York Heidelberg Berlin

Darrell C. West
Environmental Sciences Division
Oak Ridge National Laboratory
Oak Ridge, TN 37830 U.S.A.

Herman H. Shugart
Environmental Sciences Division
Oak Ridge National Laboratory
Oak Ridge, TN 37830 U.S.A.

Daniel B. Botkin
Department of Biological Sciences
and Environmental Studies Program
University of California,
Santa Barbara
Santa Barbara, CA 93106 U.S.A.

Series Editor:
David E. Reichle
Environmental Sciences Division
Oak Ridge National Laboratory
Oak Ridge, TN 37830 U.S.A.

Library of Congress Cataloging in Publication Data

Main entry under title:

Forest succession.

(Springer advanced texts in life sciences)
Papers presented at a conference held at Mountain Lake, Va., in June 1980.
Bibliography: p.
Includes index.
1. Forest ecology—Congresses. 2. Plant succession—Congresses. I. West, D. C. II. Shugart, H. H. III. Botkin, D. B. IV. Series.
QK938.F6F674 581.5'2642 81-16707
AACR2

Printed in the United States of America

9 8 7 6 5 4 3 2 1

ISBN 0-387-90597-9 Springer-Verlag New York
ISBN 3-540-90597-9 Springer-Verlag Berlin Heidelberg

Preface

Succession—nothing in plant, community, or ecosystem ecology has been so elaborated by terminology, so much reviewed, and yet so much the center of controversy. In a general sense, every ecologist uses the concept in teaching and research, but no two ecologists seem to have a unified concept of the details of succession. The word was used by Thoreau to describe, from a naturalist's point of view, the general changes observed during the transition of an old field to a forest. As data accumulated, a lengthy taxonomy of succession developed around early twentieth century ecologists such as Cooper, Clements, and Gleason. Now, nearer the end of the century, and after much discussion concerning the nature of vegetation communities, where do ecologists stand with respect to knowledge of ecological succession?

The intent of this book is not to rehash classic philosophies of succession that have emerged through the past several decades of study, but to provide a forum for ecologists to present their current research and present-day interpretation of data. To this end, we brought together a group of scientists currently studying terrestrial plant succession, who represent research experience in a broad spectrum of different ecosystem types. The results of that meeting led to this book, which presents to the reader a unique summary of contemporary research on forest succession.

The conference* was held in June 1980 at the Mountain Lake Hotel,

*We acknowledge the assistance of the National Science Foundation not only for supporting much of the research contained herein, but also for making the conference possible through support from the Ecosystem Studies Program, under Interagency Agreement Nos. DEB-77-26722 and DEB-77-25781, with the U.S. Department of Energy, under contract W-7405-eng-26 with Union Carbide Corporation. The work was also supported in part by the Office of Health and Environmental Research, U.S. Department of Energy.

Mountain Lake, Virginia. One reason for choosing this site, along with the natural beauty of the Alleghenies, was the hope that in an isolated location with minimal outside interruption substantial synthesis of ideas about succession could occur. Before the conference, we attempted to anticipate what would be the result of bringing such a diverse array of scientists to debate a controversial topic. Would the conference become bogged in semantic debates? Would the heat of disagreements generate defensiveness and divisiveness?

This trepidation was not borne out in reality. The conference participants were deeply interested in discussing one another's data. Individuals who had different viewpoints on succession were glad to have the chance to try to understand the source(s) of these differences. Scientists who felt that their data did not match the textbook definitions of succession were able to exchange views with other scientists who held similar feelings based on different ecosystems. The excitement of having so many new and old ideas, and new and old data sets about succession, was contagious.

Prose fails to capture the mood and excitement of scientific inquiry when ecologists have the glimmer of being able to synthesize an idea as important as succession. This book provides the building blocks that may lead to such a synthesis. Even if we restrict our discussion to forests, we do not yet understand succession in as thorough a way as an introductory textbook leads one to think. The conference at Mountain Lake and this book lead us, and hopefully the reader, to feel that although we do not know everything there is to know about succession, we at least possess a body of theory and empirical data that support an increasingly refined effort of analysis and documentation.

Finally, we should point out that the conference was made successful in part by the much appreciated assistance of Tom Doyle, Carolyn Henley, Joy Simmons, Tom Smith, Roger Walker, and David Weinstein. We wish to thank Forest Stearns, David Reichle, Robert Van Hook, Stanley Auerbach, Tom Callahan, Jerry Olson, and Robert O'Neill, who gave helpful suggestions regarding organization and content of the meeting and of this volume.

Contents

List of Contributors

Botkin, Daniel B. Department of Biological Sciences and Environmental Studies Program, University of California, Santa Barbara, Santa Barbara, CA 93106 U.S.A.

Brubaker, Linda B. College of Forest Resources, University of Washington, Seattle, WA 98195 U.S.A.

Christensen, Norman L. Department of Botany, Duke University, Durham, NC 27706 U.S.A.

Cooper, Arthur W. Department of Forestry, North Carolina State University, Raleigh, NC 27650 U.S.A.

Cottam, Grant Department of Botany, University of Wisconsin, Madison, WI 53705 U.S.A.

Cromack, Jr., Kermit Department of Forest Science, Oregon State University, Corvallis, OR 97331 U.S.A.

Davis, Margaret Bryan Department of Ecology and Behavioral Biology, University of Minnesota, Minneapolis, MN 55455 U.S.A.

Doyle, Thomas W. Graduate Program in Ecology, University of Tennessee, Knoxville, TN 37916 U.S.A.

Emanuel, William R. Environmental Sciences Division, Oak Ridge National Laboratory, Oak Ridge, TN 37830 U.S.A.

Fergus, I. F. CSIRO Division of Soils, Institute of Earth Resources, Brisbane, Queensland, Australia

Franklin, Jerry F. Forest Sciences Laboratory, USDA Forest Service, Oregon State University, Corvallis, OR 97331 U.S.A.

Gómez-Pompa, Arturo Instituto/Nacional de Investigaciones, Sobre Recursos Bioticos, P.O. Box 63, Xalapa, Vera Cruz, Mexico

Heinselman, Miron L. Department of Ecology and Behavioral Biology, University of Minnesota, Minneapolis, MN 55455 U.S.A.

Hemstrom, Miles A. 211 East 7th Street, Box 10607, Eugene, OR 97401

Horn, Henry S. Department of Biology, Princeton Univesity, Princeton, NJ 08544 U.S.A.

Loucks, Orie The Institute of Ecology, Indianapolis, IN 46208 U.S.A.

MacMahon, James A. Department of Biology and Ecology Center UMC53, Utah State University, Logan, UT 84322 U.S.A.

McIntosh, Robert P. Department of Biology, University of Notre Dame, Notre Dame, IN 46556 U.S.A.

Peet, Robert K. Department of Botany, University of North Carolina, Chapel Hill, NC 27514 U.S.A.

Shugart, Herman H. Environmental Sciences Division, Oak Ridge National Laboratory, Oak Ridge, TN 37830 U.S.A.

Solomon, Allen M. Environmental Sciences Division, Oak Ridge National Laboratory, Oak Ridge, TN 37830 U.S.A.

Solomon, Jean A. Environmental Sciences Division, Oak Ridge National Laboratory, Oak Ridge, TN 37830 U.S.A.

Thompson, C. H. CSIRO Division of Soils, Institute of Earth Resources, Brisbane, Queensland, Australia

Tunstall, B. R. CSIRO Division of Land Use Research, Institute of Earth Resources, Canberra, A.C.T., Australia

Van Cleve, Keith Forest Soils Laboratory, University of Alaska, Fairbanks, AK 99701 U.S.A.

Vázquez-Yanes, Carlos Depto de Botanica, Instituto de Biologia Apartado 70-233, Mexico 20, D.F.

Viereck, Leslie A. Institute of Northern Forestry, USDA Forest Service, Fairbanks, AK 99701 U.S.A.

Vitousek, Peter M. Department of Botany, University of North Carolina, Chapel Hill, NC 27514 U.S.A.

Walker, B. H. Department of Botany and Microbiology, University of Witwatersrand, 1 Jan Smuts Avenue, Johannesburg, South Africa, 2001

Walker, J. CSIRO Division of Land Use Research, Institute of Earth Resources, Canberra, A.C.T., Australia

West, Darrell C. Environmental Sciences Division, Oak Ridge National Laboratory, Oak Ridge, TN 37830 U.S.A.

White, Peter S. Uplands Field Research Laboratory, Great Smoky Mountains National Park, Gatlinburg, TN 37738 U.S.A.

Whittaker, Robert H. (deceased) Ecology and Systematics, Cornell University, Ithaca, NY 14850 U.S.A.

Woods, Kerry D. Department of Ecology and Behavioral Biology, University of Minnesota, Minneapolis, MN 55455 U.S.A.

Zedler, Paul H. Department of Biology, San Diego State University, San Diego, CA 92115 U.S.A.

Chapter 1
Introduction

Ecologists study living systems that are driven by forces of change at many spatial and temporal scales. Those interested in the change of vegetation patterns and behavior must deal with a vast array of dynamic environmental conditions. Plate tectonics theory holds that continents continuously move and change. Geomorphologists and soil scientists have long recognized the ever-changing behavior of the earth's surface. At a given location, climatic conditions change on time scales ranging from decades to millenia. Hurricanes, wildfires, and other modifying agents disturb ecological systems at varying frequencies and intensities causing change in these systems over areas of differing size. Human civilization has both directly and indirectly altered the biosphere to a great degree. Data obtained during the past few decades indicate an increase in the global concentration of atmospheric carbon dioxide, suggesting there is truly no ecosystem immune to technological influence.

Ecologists have developed a set of concepts on the nature of change in ecosystems (Fig.1.1). Many published reviews, especially those during the past ten years, have dealt with the current status of succession using a similar sequence of presentation: Clements (1916, 1928, or 1936) is usually cited; Gleason (1927) is mentioned; the review proceeds to present many of the classical works and descriptive, but elusive, plant ecology terms coined during the last few decades.

In the early 1900's, heavy emphasis on use of descriptive terms for succession grew from the accumulated knowledge of natural history. For ecologists, an interest in developing a classification scheme for ecosystems has spawned many heated discussions. In retrospect, the reasons for the arguments are understandable and can be at least partially explained. The history of any science, and possibly the reason for the beginning of

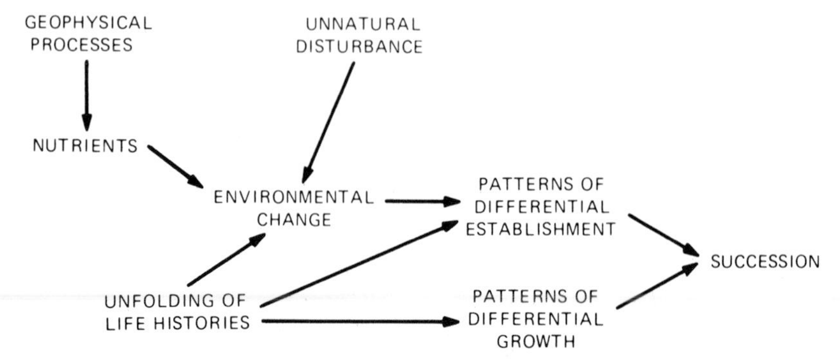

Figure 1.1. A classical concept of succession representing the major view expressed in ecology textbooks. This concept is incorporated into the modern view (Fig. 1.2) except that the classical view generally assumes disturbance as an unnatural event, whereas the modern view accepts disturbance to be a natural occurrence and recognizes that many, if not most communities, are dominated by it. Primary succession is recognized as starting with the geophysical processes while secondary succession begins after an unnatural disturbance. (Compiled and arranged by J. F. Franklin and H. S. Horn.)

all science, relates to a human desire to describe, classify, and name all the parts of any system being studied. Anyone trained in biological systematics, as most early ecologists were, would certainly be inclined to coin as many words as necessary to describe the parts and processes of an ecosystem. Further, early ecologists had before them the success of chemistry and physics in taking what was initially little more than a way of categorizing the elements, the Periodic Table, and developing theories on the basic nature of matter. Although not all elements have been discovered, the properties of the individual elements on the Periodic Table remain constant, a fact unchanged by time or space. Not so for living organisms—biological taxa arise and disappear within an environment that is always changing. Consequently, use of terms in ecology can be correct or incorrect, depending upon whether or not the context in time and space is properly and clearly defined.

As data describing a wider array of natural systems become available, studies of succession tend to concentrate less on the development of classification and terminology, and more on the associated processes and mathematical formulations of these processes. The theme throughout this book documents this shift from a qualitative to a quantitative outlook. Descriptive details of natural history are as important as they ever were, but now quantitative techniques are used and the data are subjected to analytical or predictive tests and interpretations. This later change reflects the important influence that mathematics and especially computers have had in ecology over the past few decades. From a quite general viewpoint, classical succession concepts (Fig. 1.1) evolved rather

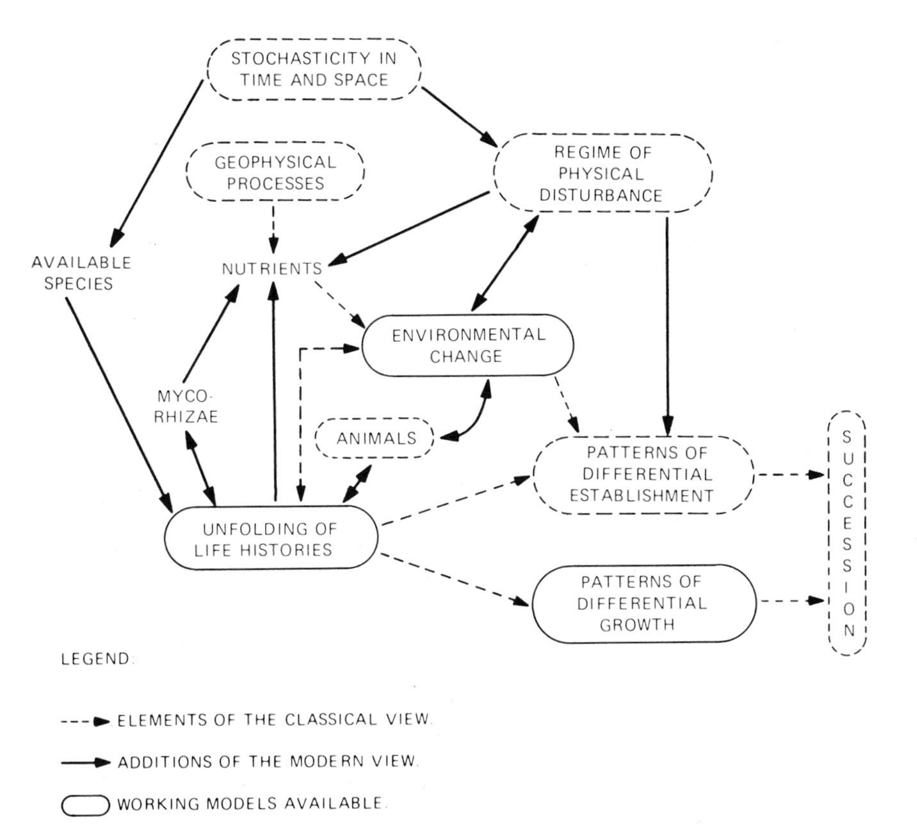

Figure 1.2. A more detailed representation of the succession concept shown in Fig. 1.1. Although not shown explicitly, all factors presented in Fig. 1.2 affect life histories when evolutionary time is considered. However, the resulting adaptations may have originated in a different environmental context from that of the present, especially in a different array of available species. (Compiled and arranged by H. S. Horn and J. F. Franklin.)

gently into what we term a modern view of succession (Fig. 1.2). The evolution is to a degree a consequence of an increased awareness of change as a strong force in natural ecosystems, and is resulting at present in the move toward more quantitative formulations of ecological processes and away from the prose-based descriptions of successional change that dominate the mood of classical theory.

It is altogether appropriate, in our opinion, that a modern theory of succession include the classic theory. Modern ecologists are provided with tools that were unavailable to ecologists developing initial concepts of succession. Ecologists who formulated classic theory drew their examples largely from North America and Europe. They based their generalizations on comparisons of relatively similar ecosystems on the two continents. In North America, in particular, regional uniformity and dynamic regularity of the ecosystems of the plains states and of western forests represented important new data to be understood. Classical succession theory (Fig. 1.1) suggested that the regularity of ecosystems in the new American wilderness, resulted in ordered communities that developed under constant prevailing climates after geological processes created new substrates. "Secondary" succession was seen as a response of ecological systems following an unnatural disturbance, particularly as influenced by man in the recent past.

Modern theory (Fig. 1.2) attempts to bring our understanding of succession into agreement with our increased awareness of change at multiple scales of time and space. The observations that lead us to classical successional theories are still valid and the new theory logically must include valid insights previously developed. However, lost from the modern theory is the comforting but misleading generality of the old. That is, as the diverse ecosystem types are explained and processes are presented, the realization emerges that it is relatively easy to form a local theory of succession but that development of a general theory with equivalent detail may be a futile objective.

Together the chapters in this book illustrate the broad range of phenomena that are necessary to consider before any unifying theory can be developed concerning succession. Variables include organisms and environments, types and regimes of disturbance, and dominant processes and the spatial and temporal scales of their dynamics (Fig. 1.2). Successional theory constructed from observations of one ecosystem invariably proves to be a special case when comparison is attempted with another system. Some common features do exist, however. Repeatable successional patterns can be defined for many common combinations of ecosystems and disturbance. Successional mechanisms, e.g., competition, facilitation, tolerance, inhibition, and secular change appear to be valid but of widely varying importance.

The objective of this book is to present contemporary (cf. Figs. 1.1 and 1.2) research regarding forest succession within the framework of *current concepts* of succession. The general order of presentation is (1) the basic concepts and issues in studies of forest succession and specific theoretical aspects and application of theory in Chapters 2–7; (2) long-term aspects of forest succession, coupling theoretical and empirical techniques in Chapters 8–11; (3) case histories of succession in selected ecosystems in Chapters 12–16; and (4) the influence of processes and disturbance on patterns and direction of vegetation behavior in Chapters 17–25. We do not attempt to present or summarize every known version and modification of succession theory. Rather, we choose examples of recent research in succession and present these studies to integrate the implications of space, time, and processes.

The book is divided into 4 sections, each with its own introduction. The first section is introduced by Loucks (Chapter 2) with a discussion of how syntheses in plant ecology studies have come to rely not only upon decades of data collection and summary, but also upon relatively new innovations in analytical and predictive tools. Chapter 3 presents the historical progress of succession studies, reviews the controversy surrounding the study and definition of vegetation dynamics, and discusses ecosystem theory and evolution to demonstrate the semantic problem that still influences the science of plant ecology. A method for using the additional variables related to Fig. 1.2 is provided in Chapters 4–7. These studies allow quantitative consideration of successional variables while employing three basic and important elements for examining succession: (1) collective or emergent properties of communities or ecosystems can be calculated or predicted on the basis of species demography; (2) functional groups, instead of species, may be considered when appropriate; and (3) hypotheses can be developed and tested.

Mathematical models are necessary for practical application of most studies of succession. For example, the predictive ability of simulation models can be of special value to applied ecology by circumventing a fundamental problem that faces anyone studying forest succession, i.e., the long-lived nature of trees. Although trees have high rates of birth and death the great potential longevity, especially for canopy trees, puts the actual observation of the mechanisms of forest dynamics beyond the realm of most investigations. The difficulties in studying tree population dynamics tend to suppress understanding of how forest ecosystems might respond to "natural" or "other" environmental disturbances. Questions have arisen regarding the effects on forests of atmospheric pollutants or climatic change, as well as about consequences of untried forest management schemes or, more specifically, forest management techniques that will ensure the survival of certain species of plants and animals. The types of modeling described in Chapters 2–7 may represent a basic approach suitable for application to real-world situations as well as to theoretical explorations of successional processes.

The studies, introduced in Chapter 8 by Brubaker and presented in Chapters 9, 10, and 11, demonstrate methods for dealing with the relatively long time scale of forest succession. Because of the long time periods during which succession occurs, these explorations consider available records of prehistoric forest dynamics. The concept of time is ever present during the discussions of these chapters. The work in Chapter 9 shows a complete change in vegetation life forms through time: a successional process going from herbs to trees to shrubs, without the benefit of an apparent change in climate during a several-thousand-year period. Chapter 10 discusses how species arrival dates can be gleaned from pollen evidence, to infer the continuing influence of a major environmental disruption, inception of the present-day climate pattern upon forest demography during thousands of years. Pollen evidence is also utilized for work described in Chapter 11, but from a single location. Using simultaneous forest stand simulations to infer forest dynamics from the pollen records, the influences derived from Chapter 9 are examined, particularly in sorting the influence of climate change from that of shifting species availability.

The chapters (13–16) introduced by Cottam (Chapter 12) support the contention that plant dynamics can be understood at the local level, but together these chapters fail to support a monolithic unifying theory of forest succession. This results from the fact that the ecosystems described in these chapters span a wide range of environmental conditions and exhibit dynamic properties at widely different temporal and spatial scales, despite the fact that all of these studies are of evergreen forests. The dynamic differences are interwoven with differences in: life histories of constituent species; size and morphology of dominant tree species; species diversity; climate; soil; and seasonality. These studies clearly show that, even without crossing the boundaries of life form, ecosystems can differ dramatically in their successional properties. One conclusion which can be derived from these chapters is that approaches to studying succession are not necessarily the same from one ecosystem to another.

The final chapters (introduced by Vitousek and White, Chapter 17) treat ecological processes, e.g., establishment and series replacement, productivity, decomposition, and their interactions with disturbance, and discuss how these factors determine the nature and course of succession. Disturbance, at varying scale, intensity, and frequency, is the initiating force for most succession (Fig. 1.1 and Fig. 1.2). Disturbances, e.g., fire in the boreal forest or chaparral, seasonal floods, or herbivory, seem to be integral to many ecosystems and the frequent actions of animal populations. If disturbance is not a process in a given system, it is a change that elicits new ecosystem dynamics from the mechanisms that are more traditionally considered ecosystem processes (e.g., element cycling, decomposition, birth, and death). Within the current view (Fig. 1.2), element cycling processes seem to be extremely important in determining the ultimate configuration of vegetation on geologically old landscapes. The point of these final chapters is not a reductionist discussion of ecological processes; it is to view the whole-system consequences of these processes.

Taken as a whole, this book presents a summary of contemporary research concepts in the science of forest succession. It brings together a wide spectrum of ideas, examples and procedures expounded by theorists and empiricists; students of tropical, savannah, temperate, and arctic regions; investigators studying parts of the system both above and below ground; practicing foresters, government researchers, and collegiate academicians. We hope that this work will help clarify ideas about succession and show the necessity of incorporating fundamental sciences with the management of the landscape.

Chapter 2

Concepts, Theory, and Models of Forest Succession

Orie Loucks

Introduction

A definitive reevaluation of the succession concept, after almost a century of study, will attract the attention of biologists and natural resource managers only if new insights of major consequence appear to be in prospect. Not only do the following five chapters warrant such attention, they reveal genuine excitement about an area rich in history and rich in current discovery. The use of recently developed analytical techniques is creating the opportunity to benefit from previous decades of data base development and conceptual innovations. The systematic interpretation of both data and concepts has led to new theory (manifested as statements of relationship, both formal and qualitative) and to new predictive tools.

These tools are becoming available at a time of considerable importance to national decision-making with respect to renewable natural resources. Air pollutant effects on vegetation, the role of biomass in national energy policy, potential species extinctions, and other resource impacts are more significant now than ever before. With the evidence now of great regional pressures on renewable resources and the environmental impact assessment process, there is a broadly based new clientele for the products of this review. The product desired, however, involves predictive capabilities with respect to succession as well as improvements in understanding of succession. Thus, studies of theory and quantitative modeling of succession are one of today's great technical and scientific challenges to field biology. In the initial chapters our goal is to present current viewpoints on theory and modeling in a way that recognizes and does credit to 100 years of preparation.

The treatment begins with Robert McIntosh summarizing what the concept of succession has been, historically, and describing the theory, or the principal hypothesis sets, implicit in our use of the term. He discusses the diverse views through which we have understood succession in both the remote and the recent past. The chapters that follow (and others throughout the volume) present the more formal structures (models) within which succession is being explored now. Diversity in concepts and applications are still evident. The models presented can be thought of as statements of the diverse relationships implicit in succession—invasions and extinctions, carbon and nutrient storage leading to site modification, and creation of conditions making fire and wind disturbance agents of initial conditions more probable—all are part of the theory of succession as we know it in 1980. The theory and models, despite their present limitations, provide paradigms and testable hypothesis sets that some field studies may address. Many of the descriptive studies in the remainder of the volume can be read in just this way. Thus, the theory and formal statements of successional relationships presented here serve as well, to determine, by the end of the volume, what questions have not been answered and what major hypotheses not yet tested, thereby framing the research of the future.

Synthesis and Summarization

It is our objective that, while seeking a broad, balanced review of successional processes, we should also seek "comprehensive state-of-the-art synthesis" or summarization. Scientific methods in ecology, have, during the development of succession as a concept, moved from largely field observation and generalization to more explicit hypothesis development and experimental manipulation of field systems for hypothesis rejection. One form of summarization may properly utilize relatively aggregated elements of succession, and abstractions such as "ideal species," for analytical work such as that presented by Henry Horn in Chapter 4. Other forms of summarization, both of data and aggregates of concepts (growth, reproduction, mortality), take the form of simulation models whose further validation requires chapters throughout the volume and additional years of study. The chapters by Botkin (Chapter 5), by Doyle (Chapter 6), and by Shugart et al. (Chapter 7) are examples of synthesis through detailed simulation studies.

Summarization in a mature science often requires major changes in the dominant paradigms of the field or innovations in the theory within which the science is understood. At one time the theory of succession was expressed adequately as verbal or prose statements of relationships then understood. Today, successional theory, although unchanged in general principles, is enriched in detail and requires expression in more formal, quantitative forms. Part of the debate in the initial chapters, and throughout the volume, therefore, concerns the extent to which formal statements of succession can or should capture the detail of our modern verbal and prose understanding. In whatever form, the statements of relationship—theory—influence how we look at our observation base, and influence our perception of the physiological and ecological properties of the species contributing to succession. Thus, the summarizations in this volume, if they are effective, are likely to further modify our information base and, therefore, the basic theory of succession.

In summarizing a science one must also recognize that it advances through differences in approach. In advancing fundamental theory in field biology, some researchers have sought to describe recurring pattern in field observations (e.g., MacArthur 1972), others to provide explanation and prediction (e.g., Bledsoe and Van Dyne 1971; Miller and Botkin 1974), and still others have sought summarization of relationships (e.g., Patten et al. 1975; Loucks 1970, 1976). The observation also is made that ecology may have no fundamental theory of its own, rather, that it derives entirely from the theory of evolution and the adaptive properties of species implicit in that theory. Formal statements of relationship exist in abundance in ecology, of course, but they appear to have originated in the chemistry or mathematics of mass transfer (e.g., acretion, population growth, or decay). In these respects, succession seems not different from much of ecology. As illustrated in Chapter 5 by Botkin and Chapter 7 by Shugart et al., succession theory, at present, can be summarized in useful predictive forms through environmental and autecological parameters expressed as part of simple acretive and demographic equations.

However, many of the chapters that follow the background and theory section of this volume present observations and data on phenomena (indeed, also theory) which is not yet incorporated in any of the more formal model structures. Species invasions as influenced by proximity to seed source or other geographic considerations (due to disturbance intensity or pleistocene recovery) obviously bear on model outcomes. The periodicity of recurring disturbance, as shown by Heinselman (1973) and the chapters by Heinselman (23), Franklin and Hemstrom (14), and Zedler (24) in this volume, also is important. The role of nutrient limitation (Vitousek and others, in later chapters) in species development or extinction is becoming quantified. All of these enrichments in detail appear to be reasonable extensions to the general statement of successional theory contained in this first section. Their very existence, however, as elements of succession not yet formally incorporated in existing models gives an indication of work yet to be done.

The theory and modeling chapters (2–7) should be thought of as extending through other parts of the volume. The entire spectrum of work on theory and modeling in succession, therefore, can be regarded as incomplete, but definable. To this extent, definition of the further research that this volume seeks to stimulate also has been accomplished.

Chapter 3
Succession and Ecological Theory

Robert P. McIntosh

Introduction

Succession has reigned as a basic concept or theory of ecology certainly since Frederick E. Clements (1916) stated as a "universal law" that "all bare places give rise to new communities except those which present the most extreme conditions of water, temperature, light, or soil." Its significance persists and was described by Eugene Odum (1969, p. 262) as critical for human society:

> The principles of ecological succession bear importantly on the relationships between man and nature. The framework of successional theory needs to be examined as a basis for resolving man's present environmental crisis.

Clements' ideas, like those of prophets generally, have been widely used and misused by his adherents and detractors; and careful rereading of his major accounts of succession (Clements 1916, 1928, 1936) is often required to distinguish his actual ideas from what they are said to be. Another advantage of rereading Clements and the writings of other early ecologists (e.g., Forbes 1880; Cooper 1926; Gleason 1927) is that it should suggest humility in claims for what is sometimes described as "new" or "recent."

E. P. Odum (1969) provided one of the best parallels to Clements' views of succession. Odum noted the similarity of succession to the development of individual organisms and converged with Clements' description of succession as: (1) an orderly process that is reasonably directional and therefore predictable; (2) resulting from modification of the physical environment by the community, i.e., "community controlled"; (3) culminating in a stabilized (climax, mature) ecosystem with homeostatic properties. Clements (1916) asserted that "succession includes both the ontogeny and the phylogeny of climax formations," and Odum (1971) also saw evolution occurring at the community level. Odum's representation of succession was described as "neo-Clementsian" by Miles (1979) and, in fact, continued the essence of the organismic, holistic concept propounded by Clements into the "new ecology" known as systems ecology (Odum 1964, 1971; Colinvaux 1973; Ghiselin 1974; Cook 1977; Golley 1977; MacMahon 1980a; McIntosh 1980a,b; Simberloff 1980).

The similarity of classical Clementsian succession and the Odum-Margalef representation of succession is dramatized by a group of widely cited articles, which found both seriously flawed (McCormick 1968; Drury and Nisbet 1971, 1973; Horn 1971, 1974, 1975a,b, 1976; Pickett 1976; Connell and Slatyer 1977). These articles reviewed traditional and modern concepts of succession, provided analyses of succession concept and theory, and some propounded "alternative" conceptualizations, models, hypotheses or theories of succession (Peet and Christensen 1980).

They share at least three characteristics:

1. They are commonly cited in recent discussions of succession as providing new insights for succession theory, although there has been little critical analysis of the substance of their interpretations, or the alternatives they offer.
2. They are explicitly critical of Clements' holistic, organismic theory of succession and of what they interpret as the successional theory of the organismic, holistic, ecosystem ecology expressed by E. P. Odum, R. Margalef, and other ecosystem ecologists.
3. The alternative models of succession proposed advocate a population-based approach emphasizing life-history attributes of organisms, the consequence of natural selection, as the essential basis of a modern theory of succession.

In an earlier paper (McIntosh 1980a) I reviewed the antithesis posed between Clementsian or ecosystem theories of succession and the alternative theories of succession suggested in these articles. It may be argued that antithesis is too strong a term. MacMahon (Chapter 18) attempts a compromise. I use the word because the articles cited above generally disavow, or question, traditional Clementsian theory of succession and also reject recent ecosystem theories of succession. Various ecosystem ecologists not only assert the appropriateness and advantages of an ecosystem concept of succession but, reciprocally, question the theoretical merit and utility of a population-centered approach to ecology and succession. The split in ecology, I suggest, was recognized in recent addresses of retiring presidents of both British and American ecological societies (Smith 1975b, MacFayden 1975), and many other ecologists have noted it. What is clear is that explanation of cause and mechanism of succession, and development of a reasonable concensus on a theory of succession require careful examination of the historical background, premises, methodology, terminology, and philosophy used in discussions of succession. Without this, continued detailed examinations of seres are not likely to resolve the obvious difficulties and lead to concensus on ecological theory in the future any more than they have in the past. The lack of concensus may be because some ecologists are not as alert as others to simple theoretical solutions. Putman (1979) wrote in a review of R. M. May's book *Theoretical Ecology*: "I was delighted to find *simple, rational explanations* for some of the practical problems that have puzzled ecologists for years; small mammal cycles, the *orderliness of succession* . . ." (italics added).

It would be unproductive to review or evaluate Clementsian succession theory, much less defend it. In the sense that it is most commonly criticized in the articles cited, it would be difficult to find a plant ecologist after 1950 who strictly adhered to it. Plant ecology always entertained varied views of succession, and, certainly after the 1940s, developed a much less tidy concept of succession than vintage Clementsianism (McIntosh 1976, 1980a). This is best seen in the publications of R. H. Whittaker (1951, 1953, 1957, 1962, 1974, 1975a), Whittaker and Woodwell (1972), and Whittaker and Levin (1977). Nevertheless, Clements' ideas have remained a favorite target of recent critics of succession concepts.

Existing differences concerning succession are in the nuances of the individualistic, evolutionary, population-centered theory, and the evolutionary ecosystem theory upon which Clements' holistic, organismic mantle has fallen. A major difficulty lies in the fact that what I have described as two poles, on the basis of some common themes, is really more elusive than the polar metaphor suggests. Like the real magnetic poles that wander, the individualistic population-evolutionary and the holistic ecosystem-evolutionary poles similarly wander. Thus, the possibility of synthesis, or even clarification, depends upon closing with two moving, and possibly diverging, targets; to adapt H. C. Cowles' phrase, it is a variable approaching two variables.

The difficulty of clearly establishing poles in ecosystem ecology is seen in a recent collection of articles on systems ecology edited by Shugart and O'Neill (1979), which includes one by Richard Levins (1974) described as a significant new direction in systems ecology. Levins reviewed three general approaches to dealing with the complexity of ecosystems. Levins (1974) described one of these as deriving from cybernetics or systems analysis which he said "attempts to measure all the links in a system, write all the equations, measure all the parameters and either solve the equations analytically or simulate the process on a computer to obtain numerical results." Another approach, distinct from the foregoing, he said, emphasizes complexity, stresses interconnection, wholeness, multiple causality and unity of structure and process. These are substantially the qualities usually claimed for systems analysis. Although Levins (1974) and Lane et al. (1974) share the holistic ideal with E. P. Odum (1969, 1977) and both are explicitly opposed to reductionist views of ecosystems, it is not clear that they fit comfortably at the same pole. Similarly, although critics of the ecosystems approach to succession converge on advocating a population-based view of succession, it is difficult to find much agreement among them on how this is to be accomplished. Inclusion of the population-based JABOWA model (Botkin et al. 1972) in the systems ecology volume also confounds a neat classification. The antipodes of holist and reductionist are not always agreed upon. Ulanowicz (1979) saw the contrast between holist and "chaoticist."

My intention in this chapter is to examine the positions concerning succession at the two poles identified in ecology, to clarify some of their historical and theoretical sources, and to begin to construct, in that way, the bridge that some ecologists have suggested is needed between them. There is a substantial third force among ecologists, less wedded to a particular theoretical position, which may provide the abutments if not the complete structure.

Ecosystem Succession

Raymond Lindeman's (1942) classic, *The Trophic-Dynamic Aspect of Ecology*, "shed new light on the dynamics of ecological succession" and provided the "metaphor" that has continued the tradition of organismic ecology into modern ecosystem ecology (Cook 1977; Ulanowicz and Kemp 1979; Gutierrez and Fey 1980; McIntosh 1980a,b; Simberloff 1980). Lindeman (1942) reviewed the history of synecological thought, discussed Tansley's (1935) then recent word ecosystem, and defined succession from the trophic-dynamic viewpoint as, "the process of development in an ecosystem, brought about primarily by the effects of the organisms on the environment and upon each other, towards a relatively stable condition of equilibrium." This definition echoed F. E. Clements' (1916) incorporation of "reaction" and "coaction" as major elements, but, like E. P. Odum's subsequent usage, differed in its emphasis on changes in productivity as the major criterion of ecosystem development as against Clements' emphasis on change in component species populations of communities.

The appeal of Lindeman's trophic-dynamic approach was, according to G. E. Hutchinson (1942, p. 417), the hope "that the final statement of the structure of a biocoenosis consists of pairs of numbers, one an integer determining the level, one a fraction determining the efficiency, may even give some hint of an undiscovered type of mathematical treatment of biological communities." Hutchinson (1942) commented that Lindeman had "come to realize . . . that the most profitable method of analysis lay in *reduction* (italics mine) of all the interrelated biological events to energetic terms." Much later, Ulanowicz and Kemp (1979) described Lindeman's trophic concept as the "first attempt at a holistic perspective of an ecosystem which met with any degree of success." They saw the elegance of the trophic concept as parallel to the gas laws in urging their own approach toward "canonical trophic aggregations":

> Just as temperature, pressure and volume allow one to characterize the incomprehensible multitude of particulate motions in a simple gas, the hope is that a small set of figures, such as trophic storages or trophic efficiencies, permit one to compare two ecosystems with overwhelmingly disparate complexities (Ulanowicz and Kemp 1979, p. 871).

Ulanowicz and Kemp (1979) converged with the hope of Hutchinson, and later ecosystem ecologists, of providing "aggregate variables" to transform community ecology into a "key component of the evolving discipline of 'macroscopic ecology.' " The difficulty of macro-ecology and its dependence on "aggregate variables," such as trophic levels and unit measures thereof, and the necessity of building conceptual bridges between macro-ecology and "processes at lower levels" persists in 1980 as one of the great challenges of ecology, according to Orians (1980). Orians (1980) commented on the limitations of using "aggregate variables" and asserted that "a predictive science of community ecology will have to be based on underlying processes," presumably derived from studies of microecology.

Systems ecology began in 1959, according to Shugart and O'Neill (1979), to develop a body of quantitative ecological theory concerning the functioning of ecosystems. The functional energetic concept of ecosystem succession anticipated by Lindeman was elaborated by ecosystem ecologists and detailed in a series of frequently cited articles by R. Margalef (1963a,b, 1968) and E. P. Odum (1960, 1962, 1968, 1969) and in several editions of Odum's well-known textbook (1971). These are of particular interest because they provide the target for recent criticisms of the ecosystems approach to succession and are commonly cited as the epitome of the viewpoint, which requires an "alternative" population-based approach to succession (Peet and Christensen 1980). Odum's (1969) review of "the strategy of ecosystem development" is a convenient base to exemplify one pole I have suggested, although subsequent ecosystems work must be cited to develop the full span of the ideas clustering at this pole. Odum noted confusion in concepts of succession and defined it as "changes brought about by biological processes *within* (italics his) the ecosystem in question" (i.e., autogenic succession) (Odum 1969). He stressed, as the core of ecosystems analysis, the bioenergetic basis for succession which had been pointed out by Lindeman (1942), H. T. Odum and Pinkerton (1955), Margalef (1963a), and Van Dyne (1966). Most specifically, Odum provided a table of 24 "trends to be expected" in succession. Margalef (1968) advanced the idea that mature ecosystems influenced the rate of maturation of younger ecosystems by exploiting the unused energy output of those systems (cf. Valiela 1971). This posed the question, also raised by some recent proponents of population-based approaches, whether single sites or ecosystems are appropriate units of study or whether the landscape "mosaic" is the proper unit of study.

Many ecologists agreed with E. P. Odum (1971) that ecosystems ecology was a "revolution" in ecology. Patten (1971) saw an "inevitable kinship" between systems science and ecology, with ecology being recast in the language of systems science (cf. Smith 1975a). A distinction between ecosystem properties and the evolutionary idiosyncracies of species is commonly encountered in the literature of ecosystem ecology. Smith (1970, p. 17–18) expressed criteria for establishing a science of ecosystems:

> If, after describing several systems, a reasonably large set of properties are found common to many, if not all, types of ecosystems, which are system properties having little to do with the evolutionary adaptations of the species, then there is a discipline called "principles of ecosystems."

The search for the "principles of ecosystems" to produce the integrated science of ecology urged by Odum (1977) has been an important part of ecology, notably in the International Biological Program (IBP). In the views of some proponents of ecosystems ecology, the models, which are the *sine qua non* of ecosystems ecology, constitute a complex hypothesis leading to a theory of ecosystems (Shugart and O'Neill 1979). The theory of ecosystems was seen by some (Patten 1975b) as a special case of system theory, which, it was said, serves as a solid philosophical base (Innis and Cheslak 1975; Patten 1975c). From the systems point of view, an ecosystem has properties which are not those of its parts but result from its organization, and, as a consequence, it must be modeled as a whole (Van Dyne 1972; Overton 1975; Van Voris et al. 1978). Webster (1979) found some ecosystem models failing in this regard and asserted that "ecosystem behavior cannot be predicted from the laws of organism behavior. This is a philosophical mistake made in many ecosystem models. We cannot expect ecosystem behavior to emerge from a set of organismic equations."

Rigler, in contrast, noted weaknesses of the trophic-level ecosystem scheme and asserted that "it is symptomatic of the sickness of modern ecology" (Rigler 1975). The limitations of other methods for dealing with succession, however, were seen by Gutierrez and Fey (1975) "as the basic motivation for the development of systems ecology" and for the utility of ecosystem models. They stated that model simulations confirmed the hypothesis that secondary successions arise from internal feedback and said their simulations produced results consistent with "those observed in nature," unaccountably citing Odum's (1969) "trends to be expected" as the basis for this observation. However, it is not clear that succession is as significant to ecosystem ecology as Gutierrez and Fey intimate. It does not appear in the index of the recent volume *Theoretical Systems Ecology* (Halfon 1979), which makes it difficult to ascertain if the book will accomplish its avowed purpose of bridging the gap between systems theory and ecologists.

Ecosystem ecologists generally focus on abiotic factors as being the ultimate "driving variables" of ecosystems (Watt 1975). One reason for the relative lack of interest of ecosystems ecologists in organisms and the equations appropriate to them was because "systems ecologists' heritage in trophic dynamics and the flow of energy through ecosystems has resulted in a concern primarily for ecosystem processes rather than for species components" (Reichle and Auerbach 1972). O'Neill (1976) asked, "Does the ecosystem change when species change?"—and answered:

> Not unless the property or measurement under study changes. It is possible for properties such as nutrient-retention time to remain constant even though species change. Further, the identity of the system remains through successional changes in species . . . There is no *a priori* reason to believe that the explanations of ecosystem phenomena are to be found by examining populations (O'Neill 1976, p. 160).

Bormann and Likens (1979a,b) suggested that some ecologists have confused modeling of species interactions, or the living fraction of ecosystems, with modeling of whole ecosystems. In their view, understanding the structure of ecosystems requires understanding of both biotic and abiotic processes and "often neither can be understood without the other" (Bormann and Likens 1979a). They commented that their emphasis on abiotic aspects of ecosystem function "might provide exceptional advantages in the study of ecosystems" (cf. Webb 1979). The key distinction between species attributes and biogeochemical processes as the basis of succession requires resolution.

Ecosystem Evolution

A striking development in current ecology is the effort to merge it with, or reduce it to, evolutionary theory and genetics. E. O. Wilson (1978) wrote, "Theoretical physics transformed chemistry, chemistry transformed cell biology and genetics, natural selection theory transformed ecology—all by stark reduction, which at first seemed inadequate to the task." The traditional view of ecologists was that ecology had a particular place in respect to evolution and adaptation (McIntosh 1980b), and many ecologists now see evolutionary thought as providing new insight for ecology and succession explained by life history "strategies." Orians (1980), for example, saw useful guidelines for constructing the "aggregate variables" of ecosystems ecology in "a rich and useful body of theory that predicts attributes of organisms from basic evolutionary principles." A reciprocal relationship was seen by Bormann and Likens (1979b), who commented that interpretation of succession is not a trivial matter because "theories of ecosystem development are beginning to play a major role in studies of evolution."

Many traditional ecologists described the ecosystem as having an evolutionary unity (Emerson 1960; Dunbar 1960; Bates 1960; Orians 1962, 1980; Ross 1962; McIntosh 1963, 1980a,b; Valentine 1968), and ecosystem ecologists commonly perpetuate this view. Odum (1969) saw ecosystem development as parallel to ecosystem evolution and commented, "It is easy to see how stepwise reciprocal selection can account for evolutionary trends toward diversity, interdependence and homeostasis at the community level" (Odum 1971). Riechle and Auerbach (1972) wrote, "The strategies of ecosystems may not always be inferred from *strategies* of species," reinforcing the distinction between ecosystem properties and species attributes noted earlier. Patten (1975b) described the "ecosystem as a coevolutionary unit"; and a basic assumption of ecosystem analysis is, according to Webster et al. (1974), that ecosystems are "units of selection and evolve from systems of lower selective value to ones of higher selective value" which, they said, "optimize utilization of essential resources."

One of the traditional objections to the analogy of the evolution of community or ecosystem and evolution of species is that the former do not share or transmit genes. Whittaker and Woodwell (1972, p. 150) asserted this position:

> There is in the community no center of control and organization, and no evolution toward a central control system . . . . Community evolution is a result of species evolution and species behavior. Community organization with its quality of "loosely ordered complexity" (Whittaker 1957) has not a really good analogy among other living systems.

The traditional population models of evolution were questioned by Wilson (1976, 1980). Wilson picked up the familiar concept that the ecosystem is an emergent supraorganismic entity as a consequence of natural selection and, like Odum, offered a model to explain it. He noted the contrast I have described between evolutionary population ecologists and ecosystem ecologists: "It seems as if ecology is divided into two schools, each inspecting a different side of the same coin and each claiming that its side is all that exists" (Wilson 1980). Wilson also cited the widespread belief in the superorganism and the lack of any "rigorous foundation" for the belief. He provided the foundation of a model of group selection in which the community selects among its members for beneficial traits for the whole and asserted: "If the superorganism concept is not a first principle of natural selection, it may still be an emergent property" (Wilson 1980, cf. Pianka 1980). Substantial differences in conception of evolution underlie theories of ecosystem and succession.

Reduction and Emergence

Clements rested his theory of succession on the assumption that the formation possessed the properties of an individual organism and the complex superorganism was widely believed to have emergent properties not inherent in its parts. Even W. S. Cooper (1926, p. 394), in most ways a critic of Clements' concepts of succession, wrote: "Assemblages of plants or of animals are subject to special laws, so that their mass action is not equivalent to the sum of the action of the component individuals." Gleason (1917) had asserted, "the phenomena of vegetation depend completely upon the phenomena of the individual" and causes of succession are a "resultant of forces" Gleason (1927). Tansley (1935) questioned measurement of the whole or its parts, "Can one in fact form any clear conception of what 'mere summation' can mean . . ." or ". . . has it any meaning at all?" The problem seen by early ecologists persisted and cropped up in F. E. Smith's (1975b) presidential address to the Ecological Society of America—"Two extreme views of ecosystems can be recognized. In one view, an ecosystem emerges as the sum of its parts." Harper (1977b) deplored the view that complex ecosystems can be described only as if they were whole organisms and called for "faith" that "the complex is no more than the sum of the components and their interactions." Webster (1979) wrote, "The higher level behavior results from a synergism. It is something more than the sum of the lower level parts." Innis (1976), like Tansley, had found difficulty in seeing "just how this summation is going to be carried out so that we may eventually study an ecosystem as the 'sum of its parts.' "

Ecological discourse about the nature of ecosystem and succession has familiarily, if peripherally, been involved with the philosophical concepts of reduction and emergence. Here ecology overlaps the traditional discussion of organicism and reduction in biology (Nagel 1952; Feibleman 1954; Rowe 1961; Grene 1967; Phillips 1970; Parascandola 1971; Ayala and Dobzhansky 1974; Ghiselin 1974; Innis 1976; Odum 1977; Hull 1978; MacFadyen 1975; Lidicker 1979; Salt 1979; Wimsatt 1980). The positions among ecologists vary widely. MacFayden (1975) supported attacks by ecologists on reductionist science as "absolutely legitimate." Eugene Odum (1977, p. 1290), the premier American spokesman for holistic ecology, stated the systems position succinctly: "We theorized that new systems properties emerge in the course of ecological development, (*read succession*) and it is these properties that largely account for the species and growth form changes that occur . . . ." An alternate theory was stated by Grime (1979, p. 154): ". . . processes of change in the structure and composition of vegetation can be interpreted as a function of the strategies of the component plant populations."

What is notably lacking in this extended discourse are clear stipulations of what is involved in sum versus whole and reduction or emergence as applied to ecology (Nagel 1952). Ecologists are divided on the appropriate use of these traditional philosophical concepts in ecology. Salt (1979) noted the frequency and importance of *emergent properties* attributed to ecosystems and called for more discriminating use of the term, which he lumped with niche as a not very meaningful "pseudocognate." The same may be said of the use of reduction in ecology. Salt distinguished what he called "collective properties," which he defined as "a summation of the proportional contribution of each of the individual species populations" (Salt 1979). According to Salt, Odum's "trends to be expected" are not emergent properties at all but are "collective properties." Salt noted the difficulty of identifying emergent properties of ecosystems and commented that some emergent properties are not clear "unless we are prepared to postulate the existence of a 'community' which is independent of the species of which it is composed" (Salt 1979). This is what some ecosystem ecologists, like O'Neill (1976) in fact, assert.

Wimsatt (1980) considered the use of reductionist versus holistic research strategies in the context of natural selection. A similarity is seen in the contrasting claims of ecosystem ecologists and population ecologists as applied to succession. Critics of holistic organismic approaches to succession urge the need for study of populations and life-history strategies (sic) as an alternative to either Clementsian or ecosystem ideas. The intimation is that examination of lower level species attributes can lead to understanding of the individual species and, these being solved, the resolution extended, in rather indeterminate ways, to the higher level community and ecosystem. The ecosystem view apparently derives from the belief that a holistic approach is required because the ecosystem has emergent properties that put their impress on the "species or and growth form changes that occur" in succession as Odum (1969) asserted. The rhetorical positions of the two groups do not usually allow for much reciprocal understanding, and there is little effective interaction between them. Here Wimsatt (1980, p. 54) has a useful caution: "Now for pragmatic as well as theoretical reasons, reduction in science is better seen as the attempt to understand the explanatory relations between different levels of phenomena, each of which is taken seriously in its own right...." If such tolerance is not achievable among ecologists, clearer exposition of hierarchical level, reduction and emergence is required.

Discussion and Questions

Although much of the renewed interest in succession stems from the heady influx of new bodies of theory into ecology, a substantial part still derives from differences in field observations and interpretation of these. However, even the most straightforward descriptions of succession are not free of theoretical premises. Whittaker and Levin (1977) noted the historical failure of unifying generalities about succession, attributing this failure to the intrinsic complexity of ecological phenomena. Their admonition that ecological theory should make allowance for this and center more on "analysis, interpretation, comparison, and modeling of cases rather than on widely applicable generalization" (Whittaker and Levin 1977) has not been heeded by ecological theorists. The often parochial concentration on efforts to produce a general unified approach to ecological systems, which permeates much of current ecology, is frequently evident in the disparate positions on succession. An important challenge to this volume is to examine these positions in the face of the admonition of Whittaker and Levin.

Virtually all current discussion of succession considers only secondary succession, which by definition follows disturbance, now widely, and unhelpfully, known as perturbation. Much of the persistent controversy surrounding succession stems from the very different starting points or pioneer states following varied kinds and degrees of disturbance, from which the seral sequence begins (White 1979). The validity or utility of an equilibrium concept, either the traditional Clementsian climax or the mature state of Margalef and Odum, is the other side of the successional coin requiring clarification. It should be clear that the term succession does *not* necessarily require the progressive development to a climax or mature state seen by certain theoretical positions. Others view it as any sequence of change, as did W. S. Cooper and H. A. Gleason, and many modern ecologists see equilibrium reached rarely, if ever.

Clements (Weaver and Clements 1938) asserted that most of an area in a state of nature was in the climax condition and disturbed areas occupied a minor portion of the regional climax because disturbances were unusual events. Much recent ecosystems modeling tacitly assumes the existence of, or trend to, an equilibrium. Bormann and Likens (1979a,b), for example, commented that their extended studies of ecosystem development of northern hardwoods forest at Hubbard Brook assumed the steady state. They wrote that their model emphasized autogenic development and deemphasized "exogenous" disturbance—"This was a necessary decision if our model was to reflect an uninterrupted sequence from the initiation of secondary development to the establishment of the steady state" (Bormann and Likens 1979a).

White (1979) reviewed the problems of assessing disturbance and described any disturbance as a tripartite continuum of frequency, predictability, and magnitude, which poses the threat of a new round of discussion of the vegetational continuum hypothesis. According to White, the traditional distinction between autogenic and allogenic is artificial, and allogenic affects are not independent of autogenic conditions. It is incumbent on ecologists to specify the types and qualities of disturbance and their role in the successional process or at least to explicitly recognize these in attributing generality to their interpretations of succession. Grime (1979) defined disturbance as "the mechanisms which limit the plant biomass by causing its partial or total destruction" by biotic or abiotic means. This harks back to Clements' (1916) definition as various degrees of denudation by the same means. Grime proposed three strategies (sic) of plants in succession based on a 2 × 2 table of two degrees of disturbance and two degrees of stress, the latter defined as extremes or absolute shortage of physical necessities. He then developed a triangular model showing succession related to various degrees of disturbance, stress, and competition. His conclusion was that changes in vegetation "can be interpreted as a function of the strategies of the component plant populations" (Grime 1979) in responding to his tripartite model. Connell (1978, 1979) shuttled between perturbation and disturbance and examined two sets of hypotheses: (1) that communities are seldom in equilibrium and disturbance is ubiquitous; (2) that communities are usually in an equilibrium state, or recovering to it, and disturbance is infrequent and small. Connell (1979) provided a 3 × 3 table of three classes of disturbance regime, based on size and frequency of disturbance, and three classes of tree species based on their relative ability to invade open or occupied sites. Connell essentially provided an elaborate description of classical secondary succession to climax in the form of an unwieldy verbal model. Connell described the evidence from studies of ecological succession as indicating that soon after a "severe disturbance" propagules of a few species arrive in the open space created by the disturbance. Harper (1977b) also saw an early stage of recovery from "disaster" as overwhelmingly controlled by dispersability of propagules. However, it is clear that the relative significance of invasion of propagules (relay floristics, *sensu* Egler, 1954) is dependent upon the characteristics of the disturbance. In Bormann and Likens (1979b) the disturbance of clear cutting was followed by an orderly process of succession based on the "buried seedbank strategy" (initial floristics, *sensu* Egler, 1954) as part of an "elaborate feedback mechanism." The problem is, of course, at what level of disturbance, as a combination of kind, intensity, frequency and area, are the various mechanisms of succession called into play. Connell (1979) described six separate hypotheses, based on two categorically opposed theories, all operating in succession depending on degree of disturbance.

The perpetual disagreement among ecologists concerning succession is commonly seen as a consequence of viewing it at different scales (Loucks 1970; Marks and Bormann 1972; Zedler and Goff 1973; Wright 1974; Cole 1977; Webb et al. 1972; Whittaker and Levin 1977). Whittaker and Levin (1977) required a topological view of climax, "variously stretched or shifted to the larger scale of the landscape." The view of the climax as a landscape pattern with varying seral stages represented in a complex mosaic or pattern continuously shifting, but retaining an overall stability, is frequently posed as an alternative to (if not an evasion of) the difficulties of resolving the perennial problems of succession. That it has neo-Clementsianism elements is seen in the statement of climax in Weaver and Clements (1938, p. 479):

> Nevertheless, all areas within the sweep of climate and climax, whether bare or denuded by man, are marked by more or less evident successional movement of communities and, hence, belong to the climax in terms of its development. In consequence, each climax consists not merely of the stable portions that represent its original mass but also of all successional areas, regardless of the kind or stage of development.

The shift of scale from seral development to climax on a presumably homogeneous site to the distribution of seral and climax states in a landscape pattern does not eliminate the problem of what are the changes, if any, on a single site. The presumed chronological relations, observed or inferred, of conditions on a single site to the various elements (stages, cells, tesserae) of the landscape mosaic are the traditional problem of succession. The shift to landscape pattern introduces a number of additional concerns (McIntosh 1980a). How large an area must be encompassed to enclose an appropriate statistical distribution of species or seral stages? What is the pattern of environment on the landscape? What is the pattern (kind, frequency, predictability, magnitude) of disturbance on the landscape? The familiar questions remain. What is the sequence on a single site, and why are some species eliminated and forced to move to other sites in the landscape mosaic? What is the mechanism of succession on a site? Do the early occupants serve as preparators of site conditions (facilitation) for later occupants? Are the evolved life history properties of the available species the explanation of the sere or does it require the supervention of emergent properties?

It may be that the two extreme theoretical poles widely evident in ecology and their interpretations of succession will persist indefinitely, like the wave and particle theories of light. The concept of succession as a collection of population processes dependent on the life history strategies (sic) of the component species as against the concept of a sere as an ecosystem process leading to the development and evolution of an emergent entity permeate the current literature. The very different interpretations of succession in the recent literature, and evident in this volume, are redolent of long familiar differences. The elaboration of ecological theory, derived from mathematical theories of evolution and populations, and the mathematical theory, derived from systems analysis and from cybernetics, makes rapproachment difficult since the methodological and philosophical premises are widely divergent. It is, no doubt, premature to assert that synthesis is not possible (Peet and Christensen 1980). Formal population theory is only about 50 years old, and formal ecosystem theory is a youngster only half that age.

An approach to a synthesis, or at least to a common platform, would be more likely occur if the groups I have suggested were less cliquish and less prone to assert their positions within the cloud of their own program. A number of important questions to be answered are:

1. Do the extended observations of succession support a general theory of a successional process? (a) That is orderly, directional and predictable? (b) That is controlled by biotic factors, i.e., autogenic? (c) That leads toward an equilibrium state in either or both its biotic or abiotic attributes?

2. Is the claim of some ecologists that successional phenomena are reducible to theories of natural selection justified? How do population theory and life history strategies explain or predict ecosystem attributes?

3. What are the emergent properties claimed to justify the ecosystems theory? If population phenomena are not additive, what is the measure of integration?

4. Does the evolution of ecosystems have any reasonable explanation in evolutionary theory? What does recent palynological evidence suggest about this claim?

5. Is the reduction of the ecosystem to trophic numbers, seen by Hutchinson (1942) and Ulanowicz and Kemp (1979) as the essence of the genius of Lindeman, holistic or simply a collective property as stated by Salt (1979)? In what sense is the ecosystem approach to succession holistic?

6. Can reasonably explicit distinctions be made between small-scale disturbance initiating the traditional serule within the community and large-scale disturbance initiating an earlier stage of a sere? Can autogenic and allogenic disturbances be clearly distinguished or are they interdependent as suggested by White (1979)?

7. Can a theory of succession be developed to incorporate a sere regularly reaching an equilibrium over an extended area as argued by Bormann and Likens (1979a) and by Franklin and Henstrom (Chapter 14) and a sere which is regularly interrupted by disturbance as described by Raup (1957) and by Heinselman (Chapter 23)? Can the achievement of equilibrium be related to a trajectory toward equilibrium?

Chapter 4

Some Causes of Variety in Patterns of Secondary Succession

Henry S. Horn

Introduction

The contrasts and paradoxes of ideas about succession are well illustrated by the history of my own ideas about succession. My introduction was the Boy Scouts' Wildlife Management Merit Badge pamphlet (Allen 1952); old fields are invaded by sun-loving species, which by their growth gradually create an environment in which only shade-tolerant species can thrive. My observation at Camp Quinapoxet in West Ridge, New Hampshire was that all the sun-loving gray birches (*Betula populifolia*) are on sandy uplands and all the shade-tolerant hemlocks (*Tsuga canadensis*) are in ravines, regardless of the ages of the stands.

When I moved to Princeton, New Jersey in 1966, I found at the Institute for Advanced Study a set of upland fields that had been abandoned at various times from 1878 to 1941 (Curtin 1978), and one patch of forest that apparently had never been plowed. The oldest tree aged so far in the unplowed wood is a beech (*Fagus grandifolia*), established before 1728 (*fide* M. C. McKitrick), and the forest floor has several generations of mounds and pits from previous wind throws. This oldest forest is dominated by beech. The successively younger woods are dominated by oaks (*Quercus*), hickories (*Carya*), tuliptree (*Liriodendron tulipifera*), and dogwood (*Cornus florida*) dating from 1878; red maple (*Acer rubrum*), black cherry (*Prunus serotina*), and tuliptree dating from about 1910; sweetgum (*Liquidambar styraciflua*) and blackgum (*Nyssa sylvatica*) dating from 1935; and gray birch and large-toothed aspen (*Populus grandidentata*) dating from 1941. In the youngest stands there are no saplings of the dominant species, though there are many partially suppressed saplings of the species that dominate older stands. The older stands have dead and dying relicts of species that dominate younger stands, along with suppressed saplings of their own species and of those that dominate still older stands. Many individual trees that are dying in the canopy are obviously being replaced by growth of the most vigorous saplings from their understory, though large openings are also invaded by new seedlings. The inferred pattern of succession (Horn 1975a) in this forest is fully consistent with the pattern idealized in the Merit Badge pamphlet (Allen 1952).

I spend most summers on a piece of abandoned farmland in Whitehall, New York. Here the pattern of succession is completely different from that in the Institute Woods in Princeton. In any old field, ash (*Fraxinus*) saplings are dominant near mature ashes in the adjacent hedgerow, pin cherry (*Prunus pensylvanica*) saplings near mature pin cherries, and likewise for black birch (*Betula lenta*), red oak (*Quercus rubra*), and sugar maple (*Acer saccharum*), among others. Furthermore, the composition of the initial species in a field appears to depend in a complicated way on the season of abandonment and the last prior treatment, in addition to details of soil, slope, and exposure.

Returning to Princeton, there is a similar pattern of patchiness and dependence on history at Stony Ford Ecological Research Preserve, Princeton University, involving red cedar (*Juniperus virginiana*), ashes, *Cornus*, and *Viburnum*, among others. Flood plains adjacent to the Research Preserve and to the Institute Woods are dominated by evenly aged cohorts of trees of several species, dating from floods or droughts (Atkin 1980). Even in the upland parts of the Institute Woods, although there is a clear pattern of early, middle, and late successional species, the particular species present at a particular site is very much affected by the proximity of parental trees and by other effects that can be broadly characterized as accidents of history.

The empirical observation of such variety in successional patterns has led reviewers to despair of unifying generalizations (e.g., Horn 1976; McIntosh 1980a). I have previously explored the case that seemed amenable to the most mechanistic biological analysis, namely the broad pattern of succession from light-requiring species to shade-tolerant ones (Horn 1971, 1975a). One stage more general than this is the representation of succession as a tree-by-tree replacement process, in which some of the replacements may be determined by the relative tolerance of the species, but others may be the result of any of a number of causes, including some that are stochastic (Horn 1975b, 1976).

This chapter presents a new model of the interaction between the spatial pattern of recruitment and the density of "safe sites" (*sensu* Harper 1977a) for the establishment of adult plants. This model suggests insights into why some successions are very much dependent on accidents of history and others are less so. Further insights come from a review of tree-by-tree replacement models with additional features from models of Connell and Slatyer (1977), taking particular note of factors that determine whether such models predict a biologically interpretable successional pattern or a historically determined patchwork. Finally, in the actual successional pattern of the upland stands in the Institute Woods in Princeton, there are elements of all of the different models reviewed.

Spatial Pattern of Recruitment and Succession

Figure 4.1 presents the first ingredient of a model of the relation between spatial pattern of recruitment and the density of safe sites (Harper 1977a,b) for the establishment of saplings. The model is an explicit extension of ideas developed by Harper (1977a, especially Chapters 2, 5, and 20). It is initially assumed that safe sites are very sparse on the ground, so that seedlings and hence saplings become established quickly and abundantly only beneath a copious rain of seeds, and become established infrequently and sparsely where the rain of seeds is scant or only moderate. Figure 4.1 shows the hypothetical pattern of recruitment to a field abandoned three years previously, all recruits being the offspring of a single, large, seed-bearing tree. In the first year of abandonment many seedlings become established beneath the tree and a short distance away because the heavy rain of seeds ensures that all of the rare safe sites receive at least one seed. Farther away from the tree, the rain of seeds in the first year is so scant that only a few safe sites are hit by seeds, and the resulting saplings are few and scattered. In the second fallow year, again many potential sites are hit near the tree, but the continuous miniature canopy of year-old saplings prevents the new seedlings from growing

Figure 4.1. Recruitment from a rain of seeds onto sparsely distributed safe sites for sapling establishment in a field that has been fallow for three years.

beyond the limit of their cotyledonary reserves. Farther from the tree, again few and scattered safe sites are hit, but many of these are unoccupied, and new seedlings become established. The same thing happens in the third fallow year; the many shaded seedlings near the tree die; the few far from the tree survive. The resulting pattern is a copse of evenly aged saplings near the seeding tree, all dating from the abandonment; and scattered saplings of a variety of ages farther from the tree.

The effect of this model on old-field succession is shown at the top of Fig. 4.2. Seed trees at opposite sides of the field beget evenly aged copses of, for example, ash and maple near the parental trees, and the middle of the field has trees of a mixture of species and of ages. The second line in

Figure 4.2. Ash (A and pointed) and maple (M and rounded) trees, seedlings, and saplings in a field with sparse safe sites for establishment (top line), and at successive stages in the growth of a woodland on a field with abundant safe sites (second, third, and fourth lines). See text for interpretation.

Fig. 4.2 shows the consequences of the same model when safe sites are much more abundant, so that even a light rain of seeds is sufficient to inoculate many sites. There are still more ash seedlings near the big ash and more maple seedlings near the big maple; however, there are appreciable numbers of ash seedlings and of maple seedlings throughout the field. The juxtaposition of ashes and maples throughout the field means that there are many opportunities for one-on-one competition for dominance as saplings. The third and fourth lines on Fig. 4.2 portray the result of this competition. In the early stages, ash dominates, owing to its higher growth rate in the open, and maple, though suppressed, persists in the understory, owing to its tolerance of shade. As the saplings grow into a forest, however, the shade-tolerant maples are more likely than the ashes to survive crowding and the subsequent thinning of the stand, and the forest loses its ash.

If safe sites are sufficiently abundant then, the forest is first dominated by rapidly growing but shade-intolerant trees, and later by slower growing and shade-tolerant trees, some of which may have persisted in suppressed form from the time of abandonment of the field, and some of which may have invaded the understory in the intervening time. The pattern is biologically interpretable as a succession that is driven by a competitive hierarchy involving reciprocity of growth rate in the open versus tolerance of shade. On the other hand, if safe sites are sparse, the forest is a patchwork of evenly aged copses near parental trees, with intervening areas filled with a mixture of species and of ages. There appears to be no succession, but rather a preemptive crazy quilt that will not change markedly in composition until the next episode of major disturbance.

Therefore, with neither structural nor qualitative change in the model, with only a change in the value of one parameter, namely the abundance of safe sites for the establishment of saplings, the prediction of the model varies from a biologically interpretable succession of dominance by different species to a patchwork of copses and individual trees whose composition is highly dependent on accidents of history.

Succession as a Tree-by-Tree Replacement Process

To model succession as a tree-by-tree replacement process, ideally one should make a map of all trees in a forest and then come back 25 or 50 years later and see which has replaced which. The data thus obtained would yield a table of replacement probabilities, which, by matrix multiplication, simulates succession for a forest of a given initial composition. In practice, estimation of the replacement probabilities must often be indirect. Using the number of saplings of various species that grow under various species in the canopy, ignoring those species for which this is not a reasonable predictor of replacement, and correcting for varying longevities, I have found that this model does indeed predict the observed species and their approximate relative abundances in the canopy of the old beech forest in the Institute Woods in Princeton, New Jersey (Horn 1975a).

My particular empirical test of this model depends on the observation that the understory of this particular forest is important in regeneration of the canopy, as Woods and Whittaker (Chapter 19) have also observed. The model itself makes no such assumption. The model assumes only that knowing what species occupies a given spot may narrow the statistical range of likely occupants of that spot in the near future, and more critically, that these probabilities of replacement depend only on what species occupies a given spot, and do not change with time other than when the occupant of a spot changes. Clearly the latter assumption will be untrue in detail, though it may be approximated when growth to the canopy is rapid and tenure in the canopy is long.

The chief virtue of this model is its simplicity and its breadth of applicability as a theoretical tool. It can be applied on a tree-by-tree level, using data like mine or those of Woods and Whittaker (Chapter 19). It can also be applied to observed transitions in composition of whole stands, as Stephens and Waggoner (1970) have done with data like those of Christensen and Peet (Chapter 15). It can even be applied to regional patterns of vegetative change (Shugart et al. 1973).

The chief weakness of this model is the difficulty of actually measuring replacements. There are very few situations in which numerical abundance of saplings in the understory is even an approximate predictor of success in attaining the canopy. Direct observation of what is replacing dead and dying trees is ideal (e.g., Barden 1980), but instances are rare for long-lived species. An intriguing, but as yet untried, possibility is to estimate transitions from the short-term results of more elaborate stand simulation models (Botkin Chapter 5; Shugart et al. Chapter 7; Solomon et al. Chapter 11). This hybrid approach combines the expensive detail of the stand simulators over the short term with the simple and somewhat blurred picture of the tree-by-tree model over the long term. It is worth noting explicitly that the difficulties of measuring replacement probabilities cannot be cited as evidence that the broad assumptions of the model are untrue or unrealistic. Even where the model works, it is intended only to be sufficient (Botkin, Chapter 5) as a caricature of reality, rather than necessary as a mechanistic explanation.

This model is not directly applicable to communities in which the dominant influence on vegetative dynamics is a recurrent pattern of large-scale disturbance, like fire (Franklin and Hemstrom Chapter 14; Heinselman Chapter 23; Van Cleve and Viereck Chapter 13; B. Walker Chapter 25; Vitousek and White Chapter 17). Similar models for the dynamics of a disturbance mosaic have been invented by Levin (Levin and Paine

1974; Levin 1976; Whittaker and Levin 1977). These models thus far have been used only metaphorically, but with promising results.

The linear replacement model is not applicable when the vegetative changes of interest are caused by secular changes over spans of time much longer than the life spans of the trees, such as climate (Brubaker Chapter 8; Davis Chapter 10; Solomon et al. Chapter 11) or nutrients in developing soil (J. Walker et al. Chapter 9). Nor is it applicable where the dynamics of species' replacement are grossly nonlinear owing to effects of animal populations (B. Walker Chapter 25), or to spatial restrictions on recruitment (e.g., the previous model in this chapter). Further critical discussions of the limitations of the linear replacement model are found in Horn (1976) and in Shugart and West (1980). Nevertheless, the model still offers insights into some of these communities through an examination of its response to violations of its basic assumptions.

When Is Secondary Succession Convergent?

I have analyzed linear tree-by-tree replacement models in detail elsewhere (Horn 1975b, 1976) and shall only list the results here and give the intuitive arguments behind them. The primary result of such models is that with successive multiplications by a given replacement matrix, the predicted community almost invariably converges on a particular stationary composition. No matter what the starting composition of the community is, eventually it reaches a fixed set of proportions of various species. Therefore, when convergence is observed, it is not necessary to postulate a biological cause for the convergence itself. The property of biological interest is not the convergence, but the structure of the replacement matrix, which is generated by the comparative life histories of the species involved. Convergence after disturbance is a statistical consequence of the existence of a replacement matrix, and the speed of convergence depends on the form of the matrix.

The critical features of the matrix that affect convergence can be summarized in a diagram of arrows showing which species can replace which other species, and which species are effective at self-replacement. A very general and rigorous proof of this assertion is found in Horn (1976, p. 198). Double-headed arrows tend to speed convergence since any two species that can replace each other can readjust their relative abundances to agree with the stationary composition within as little as one generation. Self-replacements tend to slow convergence because if a self-replacing species becomes abundant as a result of disturbance, it can maintain its abundance by self-replacement. A convergent community at its stationary composition should be a mosaic of species that are jointly connected by arrows of replacement; this is discussed more fully in the context of an excellent example by Woods and Whittaker (Chapter 19). This mosaic may include species that are typically pioneers early in succession, like the

Douglas fir (*Pseudotsuga menziesii*) that Franklin and Hemstrom (Chapter 14) find as young trees in old stands.

A particularly interesting and realistic form of nonlinearity that is easily added to the basic tree-by-tree replacement model is recruitment that is proportional to the abundance of a species at a given time (Horn 1975b). This produces multiple stationary compositions that depend upon the initial composition. Any species whose abundance is increased by disturbance also has its recruitment increased; therefore, the results of disturbance tend to perpetuate and even to augment themselves. The first model in this chapter, with rare safe sites, is a special case of this phenomenon on a fine spatial scale. A more indirect example is the alternative stable dominance of beech and tuliptree in the stand model of Shugart et al. (Chapter 7), with each species creating microsites that favor its own regeneration over that of other species.

These results can now be applied to a wide range of successional processes, which are summarized in Table 4.1.

Table 4.1. Process and convergence in secondary succession.

Process	Species' replacements	Convergence?	Chapters with examples
Chronic disturbance:			
Asynchronous & small-scale	A ⇄ C, A ⇄ B, B ⇄ C	Rapid	4,6,7,16, 18,19.
Synchronous & large scale	? → A, ? → B, ? → C	Academic	4,6,7,13, 14,18,20,21, 23,24,25.
Regeneration:			
None		Senescence	13,14,23.
Self-replacement		Slow	4,7,13,14, 24,25.
Replacement proportional to abundance (may be local & patchy)		Alternative states depend on initial state.	4,7,13,14, 15,19,21,23, 24.
Competitive hierarchy	→ A → B → C; A → C; → B; → C ↻	Gradual	4,6,7,13, 14,15,16,18, 23.
Facilitation	→ A → B → C ↻	Slow	4,13,16,19, 22,23.

Note: The list of examples is conservative, including only direct and unambiguous natural history described in this book. The inclusion of transitive interactions and plausible theoretical interpretations would result in citation of nearly every chapter for nearly every process. Species' replacements that are caused by secular changes in the physical environment are important, but are neither included in this table nor discussed in this chapter.

Chronic Disturbance

For present purposes a disturbance is chronic if it is frequent enough to end the lives of most trees. If the disturbance is asynchronous and on a small scale, that is, if scattered individual trees die and the openings thus created are invaded by any species, then any species can replace any other species by way of a disturbance. The diagram (Table 4.1) of replacements has only double-headed arrows, and convergence is rapid. This further supports the role of statistics in generating the convergence itself, because convergence is most rapid where replacing species is stochastically determined.

If disturbance is synchronous and on a large scale, then it simultaneously ends the lives of all trees in a stand. Whatever invades or regenerates immediately will persist until the next major disturbance. The stand as a whole never remains undisturbed for more than the life span of a tree, and the question of convergence is academic. The weight of current opinion suggests that most forest communities are subject to chronic synchronous large-scale disturbance (Vitousek and White Chapter 17).

Regeneration

Where regeneration does not occur, the dynamics of the stand will go from growth to thinning to senescence. As is argued above, self-replacement slows convergence, and replacement proportional to abundance produces alternative stationary communities that depend on initial conditions. Where the dispersal of seeds or other propagules is very local, the initial community itself may depend on the proximity of mature trees, and the resulting succession will be a historically determined patchwork, with different processes dominating in different places. However, as the first model in this chapter shows, this effect may be lessened on better sites.

Competitive Hierarchy

A competitive hierarchy is a list of species in order of their competitive ability, with each species able to outcompete all above it in the list, but also able to invade an open site and to hold it against all species except those below it in the list. Examples are given by Christensen and Peet (Chapter 15), Franklin and Hemstrom (Chapter 14), and Heinselman (Chapter 23). The stationary community obviously consists of the competitive dominant. The course of succession after disturbance is somewhat dependent on the competitive abilities of the initial species. Reassortment of abundances among species will occur slowly since the diagram of species' replacements contains only one-way arrows, but the arrows all point toward the competitive dominant, and species of intermediate competitive ability may be skipped. Thus successional convergence is gradual but inevitable.

Facilitation

If early species are necessary to prepare the way for later species, then convergence on the later species depends on the historical accident of which species dominate after disturbance, and the convergence is very slow since all intermediate stages must pass in order.

A Semantic Note

The competitive hierarchy corresponds roughly to what Connell and Slatyer (1977) call the tolerance model, though it also includes elements of competitive inhibition. What Connell and Slatyer call inhibition is actually complete preemption, which is likely to occur when chronic disturbance is synchronous and large scale, when there is no regeneration, or when regeneration is proportional to local abundance. The competitive hierarchy is presumably part of the mechanism behind the sorting of species in Egler's (1954) model of initial floristic composition, though Egler emphasizes differences in growth rates as causes of differential dominance, as do Drury and Nisbet (1973), Heinselman (Chapter 23), and Van Cleve and Viereck (Chapter 13). Facilitation is Connell and Slatyer's (1977) term for what Egler (1954) first clearly described as relay floristics. Facilitation is an element of Clements' (1916) notion of succession that has been attacked early and of late (McIntosh Chapter 3). However, Clements' original discussion implies a competitive hierarchy as well, and much of his original terminology can still provide a useful framework for discussion (MacMahon Chapter 18).

Reality Revisited

Returning to the successional pattern that I know best, the uplands of the Institute Woods in Princeton, I find examples of all the processes listed in Table 4.1. Figure 4.3 shows the replacements inferred among four of the commonest species. The importance of small-scale asynchronous disturbance is shown by mutual replacements among blackgum, red maple, and to a lesser extent beech. In 1976 nearly a third of the beech forest was leveled by a plow wind, and the resulting clearing has been repopulated by nearby tuliptrees. The oak, hickory, and tuliptree stand has a dogwood

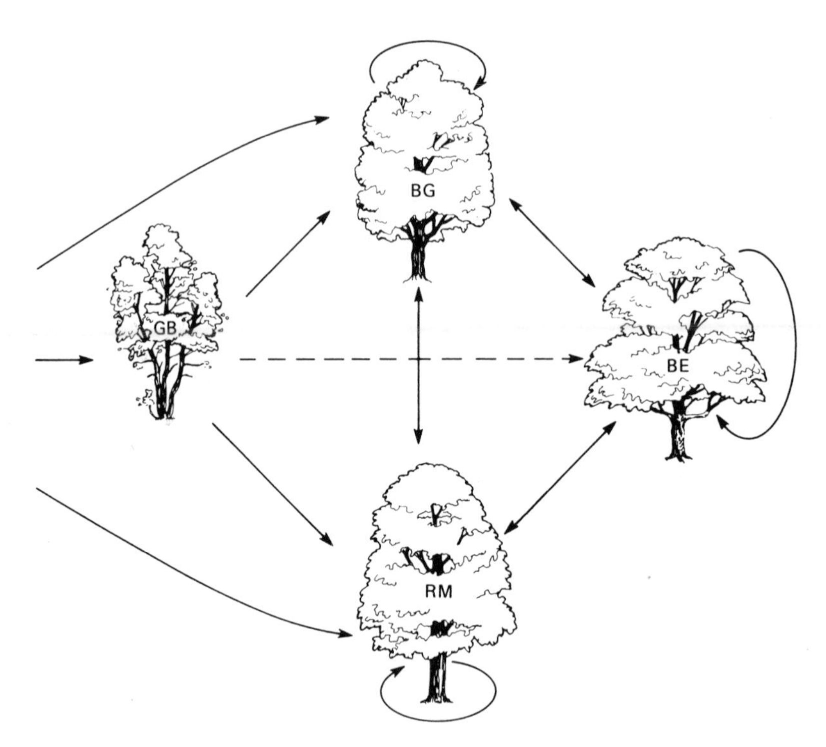

Figure 4.3. Inferred pattern of species replacements of gray birch (GB), blackgum (BG), red maple (RM), and beech (BE), all common species in the Institute Woods in Princeton, New Jersey. GB, RM, and RM invade in the open and may be replaced. BG, RM, and BE can replace themselves or each other, though replacements of BE are rare. GB can neither replace itself nor displace others. BE rarely invades beneath GB and never invades in the open.

understory beneath which there is no advanced regeneration whatever of canopy trees. Figure 4.3 shows blackgum, red maple, and beech as effectively replacing themselves. Whether forests at intermediate stages of succession are dominated by red maple, blackgum, or sweetgum seems to depend on the relative proximity of seed trees and thus on regeneration that is proportional to local abundance. There is a clear competitive hierarchy from gray birch to blackgum or red maple to beech. Finally there is some evidence the suggests facilitation since beech never invades in the open and seldom invades beneath gray birch, and thus seems to require conditions at least as crowded as those in a stand of blackgum or red maple.

Succession as the Differential Expression of Life Histories

The primacy of life histories in determining patterns of vegetative dynamics is a major theme of many chapters of this volume (Botkin, Chapter 5;

Cottam, Chapter 12; Gomez-Pompa and Vazquez-Yanes, Chapter 16; Heinselman, Chapter 23,; Shugart et al., Chapter 7; Zedler, Chapter 24). Some have gone so far as to deny any need for concepts like facilitation, inhibition, tolerance, and even succession. An alternative view that is actually complementary is that these concepts are nothing more and nothing less than statements about the comparative life histories of species that are found in association with each other. From this point of view no absolute statements may be made about any particular species and its presumed adaptation to a given successional status, just as no absolute statements should be made about adaptation in general (Lewontin 1978; Horn 1979). The only statements that should be made are comparative ones involving differences between species and differences in successional status at a given time and place. The comparative approach allows statements about adaptation to succession to be made and tested even if the species involved have not shared an ecological association over a long period of time, as Davis (Chapter 10) has shown to be the case for the very recently derived associations of the eastern deciduous forest.

Conversely the only generalizations that should be made about succession are comparative ones, and the processes listed in Table 4.1 are not mutually exclusive categories into which natural successions should be forced. Rather they form a shopping list of things for one to explore and to evaluate when comparing different successions.

Chapter 5
Causality and Succession

Daniel B. Botkin

Introduction

Like succession, *discussions* of terrestrial plant succession appear to follow a regular and repeatable pattern. An aspect of succession is observed, discussed, appears immune to test, and the discussion is abandoned only to be restarted again after some intellectual disturbance. This is particularly true for discussions of the causality of succession; we observe that some species are pioneers while others are characteristic of later stages in succession. Why is this so? Do the pioneer species prepare the way for the later ones? Do the pioneer species prevent the later ones from entering? Do the pioneer species make conditions unacceptable to future generations of their own species, and therefore create the conditions for a replacement? Or is the process merely one of time, the later successional species merely being slow growing and slow to arrive and germinate? Some recent articles suggest that we have no better understanding of these issues now than we did decades ago (McIntosh 1980a, Chapter 3). I disagree.

Ideally, in science one hopes to achieve a necessary and sufficient explanation for any phenomenon. When we observe an apparent causal sequence, like event A preceding event B, we would like to know whether there is some real connection between A and B, and whether A must always precede B (a necessary condition), or whether A can precede B, but other events, C, D, etc., can also precede B (a sufficient condition). The complexity of ecological systems and their long time scale relative the to human lifetime make it difficult to obtain direct tests of such causal sequences in succession. Without such direct tests, it seems unlikely that we can learn what is both necessary and sufficient for the occurrence of any successional stage. But all is not futile. It is still possible to obtain sufficiency. The purpose of this chapter is to consider why a direct measurement of necessary conditions is unlikely to be obtained and how a sufficient explanation can be provided through theoretical investigations begun during the last decade.

What would be required for a definitive test of the causality of succession in a forest? Let us consider a test that would provide both necessary and sufficient conditions for the causality involved in species replacement during succession. From the viewpoint of population dynamics, such a test should consider rates of birth, growth, and death, but for the sake of clarity, let us consider such a test for growth rates alone.

How would we test whether one species facilitated, inhibited, or did not affect another? At first glance, the tests seems simple, as illustrated in Fig. 5.1. Suppose that one observes that species B follows A in a successional series. We merely need to set up a control plot and two treatments. These plots would be made up of a mixture of individuals of A and B. In the first treatment, one would remove A; in the second, one would remove part of the population of B, an amount equivalent to the abundance of A removed in the first treatment. We then would measure the diameter growth of individuals of B in the control and treated plots.

What results would we expect? First, we would expect that any thinning would lead to an increase in the growth of the remaining trees. After some time, the remaining individuals of B should begin to grow faster in the treated plots than in the control plot. If removing A leads to a greater growth rate of B than removing equal numbers and amounts of B, then A inhibited B; if removing A leads to a smaller growth rate of B than removing B, then A facilitates B. If the increases are equivalent (including the possibility that the growth rates remain the same as in the unthinned stand), A has no affect on B.

However, as Connell and Slatyer (1977) have pointed out, no such test could be done easily in a forest. There are tremendous operational diffi-

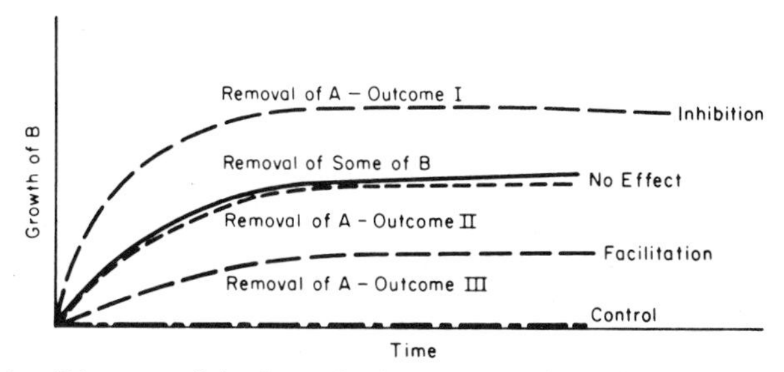

Figure 5.1. Diagram of the hypothetical results of a definitive test of the effects of one species on another during ecological succession. Suppose species A normally appears before B in a succession sequence. In the test, two treatments are set up. In one, all of A is removed. In the second, a portion of B equal to the abundance of A is removed. What might happen? If B grows better when A is removed than when a portion of B is removed, A inhibits B. If B grows more poorly with A removed (or does not appear at all), A facilitates B. If the treatments give equal results, A has no effect on B.

culties, because trees are so long lived. A truly definitive test would require an experiment of decades. For example, in the boreal forests of the Northern Hemisphere, the influence of early successional trees, such as white birch, *Betula papyrifera*, can extend for 50–100 years after a clearing. An experiment that lasted 1–3 years, even if it showed a difference in growth among treatments over that period, would not be definitive, since it is possible that such a comparatively short-term effect would disappear soon after and the treatments remain indistinguishable afterwards.

Second, there are more complex outcomes. One expects that the increment in growth resulting from thinning will extend for some time period equivalent to the time that species A would have normally occupied the forest; after that time the growth increments of the trees in the treated plots would return to the same level as those in the control plots. It is also possible, however, that the difference in the growth rates of individuals of species B in treated plots would continue for a much different time period, longer or shorter than the one observed in its normal succession, making the growth response a complex curve. That is, it is possible that there could be results that would correspond in no simple way to our original, apparently simple, hypotheses.

Third, it would be unlikely that one could find, without having planted them ten years ago, a two-species stand that would be sufficiently uniform to allow for a strict and clear test. Temporal variations in weather and spatial variations in soil, hydrology, and the distribution of trees could complicate and obscure results. For example, a series of cold years immediately following initiation of the test might lead to an obscuring of the results, so that the definitive test would require an even longer time period.

Fourth, the response of a population of one species of trees to a population of another would be affected by the age structure of each. Thus, a truly definitive test would require a span of tests with different age structures.

Fifth, tests would be required to determine whether inhibition or facilitation resulted from the presence of a unique species, or was merely a response to a class of other species. This means that the tests would have to be repeated with species other than A to learn whether A is uniquely required for B, or had a unique effect on B.

These complexities make clear that however simple such a definitive test might seem in concept, it would be almost impossible to carry out by a brute force, empirical test in the field. In some cases, a fortuitous, previously established natural or human-induced set of forest stands may allow a practical test. One could also hope that a very stable human society with consistent long-term research goals would set up such an experiment. If so, and the possibility is remote, the results lie past our lifetimes and are unavailable to us as definitive experiments for current theory.

Reviewing these problems, one's immediate response is to look to highly controlled laboratory experiments in growth chambers, where shading, nutrient limitations, or other influences could be imposed in substitution for a real species A. However, similar attempts to transpose ecological tests from the field to the laboratory have met with great difficulties. The laboratory tests seem never to provide both necessary and sufficient conditions. A failure to observe some response in the laboratory can always be attributed to the failure of the experiment to include some hitherto unknown factor. A positive result can always be reduced to a sufficient, but never to a necessary, explanation.

This discussion is not meant to imply that a definitive laboratory or field test is theoretically impossible, but to suggest that operationally our chances of conducting a definitive test of either kind is vanishingly small and can, for all practical purposes, be eliminated as a feasible part of a scientific method. This is to suggest that, in the case of the causality of succession, one is unlikely to obtain a definitive *necessary* explanation and must consider other approaches. The process of succession seems, in part, unknowable, and the questions about succession partially unresolvable. There are, I conjecture, deep implications to this limit as to what we can know about ecological succession, and I suggest that it is time these implications were carefully and precisely investigated by the means that are available. Given the limitations described above, we are forced to turn to theoretical constructs and approaches, which for the problems of succession must be examined in a mathematical model, the subject of the next part of this chapter.

Background to the Model

During the past decade, we have developed and used the JABOWA forest model to study numerous ecological processes (Botkin et al. 1972; Aber et al. 1979). In this chapter, the model is used to provide insight into the casual sequences of succession. To understand this use of the model, it is necessary to understand the assumptions and the scientific understanding implicit in the model. The approach to the development of this model was to begin with the simplest and fewest assumptions, adding complexities until all the observations about the forest modeled were accounted for—both qualitative "natural history" and quantitative published data. The required assumptions, given the great complexity of natural history and the confusion and complexity in the discussion of succession, were surprisingly few and simple. They are detailed elsewhere in several articles (Botkin et al. 1970, 1972; Botkin 1976).

Briefly, the assumptions used in the JABOWA model include the following. An individual tree of each species has an optimal or maximal growth curve, specific to that species and relating growth rate to size. This rate of growth is directly proportional to the amount of photosynthates that can be produced by the leaf tissue of the tree, and is inversely proportional to the amount of photosynthates required for the sustenance of existing nonphotosynthetic tissue. The rest of the photosynthate, according to simple algebraic formulation, goes into the production of new vegetative tissue. This optimal growth is decreased by suboptimal conditions of light, temperature, water, or any factor that a user of the model might want to add. The response to light is a monotonic curve, with an additional influence from temperature. Leaf weight is calculated from diameter. Each species has certain intrinsic population or life history characteristics, including rates of birth, growth, and death. Consistent with

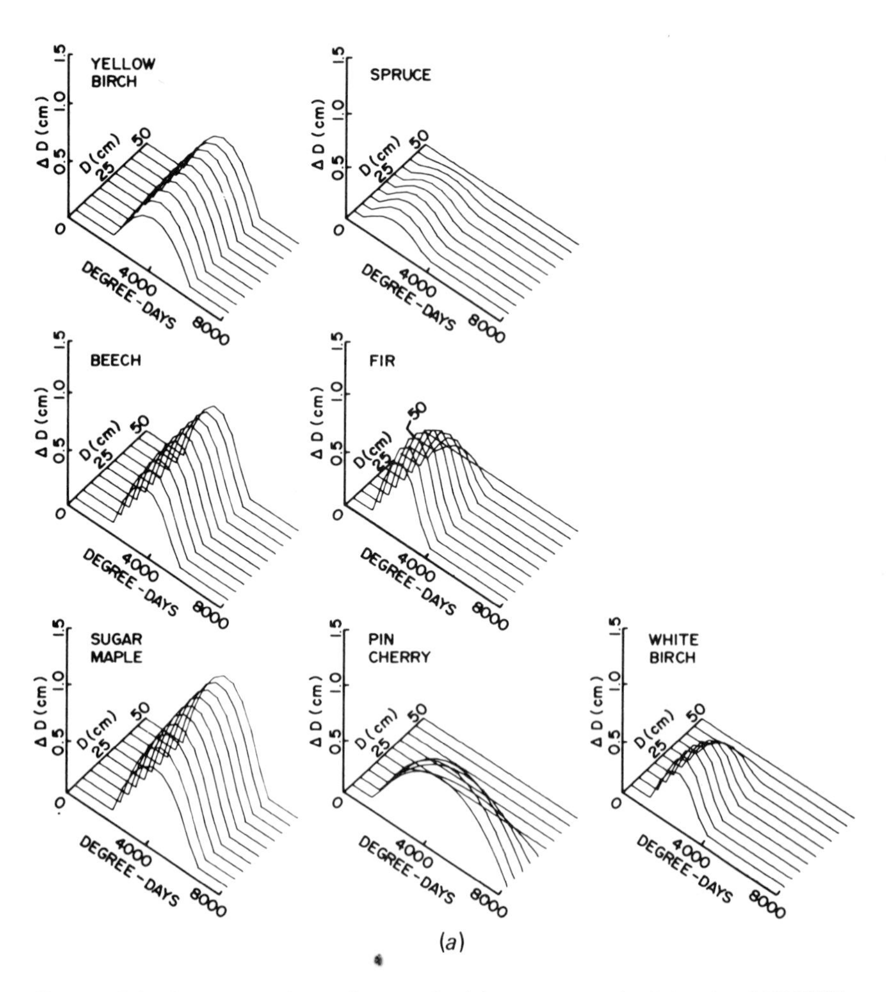

Figure 5.2. Representation of several niche axes as cited in the JABOWA model. The graphs show diameter growth as a function of (a) tree diameter and growing degree-days and (b) light and growing degree-days. In b the diameter growth represents the maximum possible for each species (from Botkin 1976).

everyone's view of natural history, pioneer species produce abundant and widely scattered seeds, grow quickly, and live a short time. Species of old-age forests are characterized by the production of fewer seeds, which germinate in less abundance, and which grow more slowly but live a longer time.

With regard to inter- and intraspecific competition, the primary assumption is that trees compete primarily for light, although some species grow much better in bright light than others, and the ones that grow well in bright light grow poorly in dim light. This well-known distinction between shade-tolerant and -intolerant species, as defined by a great deal of evidence, is the primary mover in the interaction among species in the first and second versions of this model (a third version has the competition for soil nitrogen added).

To the above assumption we must add the qualification that the

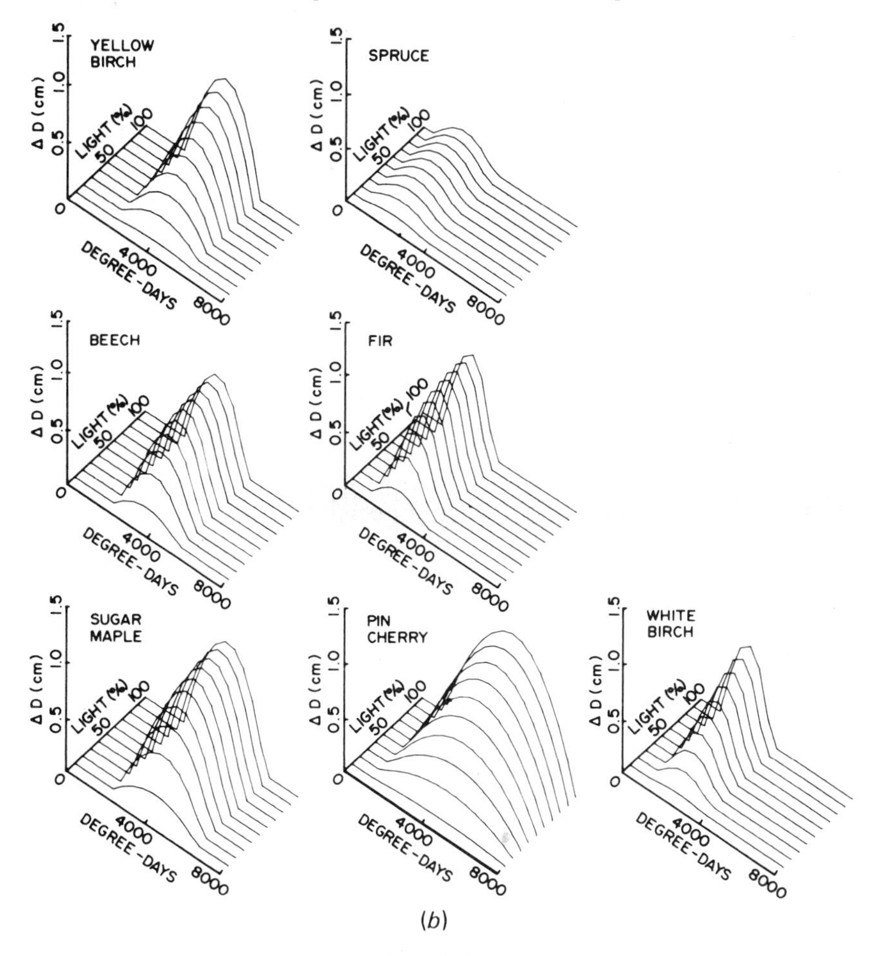

Fig. 5.2b

pioneer species have seeds which sprout and grow rapidly under bright light, but that this may be affected by soil and soil moisture conditions. Seeds of these species are sensitive to light, and their germination is suppressed by the high far-red/red light ratio found in deep forest shade. Seeds of late-successional species, particularly two used in the original version of the model (*Fagus grandifolia* and *Acer saccharum*), have been shown to be unaffected by light (Schopmeyr 1974). As concerns mortality, we assume that a tree must grow well and must continue to grow if it is to survive. A poorly growing tree becomes subject much more readily to disease and death, but there are certain extrinsic and random events such as fire and windstorms that could cause the death of any tree, healthy or not.

These assumptions were set down in an explicit fashion. In some sense it is unfortunate that this had to involve the use of computers, because during the past decade ecologists in the United States have been substantially separated into those who do and those who do not. Although a very long discussion of succession has occurred in ecology throughout the twentieth century (McIntosh 1980a), some believe that we have learned little in the last 80 years and have merely replaced one term with another. On the contrary, I believe that in the last ten years something of value has been learned from the careful formulation of the JABOWA model.

First, this model, which was described briefly above, serves in an integrating role—as a way of summarizing and bringing together the natural history, physiological, and ecological information that has been gathered over many years by many people. One of the attributes of ecology that is appealing is the natural history detail. Often our theoretical approaches, in an attempt to seek simplicity, omit what is most interesting to us and what originally interested us in the subject. The problems and implications of this have been discussed elsewhere (Botkin et al. 1979). Work in ecology often seems to lead to many diverse approaches. Here we perceive a method to bring these approaches together.

Within the field of ecology there are several schools that tend to be considered as distinct and to be pursued separately by different groups. These include the study of population dynamics, the pursuit of niche theory, the study of succession, the study of communities, and the study of ecosystems. An interesting advantage of the simple set of assumptions that I have reviewed and make up the JABOWA model is that they are easily translated in quite a formal fashion into the views of each of these schools. For example, the assumptions can be viewed as assertions of an explicit and quantitative representation of a multidimensional niche for tree species (Fig. 5.2). They show that the niche of a species, or at least several axes of the niche, can be represented simply. Here we see represented light and temperature and the response in terms of the diameter growth of six tree species. These two axes are most important, and along with water, soil characteristics, and in some cases the nitrogen in the soil, provide an entirely sufficient set, as far as our observations allow us to pursue succession, to account for all that we observe about succession.

We can also translate these assertions into standard population dynamics. The assumptions relate to the population dynamics of trees. They concern birth rates, growth rates, death rates, and age structure, as modified by competition among individuals of different species for light, temperature, and moisture. The original version of the model involved only those three environmental factors. However, it is a simple matter, when data are available, to add new axes, and the response of the growth curves, as shown in Fig. 5.3, have been added to a more recent version in order to consider the effects of nutrient removal from forest soils on forest growth (Aber et al. 1978, 1979).

As will be seen in the next section, the model described above is now relatively verified, or well verified, particularly when compared with many other models or theories in ecology. It is, therefore, possible to examine the model and ask what it implies about the causality of succession.

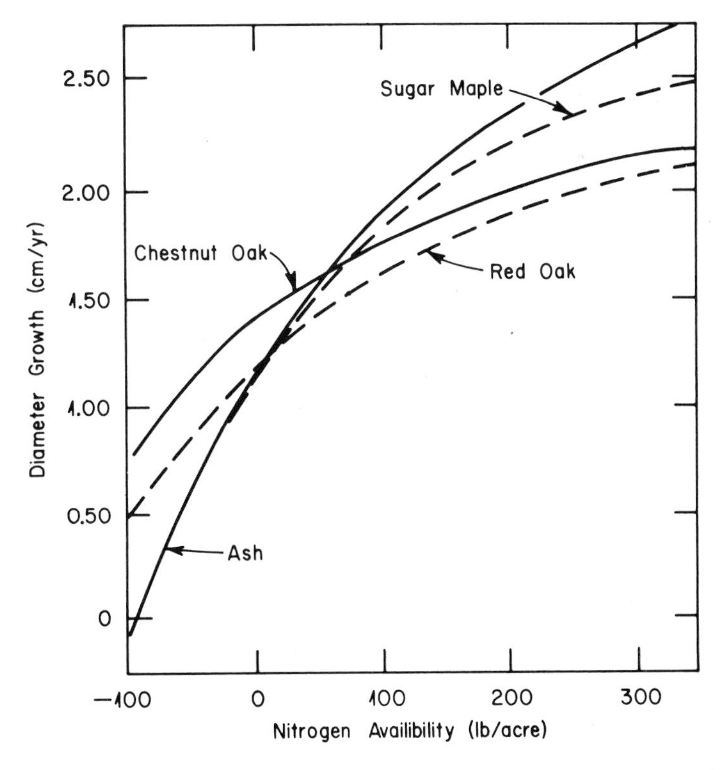

Figure 5.3. Growth responses of several tree species of the eastern deciduous forests of North America to nitrogen fertilization (data from Mitchell and Chandler 1939).

For the condition in which it is verified, the model provides a sufficient explanation; that is, it provides a set of assumptions that are sufficient to account for successional patterns of species replacement. The model is not a necessary explanation because it in no way eliminates the possibility that a model developed from a completely different set of assumptions could give equivalent results. The JABOWA model provides a challenge to empiricists to test whether contrary sets of assumptions lead to similar results. It also provides a challenge to theoreticians to investigate whether alternative sets of assumptions, made explicit in a model, give equivalent results.

In this chapter I shall proceed to review the verification, including the presentation of some new evidence for the verification of the model. With this background, I consider the causality implied by the model for species replacement and indicate what this suggests to us about succession.

Verification

I have just asserted that the JABOWA forest model is well verified for an ecological model. What do I mean by that statement? Verification for ecological theories and models can never be complete (O'Neill 1976). The model is general for forests, at least as indicated by the successful modifications of the model, making it applicable to forests of the southern Appalachians in the United States and eucalyptus forests in Australia (as indicated in Shugart et al. Chapter 7). It appears to be *realistic* in that it qualitatively reproduces all of the major characteristics known for forest succession. This qualitative realism of the model has been described elsewhere and need only be briefly summarized here (Botkin et al. 1970, 1972; Aber et al. 1978, 1979). It includes realistic representation of the competition among trees, the process of succession under a variety of climatic conditions in terms of the dominance of specific tree species, the change in biomass and nutrient content over time, and the effects of fertilization and disturbance on forests under a variety of soils and climate. Past discussions of the model have not included attempts to verify the model with regard to some details of the growth of forest trees on specific sites. A complete verification would require a real time series for real plots, including measures of the growth of each tree and the local rainfall and temperature regions. No such time series existed.

Lacking a real time series, a partial verification was attempted. The method of verification was as follows. The model was originally designed for the eastern deciduous and boreal forests of New England, for which parameters were devised from data independent of those used in the verification here. For the detailed verification, plots were chosen in the White Mountains of New Hampshire to represent a variety of soils, slope, elevation, aspect, and successional stage. Within these large areas points were selected at random. At these points 10-m^2 plots were established. All trees on the plots that were saplings or larger were measured for height and diameter, and the trees were aged from growth rings. Dead trees were recorded, seedlings subsampled, and other information including the soil depth, percentage of rock on the soil surface, the height of the lowest living branch, and the width of the canopy of every tree recorded. Although more than 50 plots were originally located, in many cases inevitably some of the data were missing. Some trees were rotten in the center, and cores were not successfully obtained. The growth rates of all trees were obtainable on only a few plots. This was taken as a minimal criterion for an attempt at detailed verification.

The last 10 years of growth for every tree on a plot was measured, and and the diameter of every living tree on the plot 10 years before was calculated by subtracting the last 10-year diameter growth from the current diameter. This provided a partial description of the plot as it looked 10 years in the past. Verification was carried out by using the reconstructed status of the plots 10 years ago as the initial condition. The JABOWA model was used to grow the plot forward in time, and the results compared at the end of the tenth year with the observed diameters.

This is clearly only a partial verification. Some trees that died and disappeared, or were present as rotting logs on the day of the sample, were not counted, and their effect on succession and competition could not be determined. This would be a particular problem for small trees. However, these omissions would tend to increase the errors in the estimated growth and decrease the likelihood of the success of the verification. Furthermore, no plot had its own weather records, and records from a single station in the Hubbard Brook Forest were used. Therefore, any agreement between the model and the data would be an underestimate of the precision of the model for a complete verification.

Results of the Verification: Individual Tree Growth

Prior to beginning the verifications we knew that the parameter data base for one species, balsam fir, was more restricted than those for other species, in that the balsam fir parameters were derived from a much smaller range of size classes. The plots utilized included some from early and late-successional stages, plots on almost level ground and steep slopes, plots facing southwest, northwest, and southeast, and plots with thin and deep soils.

The relative importance of species can be compared for these verification plots through changes in total basal area by species (Table 5.1). In

Table 5.1. Observed and predicted changes in basal area of six plots over a 10-year period.

A. *Plot 6*: Northeast aspect and slope of 18%. Hardwoods and softwoods (particularly sugar maple) were regenerating. Main species red spruce, balsam fir, sugar maple, white birch; other species scattered throughout plot. Average soil depth was 20 cm; maximum soil depth was 70 cm.

B. *Plot 8*: Southwest aspect and slope of 35%. Dominated by sugar maple, yellow birch, and red spruce. Two beech were located just south of plot with thick regeneration beneath. Balsam fir, white birch, mountain maple, and striped maple were also present. Average soil depth was 20 cm; maximum soil depth was 70 cm.

C. *Plot 10*: A south-facing slope (35%) with numerous large trees (mainly spruce and yellow birch) and lots of seedlings and small saplings (spruce, beech, and sugar maple). No yellow birch regeneration was seen. Fir was uncommon on plot. Soil was very rocky, with depth ranging from 10 to 50 cm.

D. *Plot 20:* A slope (25%) with a southern aspect at the top of a knoll. Red spruce and white birch were the dominant species, although several others were moderately common. Regeneration of fir and striped maple was common; that of birch, fairly common; and that of spruce, beech, and mountain maple, uncommon. Soil depth ranged from 15 cm to 1m, with few rocks.

E. *Plot 21*: Plot with slope of 5%, dominated by red spruce, with balsam fir also present in fair numbers and occasional white birch. Canopy was fairly open (many dead stems on the ground, apparently blown over after having succumbed to suppression). Regeneration was abundant - mostly fir, but also spruce, white birch. Red maple seedlings common - seed from vicinity of plot 22 (400 cm). Soil depth ranged from 20 to 80 cm; no rocks.

F. *Plot 31*: Plot on a small level knoll with southern aspect. Few rocks; soil depth ranged from 5 to 30 cm. Beech was dominant and was also the only species which showed good regeneration. Spruce saplings were common, but all large spruce were dead. Pin cherry and yellow birch appeared to be drying out. Some yellow birch regeneration (small seedlings) on rotting spruce log.

JABOWA validation run (deterministic)

Plot number	Elevation (m)	Soil depth (m)	Rock (%)	Growing degree-days	Evapotranspiration index
6	546	0.2	5	3188.2	235.4
8	838	0.2	4	2593.9	232.6
10	506	0.3	15	3268.6	303.1
20	664	0.3	2	2946.8	336.3
21	625	0.4	0	3027.2	370.9
31	686	0.1	1	2903.4	180.4

Species	Total basal area year 1 (cm^2)	Total basal area year 11 (cm^2)	Observed basal area year 11 (cm^2)
Plot 6			
White birch	873.0	1136.1	1075.8
Yellow birch	135.1	167.5	169.7
Balsam fir	957.9	1466.8	1044.6
Mountain maple	27.4	53.4	31.4
Sugar maple	712.6	1082.3	939.3
Red spruce	649.2	718.4	776.5
Totals:	3355.6	4624.7	4037.3
Plot 8			
Yellow birch	679.6	818.9	820.3
Balsam fir	2.5	45.5	2.5
Mountain maple	10.0	16.2	15.2
Sugar maple	870.4	1130.3	1151.8
Red spruce	153.6	228.0	188.6
Totals:	1716.3	2239.1	2178.5
Plot 10			
Beech	116.3	215.2	169.4
Yellow birch	829.7	1013.4	1008.9
Balsam fir	26.4	66.9	31.1
Sugar maple	485.4	675.5	606.9
Red spruce	1889.6	2025.2	2063.8
Totals:	3347.3	3996.4	3380.3
Plot 20			
White birch	801.9	1304.3	1106.3
Yellow birch	321.7	387.3	466.2
Pin cherry	419.5	595.1	647.8
Balsam fir	447.3	904.4	597.4
Red spruce	1197.9	1465.7	1574.2
Totals:	3188.5	4657.0	4392.0
Plot 21			
White birch	214.4	387.9	257.8
Balsam fir	356.8	686.5	740.1
Red spruce	3057.6	3551.0	3726.4
Totals:	3628.9	4625.5	4724.4
Plot 31			
Beech	1907.1	2505.7	2827.4
Yellow birch	82.5	118.6	133.4
Pin cherry	108.2	173.7	193.9
Balsam fir	4.9	39.6	7.5
Striped maple	13.4	23.9	51.5
Red spruce	104.1	146.8	149.9
Totals:	2220.6	3008.6	3363.6

most cases, the estimate of basal area is within 10% of that observed. For example, in a low elevation site (Plot 6, elevation 546 m above sea level), sugar maple contributed a basal area of 713 cm^2 at the beginning, was observed to contribute 939 cm^2 10 years later, and was predicted to contribute 1082. Yellow birch began at 135 cm^2 of basal area (calculated from diameter at breast height-dbh), was observed at 169 in 10 years, and was predicted to reach 167. Red spruce began at 649 cm^2, was observed to reach 776, and was predicted to reach 718. White birch began at 873cm^2, reached 1076, and was predicted to reach 1136. The total basal area on the plot began at 3356 cm^2, was observed to reach 4037 in 10 years, and was predicted to reach 4624. Similar results were obtained for a high elevation of 838 m. Again, the predicted basal area by species was generally within 10%, but as was observed previously, the projections were less accurate for balsam fir than any other species. Sugar maple was projected to reach 1130 cm^2 and actually reached 1152. Balsam fir was projected to reach 45.5 cm^2, and only a small single balsam fir of 2.5 cm^2 was found. However, in this case the original balsam fir tree had died and was not counted, and another sapling was found on the plot.

In an early successional area (Plot 20), the model gives comparable results. Basal area for yellow birch, pin cherry, spruce, and white birch are accurately projected, but basal area for balsam fir is much less accurately projected. The total basal area projections are within 10% of the observed basal area.

The estimated growth of individual trees on the plots also can be compared with the observed growth and these are presented for three plots (Table 5.2). One would expect this estimation to be subject to the greatest error, because no averaging was allowed and because the effect of competitors that died during the last 10 years (and were not counted) would magnify errors. In comparing initial and final sizes, the results seem encouraging, as the predicted final dbh measurements are within 10% or 20% of the observed values. For example, on a plot at an elevation of 546 m with a shallow soil (Table 5.2), only early and late-successional species are found, including sugar maple, yellow birch, mountain maple, white birch, balsam fir, and red spruce. The model reproduces the growth of most trees over the 10-year period within 10%. The dominant large white birch tree was observed to reach 26.4 cm after 10 years and was predicted to reach 26.2. A large sugar maple was observed to reach 19.7 m and was predicted to reach 22.4. Another large sugar maple was observed to reach 20.8m, having been predicted to reach 22.4. Small trees were also predicted with reasonable accuracy. A small sugar maple that grew from 5.2 to 8.7cm was predicted to have reached 8.2. Interestingly, the balsam fir projections are again subject to the largest error.

In another plot at a higher elevation (838m), with an equivalently shallow soil (Plot 8, Table 5.2), similar results were obtained for individual tree diameters. A sugar maple tree that began at 11.7 cm and grew to 14.1 was predicted to grow to 13.4. A yellow birch that began at 22.2 cm and was predicted to grow to 24.3 grew to 23.2. A balsam fir that began at 1.8 and was observed to grow to 7.8 was predicted to grow to 7.6.

A third plot, which contained the early successional species pin cherry, white birch, and yellow birch as well as balsam fir and red spruce, shows equivalent projections (Table 5.2). For example, a pin cherry that began the period at 17 m and grew to 20.4 was predicted to grow to 20.0. Another pin cherry observed to be 8.2 cm grew in 10 years to 11.5 and was projected to grow to 10.2.

Although comparison of predicted and observed final sizes of individual trees gives good agreement, there is another way to approach the analysis: to compare predicted and observed changes in diameter. Since over a ten-year period trees grow only a few centimeters, and since many data are unavailable for the plots, including the weather on the sites, this is a stringent test. On this basis the model predicts on the average only within 100% of the observed. Although the concept of the test appeared initially straightforward, in actuality the attempt to use sites without actual annual measures of weather and tree conditions suggests that the tests are insufficient on the basis of changes in diameter. The model reproduces longer term forest dynamics realistically, but its precision for short time spans remains to be determined.

Implications for Analysis of Causality in Succession

Given a model verified to this extent, one can then consider its assumptions concerning the causality of succession. The model represents a sufficient set of assumptions to the degree it is verified. In pursuing the discussion of causality, we shall use the division of the ideas in such causality as discussed by Connell and Slatyer (1977).

Connell and Slatyer divide the possible explanations for the causality of succession into three basic types: facilitation, tolerance, and inhibition. In the facilitation type, the early successional species prepare the way for the later ones. In the tolerance type, later succession species are more tolerant of the low resources that occur later in succession. In inhibition, the early successional species prevent the occurrence of the later ones, which can only enter when the earlier ones have passed through their life cycle. Connell and Slatyer discuss modifications of these types following a disturbance. To these we must add a fourth, a null hypothesis: that the temporal change is merely a consequence of the individual life history charcteristics of the trees.

To examine which of these the model supports (or fails to support), one can remove early successional species from the model and examine the the responses. The simulations were conducted by utilizing a plot with an elevation of 972 m and a deep soil (2 m) and zero slope. This places the plot within the boreal forest, which has fewer species and can be used

Table 5.2. Observed and predicted growth of individual tree diameters for three of the plots given in Table 5.1. (For an explanation, see text and Table 5.1.)

JABOWA validation run (deterministic)

Plot number	Elevation (m)	Soil depth (m)	Rock (%)	Growing degree-days	Evapotranspiration index
6	546	0.2	5	3188.2	235.4

Species	Tree diameters by species (cm) Observed year 1	Observed year 11	Predicted year 11
White birch	23.2	26.4	26.2
	10.7	13.9	13.0
	20.3	21.9	23.3
Yellow birch	13.1	14.7	15.1
Balsam fir	3.4	4.9	7.3
	29.7	30.4	33.7
Mountain maple	2.8	3.6	4.8
Sugar maple	18.2	19.7	22.4
	18.2	20.8	22.4
	5.2	8.7	8.2
	14.0	15.5	17.8
	4.3	7.7	7.1
Red spruce	26.7	29.1	27.9
	2.5	4.1	3.3
	2.9	2.9	3.6
	10.1	10.8	11.1

JABOWA validation run (deterministic)

Plot number	Elevation (m)	Soil depth (m)	Rock (%)	Growing degree-days	Evapotranspiration index
8	838	0.2	4	2593.9	232.6

Species	Tree diameters by species (cm) Observed year 1	Observed year 11	Predicted year 11
Yellow birch	22.2	23.2	24.3
	19.3	22.5	21.2
Balsam fir	1.8	7.8	7.6
Mountain maple	3.5	4.4	4.5
Sugar maple	11.7	14.1	13.4
	3.7	5.1	4.9
	12.4	14.1	14.1
	1.8	2.9	2.8
	10.0	12.8	11.6
	8.2	8.6	9.7
	4.5	5.7	5.8
	18.5	20.1	20.3

	.6	1.7	1.2
	12.4	14.0	14.1
	10.5	12.7	12.1
Red spruce	3.5	4.8	5.1
	2.3	3.5	3.7
	5.3	6.2	7.0
	12.1	12.9	14.1

JABOWA validation run (deterministic)

Plot number	Elevation (m)	Soil depth (m)	Rock (%)	Growing degree-days	Evapotranspiration index
20	664	0.3	2	2946.8	336.3

Tree diameters by species (cm)

Species	Observed year 1	Observed year 11	Predicted year 11
White birch	16.0	20.4	21.2
	14.3	18.5	19.0
	23.6	25.5	29.1
Yellow birch	13.3	16.2	14.6
	15.2	18.2	16.6
Pin cherry	17.0	20.4	20.0
	8.2	11.5	10.2
	13.2	16.5	15.8
Balsam fir	9.7	11.1	14.7
	11.3	12.3	17.1
	18.6	22.1	25.2
Red spruce	5.2	6.3	6.0
	12.0	14.2	13.1
	6.4	7.7	7.3
	5.5	7.7	6.4
	8.3	10.5	9.3
	2.5	6.9	3.2
	2.3	3.1	3.0
	7.0	8.8	8.0
	18.3	19.4	19.9
	12.4	15.0	13.7

to demonstrate the principles in a simpler manner. [For an explanation of the climatic regime, see Botkin et al. (1972).] These forests contain choke cherry (*Prunus virginiana*), white birch (*Betula papyrifera*), red spruce (*Picea rubens*), and balsam fir (*Abies balsamea*). Figure 5.4a illustrates the "normal" process of succession projected by the model and succession in a forest where the pioneer birch and cherry are suppressed so that only the late-successional fir and spruce remain. As one would expect, the model projects that the removal of the early succession species reduces greatly the total increase of the basal area in the first century. After that, the "control" and "treated" forests are indistinguishable.

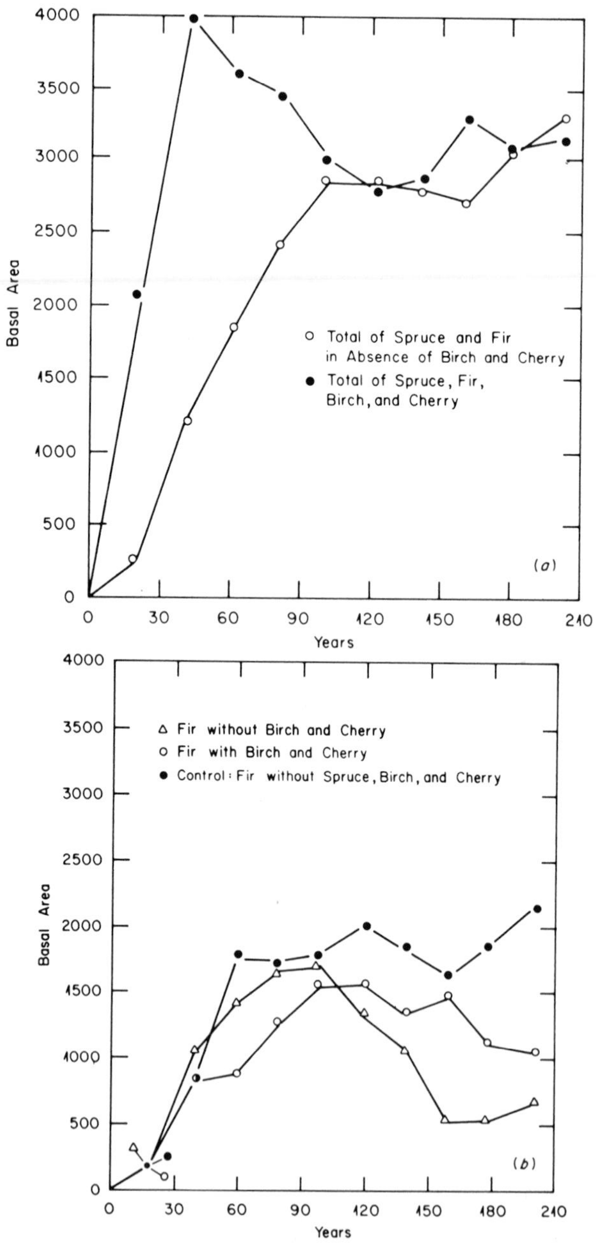

Figure 5.4. The response of the JABOWA version II simulation forest model to the removal of one or more species. The model is operated under a constant climate characteristic of the boreal forest zone in New Hampshire, found at elevations between 2700 ft (900 m) and 4500 ft (1500 m). Each point is the mean of ten trials that begin with the same initial conditions. (a) The total basal area; (b) the basal area, for fir; (c) for spruce; and (d) for white birch.

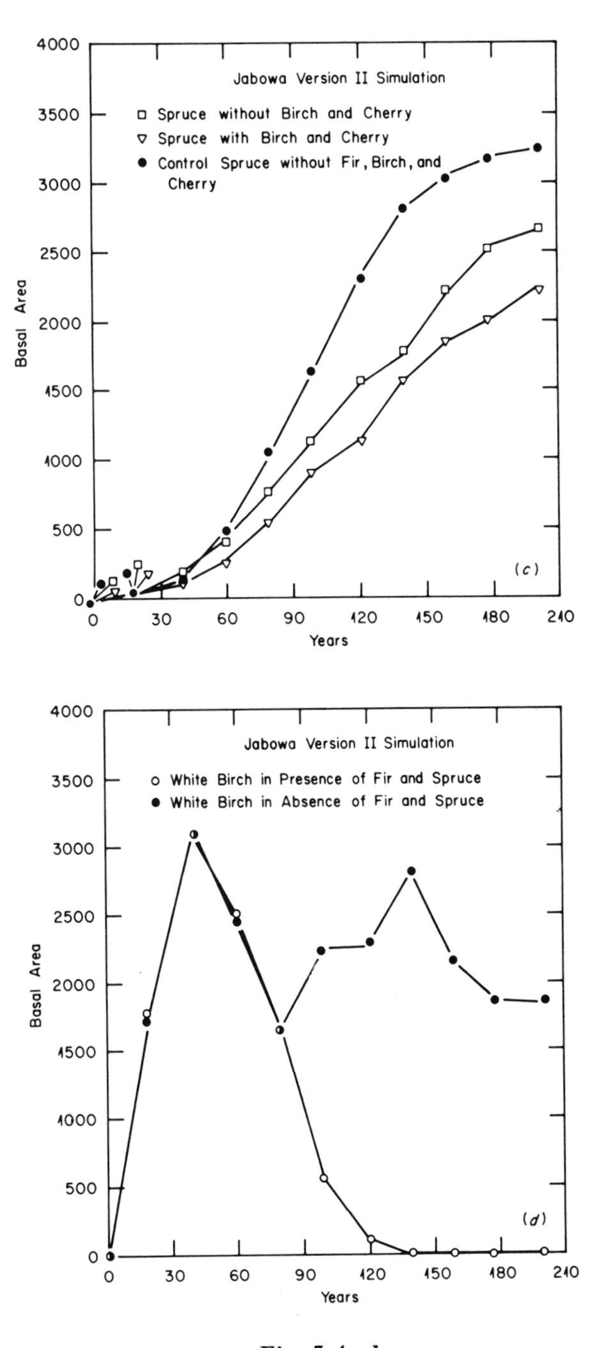
Jabowa Version II Simulation
Spruce without Birch and Cherry
Spruce with Birch and Cherry
Control Spruce without Fir, Birch, and Cherry
Basal Area
Years
(c)
Jabowa Version II Simulation
White Birch in Presence of Fir and Spruce
White Birch in Absence of Fir and Spruce
Basal Area
Years
(d)

Fig. 5.4c,d

Figure 5.4b illustrates the projected growth, according to the model, of balsam fir during the first 200 years following clearing for simulated plots where balsam fir occurs with its early successional competitors, birch and cherry, and where these are suppressed. When the early successional competitors are suppressed, the *quantity* of balsam fir, as indicated by its basal area, increases significantly but the *timing* of the increase to a maximum basal area does not change. Balsam fir in both cases achieves its maximum basal area in the first 100 years. Afterward, its contribution to the basal area of the forest decreases as it competes with spruce (a slower growing species, and the most slowly growing of all trees in the forest).

The effects of the removal of earlier succession competitors on spruce are shown in Fig. 5.4c. The results are similar in kind to those for fir. When the early successional species, birch and cherry, are removed, the quantity of spruce trees (as indicated by basal area) increases. When all other species are removed, including fir, the quantity of spruce increases even more rapidly. Yet the timing of the increase in spruce does not change with the removal of other species.

Finally, Fig. 5.4d illustrates the boreal forest composed only of early successional species, birch and cherry. The forest undergoes oscillations, with birches coming in, growing rapidly, then decreasing in abundance as the large individuals die and the forest floor remains too dark for regeneration. Finally a point is reached, after about 90 years, in which the death of mature birch trees opens up the forest sufficiently for a new regeneration of birch. An increase in basal area occurs again.

What do these results imply about the causality of succession? They make clear that qualitatively the role of a species is determined by its intrinsic life history and characteristics, and that quantitatively the importance of a species is affected by inhibition competition and tolerance. The inhibition competition is for light. The tolerance also relates to light. Facilitation is not involved.

The model makes use of the population life history characteristics of individual tree species, and the importance of the life history characteristics is made clear in the figures presented here. Spruce, the slowest growing of all species, increases in basal area more slowly than any other, regardless of the presence or absence of competitors. Fir, the next most slowly growing species and second in shade tolerance, accumulates slightly faster than spruce, but again the accumulation of basal area is controlled by the fundamental life history characteristics of the species. Only the quantity is affected by competition. The model tells us that inhibition competition and tolerance, coupled with intrinsic life history characteristics of species provide a sufficient set of mechanisms to reproduce forest succession, without an explicit facilitation mechanism. Tolerance is involved in that the later successional species are tolerant of low light conditions, and the early successional ones are less tolerant. Tolerance, inhibition, and life history characteristics together are sufficient to account for precision, generality, and realism in succession.

These results provide a challenge to empiricists and theoreticians to test whether alternative sets of assumptions can reproduce equally well the processes and events that take place during succession. We have learned something about succession. During the last decade a model, empirical tests of its projections, and an investigation of its assumptions, led us to know what is sufficient to explain one of the oldest, most interesting, and central questions in ecology.

Chapter 6

The Role of Disturbance in the Gap Dynamics of a Montane Rain Forest: An Application of a Tropical Forest Succession Model

Thomas W. Doyle

Introduction

Most discussions of forest disturbances have focused on large-scale phenomena (e.g., fire, volcanic eruptions) and man-induced changes (e.g., abandoned fields, cleared forest lands). Increasing attention, however, is being focused on the important influence of minor forest disturbances such as tree falls and canopy gaps on the composition, structure, and dynamics of forest communities. The relative importance of major and minor disturbances is, however, not yet clear. Observations in the wet tropics indicate that natural forest communities are composed of a "mosaic" of patches in different stages of successional and/or compositional maturity (Aubreville 1938; Richards 1952; Whitmore 1975, 1978). Watt (1947) interpreted such forest mosaics as a spatiotemporal patterning of phases linked in a regular cyclic succession. Each patch is thought to originate as a canopy gap formed by a localized disturbance such as tree falls or landslides. The degree of patchiness reflects the magnitude and periodicity of these perturbations (Whittaker and Levin 1977). High species richness, niche specialization, and chance factors also contribute to this dynamic patchwork in tropical rain forests (Knight 1975).

The pattern of forest disturbance governs the rate and recovery process of gap formation and replacement. The type, frequency, and extent of disturbance is determined not only by the environment but also by the nature of the forest system and its component species. More important than the result of any one disturbance type or single event is the community response to the cumulative effect of all the kinds and scales of disturbances perturbing a forest system.

Natural extrinsic disturbances, including cyclonic storms (Webb 1958; Whitmore 1974), wildfires (Budowski 1959, 1965), and geophysical eruptions (Beard 1945, 1976), are formidable forces in tropical regions, creating gap openings, and/or in some cases extensive forest destruction. In the West Indies, heavy winds and rainfall associated with tropical storms cause frequent blowdowns and landslides (Beard 1945; Wadsworth and Englerth 1959). Aside from major disturbances, senescence, suppression, disease, and damage by animals or other tree falls are principal elements contributing to tree death and gap occurrence. Superficial root systems and saturated soils, common in tropical wet forests, foster landslide and wind throws (Foster 1980). The weight of epiphytes has also been suggested as a major factor of tree fall in tropical forests (Strong 1977).

The regularity and severity of forest disturbances are major determinants driving the structural and compositional dynamics of the forest community. In a lowland rain forest in Costa Rica, Hartshorn (1978) found a higher frequency of natural tree falls than recorded for either dry tropical or extratropical forests. He attributed the low density of large trees in tropical wet forests to the high occurrence of tree falls. Many rain forest species are dependent on gap occurrence for regeneration, and consequently for their continued persistence in mature forest (Jones 1945; van Steenis 1958). This article presents an application of a tropical forest succession model, FORICO (Doyle 1980), that was developed to investigate the ecological role of major and minor disturbances on a montane rain forest in Puerto Rico, where localized perturbations from tropical storms are prevalent. A comparison of hurricane and nonhurricane simulations provides the basis for assessing the importance of disturbance on this forest system.

Description of the FORICO Model

The FORICO model is a detailed simulator of natural gap succession in the lower montane rain forest of the Luquillo Experimental Forest in northeastern Puerto Rico. Floristically, this lower mountain forest has been labeled the "*Dacryodes-Sloanea*" association (Beard 1944, 1955) throughout the Lesser Antilles; locally it is called the "tabonuco" forest type (Wadsworth 1951, 1957), after the dominant tabonuco tree, *Dacryodes excelsa*.

The FORICO model was constructed by modifying the FORET model developed by Shugart and West (1977) for southern Appalachia to conform to the characteristics of the tabonuco forest and its component species. This model is one of many gap models (Shugart and West 1980;

Shugart et al., Chapter 7; Botkin et al. 1972) designed to simulate the successional dynamics and competitive interrelations among individual trees on a spatial scale that approximates a large canopy gap (ca. 0.01–0.1 ha). In contrast to other forest model types, matrix tree models (Horn Chapter 4), for example, constitute a repetitive matrix multiplication of tree replacement probablities to predict forest succession for stands of unrestricted size. Also, parameter estimations for gap models are less dependent on successional field data than on observed biological and physical characteristics of individual species and model forest. The simulated gap of the FORICO model represents a 1/30-ha area. This gap is assumed to occur within an otherwise intact forest that provides an adequate and equitable seed source from all species. Key autecological parameters of 36 common tree species are used as input for the model (Table 6.1). Model processes, mainly tree birth, growth, and death, are expressed as functional relationships computed in yearly intervals. A flow diagram of model subroutines illustrates the order and function of these processes and support programs (Fig. 6.1). A general summary of the design and mechanics of FORICO and similar gap models by major subroutines follows.

The BIRTH subroutine (Fig. 6.1) determines the species and number of new seedlings that are established on the plot each year. Eligibility of model species for recruitment is based on their silvicultural characteristics and regenerative requirements for successful establishment. These vary between models, but consider such species-specific factors as dispersal mechanisms, dormancy periods, grazing palatability, shade tolerance, and site requirements. Selection of species and seedling numbers is modeled as a stochastic process. Each new tree is assigned a randomly determined seedling size. At the initiation of a simulation trial or following a significant turnover or disturbance, the model continues to seed in new trees until a minimum leaf area or biomass level is attained.

The GROW subroutine (Fig. 6.1) calculates and increments the diameter growth of each tree. Under optimum conditions, annual tree growth is directly related to the amount of sunlight that the tree receives in leaf area times the energetic cost of maintaining its size. Species' maximum diameter and height figures are used to derive coefficients B_2 and B_3 (Table 6.1) from a parabolic expression (Ker and Smith 1955) for estimating tree height from tree diameter. Tree size, B_2 and B_3, and the growth parameter G (Table 6.1) are used to calculate the maximum expected diameter increment for each tree. The species growth constant, G, is approximated for each species as either a function of its maximum growth rate or of its maximum age. To determine actual tree growth, however, the optimum growth increment is reduced by growth-inhibiting factors, namely shading, competition, climate, and other environmental stresses as they apply. Taking into account the stand structural profile, the model determines the vertical distribution of tree heights in 0.1-m units and the leaf area of each tree. The sum leaf area above each tree, obtained from adding the leaf area of all taller trees, is then used to estimate the amount of light available to each tree for growth due to shading. The distribution of photosynthetic light through the stand canopy is determined from a negative exponential function defining the relationship between leaf area and light extinction. After determining the amount of light available to each tree, the model computes the growth reduction associated with the decreased photosynthetic potential under shaded conditions. Light saturation and compensation values for species in general tolerance classes (Tol, Table 6.1), are used to generate photosynthesis curves that define the shade effect on a scale from 0 to 1. The tallest tree, for example, receives full sun exposure, and therefore computes a value of 1, indicating no reduction in growth. Competition for space and nutrients also limits potential tree growth. The model employs a linear function that proportionally reduces tree growth due to crowding as the total stand biomass approaches a maximum biomass level recorded for the model forest. This function is likewise scaled from 0 to 1, whereby an open stand would have a value of 1. Climate is another important factor influencing growth. Each model treats climate in one of two ways, depending on the derivation of the growth constant G. If G is calculated from actual field measurements of species growth, as in the FORICO model, then a climate factor is intrinsically incorporated in the optimal growth increment. If, on the other hand, G is approximated from maximum age figures, then climate is modeled as a function of upper and lower temperature tolerances based on species' distribution ranges. Each year a random degree-day value is computed from a normal distribution for the particular geographic location of the model forest. This is compared with the degree-day optimum and limits for each species, which in effect scales the growth reduction factor from 0 to 1.

The KILL subroutine (Fig. 6.1) eliminates trees from the simulated stand as a stochastic function of age- and growth-related deaths. Each tree is assumed to have an intrinsic mortality rate such that under normal conditions only 1–2% of all seedlings in a cohort ever live to reach their life expectancy. The probability of a tree dying in any one year is expressed as a negative exponential of the above survival percentage and the species' maximum age. A random number is generated and compared with this probability to determine whether each particular tree survives or dies in that year. Suppressed individuals that fail to maintain a minimal growth rate are subjected to an additional probability of dying. This additional mortality factor has the effect of allowing only 1% of a suppressed cohort to survive 10 years.

Subroutine HURCAN (Fig. 6.1) of the FORICO model stochastically mimics cyclone storm effects common to West Indian forests. Wadsworth and Englerth (1959) reported from historical records that 50 hurricanes have passed over Puerto Rico over the last 450 years. This indicates a re-

Table 6.1. Species list and autecological parameters used in the FORICO model.[a]

Species name[b]	D_{max}[b]	H_{max}[b]	B_2	B_3	Tol[c]	Age[d]	G[e]	S_{min}	S_{max}
Alchornea latifolia	46	1524	60.30	0.6554	2	58	240.19	0.05	0.12
Alchorneopsis portoricensis	46	1524	60.30	0.6554	2	56	246.40	0.05	0.12
Buchenavia capitata	122	2438	37.72	0.1545	2	120	182.85	0.05	0.12
Byrsonima coriacea	46	1829	73.56	0.7996	3	46	202.78	0.12	1.00
Casearia arborea	15	914	103.60	3.4533	3	47	175.37	0.12	1.00
Casearia sylvestris	10	457	64.00	3.2000	2	44	99.00	0.05	0.12
Cecropia peltata	61	2134	65.47	0.5366	3	50	271.12	0.12	1.00
Cordia boringuensis	13	610	72.76	2.7988	1	43	131.99	0.004	0.75
Cyathea arborea	13	914	119.53	4.5976	3	17	461.89	0.12	1.00
Dacryodes excelsa	152	3048	38.30	0.1259	1	205	132.34	0.004	0.05
Didymopanax morototoni	46	1829	73.56	0.7996	3	52	267.65	0.12	1.00
Drypetes glauca	15	914	103.60	3.4533	1	56	147.19	0.004	0.05
Eugenia stahlii	30	1829	112.80	1.8800	1	103	154.85	0.004	0.05
Euterpe globosa	20	1524	138.70	3.4675	1	22	587.34	0.004	1.00
Guarea ramiflora	15	762	83.33	2.7777	1	41	169.56	0.05	0.12
Guarea teichiloides	91	2286	47.23	1.2095	2	121	169.56	0.004	0.05
Hirtella rugosa	8	610	118.25	7.3906	1	32	164.62	0.004	0.05
Homalium racemosum	61	2134	65.47	0.5366	2	107	178.41	0.05	0.12
Inga laurina	46	2134	86.82	0.9437	1	75	229.10	0.05	0.12
Inga vera	46	1524	60.30	0.6554	2	65	271.38	0.05	0.12
Laetia procera	30	2286	143.26	2.3877	3	86	228.60	0.05	0.12
Linociera domingensis	30	1829	112.80	1.8800	1	65	247.67	0.05	0.12
Manilkara bidentata	122	3048	47.72	0.1955	1	157	171.76	0.004	0.05
Matayba domingensis	46	1829	73.56	0.7996	1	70	232.60	0.004	0.05
Miconia prasina	10	762	125.00	6.2500	2	26	257.17	0.05	1.00
Miconia tetrandra	30	614	92.46	1.5411	1	53	257.17	0.05	0.12

Ocotea leucoxylon	25	1524	110.96	2.2192	2	35	387.09	0.05	0.12
Ocotea moschata	76	2438	60.55	0.3983	1	76	57.03	0.004	0.05
Ormosia krugii	61	1829	55.47	0.4547	2	84	197.89	0.05	0.12
Palicourea riparia	8	457	80.00	5.0000	2	29	142.81	0.05	1.00
Psychotria berteriana	10	610	94.60	4.7300	3	28	195.68	0.05	1.00
Sapium laurocerasus	61	1829	55.47	0.4547	2	167	99.00	0.05	0.12
Sloanea berteriana	91	3048	63.97	0.3515	1	118	226.08	0.004	0.05
Tabebuia heterohylla	46	1524	73.56	0.7996	3	79	206.75	0.05	0.12
Tetragastris balsamifera	46	2438	100.04	1.0874	1	169	125.87	0.004	0.05
Trichilia pallida	15	914	103.60	3.4533	1	254	32.23	0.004	0.05

[a]D_{max} - maximum dbh attained by species; H_{max} - maxim um height attained by species; B_2, B_3 - coefficients relating height to dbh; Tol - shade tolerance class (see text); Age-maximum known age attained by species; G - derived growth constant (see text); S_{min},S_{max} - minimum and maximum light levels within which recruitment can occur.

[b]Taken from Little and Wadsworth (1964).

[c]Tolerance classes for each species drawn from Smith (1970).

[d]Computed from a supplemental program using the G constant and maximum size figures.

[e]Derived from tree growth data provided by the U.S.D.A. Forest Service Institute of Tropical Forestry (see Crow and Weaver 1977).

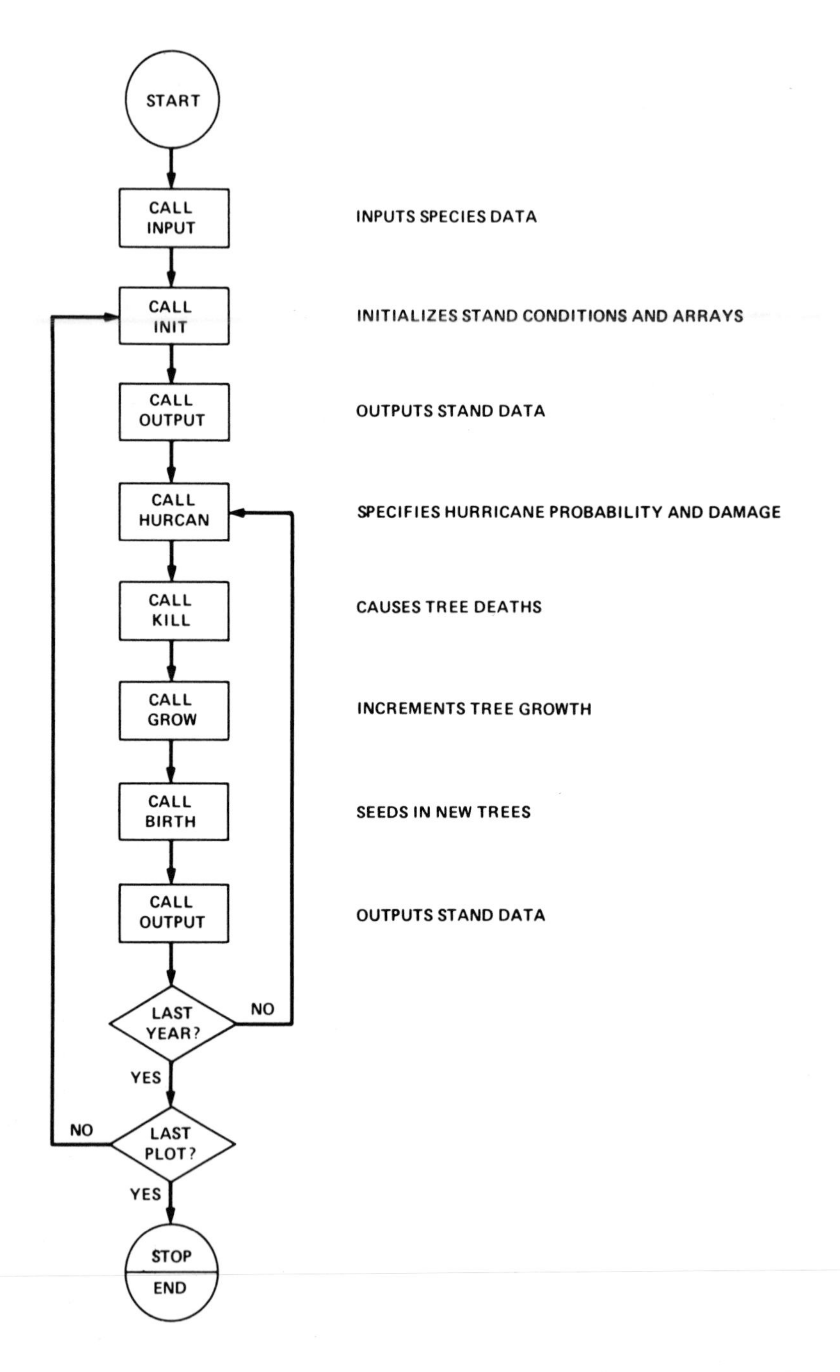

Figure 6.1. Flow diagram of calls to subroutines in the FORICO model. The general function of each subroutine is listed beside the subroutine named in the box.

turn frequency of one hurricane every nine years, and thus the model uses 0.111 as the probability that a hurricane will occur in any one year of simulation. Quantitative data do not exist on the impact of historic storms in the study area. In the simulations reported here, it was assumed that hurricanes removed, on the average, 0–10% of the stand density. In each hurricane year, a random number between 0 and 0.1 was used as the probability of stem removal for that year. Individual stems were then stochastically selected for removal, based on that probability.

Support subroutines that facilitate model operation from the programming aspect, rather than from a biological end, are termed utility subroutines. Subroutine INPUT (Fig. 6.1) calls in the species list and model parameters in Table 6.1, and stores this information in preassigned arrays. The INIT subroutine (Fig. 6.1) initializes stand conditions and various storage and counting arrays. The OUTPUT subroutine (Fig. 6.1) stores model results after each year to average and output stand and species data at the end of the simulation.

Model Forest and Species Dynamics

Long-term stand projections and species responses are presented to illustrate what might be considered typical model results. These simulations represent a 500-year successional sequence of continuous forest growth and development beginning from an open gap with no standing trees present.

The general pattern in model simulation of the average stand dynamics for leaf area, biomass, and stem density is shown in Fig. 6.2. Leaf area increases sharply through the first 50 years of gap recolonization to its highest level of 7.8 m^2/m^2 (Fig. 6.2a). A stand thinning and canopy stratification becomes evident as the average leaf area declines about a stand equilibrium ranging around 6.5 m^2/m^2. Several studies conducted in the tabonuco forest have reported comparable leaf area averages of 6.4 m^2/m^2 (Odum et al. 1963), 7.3 m^2/m^2 (Odum 1962), and 6.6 m^2/m^2 (Jordan 1969). Stand biomass displays a classic trend of forest development through year 125, after which it stabilizes and fluctuates between 175 and 250 metric tons per hectare (Fig. 6.2b). This range is compatible with above-ground biomass figures of 150, 200, and 245 metric tons per hectare recorded by Ovington and Olson (1970) for three study sites in the tabonuco forest. The total number of stems above 1 cm in diameter for an average plot decreases from a high-density stand in the intial years to about 100 trees at year 50 (Fig. 6.2c). Stem density remains relatively constant to the end of the simulation, averaging about 110 trees per plot.

Model simulation of successional dynamics for several dominant species in the tabonuco forest is shown in Fig. 6.3. The predicted species response through succession is noted by the change in species biomass.

These response patterns are analyzed according to the species observed successional role.

Cecropia peltata and *Didymopanax morototoni*, Fig. 6.3a and b respectively, are typically characterized as "early secondary" or "pioneer" species due to their early colonization and establishment of newly opened forest sites. This group of species is generally short-lived-with life spans rarely exceeding 50 years. Also, they are mostly shade intolerant, and

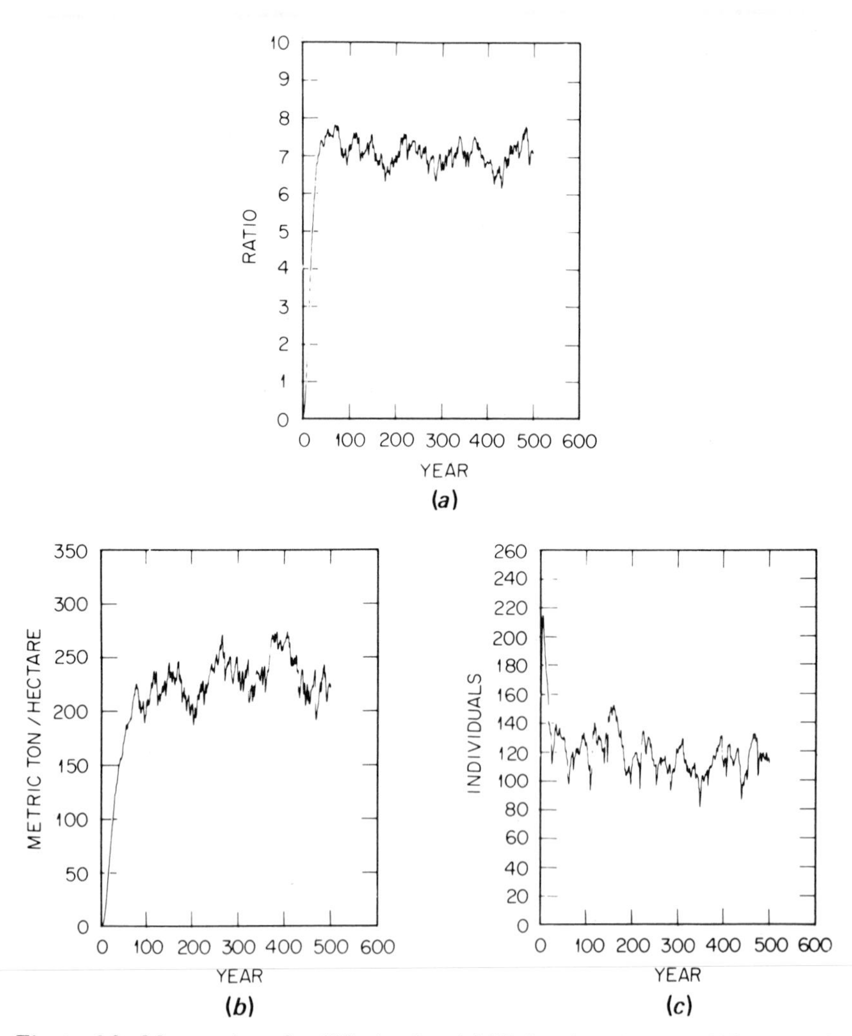

Figure 6.2. Mean values for 120 simulated 1/30-ha plots through 500 years of model simulation: (a) leaf area index (m^2 of leaves/m^2 of land area); (b) total aboveground biomass (metric ton/ha); (c) total number of trees (individual trees greater than 1.3 cm diameter at breast height, or dbh).

thus do not persist beneath a closed canopy. The successional patterns of these species peak around 15–30 years after initial site invasion (Fig. 6.3a and b). By year 60, canopy displacement of aged and overshadowed pioneer species by shade-tolerant, late-successional species is nearly complete. However, low biomass levels are maintained over the remaining years through random occurrences of sizable gaps, which allow the repeated establishment of these species in mature forest.

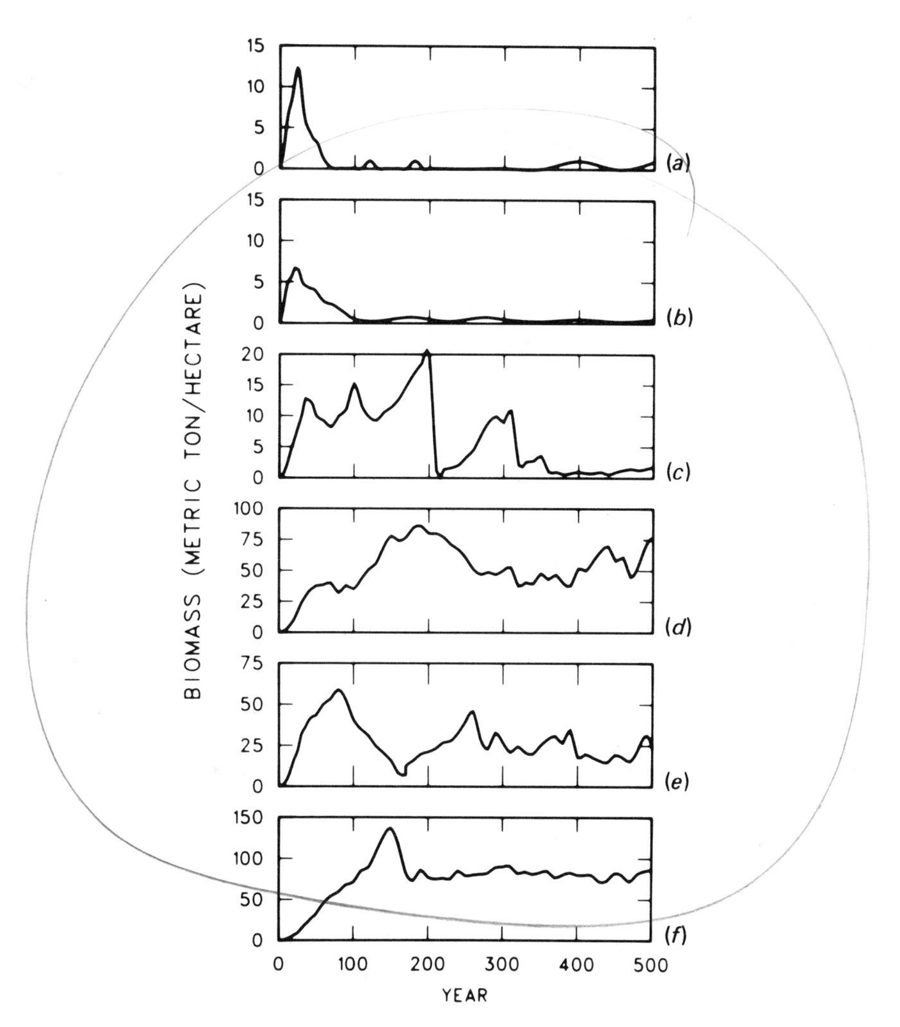

Figure 6.3. Successional dynamics of six dominant species of the tabonuco forest as simulated by the FORICO model: (a) *Cecropia peltata*; (b) *Didymopanax morototoni*; (c) *Buchenavia capitata*; (d) *Manilkara bidentata*; (e) *Sloanea berteriana*; (f) *Dacryodes excelsa*.

Figure 6.3c displays the simulated dynamics of a "late-secondary" species, *Buchenavia capitata*, which typifies a fairly long-lived species of intermediate shade tolerance. This particular species shows a stepwise increase in biomass over the 0–200-year range, followed by a sharp decline and another notable peak at year 300 (Fig. 6.3c). The last 200 years exhibit a rather flat response of low biomass values. The overall successional pattern of this species appears to share the fast growth inclines of pioneer species along with the long-lived behavior of late-successional species.

The remaining examples, *Manilkara bidentata*, *Sloanea berteriana*, and *Dacryodes excelsa*, Fig. 6.3d, e, and f respectively, are the primary dominants of the tabonuco forest, comprising more than 50% of the total biomass. These species represent the long-lived, shade-tolerant, and late-successional species often called "primary," "climax," or "mature phase" species. This group displays a more steadily increasing biomass trend that peaks from 100 to 200 years into the simulation. Submaximal levels are then maintained through the remainder of the succession. The rates of increase and maximum biomass levels achieved vary markedly between species. However, their much higher biomass values over the species of other successional roles substantiates their overwhelming representation in this forest system. The successional dynamics of these primary species reflect a pattern of codominance among these three species.

Model Testing and Validation

Structural and compositional agreement between simulated results and field observations provides the basis for model validation. The FORICO model was initially tested for its ability to duplicate the stand density characteristics of young and mature aged stands. This test involved the statistical comparison of density-diameter distributions between model stands and forest plots of like successional age.

Density-diameter distributions represent the number of trees apportioned in defined diameter size classes over the diameter range of a stand. From one year to the next, individual trees may grow into larger size classes or remain in the same one, provided tree death or disturbance does not occur. Therefore, these distributions are not static and are subject to continual change as governed by rates of growth and mortality. The shape and position of these curves relative to each other can also indicate the successional maturity and dynamics of a stand. Figure 6.4a and b shows the observed and predicted density-diameter distributions of stands, 9 and 150 years of age, respectively. Field data used in this part of the study was provided by the USDA Forest Service Institute of Tropical Forestry in Rio Piedras, Puerto Rico.

The stem density values from both simulated and actual independent data were converted into cumulative percentages of the total stand density for successive diameter classes from smallest to largest. This results in cumulative frequency distributions which can be statistically compared by using the Kolmogorov-Smirnov two-sample test (Siegel 1956). The diameter distributions for both stand comparisons were found not significantly different at the 0.1 level. The model therefore seems capable of balancing recruitment and mortality rates for each size class, and compensating for the differences in these rates for young and mature aged stands. The structural agreement in tree size and number for variable aged stands verifies the ability of the model to mimic the successional dynamics of forest structure in the tabonuco forest.

To test for compositional conformity, model results were validated against a 2-ha sample of forest data (see Table 2 in Wadsworth 1970) by correlating the numerical ranking of species by abundance. Model data were extracted from each stand simulation at 150 years to compare with the estimated age of the sample forest data. The data sets included species

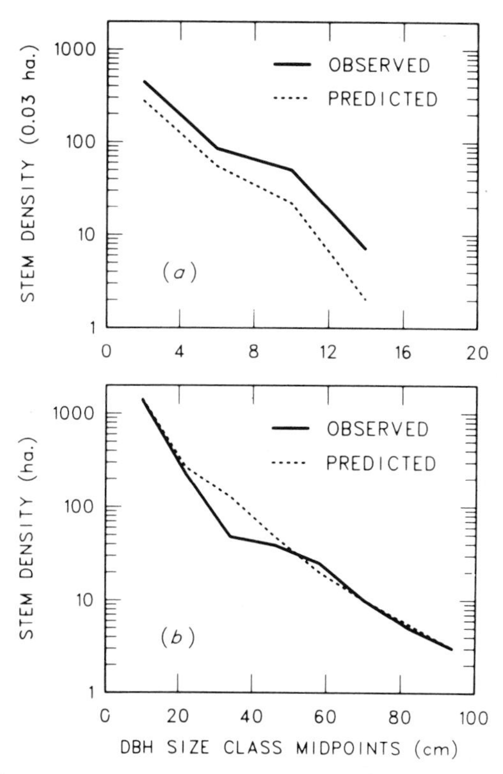

Figure 6.4. Density-diameter distributions between simulated results and field data from stands of different ages: (a) 9-year-old stand; (b) 150-year-old stand.

names, number of stems per species, and ranking of each species by stem count. Using a Spearman rank correlation test (Siegel 1956), an r_S value of 0.66 indicated that the observed and predicted species rankings were significantly correlated to the 0.05 level.

Simulated Disturbance Patterns and Forest Responses

Simulations without hurricane effects were used to further confirm the validity of the model. Model simulations excluding hurricane disturbance produced a compositional assemblage that did not correlate significantly with the same set of forest observations. This finding not only substantiated model performance under a hurricane disturbance regime, but also prompted a further evaluation of the role of disturbance in the gap dynamics of this Puerto Rican rain forest.

Model results from hurricane and nonhurricane simulations were compared to investigate effects of different disturbance regimes on structure, composition, and diversity of forest communities. Figure 6.5 compares dominance-diversity curves from field observation and model simulations with and without hurricane effects. These curves were drawn from the same data sets generated to validate the model with respect to forest composition.

The dominance-diversity curve from the hurricane simulation converges with the curve of field data from the tabonuco forest, while the curve without hurricane effects differs significantly. In the absence of disturbance, the model predicts a forest community lower in species number and dominated mostly by late-successional species. In actuality, species dominance in the tabonuco forest is shared by primary and pioneer species. However, in the nonhurricane simulation primary and late-successional species accrue greater dominance because low disturbance frequencies favor species with a shade-tolerant reproductive strategy; hence, these species show classic self-replacement. Furthermore, the rankings of less common shade-tolerant species improve because of the loss of early successional species, but their numbers remain relatively unchanged. Despite the fact that gaps are generated in the nonhurricane treatments by normal tree mortality, sizable openings are too few and far between to allow sufficient regeneration by pioneer species. Without frequent, large gaps, gap-dependent species risk local extinction.

Evidence from the dominance-diversity curves in Fig. 6.5 reveals that hurricane disturbance may not only influence the compositional makeup and total species number of mature tabonuco forest, but also species diversity as determined by the proportions of abundance or biomass shared among species. Figure 6.6 features the difference in the forest diversity pattern between hurricane and nonhurricane simulations over a 500-year succession. Diversity, in this case, is a normalized index of shared dominance determined from the Shannon formula for evenness. Index values range from 0 to 1.0, where 0 represents a single-dominant stand and 1.0 signifies an equal distribution of the stand biomass among all species. Both simulations share almost identical diversity values from 0 to 60 years in succession (Fig. 6.6). At year 60, however, these diversity patterns diverge and follow steady declines through year 200 before reaching an equilibrium condition. The average dominance index of these patterns differs most through the latter years of succession. Between years 200 and 500, the nonhurricane simulation has an average dominance index of 0.38 in contrast to 0.58 for the simulation including disturbance effects from hurricanes. Diversity values from independent data of mature tabonuco forest ranged between 0.50 and 0.56, further confirming the importance of disturbance in the composition and succession of this tropical forest system.

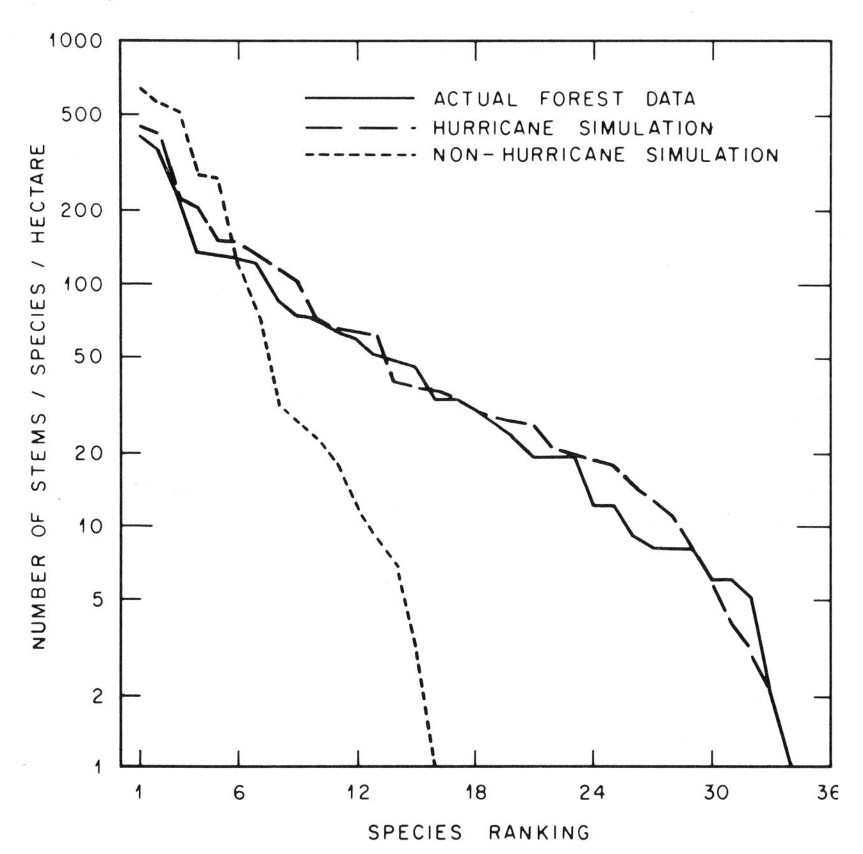

Figure 6.5. Dominance-diversity curves for model simulation with and without hurricane effects and tabonuco forest data. Species rank ranges from most abundant (1) to least abundant (36). Species abundance is represented as the total number of stems (above 4 cm dbh) per hectare.

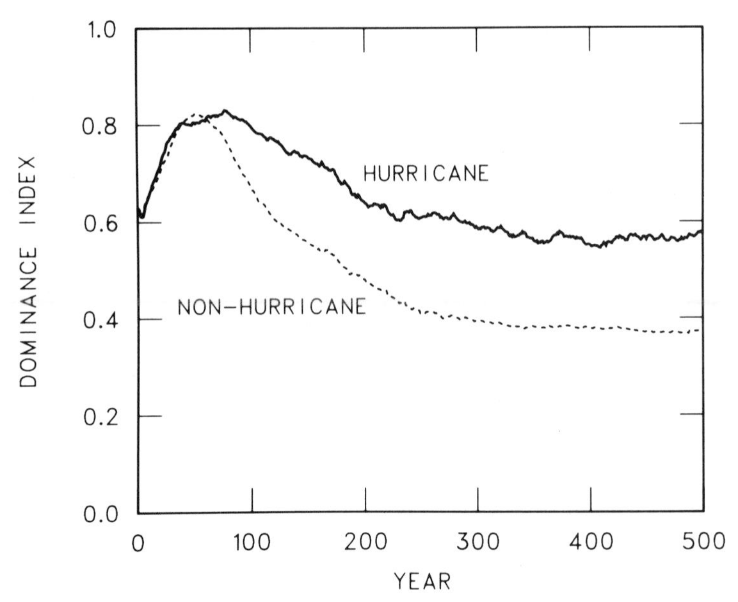

Figure 6.6. Forest diversity pattern of a 500-year succession from model simulations with and without hurricane disturbance.

Although the species diversity of the tabonuco forest seems related to hurricane disturbance, it is not known how hurricane frequency and magnitude influence this diversity. Figure 6.7 illusrates the disturbance-diversity relationship under varying disturbance frequencies and magnitudes composed from 40 different model simulations. A disturbance frequency of 0.1 represents the probability that a hurricane will occur once every 10 years. The magnitude of disturbance simulates the percentage loss of stand density such that a probability of 0.1 indicates the removal of 1 in every 10 trees from the stand. The contour lines signify the plane on which the associated dominance index value passes. Each diversity value denotes the average dominance index over an entire 500-year succession. Results clearly demonstrate that different disturbance regimes can indeed affect overall forest diversity. It might otherwise be stated that species diversity varies according to the patterns of disturbance. Intermediate disturbance levels appear to generate the highest diversity values, while disturbance extremes share the lowest values. Though not apparent from this disturbance-diversity figure, each contour corresponds to a shift in the community structure and composition of this Puerto Rican rain forest, such that highly disturbed assemblages are dense stands dominated by secondary species, and relatively undisturbed stands are composed primarily of late- successional species.

Model trials including hurricane disturbance reflect the compositional variation and species diversity characteristic of the tabonuco forest. Unlike the nonhurricane simulation, model results with hurricane effects exhibited statistical agreement with forest data in both structure and composition. Model output shows strikingly accurate figures of ranking and abundance among the forest dominants, including primary species, *Dacryodes excelsa* and *Sloanea berteriana*, and pioneer species, *Cecropia peltata* and *Didymopanax morototoni.* It is evident then, that an alteration in the disturbance pattern causes a shift in both the shape and position of the dominance-diversity curve as well as species ranking and diversity figures. This result indicates that a variation in the disturbance regime could alter the successional pattern and subsequent composition, structure, and diversity of the tabonuco forest.

Gap Dynamics and the Role of Disturbance

Current diversity theory suggests that forest systems subject to chronic disturbance tend to support comparatively richer species assemblages. Connell (1978) used this argument to explain the high tree diversity of

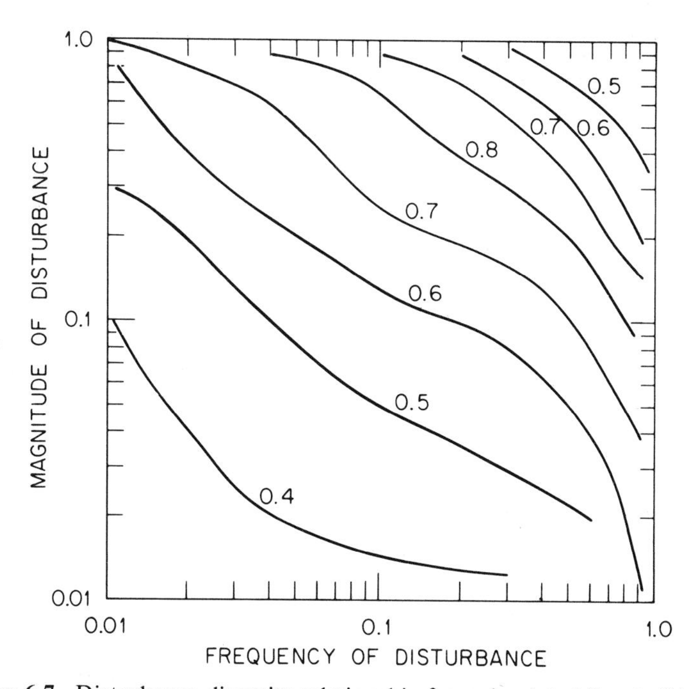

Figure 6.7. Disturbance-diversity relationship for a simulated Puerto Rican rain forest under varied disturbance regimes as determined by hurricane disturbance frequency and magnitude.

tropical rain forests. A comparison of hurricane and nonhurricane simulations from the FORICO model substantiated the important role of disturbance in maintaining the compositional variation and species abundance of a montane rain forest in Puerto Rico. The simulation without hurricane effects showed a lower species number than runs including disturbance effects. Using a linear Markov model of succession in a temperate forest, Horn (1975b) reported similar results of higher diversity at intermediate levels of disturbance than at either high or low extremes of disturbance.

The structural aspects of forests may also be affected differentially by disturbance. Forest sites prone to windthrow consequently support fewer large stems. Hartshorn (1978) explained the low density of large trees (ca. 1 m dbh) in tropical wet forests, based on their dynamic nature and high frequency of treefall. Model behavior and output from hurricane and nonhurricane simulations similarly confirm this expected trend of stand density and size distribution changes with the changes in the disturbance regime.

The differences in community structure, composition, and diversity between hurricane and nonhurricane simulations further suggest that certain floristic associations are likely to predominate under particular disturbance patterns. Crow and Grigal (1980) identified eight major floristic groups in the tabonuco forest that correlated best with topographic position and disturbance history. Under a no-disturbance regime, canopy gaps were not frequent or sizable enough to allow early successional species to maintain canopy positions. This result emphasizes the importance of disturbance frequency and intensity for the continued persistence of gap-dependent species. It also shows an altered successional pattern in the absence of disturbance that favors a self-replacing community of primary species. These findings support the notion that variations in the disturbance regime from one forest locale to another plays a major role in determining community type and species persistence.

Models and the Study of Forest Disturbance

Forest simulation models can be effectively used to examine the effect of disturbance on community structure, species rankings, and diversity. In the case reported here, a tropical forest simulator, FORICO, was used to model the effects of hurricane frequency and severity on the lower montane rain forest of Puerto Rico. A comparison of hurricane and nonhurricane simulations confirms the important role that disturbance has in maintaining compositional diversity and structural characteristics of this forest system. Variations in disturbance regime could alter the successional pattern and subsequent composition, structure, and diversity of rain forest in Puerto Rico. This result supports other studies of the role of natural disturbance in tropical forests. Perhaps the greatest advantage of forest simulators is that they allow exploration of the consequences of variation in disturbance regime.

Acknowledgments

This research was supported by National Science Foundation's Ecosystem Studies Program, under Interagency Agreement Nos. DEB-77-26722 and DEB-77-25781, with the U.S. Department of Energy, under contract W-7405-eng-26 with Union Carbide Corporation. Publication No. 1768, Environmental Sciences Division, Oak Ridge National Laboratory.

Chapter 7

Patterns and Dynamics of Forests: An Application of Simulation Models

Herman H. Shugart, Darrell C. West, and W. R. Emanuel

Introduction

Studies of forest succession characteristically rely heavily on inference to order the spatial pattern of forest stands (usually on a regularly disturbed landscape) into a temporal pattern. The inferential structure used to arrange these data is, in a sense, a "theory of succession," and the relation between the artificially ordered successional pattern and the ordering theory are, of necessity, tautological. This circularity persists regardless of the particular theory of succession (monoclimax, polyclimax, multiple paths, etc.) to which an ecologist happens to subscribe. The fact that both the observation data sets and the theory may both be quite elaborate makes it difficult to test the theories of succession. In this chapter, we shall take an alternate approach to studying successional dynamics. Instead of a direct and long-term observation of vegetation plots or a synthetic ordering of several plots, we shall use a set of reasonably detailed computer models of forest dynamics to explore system-level responses of different ecosystems. This simulation technique does not rely upon inferences made regarding temporal and spatial scales as do empirical studies of vegetation dynamics.

During the past decade, forest succession models have become a valuable tool for exploring the long-term consequences of theories about the fundamental mechanisms underlying forest dynamics (Shugart and West 1980). In this chapter models are used as a unifying framework for synthesis of ecological processes operating on individual trees to inspect ecosystem dynamics. Detailed descriptions of each of these models are available in the scientific literature, but we shall provide a brief overall description of the models below. The four models are all derived from a similar modeling approach, and each incorporates species characteristics of regeneration, growth, and mortality pattern. The models are:

1. The FORET model (Shugart and West 1977), which simulates dynamics of 33 tree species typical of southern Appalachian deciduous forests.
2. The BRIND model (Noble et al. 1980b; Shugart and Noble 1981), which simulates the dynamics of 18 arborescent species in the *Eucalyptus*-dominated forests of the Australian Alps.
3. The FORAR model (Mielke et al. 1978), which simulates 33 tree species found on upland sites in the mixed oak (*Quercus*)-pine (*Pinus*) forests of Arkansas.
4. The KIAMBRAM model (Shugart et al. 1980a; Shugart et al. 1981), which simulates the dynamics of 125 species found in the subtropical rainforest in the vicinity of the New South Wales-Queensland border.

Model Description

The four models each contain certain mathematical functions in common (summarized in Table 7.1), but each model has some unique algorithms treating special features of the modeled forest (summarized in Table 7.2). The models are all derived from a similar modeling philosophy—*viz.*, that of simulating a forest stand of some specified size by annually computing the growth in diameter at breast height (dbh) of each individual tree in the stand (Table 7.1). Growth of trees is decreased from a species-specific optimum growth for a given diameter depending on:

1. Case-specific responses to environmental conditions (e.g., temperature, drought, altitude).
2. Shading from other trees computed as a geometric explicit competition function based on tree height and light profiles (Table7.1) in the simulated forest.
3. Crowding from other trees computed as a reduction in growth potential based on plot biomass or basal area. The crowding effect is designed to mimic competition for utilization of some limiting nutrient or water.

Trees are killed on a tree-by-tree basis as a stochastic function of their species-specific expected longevity. Trees that are suppressed and are growing very slowly are exposed to an increased probability of mortality. Fire-damaged trees are exposed to increased mortality in a fashion analogous to suppressed trees. Trees may also be killed by wildfire and by having other larger trees fall on them (Table 7.2).

Table 7.1. Model equations common to each of the four simulation models.

Process	Equations
Growth of each tree under optimal conditions[a]	$\frac{d(D^2H)}{dt} = RLA(1 - \frac{DH}{D_{max}H_{max}})$ R = growth rate parameter, LA = leaf area of tree, D = diameter at breast height, H = height of tree, D_{max} = maximum diameter for a particular species, H_{max} = maximum height for a particular species, D^2H = index of tree volume.
Height/diameter relation[b]	$H = 137 + b_2D - b_3D^2$; $b_2 = 2(H_{max} - 137)/D_{max}$, $b_3 = (M_{max} - 137)/D^2_{max}$, b_2 and b_3 determined by setting $H = H_{max}$ and $dH/dt = 0$ when $D = D_{max}$.
Extinction of light as a function of leaf area in forest canopies[c]	$Q(h) = Q_o \exp(-k \int_h LA(h')\, dh')$ LA(h′) = distribution of leaf area as a function of height Q_o = incident radiation, Q(h) = radiation at height (h), K = constant[c]
Reduction of photosynthesis due to shading[d]	Various empirical equations fitted to light-photosynthesis curves for shade-tolerant or shade-intolerant species found in each forest. These equations are used to reduce the magnitude of the growth equation (above) for shaded trees.
Crowding effects related to stand biomass (competition)[a]	S(BAR) = 1 − BAR/SOILQ, BAR = total biomass (basal area) of simulated stand, SOILQ = maximum biomass (basal area) recorded.
Intrinsic tree mortality[a]	$p = 1 - (1 - \epsilon)^n$, p = probablilty of mortality at year n is chosen such that p = 0.99 when n = AGE_{max} (the maximum age for the species).
Mortality of trees with suppressed growth[a]	If growth is less than a critical value for the species, p = 0.368.

[a]Botkin et al. 1972.
[b]Ker and Smith 1955.
[c]Kasanga and Monsi 1954, Loomis et al. 1967, Perry et al. 1969.
[d]Kramer and Kozlowski 1960.

Table 7.2. Biotic and abiotic factors that modify annual growth rates of individual trees, probabilities of establishment for a given tree species in a given year, or annual probablities of mortality for individual trees for each of four succesion models.

Model (Ecosystem):	Growth rate modifiers					Modifiers of establishment probabilities											Modifiers of mortality probabilities			
	Crowding	Drought	Fire damage	Temperature	Shading	Deer browse	Dispersal	Fire	Leaf litter	Lignotubers	Mineral soil	Phenology	Raised site	Seed source	Shading	Strangler	Fire	Harvest	Lack of growth	Tree fall
BRIND (Australian *Eucalyptus* forest)	×		×	×	×		×	×		×				×	×		×		×	
FORAR (Arkansas mixed pine-oak forest)	×			×	×	×	×		×		×				×			×	×	
FORET (Tennessee Appalachian hardwood forest)	×	×		×	×	×	×		×	×					×			×	×	
KIAMBRAM (Australian subtropical ranforest)	×				×				×		×	×	×	×	×	×		×	×	×

The establishment of small trees in the typical gap model is simulated by allowing small trees to be recruited into the simulated stand as a stochastic function of stand site conditions (Table 7.2). Small trees are usually initiated assuming a dbh of approximately 2 cm. In some of the models, seed storage or the survival of special reproductive organs (i.e., lignotuberous sprouting in genus *Eucalyptus*) is also computed. Vegetative sprouting from individual trees of appropriate size and species is also included. One of the models (the KIAMBRAM model, Shugart et al. 1980a) also simulates the epiphytic, strangling habit of certain tropical figs (*Ficus* spp.).

The models taken as a set incorporate 11 ecological factors that can alter the regeneration of young trees. Some of these factors are related to conditions needed for dispersal of certain species to a potential germination site. Such factors (Dispersal, Table 7.2) include a suite of relatively straightforward model algorithms for eliminating species from eligibility for regeneration in years when a species-specific dispersal agent is absent. Several of the factors involve the state of potential germination sites in a given year (has the site been burned, flooded? is there mineral soil present? etc.) which might serve to eliminate or amplify a given species' reproduction. Other factors are concerned with seedling mortality occurring between germination and the time when the seedlings are large enough to be tallied in the model. Such sources of mortality are considered as a part of the regeneration process (e.g., deer browse).

Output from the model resembles a field data sheet from a sample plot (i.e., diameters of each tree and the species of each tree). In general, the model is run for many replications to obtain the mean of the overall forest response, and the simulation results are amenable to the same data reductions as might be applied to actual field data.

Model Testing

Because of the level of detail in the model output there are several kinds of detailed model-data comparisons that can be used to test the models. All simulations presented here are averaged over 50 simulated plots starting from an open stand (with no trees). Model validation is often central to the development of useful models of any ecosystem. While some theoretical applications use models in ways in which the testing of the models is not particularly germane, in most cases the utility of a model's predictions is often gauged by the veracity of the model. This is perhaps clearer when we compare ecological models with another projection technique, statistical regression. In developing regression equations, one adopts the position that the future behavior of a system is best determined by using simple models to *interpolate* between actual observations on the system. In applying systems models, the stated objective is to use a realistic model to *extrapolate* the behavior of the ecosystem outside the limits of observation. Extrapolation from statistical models is notoriously unreliable and any extrapolative methodology is of its basic nature difficult.

Model testing can be divided into three types of procedures (Shugart and West 1980):

1. Verification procedures: in which a model is tested on whether it can be made consistent with some set of observations. The model structure and/or parameters are sometimes altered as one attempts to make the model agree with the observations. In this sense, linear regression is a verification procedure in which simple models are forced to have a maximum agreement with observations by changing the parameters over an infinite range.
2. Validation procedures: in which a model is tested on its agreement with a set of observations that are independent from those observations used to structure the model and estimate its parameters. The degree of independence between the model and the observations prior to the validation test can be as important as the degree of agreement between the model and the observations following the test.
3. Application procedures: in which results from a model are judged as to their utility in providing insights into either basic or applied problems of interest. The applications of models often involve some of the most powerful tests on the models. Application also provides hypothetical model predictions with a high likelihood of being tested by future observations or experiments.

The four models have been tested with several different verification, validation, and application procedures (Table 7.3). The results from these tests indicate that the behavior of the models strongly resembles that of real forest ecosystems.

Patterns of Compositional Dynamics Simulated by the Four Models

Changes in stand composition, expressed as the dynamics of proportions of biomass associated with different species, for each of the four models (Fig. 7.1) can be compared directly with accounts of inferred patterns of replacement in the forests.

Table 7.3. Brief descriptions and testing of the four models.

Model	Verification
FORET (Shugart and West 1977) Model simulates dynamics of 33 tree species typical of Southern Appalachian deciduous forests. Nominal locations of simulated stands are on moist sites in Anderson Co., Tennessee.	1. Prediction of composition of post-chestnut-blight forests.
BRIND (Noble et al. 1980b; Shugart and Noble 1981) Model simulates dynamics of 18 arborescent species found on southeast-facing slopes from 900 to 2400-m elevations in the Brindabella Range, Australian Capital Territory. Nominal locations of simulated stands are in the Cotter catchment near Canberra, A.C.T. All forests are dominated by *Eucalyptus.*	1. Prediction of qualitative pattern of tree species replacement at at 1600-m elevations in the Brindabella Range under different wildfire frequencies.
FORAR (Mielke et al. 1977; Mielke et al. 1978) Model simulates 33 species found on upland sites in Union Co., Arkansas. Forest are of a mixed oak-pine type.	1. Predicition of forest composition based on an 1859 reconnaissance (Owen 1860) of southern Arkansas. 2. Prediction of yield tables for loblolly pine (*Pinus taeda*) (L.K.Mann,pers. comm. 1980)
KIAMBRAM (Shugart et al. 1980a; Shugart et al. 1981) Model simulates the dynamics of 125 species found in the subtropical rain forest in the vicinity of the New South Wales-Queensland border. Nominal locations of the simulated stands are Wiangaree State Forest, New South Wales.	1. Comparisons with composition of stands of known age in Lamington National Park, Queensland.

Validation	Application
1. Prediction of composition of pre-chestnut-blight forests.	1. Prediction of chronic atmospheric pollitant effects on forests of Southeastern U.S. (West et al. 1980; McLaughlin et al. 1978). 2. Reconstruction of 16,000-year pollen chronology under changing climatic conditions (Solomon et al. 1980). 3. Prediction of habitat dynamics for nongame birds following harvest of select hardwood trees. (Smith et al. 1981).
1. Prediction of fire-succession response of communities at different altitudes in the Brindabella Range. 2. Prediction of altitudiual zonation of forests from 900 to 2400-m in the Brindabella Range. 3. Prediction of mean dbh, basal area and stocking density for alpine ash (*Eucalyptus delegatensis*) stands.	1. Model is currently being tested for its utility in managing the watersheds supplying Canberra, A.C.T. (I. R. Noble 1978, pers. comm.).
1. Prediction of abundances of trees by species on uplands in southern Arkansas. 2. Prediction of density by diameter class of upland forests in southern Arkansas.	1. Design of tree management schemes for providing habitat for the rare and endangered bird, the redcockaded woodpecker (*Dendrocopus borealis*). (Mielke et al. 1978; also L.K. Mann 1980, pers. comm.).
1. Prediction of abundances and basal areas of trees in mature forests at Wiangaree State Forest, N.S.W.	1. Evaluation of timber harvest schemes for Australian subtropical rain forest (Shugart et al. 1981).

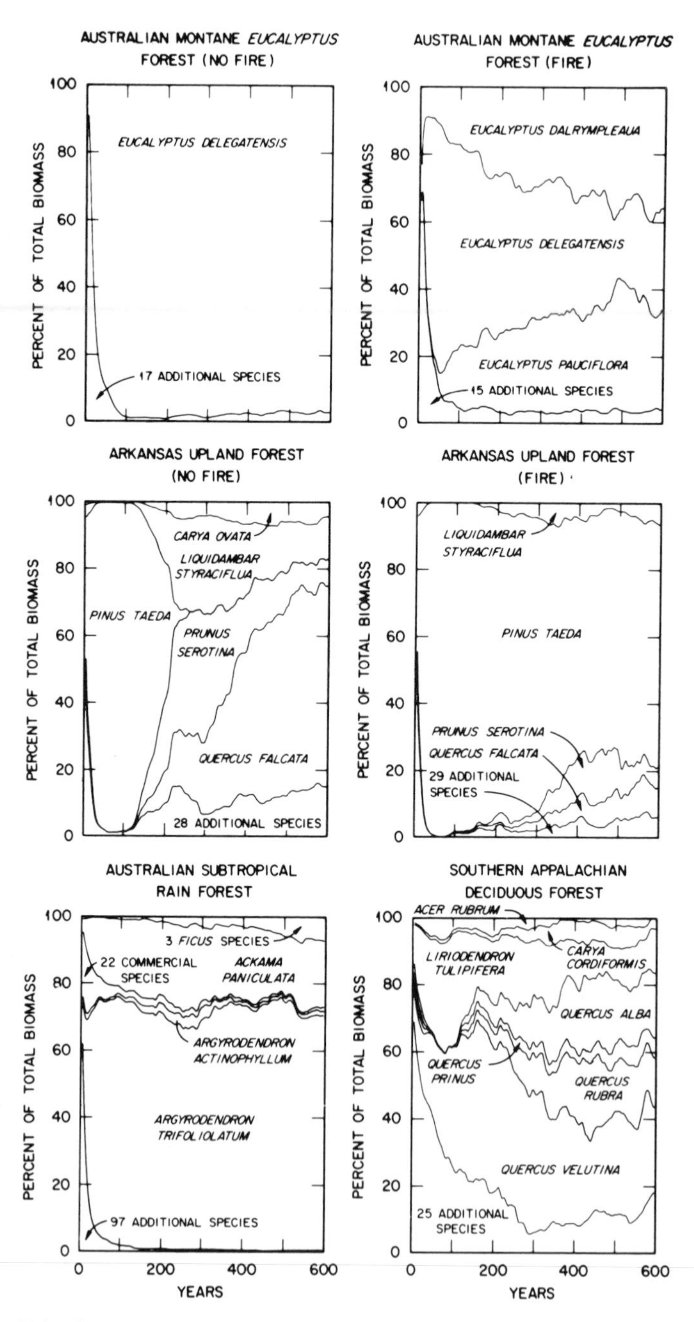

Figure 7.1. Compositional dynamics of succession models of four different ecosystems. Each graph is the average response for 50 simulated forest stands for 600 years. Each simulated run is started from an open stand with no trees (except for the simulations for the Australian montane *Eucalyptus* forest, which are initiated with four small trees of all species to initiate the seed banks). The area associated with each species on each graph indicates the proportion of the stands' total biomass associated with that species.

Australian Montane Eucalyptus Forests

The BRIND model has been used to simulate several forests associated with altitudinal zonations found in the Brindabella Range (see Table 3 in Shugart and Noble 1981). Work described here is for simulations of the forests at the 1600-m elevation: the alpine ash (*Eucalyptus delegatensis*) zone (Costin 1954; Pryor and Moore 1954; Anon. 1973). Two simulation cases were run for this forest: one with no wildfire and the other with a fire return probability of 0.020/y (Anon. 1973). Fire intensity is simulated as a function of climatic conditions and fuel loads using climate records for the area (Anon. 1973) combined with fire equations (Noble et al. 1980a) derived from the empirical fire meters of McArthur (1967) and Luke and McArthur (1978).

Without fire in the model, *E. delegatensis*, a fast-growing, tall eucalypt, completely dominates the stands and comprises almost 98% of the stands' biomass by year 100 (Fig. 7.1). Of the remaining biomass, most is found to be associated with the understory composite shrubby trees, *Olearia argophylla* and *Bedfordia salicina*. The dominance of *E. delegatensis* can be attributed to its superior growth rate relative to any of its possible competitors and to its potential height (61m, Hall et al. 1970). *E. delegatensis* is quite sensitive to fire, and when fire is added to the model the dynamics are greatly altered.

With periodic fire (50-year return frequency) (Fig. 7.1), there is a systematic reduction of *E. delegatensis* over the 600-year period of the simulation. *E. delegatensis* regenerates from seeds that are released after a fire that also kills the mature trees. If two fires occur close enough in time (ca. 20 years apart), there are no seed-bearing trees following the second fire and the species is eliminated for lack of a seed source. As *E. delegatensis* declines with recurrent fire, the mountain gum (*E. dalrympleana*) and the snow gum (*E. pauciflora*) become more dominant. Both of the latter species are quite fire tolerant. Both have the ecological attributes that confer increased fire survival (e.g., lignotuberous resprouting, epicormic buds, ability to survive total leaf scorch).

The responses of the model in Fig. 7.1 [as well as more extensive experiments with different fire probabilities and at different altitudes in Shugart and Noble (1981)] are consistent with replacement patterns of vegetation in the Brindabella Range in particular (Costin 1954; Pryor and Moore 1954; Anon. 1973) and the wet schlerophyll forests of southeastern Australia and Tasmania in general (Ashton 1979). Both the BRIND model and the forests of the region appear to have multiple stable vegetation types when regularly perturbed by fire.

Arkansas Upland Forest

Without fire, FORAR (Mielke et al. 1978) model simulations of the mixed oak-pine forests found in the uplands of south Arkansas (Fig. 7.1) demonstrate a pattern, agreeing with actual field studies, of successional replacement even when the response is averaged over 50 stands. For the first 250 years of the simulation, loblolly pine (*Pinus taeda*) dominates the forest stands. As these stands break up, with mortality processes working on the mature pine canopy, the understory hardwoods, in particular sweetgum (*Liquidambar styraciflua*) and black cherry (*Prunus serotina*), fill the forest gaps. Southern red oak (*Quercus falcata*) dominates in hardwood stands by year 600. These dynamics are in agreement with forest types found in south Arkansas and are generally the pattern expected in nonburned mixed pine-hardwood forests throughout the southeast (Oosting 1942; Johnston and Odum 1956), particularly in abandoned old-fields.

With periodic fire (mean return frequency of 24 years corresponding to the apparent fire occurrence rate for south Arkansas), the pattern of compositional dynamics (Fig.7.1) is altered radically. *Pinus taeda* dominates the stands for the entire 600-year simulation. The hardwood species that dominated the older unburned plots are considerably reduced in abundance and are found only in stands that are fortunate enough to only have light fires or no fires for a long period (owing to the stochastic nature of the wildlife generating function). This result is consistent with older observations of extensive (fire-maintained) pine lands in south Arkansas (M. Leo Lesquereux in Owen 1860).

Australian Subtropical Rainforest

The initial dynamics of the KIAMBRAM model (Shugart et al. 1981) correspond to those actually observed by Webb et al. (1972) and Hopkins et al. (1977) in subtropical rain forests in the vicinity of the Queensland-New South Wales border. Following a brief period (ca. 15 years) that under close inspection includes the increase and demise of pioneer species (e.g., *Dubosia myoporoides, Trema aspera*), both the simulations and the forest (Webb et al. 1972; Hopkins et al. 1977) have a short-lived appearance of "early secondary" or "nomad" species (van Steenis 1958) between years 30 and 60. Typical species in this period would be *Acacia melanoxylon* or *Polyscias elegans.* The decrease in the dominance of these pioneer and early secondary species is indicated by the decrease in the "97 other species" category shown in Fig. 7.1. For a brief period, "late-secondary" species (Hopkins et al. 1977), including many of the "22 commercial species" shown in Fig. 7.1, have a period of abundance. The simulated mature forest is dominated by *Ackama paniculata*, *Argyrodendron* spp., and *Ficus* spp. The composition of the mature rainforest shows a fairly striking similarity to that found in the Wiangaree State Forest in New South Wales (Burgess et al. 1975).

Southern Appalachian Deciduous Forest

Results from the FORET model (Shugart and West 1977) simulating the southern Appalachian deciduous forest (Fig. 7.1) are typified by the development of a mixed oak (*Quercus* spp.) forest type and by the persistence of the intolerant-species tulip poplar (*Liriodendron tulipifera*) by means of gap-phase replacement (Watt 1947). The behavior of the model compares favorably with records by Foster and Ashe (1908) regarding forest composition before the chestnut blight (Table 7.3). The mature forest composition resembles historical accounts for forests of the southern Appalachians (Sargent 1884; Ashe 1897; Pinchot 1906, 1907; Greeley and Ashe 1907; Hall 1907; Ashe 1911).

Structural Similarities in Ecosystem Dynamics

The patterns of compositional dynamics shown in Fig. 7.1 clearly demonstrate some striking differences in pattern. Nonetheless, certain patterns of dynamics tend to be quite consistent over all of the 600-year simulated successions and seem to stem from the dynamics of the forest population structure (Goff and West 1974). The form of the density-diameter function (a frequency histogram of number of stems plotted in equal intervals of diameter) forms an inverted j-shaped curve which approximates a negative exponential series. This was first described mathematically by deLiocourt (1898), and has been observed in a number of mature forests (Goodman 1930; Hough 1932; Meyer and Stevenson 1943; Meyer 1952). Several authors noted that the diameter-density function for the stable structure of a stand may be a rotated sigmoid form (Knuchel 1953; Leak 1964; Schmelz and Lindsey 1965; Bailey and Dell 1973; Goff and West 1974) and explained this shape as a consequence of high-understory tree mortality, lowered mortality in vigorously growing understory trees, and higher mortality in senescent overstory trees.

When plots are initiated from an open stand (as has been done in the present simulations), there is an even-aged structure to the forest with a modal diameter-density curve. The transition from a bell-shaped diameter-density curve to a rotated sigmoidal curve has consequences on the shape of the biomass dynamics of stands. Figure 7.2 shows the response of total stand biomass from three selected examples from the six cases shown in Fig. 7.1. These cases are selected to demonstrate the range of responses that we observed in these simulations. Invariably, in the model there is a rapid increase in stand biomass over the first 100 years of succession to some local maximum. The maximum biomass is obtained when the relatively even-aged cohort created by the initial recruitment into the open stand used to start the simulations reaches its maximum

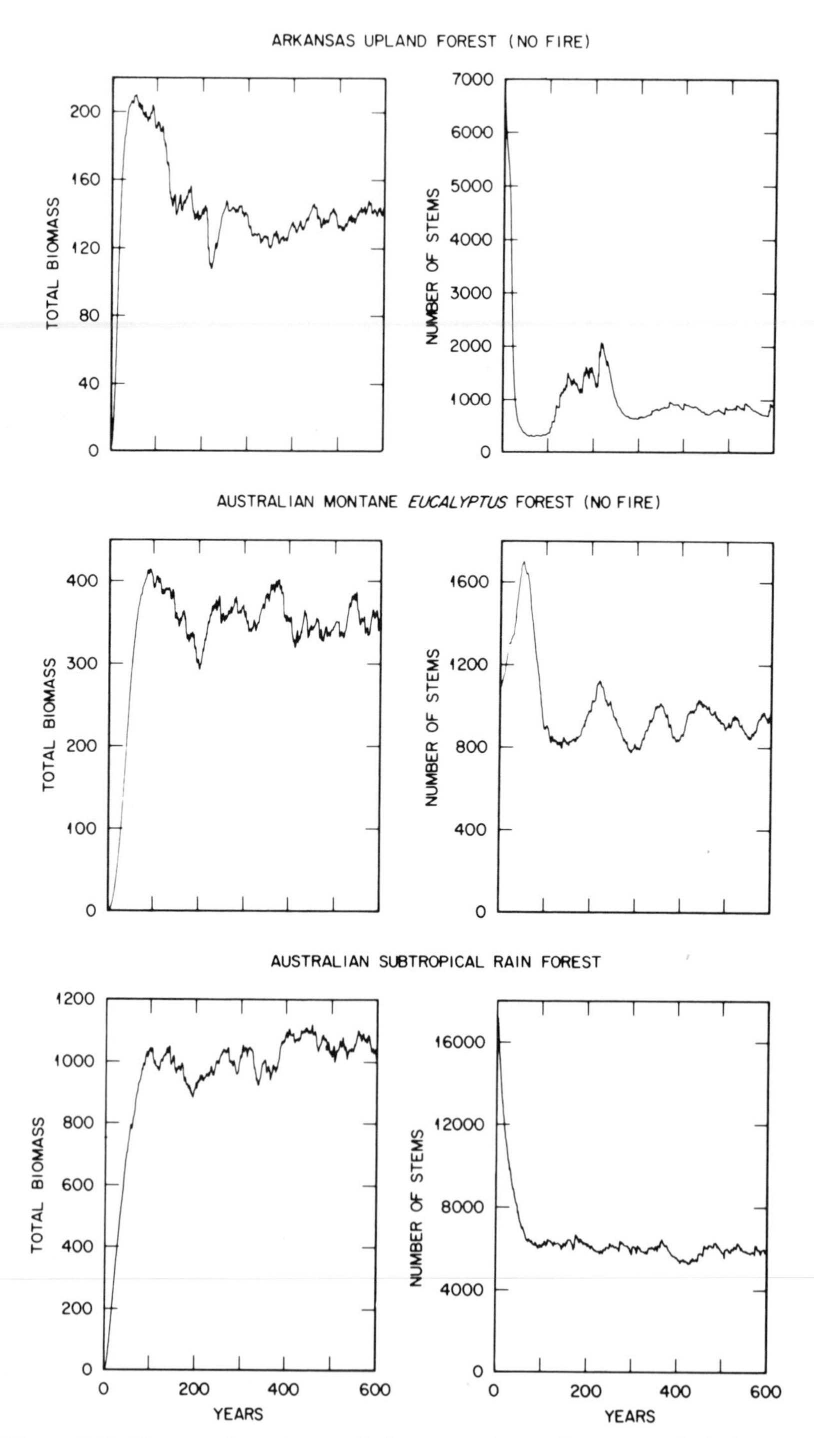

Figure 7.2. Biomass (metric tons/ha) and numbers (individuals/ha) for three selected succession simulations over 600 years.

development. As this initial cohort dies, the canopy breaks up and the forest takes on a mixed-sized structure. In forests in which there is a pronounced species change associated with the canopy breakup, the drop in biomass following the maximum can be very large (e.g., Arkansas upland forest—no fire, Figs. 7.1 and 7.2). In other forests (e.g., Australian subtropical rain forest, Figs. 7.1 and 7.2), the drop in biomass is not nearly so pronounced.

Related to the biomass dynamics of the simulation runs, there is a tendency in all of the models for numbers of stems (Fig. 7.2) to have a reciprocal relationship with biomass. In some forests the canopy breakup (e.g., Arkansas upland forests, Fig. 7.2) is so pronounced that there is an intensified period of recruitment at about 200 years in the simulations. In fast-growing Australian subtropical rain forests, the replacement of missing canopy trees is rapid and the numbers of stems on the stands rapidly seek an equilibrium (Fig. 7.2). In Australian montane *Eucalyptus* forests with no fire (Fig. 7.2), the transient response of numbers of trees oscillates with decreasing amplitude as the forest slowly develops an all-age character. The *Eucalyptus* forest has these pronounced oscillations in its numbers curve primarily because the forest is dominated by a single species (*E. delegatensis*, see Fig. 7.1) and the lengths of time for the forest structure to develop a mixed-sized distribution is over two or three generations of this species (400–700 years).

The pattern of numbers and the biomass dynamics in all of these forests when taken at the ecosystem level have certain unifying dynamic attributes. These attributes are the product of a complex interaction of population age structure, the orders of magnitude of size that individual trees can have during their life span, the physical interactions between size and light, and several other features of tree competition involving geometry. These unifying dynamic attributes can be seen to persist in the face of radically different patterns of composition dynamics.

Compositional Dissimilarities in Ecosystem Dynamics

It is clear from Fig. 7.1 that the pattern of species changes through succession can have some pronounced differences when compared across forest ecosystems. There are even differences in the response from different forests to the same sort of disturbance. For example, fire in the Australian *Eucalyptus* forest (Fig. 7.1) tends to transform the composition to a more diverse mix of three co-dominant species. In Arkansas, fire has the opposite effect, with a shift to a pine dominant with fire. When we compute the familiar diversity index based on biomass and then normalize this index between 0 and 1 by dividing by the log of the number of (H_{max},

Pielou's J′), the resultant index can be taken as an indication of dominance with a value near zero in cases in which the stand biomass is all associated with one species and a value of one when the biomass in the stand is evenly distributed across all species (Fig. 7.3). This index takes on a maximum value at the beginning of each simulation (because these simulation runs are started with all species present or able to seed in to the bare plot). Following this initial maxima the shape of the dynamic (Fig. 7.3) can equilibrate, rise, undershoot, and then equilibrate, or rise and then decline. When the patterns of the index (Fig. 7.3) are compared with Fig.7.1, it is obvious that the response of the index reflects the compositional dynamics accurately. The responses of other indices of diversity and dominance based on either numbers or biomass do *not* show the regular across-ecosystem pattern of response with succession that we found in the biomass and number-of-stems responses of the models. It appears, in the present cases, at least, that species compositional pattern can be considered a within-ecosystem not an across-ecosystem phenomenon.

Species Roles in Ecosystem Dynamics

One of the principal reasons for our initial discussion to attempt to model a number of forest ecosystems with the same modeling paradigm was to determine comparable "roles" of various tree species across ecosystems. It was our hope to develop these comparisons with some degree of formalism. The formalism has been difficult to develop, although consistencies among species attributes, at least in a general sense, are quite apparent. An example is the frequently occurring combination of shade intolerance, pronounced apical dominance, and relatively high mechanical strength of wood per unit mass found in trees that persist in mature forests by virtue of gap-phase replacement. The classification of these patterns into clusters of adaptive traits found in the species of various forest ecosystems may only be possible at a qualitative level, and the work of Noble and Slatyer (1980) may be a logical step in this direction.

We noted several features of individual tree species that are manifest as changes in ecosystem behavior in which these species are important:

1. Certain trees in some ecosystems are able to so completely dominate other species that the ecosystem dynamics (particularly the compositional changes) can be explained as a consequence of these single key species. In the Arkansas upland forest such a species can be loblolly pine (*P. taeda*). This species can outgrow and outcompete all of the others in Arkansas uplands but it does not regenerate well except in open stands. The result is an even-aged cohort of pines that forms a discrete and easily observed successional stage following disturbance. In the Australian *Eucalyptus* forest, alpine ash (*E. delegatensis*) is similarly able to perform as a

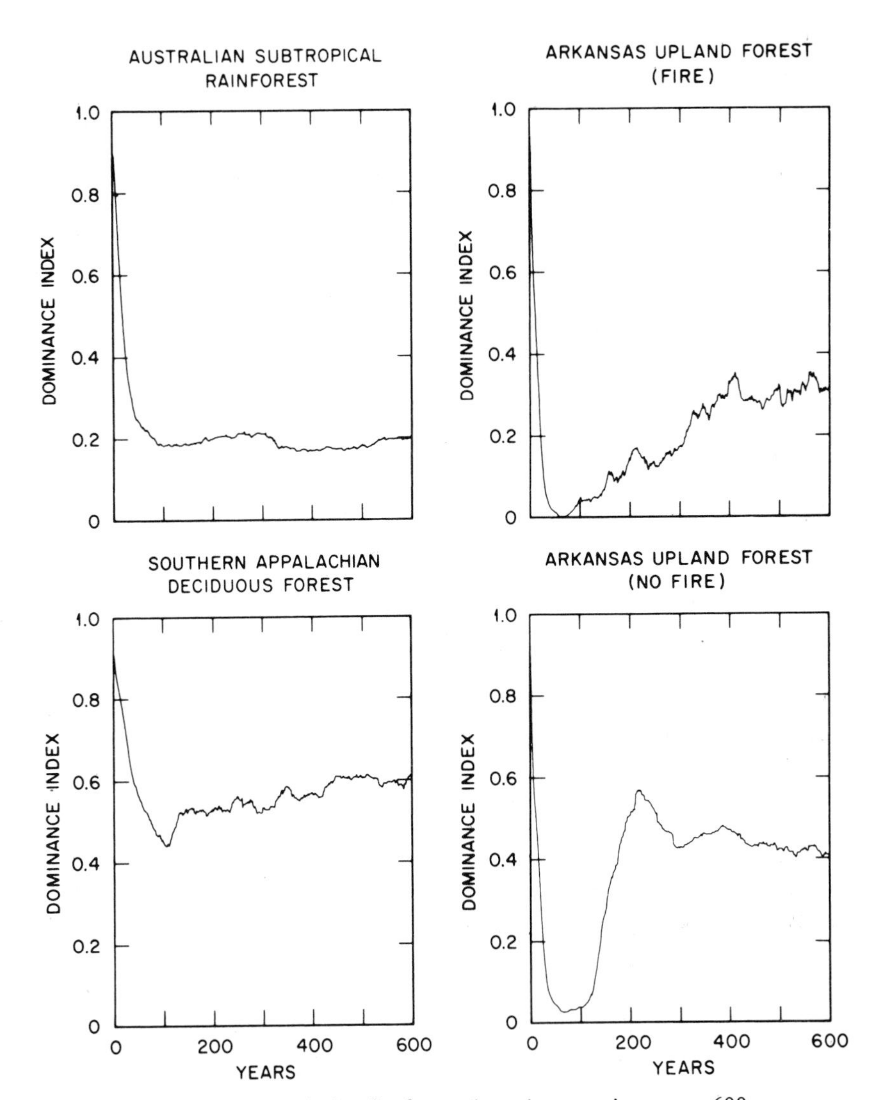

Figure 7.3. A dominance index for four selected successions over 600 years.

"super" tree. In this case, the Achilles heel of the alpine ash is from recurrent wildfires with a short time interval between fires.

2. The sprouting habit can serve to make a species difficult to eliminate and always places the species in a position to capitalize on a favorable set of environmental conditions. At the ecosystem level the presence of a fast-growing "sprouter" is manifest as a tendency for gaps to close rapidly.

3. Some species that usually have great height, that are shade intolerant, and that maintain their persistence in the mature forest by gap-

phase replacement are able to combine attributes that allow a positive feedback between catastrophic disturbance with mortality and gap regeneration from this disturbance. Such species might include tulip poplar (*Liriodendron tulipifera*) and Australian cedar (*Toona australis*). The system-level effect of the prescene of these species is to increase the variance of the biomass response through succession.

4. Shade-tolerant species tend to reinforce an all-age (all-size) structure on the forest and tend to reduce the variance of the biomass response through time.

5. Certain species can capitalize on specific structural aspects of a forest and their abundance is increased when these structures are generated. For example, in the KIAMBRAM model, stranglers (*Ficus* spp.) are most successful with victims of canopy height and sufficient vigor to allow the strangler's establishment as a tree. Such trees occur in greater abundance in the 150–250-year range into the successional sequence.

6. Other researchers have identified species' role in tropical systems in particular, and we see such roles as "nomad" species (van Steenis 1958) having a system-level effect of speeding canopy closure in gaps.

An important result that we noticed in our work with the FORET model is the tendency for species of different roles to compete by altering the dynamic nature of the ecosystem to each species' own benefit. This sort of competition is one of changing the rules of the competition as opposed to competing within some fixed set of rules. We first noticed this effect when considering the frequency components of the time response of the FORET model (Emanuel et al. 1978a,b) with and without American chestnut (*Castanea dentata*). Each significant peak in the power spectral density (Fig. 7.4) corresponds to a cyclic phenomenon in the model forest stand dynamics. The cyclic behavior is not necessarily the result of a system periodicity (e.g., a limit cycle) but is identifiable as a component of the time series of interest (viz., total biomass). In each of the time series there is an important cyclic component with a period of 200–250 years. This component appears to coincide with the average longevity of a dominant canopy tree.

The expected pattern of changes through time in a forest stand (a 1/12-ha plot of land) is for a few trees or even just one tree to grow above the other trees. Capitalizing on the positive feedback mechanism of the size of trees being directly related to the ability to capture light and nutrients, and this in turn being directly related to growth, dominant trees tend to become more and more dominant until they die. When death of a superdominant tree occurs, the cycle of reestablishment and attainment of dominance is reinitiated. It is this intrinsic cycling that, excited by the random mortalities of large trees, creates the strong 200–250-year period component that is evident in the power spectral density of each case (Fig. 7.4). We also know from the biology of the tree species involved that the tulip poplar (*Liriodendron tulipifera*), because of its rapid growth

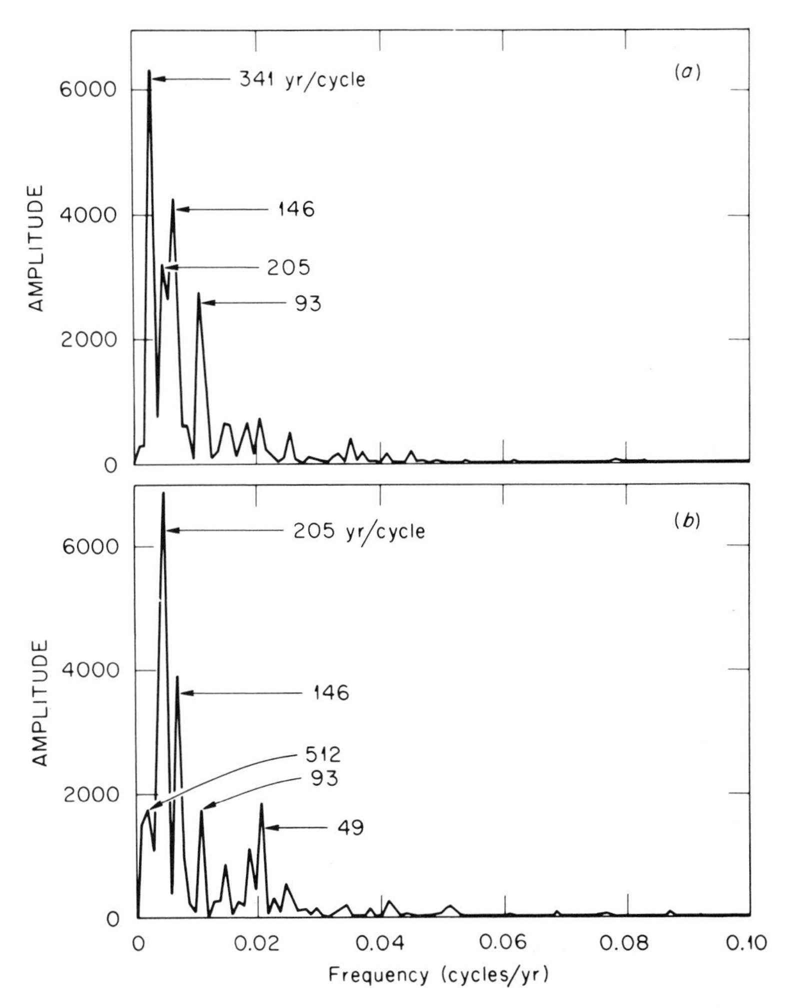

Figure 7.4. Power spectral density for the total biomass variable for each case: (a) base case; (b) American chestnut added as a viable species.

rate and its ability to grow to great size (5m in diameter, 70m in height), tends to amplify the cycle created by the growth, establishment, and death of a canopy tree. Chestnut (*Castanea dentata*), because of certain unique attributes in its biology (largely because it is a rapidly growing, tolerant tree that reproduces vegetatively), tends to suppress this cycle. The variations from case to case in the length of this period are attributable to differences in the biology of the important species.

It is important to note that the cyclic behavior observed from examination of the power spectral density is present in the total biomass time series averaged over 100 model stands. This implies a fair degree of uniform phase relationship among the stands. To some extent, this results from the choice of a open stand as an initial condition for all simulations. However, if the frequency component structure of the total biomass time series were not generally the same for each stand, any initial phase relationship would not be preserved.

We attribute periodicities that are evident in the base case and in the case with chestnut included as a viable species (the increased spectral richness evident in Fig. 7.4) to the diversity of growth responses associated with a relatively large number of important species in the system. The tendency of the 250-year cycle to be suppressed by chestnut is apparent in Fig. 7.4b, as is the tendency for longer periods to appear important in the time trace. We associate these longer periodicities with multiple-generation replacement patterns that are clearly system-level responses in that they occur at periods significantly greater than those associated with the growth replacement process of the species considered.

Having found that the introduction of a single species was sufficient to alter the frequency dynamics of a forest, we conducted a model experiment to observe how two species with differing roles would replace one another in response to a gradient that slowly and systematically changes the competitive advantage (Shugart et al. 1980b). This involves changing the climate (growing degree-day parameter, D_e in the FORET model indicated by the temperature growth modifier in Table 7.2) in a hypothetical forest composed of beech (*Fagus grandifolia*) and tulip poplar.

Figure 7.5 shows the percentage of the total stand biomass in yellow poplar (averaged over five simulated forest stands) as the number of growing degree-days (GDD) is changed at a rate of 1.0 GDD per year from 3800 to 5300 GDD (or vice versa). In each case, the model stands were allowed several hundred years for equilibration before the change in D_e was introduced. The change of one GDD from one year to the next would correspond to a change in the mean temperature during the growing season of ~1/93°C, a virtually undetectable change given typical year-to-year variations in temperature.

Both for increasing and decreasing D_e, the cool (smaller D_e) forests are dominated by beech and the warm forests are dominated by yellow poplar. Under the conditions of slowly changing D_e, these dominance relationships change rather abruptly. The value of D_e at which the abrupt change in dominance occurs depends on whether D_e is increasing or decreasing.

The model behavior that we observed seems to be related to the different dynamics associated with the silvics of the two tree species. The model beech-dominated forests tend to have two or three canopy trees (on a 1/12-ha plot) with a well-developed subcanopy and with a more or less exponential diameter-density distribution on a given stand. Following the death of a canopy tree, there is recruitment from the subordinate trees. The biomass of the stand does not show great variation from year to

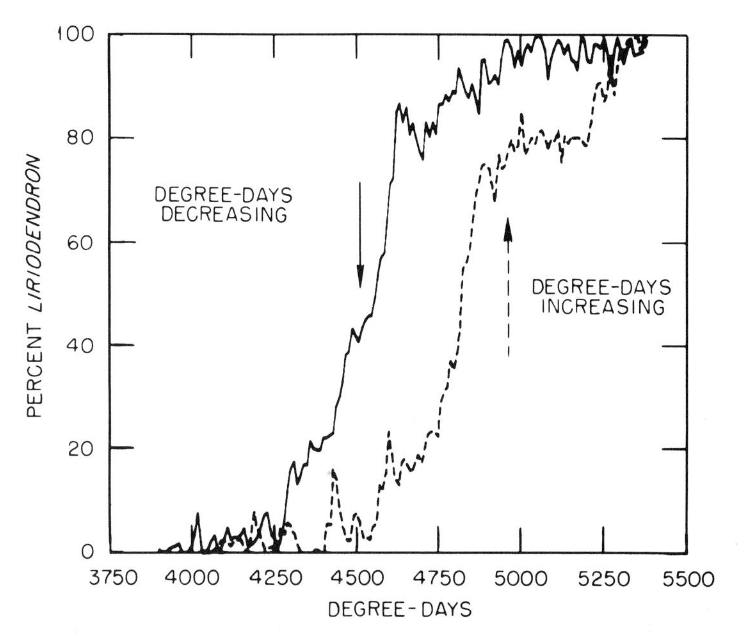

Figure 7.5. Percentage of total stand biomass attributable to yellow poplar under gradually changing D_e values. One curve (dashed line) illustrates the mean value of five simulated 1/12-ha forest stands as the annual value of D_e is increased from 3800 to 5300 GDD at a rate of 1.0 GDD/y (following an initial equilibrium period of several hundred years). The solid line illustrates the same conditions except the annual value is decreased from 5300 to 3800 GDD.

year even in the face of stochastic mortality of canopy trees. This sort of dynamics is likely to be a product of the shade-tolerant, slow-growing (relative to yellow poplar), and smaller-sized silvics of beech.

There are also indications of evidence for such hysteretic behavior in natural ecosystems. This evidence involves cases in which natural communities appear to have more than one potential equilibrium state at equivalent points in space. There has been considerable debate on the existence of multiple equilibrium states in plant communities (Tansley 1920; Nichols 1923; Clements 1936; Gleason 1939) that is to some extent semantic (Cain 1939). But there are several cases of communities near extremes of their natural distribution being replaced by a different community, following a disturbance. For example, Eric the Red's estate at Qagssiavssuk, Greenland, has not been restored, even after 400–500 years, to the birch forest that was cleared by Norsemen (Polunin 1937). Similarly, Marie-Victorin (1929) notes that the spruce forest on the Gulf of St. Lawrence is maintained only by virtue of mutual protection among the trees, and if this forest is destroyed over a large area, it is not able to reconstitute itself. Griggs (1946) noted that once forests are removed from the higher zones on Mount Washington, reforestation does not occur. Griggs supposed that the occurrence of these high-altitude forests, in an area where they will not presently regenerate from a nonforested habitat, resulted from to their establishment during more conducive climatic conditions. A subsequent climatic change to a less favorable condition has not destroyed the forest because of its vegetative method of reproduction and maintenance. This explanation indicates that a hysteresis effect can be involved in montane plant community zonation.

Conclusions

Our investigations identified several species adaptations (in the forms of patterns of model parameters associated with a particular species) that can be manifest as ecosystem behavior when these species are important. We also noted ecosystem behavior that clearly is emergent at the system level and is not a simple consequence of the species that comprise the ecosystem. The hysteretic behavior that we noted is a clear, albeit theoretical, result of the latter kind, as is the overshoot behavior of the ecosystem biomass curve. In our opinion, a fruitful line of ecological investigation lies in the simultaneous consideration of individual attributes in the context of ecosystem effects. While such investigations are difficult, the use of computer models of succession form a basis of a theory of community dynamics that can be tested against observations, that can be manipulated to produce testable hypotheses, and that can provide long-term predictions on ecosystem behavior.

Acknowledgments

This research was supported by National Science Foundation's Ecosystem Studies Program, under Interagency Agreement Nos. DEB-77-26722 and DEB-77-25781, with the U.S. Department of Energy, under contract W-7405-eng-26 with Union Carbide Corporation. Publication No. 1691, Environmental Sciences Division, Oak Ridge National Laboratory.

Chapter 8

Long-Term Forest Dynamics

Linda B. Brubaker

Introduction

Forest succession is a lengthy process compared to the life span of human beings. Secondary succession in hardwood forests of eastern North America and Europe, for example, requires several hundred years to reach a theoretical end point (Hough and Forbes 1943; Jones 1945; Raup 1957; White 1979). Some conifer forests of western North America may not reach climax for more than 1000 years owing to the dominance of long-lived early successional species (Franklin and Dyrness 1973; Franklin and Hemstrom Chapter 14). Primary succession and vegetation adjustments to climatic change (Wright 1971; Davis 1969a, Chapter 10) occur over even longer time periods than these.

Ecologists must, therefore, adopt indirect methods to study successional changes. One indirect method involves the use of chronosequences, sets of different-aged communities that are assumed to represent stages in plant succession on a single plot of ground (Drury and Nisbet 1973). Chronosequences theoretically assess the effect of time by holding all other variables constant. Another approach is the reconstruction of vegetation changes from plant remains. For relatively short time periods (e.g., 100 years), forest histories can be reconstructed from the growth rings and stem positions of living and dead trees (Stephens 1956; Henry and Swan 1974; Oliver and Stephens 1977). For longer periods, the approach typically involves the analysis of pollen grains preserved in lake and bog sediments (Faegri and Iverson 1975; Davis 1969b). The most recent method for investigating forest succession is simulation modeling (Botkin et al. 1972; Shugart and West 1977). Based on information on species' life histories and responses to physical factors and stochastic events, simulation models can provide information about successional changes in numerous components of forest ecosystems.

These three approaches for studying succession are represented by the next three chapters. In each chapter, factors causing very long-term changes in forests, those which span several millenia or more, are considered. While such changes may be mediated by disturbance events, the role of disturbance as it relates to secondary succession is not the primary focus of these discussions. This chapter discusses some of the assumptions of chronosequence, reconstruction, and simulation approaches to the study of long-term forest changes and examines some of the contributions such studies have made to current ideas about succession and climax.

Methods of Study

Chronosequences

Chronosequences have been research tools in ecology from the inception of the field. For example, Cowles (1899) described the vegetation on a series of dunes along the shores of Lake Michigan, U.S.A., and, in so doing, set the foundation for Clements' theories of succession and climax (Clements 1916). Since that time, vegetation chronosequences have been used repeatedly to describe community and ecosystem changes during primary and secondary succession. Most of the articles in this volume, for example, are based on data from chronosequences. Since various attributes (e.g., species abundances, net primary production, soil nutrients) of the communities making up chronosequences can be directly measured, very precise information can be obtained, and conclusions need not rely heavily on the more indirect reasoning involved in reconstruction and simulation techniques. Chronosequences have become important recently, furthermore, for the verification of forest simulation models (Botkin et al. 1972; Shugart and West 1977).

Soil scientists have also adopted the chronosequence approach to examine the processes and rates of soil development. Jenny (1941) first described the underlying assumptions of this approach, stating that soil properties are a function of the (1) age of the soil, (2) climatic environment, (3) plants and microbial organisms, (4) topography, and (5) parent material. Theoretically, a chronosequence would vary soil age and hold the other four variables constant. It is hard to imagine, however, a situation in which all other variables could be strictly constant. For example, soils initiated at different times in the past have not passed through the same sequence of climatic variations because climates vary uniquely through time. Unique climatic sequences would therefore lead to different vegetation compositions on similarly aged soils. Information from historical or paleoecological records makes it possible to assess the importance of these climatic and vegetation variations on the soil processes under study. If climatic and vegetation shifts are found to be relatively minor, chronosequences can be successfully applied.

The study of soil chronosequences is also limited by the availability of accurately dated land surfaces of similar topography and parent material. Glacial deposits, dune systems, mud flows, and raised terraces provide useful surfaces, but these types of deposits are rare in many parts of the world.

The power of chronosequences for studying long-term successional processes is well illustrated in the study by Walker et al. (Chapter 9). From the analysis of soils, nutrients, biomass, and forest vegetation across the Cooloola sand dunes in east coastal Australia, these authors have been able to show that the composition and structure of forest communities and their recovery following disturbance are dictated primarily by soil changes that have taken place over several hundreds of thousands of years.

Vegetation Reconstruction

Pollen grains in lake and bog sediments provide continuous records of vegetation change over time (Faegri and Iversen 1975; Davis 1969b). The interpretation of pollen frequencies in terms of vegetation is complicated, however, by the differential pollen production and dispersal rates of plant species involved and the difficulty of distinguishing many taxa at the species level.

While early pollen studies interpreted vegetation in very general terms, recent work has greatly improved the precision of palynological inferences. One improvement is the use of absolute pollen influx (pollen grains accumulating per square centimeter of sediment per year) rather than relative pollen percentages (Davis 1969b). Unlike pollen percentages, the pollen influx of any one species is independent of the abundance of pollen from other species. The relationship between pollen influx values and actual plant abundances is still quantitatively different for every species, however, depending on their pollen production and dispersal characteristics (Davis et al. 1973). This relationship may also vary between sites owing to sedimentation processes unique to each bog or lake (Davis et al. 1973).

Several recent studies attest to the validity of pollen records from small lakes (Webb 1974a,b,; Brubaker 1975) and localized organic deposits (Iversen 1969; Andersen 1970) for examining the long-term dynamics of individual forest communities. The accuracy of pollen records has also increased in recent years by the identification of plant macrofossils (leaves, fruits, seeds), which provides conclusive evidence that a species occurred in the immediate vicinity of a collecting basin (Birks 1973; Watts 1967). Finally, in the past few years numerous species-specific identifications have been made possible by the careful measurement of pollen grain dimensions (Birks and Pegler 1979) and by the use of scanning electron microscopy (Martin 1969). All of these techniques have improved the ability of palynologists to interpret past species abundances and distributions.

The time resolution of pollen records (decades to centuries) is appropriate for studying the population dynamics of tree species. No other data source provides continuous information about changes in tree abundances in vegetation unaffected by humans. Such information is necessary for evaluating conceptual and mathematical models of tree population and community dynamics. Davis (Chapter 10) and Solomon et al. (Chapter 11) demonstrate that pollen studies provide unique information for evaluating traditional ecological concepts. They have examined the climax concept from evidence on the long-term stability and climate-dependence of forest associations in eastern North America.

Simulation Models

Forest simulation modeling and pollen investigations have recently been combined to examine processes of forest succession (Solomon et al. 1980). Comparisons between model output and pollen records can be made because these approaches deal with comparable time scales of forest change. In addition, both consider the composite behavior of trees in a number of stands at different stages in secondary succession.

The most compelling reason for combining pollen and simulation studies is to test hypotheses about causes of vegetation change. Since model structures are statements of species interactions and environmental responses that bring about vegetation change, one can alter model structures (or parameters) to identify which factors were most important in causing the vegetation changes revealed by pollen records. Solomon et al. (Chapter 11) illustrate this approach. In one of their simulations, for example, they tested the hypothesis that responses to immigrating species rather than to climate caused the forest changes recorded in a 16,000-year-long pollen record from central Tennessee, U.S.A. (This test was accomplished by holding the climatic driving variables of the model constant and varying the species available for establishment.) Forest simulation models can similarly be used to examine the independent or combined roles of climate, competition, soil leaching, and disturbance in the vegetation changes revealed by pollen records.

The use of models to test hypotheses of past vegetation change relies on accurate model representation of the biological and physical interactions in the ecosystem. Models that have been verified and validated to the extent of JABOWA and FORET appear to be suitable for use in conjunction with pollen studies. It is possible, of course, that existing models cannot be altered to simulate some of the vegetation changes in pollen records, or that the necessary alterations are biologically unreasonable. In such cases, the structure of the model, the interpretation of the pollen record, or the hypothesis being tested may be incorrect. This should lead to a reevaluation and improvement of the assumptions involved and an increase in our understanding of the system under study.

Simulation results may also clarify some of the taxonomic uncertainty in pollen records (Solomon et al. 1980). Sometimes species with contrasting ecologies but similar pollen morphology must be lumped into the same category for the interpretation of pollen records. Simulation models, however, describe the abundance of individual species, and thus provide important taxonomic clues for the interpretation of pollen records.

Implications of Results

Physical and biological evidence of past environments shows that the earth's climate is continuously changing (Lamb 1972, 1977). Some variations, such as glacial epochs, span more than 100,000 years, but others last only a few years, decades, or centuries. While climatic changes are better documented than other types of environmental changes, nonclimatic changes (e.g., soil development) may also be very important influences on forest systems.

Forest ecologists have not rigorously examined the effects of changing environments on successional processes. Such considerations can, however, become an important priority of future research because information of past environments is now increasing to a level where it can be integrated into studies of forest succession. The articles in this section demonstrate that new insights can come from this type of integration. Some of the implications of these and other studies to the population dynamics and evolution of tree species are discussed below.

Interspecific Adaptations

In eastern North America, many tree species that occur in the same stands today did not grow together in the past, and vice versa (Davis 1976, Chapter 10). Many present-day boreal and north-temperate forest communities, in particular, did not exist 20,000 years ago. The physical and competitive environments have certainly changed too rapidly to permit close coevolutionary relationships to develop among those species. In those forests, we should not expect that two coexisting species display genetic adaptations specifically to each other, even though we consistently see them together in current forest communities. Rather, we should expect that they interact in general ways or that they can interact in a particular way with several different species, including those with which they do not occur today.

Climax Communities

Davis (Chapter 10) emphasizes that American and European forests have changed repeatedly during the Quaternary Period (1.7 million years ago to the present), as tree species adjusted their ranges individually to alternating glacial and interglacial conditions. Throughout this period, tree species occurred in loose associations rather than as tight complexes that respond as units to environmental change. The postglacial migration patterns of individual tree species in eastern North America also suggest that many species have not been in equilibrium with climate during the past 14,000 years. As a consequence, some vegetation changes during that period resulted from the delayed immigration of species following glacial retreat. These ideas strongly refute the traditional concept of climax.

Solomon et al. (Chapter 11), on the other hand, argue that pollen records from unglaciated central Tennessee, U.S.A., support certain aspects of traditional climax theory. According to their simulation model-versus-pollen comparisons, tree populations in central Tennessee have been in equilibrium with prevailing climates throughout the past 16,000 years. The concept of climax as a climatically controlled regional vegetation, therefore, appears to be valid in this area.

The apparent conflict between these two chapters, as Solomon points out, occurs because Tennessee is near the glacial refugia of most of the tree species involved (Delcourt 1979), while most of the areas discussed by Davis lie far to the north. Since differential species migration rates will cause vegetation changes to be out of balance with climate as distance from the glacial refugia increases, pollen records from north of the glacial border should be expected to emphasize the existence of disequilibrium conditions within the vegetation.

These two chapters together put certain limitations on the climax concept. Neither supports the idea that forests are made up of stable species combinations with attributes of superorganisms. Even in the relatively stable, unglaciated landscape of central Tennessee, for example, many of the postglacial forests have no exact modern counterparts. The idea of a regional vegetation in general equilibrium with the climate appears to be valid in certain areas where proximity to seed source allows species to adjust quickly to changes in the prevailing climate, but not reasonable in instances when the environmental changes are very rapid and distances to seed source are great.

Population Age Structures

Long-term changes in the climatic and competitive environments of trees may prevent the establishment of stationary age distributions. Physical evidence of climatic change during the past 20,000 years, such as $^{18}O/^{16}O$ ratio changes in ice cores, indicates that temperatures have rarely remained stable for more than 3000 years (Lamb 1972, 1977). Pollen diagrams from north temperate areas similarly show that forest compositions and, therefore, the competitive interactions of trees have seldom remained constant for more than 2000–3000 years in a given area.

The potential effect of such environmental changes on the age structures of forest trees can be evaluated by considering results of recent ecological studies. Mathematical and experimental analyses of population growth and competition typically encompass periods of 5–10 generations of the species involved (Colinvaux 1973, Pianka 1978). For long-lived trees, such as Douglas fir (*Pseudotsuga menziesii*) and western red cedar (*Thuja plicata*) (Franklin and Hemstrom Chapter 14), this period could span a few thousand years. Assuming a mean generation length of approximately 150 years for hardwoods and 300 years for conifers, pollen records suggest that there have been few opportunities for long-term stationary age distributions to develop. Climax, in the strict sense of an assemblage of trees with stationary age distributions, is improbable, therefore, given our current knowledge of rates of environmental change. Stationary age distributions are even less likely, of course, because the effects of disturbance events are superimposed on those of long-term changes in the physical environment.

Appropriateness of Twentieth Century Environmental Data

Climatologists have recently asserted that the twentieth century has been warmer than any other century in the past 1000 years (Lamb 1972; Bryson and Hare 1974). Temperature estimates from $^{18}O/^{16}O$ measurements of ice cores for the past 1000 years (Fig. 8.1) clearly show the anomalous nature of twentieth century temperatures (Bergthorsson 1962). Though shorter in duration, instrumental temperature data indicate that the twentieth century warming trend has been widespread (Brinkman 1976).

This recent shift in climate makes it difficult to identify environmental factors that have been important in determining the composition and distribution of existing forests. For example, if climate-related factors strongly affect seed germination and juvenile tree mortality rates, pretwentieth century conditions may have been more important than recent conditions in determining the species composition of some forests.

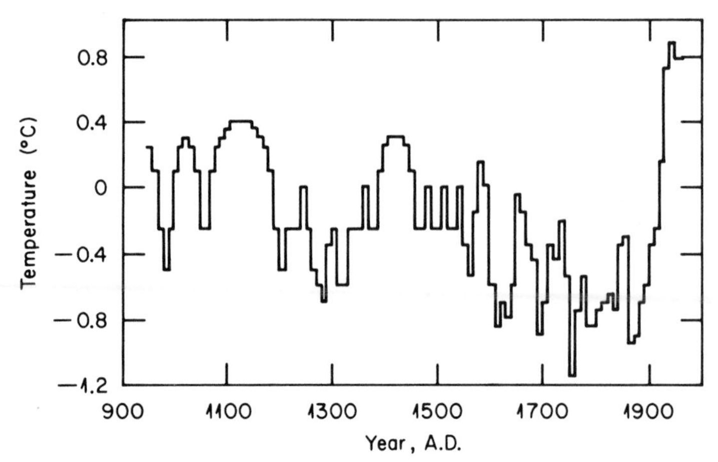

Figure 8.1. Running 20-year means of mean annual temperture in Iceland derived primarily from sea ice data (from Bergthorsson 1962).

Unfortunately, long-term instrumental weather records were not systematically kept in most parts of the world before this century (Lamb 1972, 1977). As a result, many forest ecologists are forced to work with weather data biased toward the anomalously warm twentieth century. The problem is particularly acute in the United States and Canada, where national weather data have been summarized for a standard period of 1931–1960 (United States Geological Survey 1965, Canadian Geological Survey 1965), the warmest decades of the twentieth century.

In the absence of abundant data on recent climatic change, there is naturally a tendency to ignore the potential effect of such changes on succession processes. The evidence for a significant climatic shift during the past 100–200 years is so compelling, however, that empirical and simulation studies of succession should evaluate the appropriateness of twentieth century environmental data for their investigations. The following paragraphs present one example to illustrate the misleading effect of using climatic data from the standard 1931–1960 period in forest simulation models.

Existing simulation models (Botkin et al. 1972; Shugart and West 1977) calculate the growth rates for each tree species as a function of the total growing degree-days per year (GDD) for a given site in relation to GDD's at the northern and southern limits of the species' distribution. GDD values are calculated from mean July and January temperatures published by the United States Geological Survey (1965) and the Atmospheric Environmental Service of Canada (Downsview, Ontario) for the standard 1931–1960 period. Long-term records of July and January temperatures from several parts of the world (Fig. 8.2) show that 1931–1960 values are often different from those of the preceding century. In such cases, the less anomalous eighteenth or nineteenth century data would be

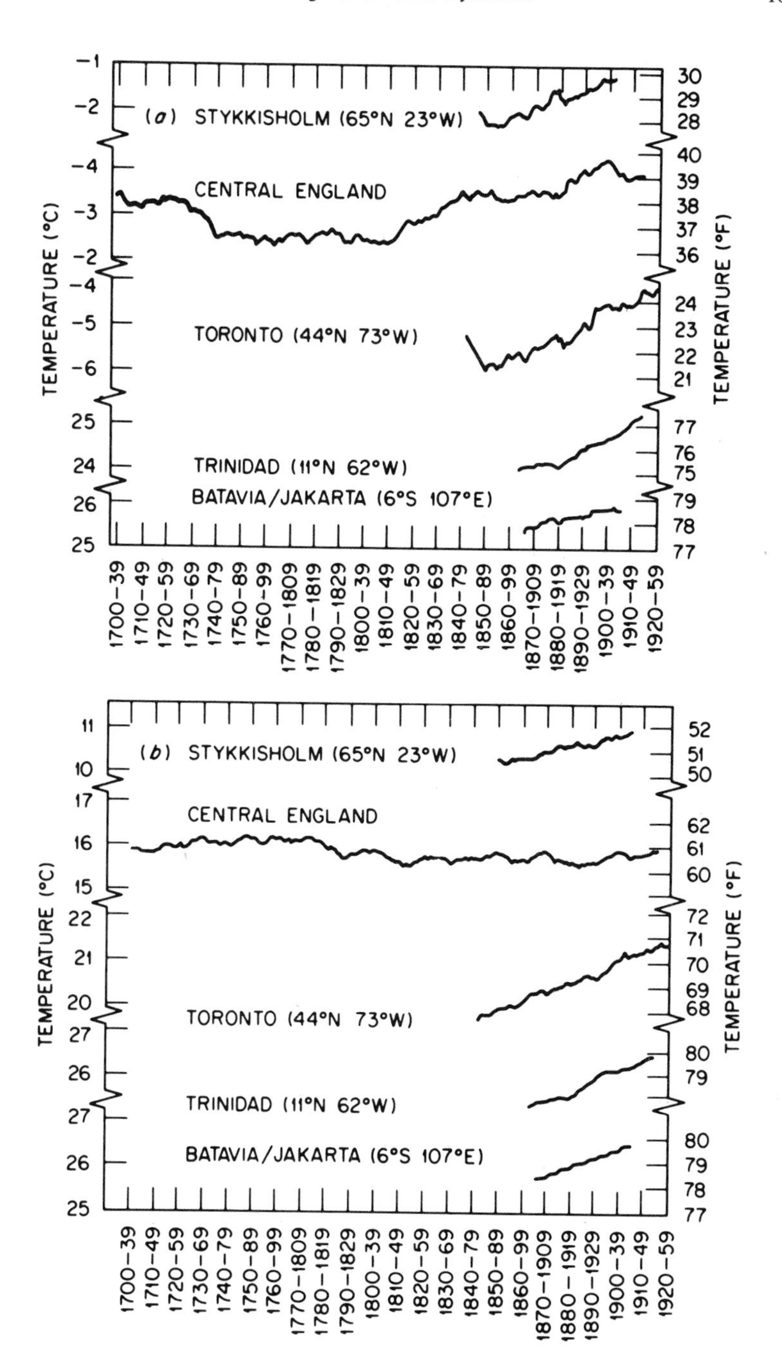

Figure 8.2. Average temperatures at selected places in different latitudes. Forty-year running means: (a) January; (b) July (from Lamb 1977).

better for calculating GDD values used to simulate the development of existing mature forest stands.

When GDD's are calculated using temperature data from Toronto, Canada (Fig. 8.2) in the standard equation of simulation models, the GDD values increase by approximately 500 GDD from the mid nineteenth to mid twentieth centuries (Fig. 8.3). Increases in GDD of this magnitude may significantly change the growth, reproductive, or competitive ability of some trees, especially those near their distribution limit. Solomon et al. (Chapter 11), for example, suggest that a 250 GDD increase caused the disappearance of spruce from central Tennessee 13,000 years ago. The Toronto example suggests that ecologists in eastern Canada should question the suitability of twentieth century climatic data for modeling or otherwise explaining successional sequences that began 100 or more years ago.

Instrumental weather records (Kutzbach 1970; Van Loon and Williams 1976) and tree ring evidence (Fritts et al. 1979) of past climatic variations indicate that climatic conditions do not vary uniformly in magnitude and direction across long distances. As a result, one cannot generalize that GDD changes like those in Toronto have occurred everywhere. The GDDs calculated from the London data in Fig. 8.2, for example, changed only about 100 GDD during the past century. To assess the importance of recent climatic changes to a given study, ecologists must, therefore, seek out long climatic records close to the investigation site. Even if early records are sporatic, they may provide a more representative picture of long-term climatic conditions.

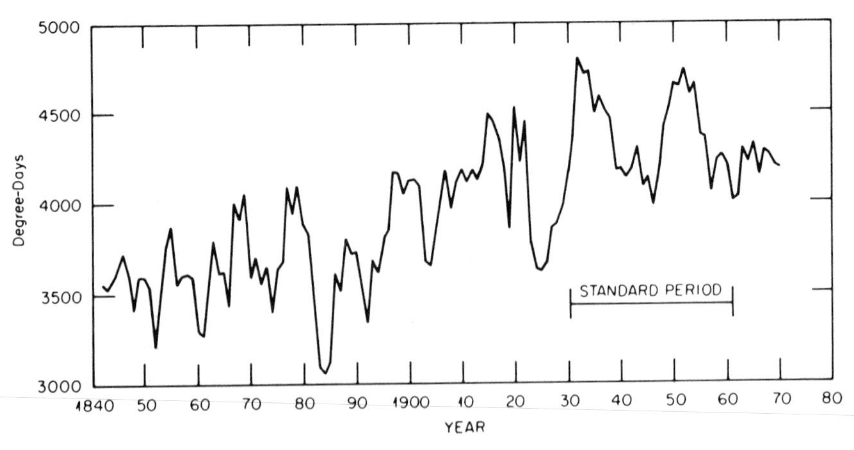

Figure 8.3. Three-year running average of GDD calculated from Toronto, Canada, mean July and mean January temperature records. The standard period of data summarization is indicated. Average values and variances during this period are typically used as parameters in simulation model runs.

Effects of Soil-Forming Processes

Soil-forming processes are a combination of many individual reactions taking place at different rates within the soil system. Some reactions, e.g., oxidation-reduction reactions, equilibrate rapidly to environmental changes, while others, e.g., the downward movement of clays, proceed slowly in a given direction, even under oscillating environmental conditions (Yaalon 1971). The longer term soil reactions can theoretically act as a driving force to bring about vegetation change under otherwise constant conditions. However, evidence of long-term reactions and their effects on vegetation is difficult to obtain because well-dated chronosequences are necessary for studying them.

Despite this dilemma, several convincing studies exist which suggest that long-term, irreversible soil processes are widespread and have led to significant vegetation changes in large portions of the world (Yaalon 1971). Soil-forming processes, for example, have induced peat land formation in southeast Alaska and perhaps other boreal regions (Ugolini and Mann 1979). The long-term leaching of nutrients in old soils of some temperate and subtropical regions have produced infertile soils that support only sparse vegetation, even though the prevailing climate favors rapid tree growth and large biomass accumulation on young soils (Westman and Whittaker 1975; Walker et al. Chapter 9). In semiarid regions, the accumulation of calcium carbonate and silica in the soil can proceed to a point where calcareous and/or siliceous deposits inhibit the growth of many plant species across large areas. The effect of long-term soil forming processes on forest structure and composition is dramatically evident in areas such as New Zealand and Australia, where soils of greatly different ages are interspersed on the landscape (Gibbs 1971; Walker et al. Chapter 9). In such areas vegetation mosaics conform to age-determined properties of underlying soils.

The Cooloola chronosequence described by Walker et al. (Chapter 9) has special relevance to theories of forest succession because it demonstrates that patterns of ecosystem change following disturbance can also be controlled by the soil developmental stage of a site. In the very young soils of the Cooloola dune system, nutrient inputs through mineral weathering exceed nutrient losses through leaching, and excess nutrients accumulate within the organic components of the ecosystem. On intermediate-aged soils most nutrients are cycled between the soil and the vegetation, with minor amounts lost through leaching. This stage of soil development is maintained for very long periods because plants are able to take up and recycle nutrients from the B horizon (zone of accumulation) of the soil. However, continued leaching eventually lowers the B horizon below the rooting zone, and the vegetation can no longer take advantage of this nutrient store. As a result, vegetation biomass and productivity decrease.

According to Walker et al. (Chapter 9), the characteristics of secondary succession in this sequence depend on the soil-forming stage of a given site. Disturbed vegetation will return to a preexisting state, the traditional view of progressive succession, only at young and intermediate soil-developmental stages when nutrient inputs compensate for nutrient losses. A retrogressive sequence, in which biomass levels are permanently reduced and the species composition permanently altered, is characteristic of secondary succession on very old soils where nutrient losses predominate.

These authors suggest that the processes documented at Cooloola are general features of soil development over very long periods and can be found on different-aged soils throughout the world. Since most of the earth's land surface is relatively old, their findings suggest that retrogressive successional changes may be more widespread than progressive changes. In this example, we are again faced with the possibility that the traditional concept of progressive succession has predominated in the ecological literature because most studies have been carried out in the relatively young landscapes of North America and Europe. Evidence from other parts of the world must be examined to evaluate the variation that can exist in secondary successional sequences and the role of soil-forming processes in bringing about that variation.

Conclusions

Our understanding of forest succession is constrained by the fact that we can witness only a small portion of the entire process. Indirect means such as chronosequences, pollen records, and simulation models permit the study of succession at time scales that are more appropriate for the changes taking place. The following three chapters demonstrate that such methods provide important data for evaluating traditional concepts of succession and climax. These studies suggest, furthermore, that changes in soil factors, competition, and climate are continuously modifying the composition of the earth's forests. If we are to understand the evolutionary and ecological characteristics of tree species, we must evaluate the influence of these long-term environmental changes.

Chapter 9

Plant Succession and Soil Development in Coastal Sand Dunes of Subtropical Eastern Australia

J. Walker, C. H. Thompson, I. F. Fergus, and B. R. Tunstall

Introduction

Progressive plant succession to a climax has developed as a central ecological concept (Cowles 1899; Clements 1916), and forms an important basis for predicting changes (over moderate time scales) in primary and secondary plant succession. Although the concept of progressive succession leading to a stable, self-perpetuating climax has been questioned (Whittaker 1953, 1974), the climax concept has guided much of our thinking in the management of indigenous forests and in the prediction of forest development following various forms of disturbance in different environments.

Observations that appear to validate the climax theory have largely been carried out on relatively fertile soil sequences in postglacial landscapes of the northern hemisphere (Crocker and Major 1955; Olson 1958). The effects of prolonged soil weathering are absent in such areas and field evidence is biased toward the recognition of progressive succession stages to a climax since insufficient time has elapsed for retrogressive changes to be important. Nevertheless, because most of the earth's surface is unglaciated and, therefore, relatively old, it is of considerable scientific importance to evaluate the applicability of the climax concept to successional sequences in landscapes subjected to prolonged periods of soil weathering.

Successional sequences that stretch beyond recorded history pose a basic problem in demonstrating succession. Drury and Nisbet (1973) state that in this situation, "it is necessary to assume a homology between spatial sequences of zones of vegetation visible at one time in the landscape and a long-term sequence of vegetation types on a single site."

A similar assumption has been made in pollen analytical studies (Walker 1971) because of "a reliance on the parallelism of pollen curves to substantiate supposedly equivalent climate sequences in distant parts of the world . . . ." The problem in also encountered by pedologists in the interpretation of soil chronosequences (Jenny 1941; Stevens and Walker 1967).

The large sand dune systems on the subtropical coast of Queensland may provide an opportunity to validate the climax concept experimentally and possibly by direct observation. The soils have low nutrient reserves, and weathering, soil development, and forest growth have apparently proceeded without interruption by the last glaciation and probably for a much longer period before that. Six of these dune systems form a relative age sequence extending through the late Quaternary Period (Thompson 1980). Each dune system supports different sclerophyll forest communities ranging from dwarf woodlands and grassy forests to extremely tall layered woodlands and forests, varying in height from 3 to 50 m. Depth of weathering and soil profile development also change with age of dune system from < 1 m in the youngest to > 20 m in the oldest system.

This study of soil-vegetation relationships forms part of a much more extensive investigation by CSIRO Division of Soils into the soil and landscape processes controlling the stability of vegetated coastal dunes in the subtropics. In this chapter we summarize the main soil features, nutrient inputs, and forest vegetation of the dune systems and examine both progressive and retrogressive phases of forest succession evident at Cooloola.

The Sand-Dune Systems of the Subtropical Coast

Five large accumulations of windblown sands occur as island and mainland deposits along the east coast of Australia between 25 and 28° south latitude (Coaldrake 1962; Thompson 1975). The largest mainland deposit, the Cooloola sand mass (lat. 153°E, 26°S), located 50 km northeast of Gympie, Queensland, extends 40 km along the coast and 10 km inland (Fig. 9.1). It consists of dominantly quartz sands deposited in overlapping dune systems, which form large compound hills rising to elevations of up to 225 m. Large areas of six of these dune systems have not been covered by younger deposits and form a relative age sequence extending through the late Quaternary Period.

Sands of the Cooloola dune systems consist almost entirely of quartz grains but include some feldspar and heavy minerals such as ilmenite, rutile, zircon, and monazite. Because most of the sand grains are coated with sesquioxide, they form dominantly yellow and yellowish-brown beds, but various shades of white, red, and black sands also occur. Shell fragments are insignificant even near the coast, and calcium carbonate cementation is absent in the main sand mass. Cementation by iron and/or organic matter to form consolidated bands of varying thickness and hardness is relatively common, however.

Description of Dune Systems

Two distinct series of dunes younger than Tertiary were initially postulated by Ball (1924) and Coaldrake (1962). These authors discussed their possible origins and ages, concluding that they were the result of intermittent dune building during the Pleistocene. However, more than two series of dunes were identified during reconnaissance soil mapping, and new units were recognized to accommodate variations in both dune geomorphology and soil development (Thompson 1975; Thompson and Ward 1975). Later, Ward et al. (1979) found that two units were inadequate to explain stratigraphical differences in a sea cliff section.

The Cooloola sand mass is now seen as a series of overlapping, Quaternary dune systems in which eight periods of dune building can be recognized in the deposits above sea level (Thompson and Moore, 1980). Although the absolute age of these systems is unknown, large parts of six of them have been continuously exposed to subaerial weathering since deposition (Figs. 9.2 and 9.3). These dunes form a chronosequence showing increasing weathering, soil formation, and erosion (Thompson 1980).

When a sand dune is vegetated, aeolian erosion and deposition cease and the dune sands are exposed to weathering and water erosion. These processes lead to soil profile development and modifications to dune

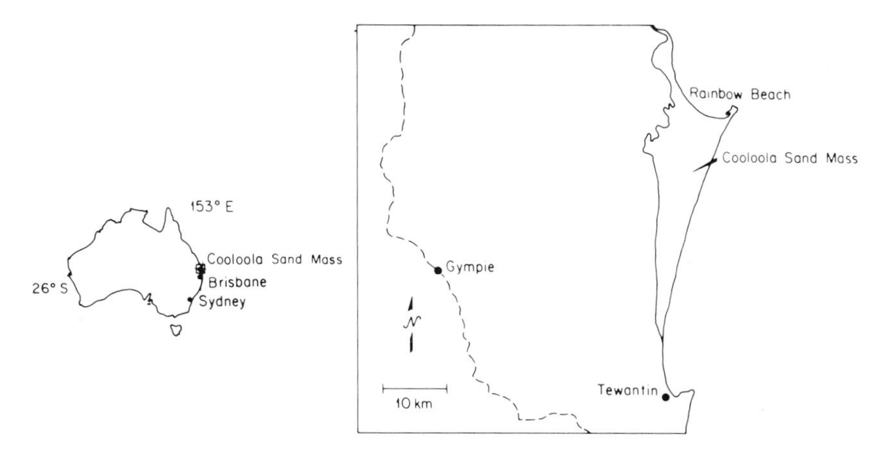

Figure 9.1. Location of the Cooloola sand mass.

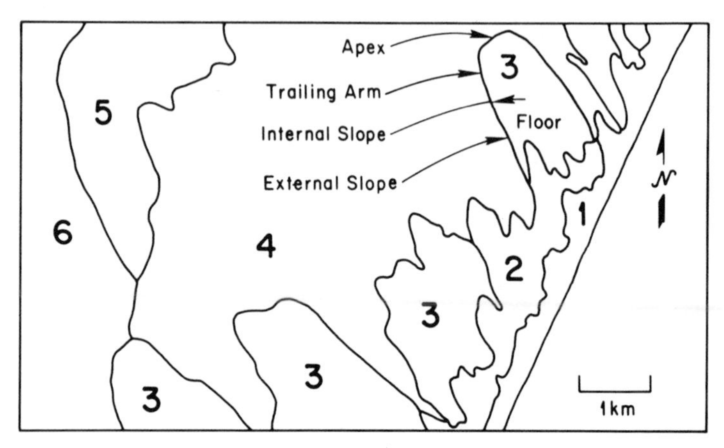

Figure 9.2. Interpretation of dune systems from an air photograph of part of the Cooloola sand mass (Thompson 1980). The parabolic shapes are not drawn in detail; the geomorphic terms used are shown for a large parabolic dune of system 3.

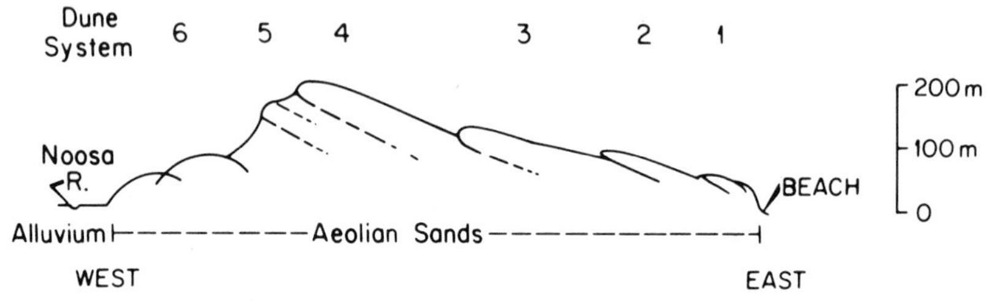

Figure 9.3. Schematic representation of overlapping sand dune systems at Cooloola, Queensland, forming a relative age sequence (after Thompson and Moore 1980).

shape. The four youngest dune systems recognized at Cooloola (1, 2, 3, 4 of Fig. 9.2) are made up of U- and V-shaped (parabolic) dunes open to the formative southeast winds. This is the normal shape of blow-out dunes formed by onshore winds blowing sand into humid vegetated lands, particularly those supporting woodlands and forests.

The vegetation on either side of an area of drifting sand traps some sand, building roughly parallel sand ridges, the trailing arms, extending upwind behind the advancing dune apex and forming a U- or V-shaped dune with a concave *floor* between the trailing arms. The external slopes of the dune are steep and unstable and highly susceptible to water erosion. In comparison, the erosion of the crest of the trailing arms is rather slow, and the dune floor, near the apex, is protected by the dune shape.

Relative Age of the Dune Systems

No materials have been dated from ground surfaces underlying any of the dune systems at Cooloola. Radiocarbon dates from wood samples in basal beds of the sand mass, near sea level (Coaldrake 1962) and in sandy alluvium derived from the sand dunes (Polach 1978) are >45,000 years b.p.

Samples of an organic B horizon (humus podzol, Stace et al. 1968) and of charcoal embedded in nearby sand, both exposed in sea cliffs 80 m above the beach, have yielded ages of 30,300 ± 800 and 39,000 ± 3000 b.p., respectively (Coaldrake 1962). These materials both overlie beds of colored sands that stratigraphically under lie dune system 5. The location of this sampling site is poorly recorded, and the dune system it represents has not been identified; however, it is likely to lie at the base of dune system 4.

Although absolute ages are not available for the dune systems, they can be put in a relative age sequence and some deductions may be made of the magnitude of the periods involved from the degree of erosion and depth of weathering each dune system has experienced. The six sand dune systems exposed at Cooloola are believed to have been formed during episodic periods of dune building, which are probably related to the oscillating sea levels of the Quaternary Period. The preservation of initial dune shape and the limited depth of soil development in systems 1, 2, and 3 (see Fig. 9.2) show that they are relatively young and that the intervals separating their deposition are short; therefore, they are probably of Holocene age and can probably be correlated with the transgressive parabolic dunes of the Outer Barrier sands of New South Wales, which have been deposited since the sea reached its present position about 6000 years ago (Thom and Chappell 1975).

Dune systems 3 and 4 are separated by a much longer interval of weathering and erosion in which the trailing arms of system 4 have been breached in many places and the dune floors eroded; soil weathering on the protected remnants of trailing arms and dune floors has reached depths of 6–10 m. Dune system 4 is regarded as being of Pleistocene age.

Dune system 5 is even more extensively eroded, and only remnants of what were possibly the trailing arms persist as high sand ridges. The dune floors have been extensively weathered and in places carry small running streams. Soil profile development on the high ridges usually exceeds 12 m. Dune system 5 is certainly older than Holocene and may be equivalent to the parabolic dune remnants associated with parts of the Inner Barrier of new South Wales, which is regarded as high sea level deposits of the last interglacial and dated between 112,000 and 143,000 years ago (Marshall and Thom 1976).

Dune system 6 is clearly much older. It has been so strongly modified by erosion and weathering that no indications of initial dune shape or orientation have been found. It is characterized by broad whale-back sand hills separated by shallow open valleys with small running streams. The dunes of this system have been severely eroded, and there is little doubt that erosion has reduced the thickness of the soil weathering column; weathering profiles >15 m deep are common, and at some sites soil weathering extends to well beyond 20 m.

Sand grain sizes are similar across all of the dune systems (Ross 1977), but there is a higher proportion of weathered grains in dune system 6 (Little et al. 1978). This indicates that all the sands have had a similar origin and have arrived by similar means. Differences in sand hill shape and depth of soil development therefore are probably the result of weathering over a very long period.

In one place where the deeply weathered sands extend to the coast, a fossil beach rising to 3 m above the present ocean occurs at their base and is enveloped by the cemented organic B horizon of a giant humus podzol similar to profiles seen elsewhere in dune system 6. It has been suggested (Ward et al. 1979) that the fossil beach represents deposition during the warmest part of the Pleistocene (400,000 b.p.). If so, the deposition of dune system 6 possibly dates from this period.

Present and Past Climates

The Present Climate

The climate of the Cooloola sand dunes is influenced by continental weather systems responsible for cool season rains in the south and warm season rains in the north. The mean annual rainfall of about 1500 mm is characterized by a summer maximum and a high degree of monthly variability. The wettest months, January to March, have a mean monthly rainfall of about 200 mm, while the mean for each of the driest months— August and September—is near 60 mm (Anon. 1975). Short-term measurements over 2.5 years show little difference in rainfall between the coast and inland margin, or between different elevations (P. J. Ross 1980, pers. comm.).

Mean monthly minimum and maximum temperatures for the hottest and coldest months are 20 and 27°C in January and 13 and 19°C in July (Anon. 1968, 1975). Precipitation usually exceeds the calculated evaporation during the first six months of the year (Fig. 9.4). Late winter and early spring are usually drier, and moisture stress occurs in the native vegetation on the drier sites during the period. However, thunderstorms during the spring and early summer often provide sufficient rain to recharge the soil and initiate leaching. The soils are very permeable, and even during high-intensity falls there is little runoff from the sand mass; areas of water-repellent sands can cause local redistribution of water (B. J. Bridge 1980, pers. comm.).

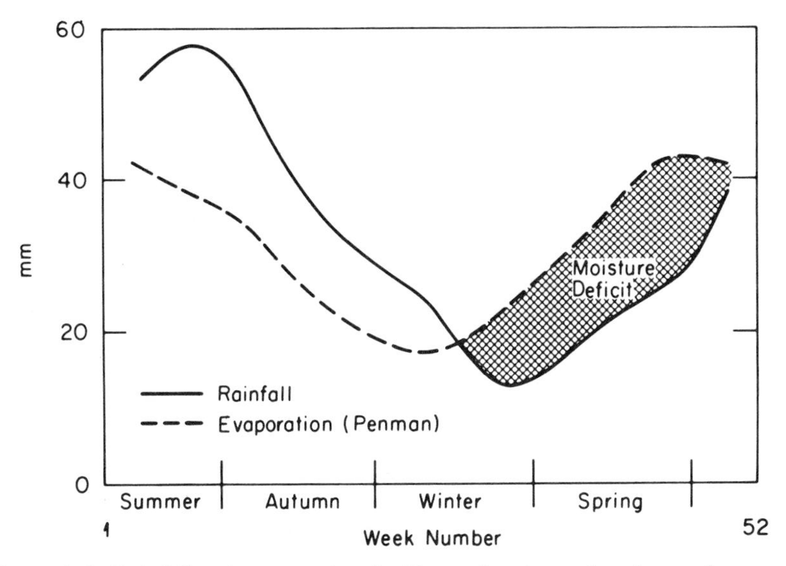

Figure 9.4. Rainfall and evaporation for Tewantin, situated at the southern end of the Cooloola sand mass.

Climates in Subtropical Queensland

Information about the late Quaternary climates of Australia has recently been reviewed (Bowler et al. 1976) using evidence from both southern and northern Australia and from New Guinea. Generally cooler and drier conditions prevailed 60,000–40,000 b.p. This was followed by a period of highly effective precipitation, which lead to widespread fluvial and lacustrine activity in southern Australia 40,000–25,000 b.p. Northern evidence, however, favors increasingly dry conditions during this period. Lower temperatures and drier conditions induced by the last glaciation from 22,000 to 15,000 b.p. were followed by relatively stable conditions similar to the present climate over the last 10,000 years, although there is some evidence for slightly higher temperatures and rainfall 8000–5000 b.p. and a slightly drier period 5000–2000 b.p.

Although variations in climate must have occurred along the subtropical coast during the Quaternary Period, no evidence has been found of marked climatic changes leading to total loss of vegetative cover. Questions remain, therefore, as to the magnitude of the climatic changes in this area and their effects on the coastal sand dune systems. Coaldrake (1968) has suggested that areas along the subtropical coast, lying within the overlap of the tropical and southern weather systems, have not been seriously affected by climatic changes. This rests on the premise that the estimated decrease in mean annual temperature (< 5°C) associated with glaciation would probably induce northerly movement of the southern anticyclone weather systems and so increase the winter rainfall to these areas, thereby compensating for the lower summer rain. Some of the lakes in the sand mass and in the adjoining coastal plain have been higher and larger in the past, but it is uncertain whether these reflect wetter conditions or higher sea levels.

Perhaps the best indirect evidence of a relatively stable climate during the Late Quaternary Period is indicated by the parabolic shapes of the dunes in systems 1–4. Since the parabolic shape and raised trailing arms and apices of the dunes are a consequence of sands blown into vegetated landscapes, it follows that the sand mass has been vegetated since before the deposition of system 4 began. While the absolute age of this system is unknown, the degree of erosion implies considerable age, older than Holocene.

In addition, the general lack of dune shapes characteristic of regional areas of bare sand and the evidence of different periods of fluviatile erosion lend support to the view that the sand mass has supported vegetation continuously since well before the last glaciation. Although the sea level has been considerably lower than that at present during most of this period, it seems that the lowering of water tables did not result in the complete loss of vegetative cover and that precipitation during this period remained effective.

Changes in atmospheric accession of nutrients certainly occurred with the changing position of the coastline and variations in wind direction and strength. Even so, the on hand evidence discussed above leads us to conclude that the sand mass carried vegetated cover throughout the last glaciation and probably for a considerable period preceding it. The effects of climatic change associated with the last glaciation appear to have been oscillations in some of the present vegetation formation boundaries.

Relationships between Soils, Nutrients, and Vegetation

Soil development in quartzose sand dunes in the subtropics is largely dependent on moisture regime, rate of erosion, inputs of organic constituents from the vegetation, and period of exposure to weathering. Since these factors vary among the different geomorphic components of parabolic dunes, different parts of a single dune show marked differences in intensity and depth of profile development. These differences on a single dune are usually reflected in local mosaic patterns in the vegetation. As the purpose of this study was to investigate the changing relationships between vegetation and soil profile development with relative dune age, it was essential to examine equivalent geomorphic sites in the different dune systems. The dune floors and broad crests of the trailing arms are

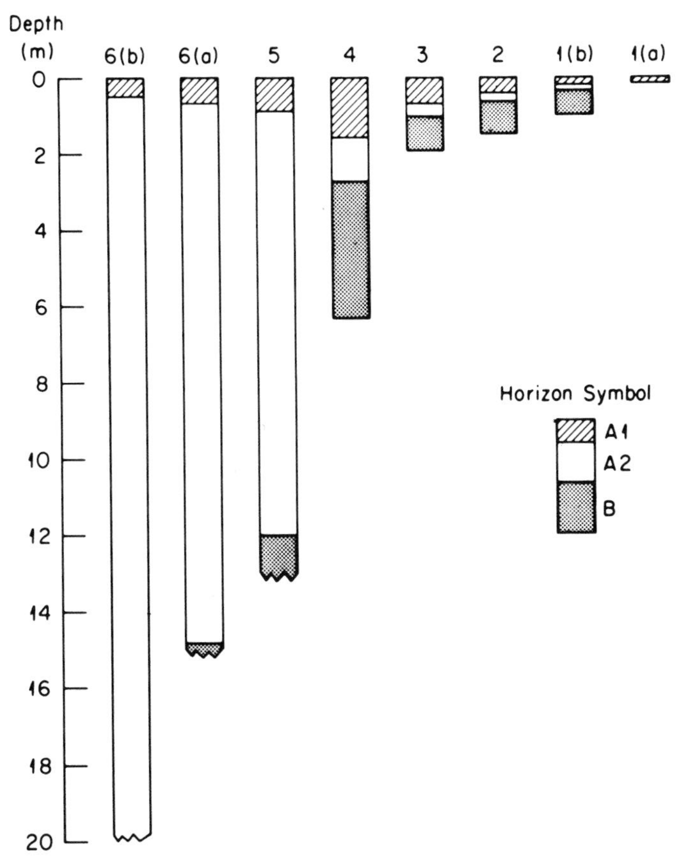

Figure 9.5. Thickness of soil horizons and depth of soil development on the dune systems of increasing age (after Thompson and Moore 1980).

least affected by water erosion and deposition. Dune floors are poorly developed in small V-shaped dunes, and at Cooloola they have not survived erosion beyond dune system 4. Therefore, sites on trailing arm crests have been used in our study, even though soil profile development is shallower than the maximum for each of the systems because of some erosion at these sites. In this study, 54 trailing arm sites were sampled, with a minimum of four sites in any one dune system or subdivision of it, for example, 1a, 1b, 6a, 6b (Fig. 9.5).

Soils

The soil profiles at Cooloola were described and identified from hand auger drilling through the solum, except at the very deep soil sites, where it was possible to drill only into the B horizons; the maximum drilling depth was 20 m.

Soils with the profile morphology of subtropical podzols have developed on freely drained sites on all but the very youngest dunes. These soils are characterized by very thin, discontinuous zero horizons; thick, dark, organic-enriched A1 horizons; thin to extremely thick bleached sand A2 horizons, and B horizons of accumulated organic, iron and aluminum compounds (Thompson and Hubble 1979). Soil development on the youngest dunes is restricted to organic enrichment of the top few centimeters, and these are classed as siliceous sands (Stace et al. 1968). Subtropical humus podzols having B horizons rich by organic and aluminum compounds but low in iron have formed where water tables reach the soil weathering zone, such as in interdune corridors. Most of the subtropical podzols have been placed as quartzipsamments in soil taxonomy (Soil Survey Staff 1975) because the soft and friable organic B horizons fail to meet either the field, micromorphological, or chemical requirements for spodic horizons or are too deep for spodosols (Thompson and Hubble 1979; Brewer and Thompson 1979). In addition, some profiles are classed as troporthods.

The mean thickness of the A1 horizon, mean depth to B2 horizon, and, where determined, mean depth of soil profile development are shown in Fig. 9.5. Soil profiles across the dune systems show progressive increases in depth and intensity of podzol development with increasing age; profile depth increases from <1 m on young dunes to >20 m in the oldest dunes. With increasing age both A and B horizons become progressively thicker, with a marked increase in the thickness of A2 relative to A1 and B horizons (Thompson and Hubble 1979). This age sequence exhibits an increasing thickness of the A1 horizon through dune systems 1, 2, 3, and 4, and a decline through systems 5 and 6 (6a and 6b). Soil formation lead removed sesquioxide coatings from the quartz grains in both A1 and A2 horizons (Brewer and Thompson 1979), with the only detectable difference between the two being the addition of organic matter to the A1 horizon.

Vegetation

The following data were collected at each sample site (ca. 1000 m^2): height and crown cover of each recognizable tree or shrub layer (using the crown ratio technique of Walker and Tunstall 1978); total number of species present, density of each woody species, height and percent age of cover of ground layer species. The total number of species identified (236) probably represents 95% of the species present of freely drained sites at Cooloola (A. G. Harrold 1980, pers. comm.).

The species data (binary data) for all sample sites (54 sites and 236 species) were subjected to a divisive monothetic numerical classification, which contains a reallocation procedure (Weir 1972). The program uses an information statistic for group recognition. Data on species presence and absence were perferred to density or cover data because the dune systems have been subjected to selective logging for hardwood timbers over the last 80 years and have experienced fires of varying intensity at intervals of 3–10 years since the end of the ninteenth century (Hawkins 1975). We have assumed that species presence is less affected by such disturbances than plant abundance variables, in particular species density and cover.

Floristic Groups

The classification is shown in Fig. 9.6. Abbreviated species lists for the groups are given in Table 9.1. The discrete groupings indicate that the different dune systems support different species assemblages, and the major floristic subdivision, Groups 0–5 versus Groups 6–10 in Fig. 9.6, is between "younger" and "older" dune systems. The major floristic groups within dune system 1, 4, and 5 can be divided into smaller groups, which appear to be related to the depth of soil weathering. For dune system 4, the split on each side of the major dendrogram subdivision is associated with the depth of the soil profile; the floristic groups with shallower or deeper soil profiles are allied to the younger or older dune systems, respectively. The number of tree species and the total plant species present in each dune system increases from system 1 to system 4 or 5 and then declines. It appears, therefore, that the depth of soil weathering is related to vegetation succession to the extent that successional changes are reflected in species composition.

An ordination of the 54 trailing arm sites (Fig. 9.7), using the same similarity matrix as the classificatory analysis, gives an indication of the floristic affinities across the dune systems. Dune systems 3 and 4 occupy roughly the same space on vectors I and II, and dune systems 1 and 2 are markedly different from systems 5 and 6. Vector I shows a sequential floristic change from dune system 1 to system 4; vector II shows floristic changes (termed retrogressive changes later in this chapter) in dune systems 5 and 6 away from dune systems 3 and 4.

The structural features of the vegetation on the trailing arms of the dune systems for the recognized floristic groups are shown in Table 9.2. The heights of the upper and lower vegetation layers increase from system 1 to system 4 and then decline through systems 5 and 6. The crown cover of the vegetation is very low at the colonizing stage of dune system 1a, but is high and relatively constant for dune systems 1b–3. The maximum crown cover occurs at dune system 4 and is followed by a rapid decline to the older dune systems 5 and 6. The older dune systems have little tree crown cover. The sequence of increasing vegetation height and cover from dune systems 1 and 4 is followed by a decline through systems 5 and 6.

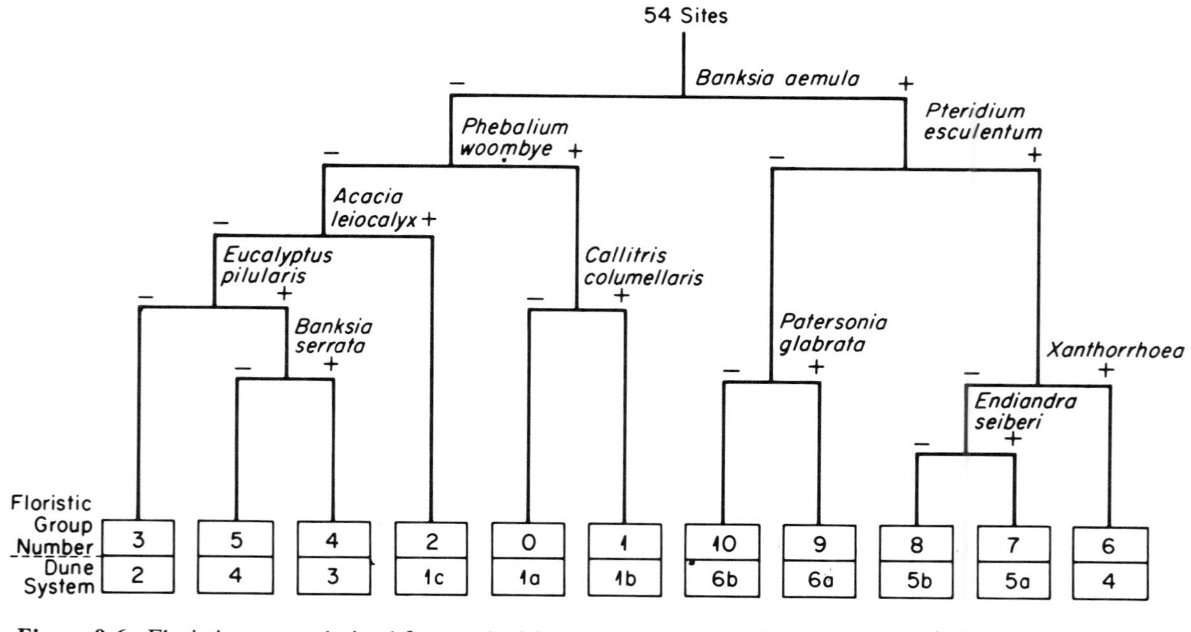

Figure 9.6. Floristic groups derived from a devisive monothetic classificatory analysis (species listed in Table 9.1).

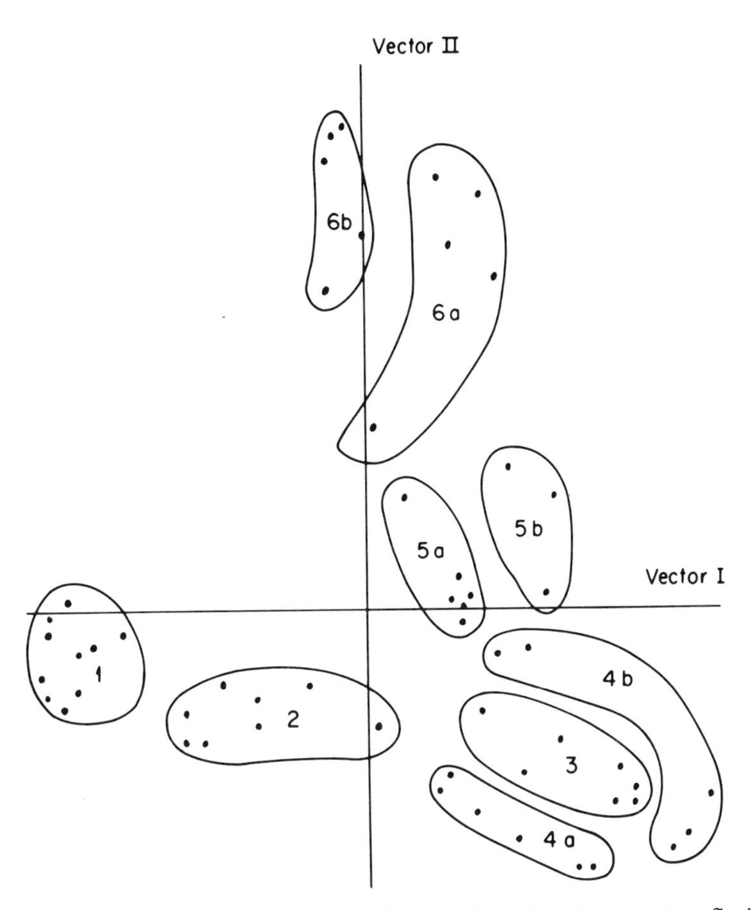

Figure 9.7. The site ordination using floristic data: the dune system-floristic group classes are superimposed (see Fig. 9.6).

Table 9.1. The common species in each dune system, defined by the numerical analysis DIVINFRE.[a]

Species	Classification	Group number in dendrogram										
		10	9	8	7	6	5	4	3	2	1	0
Angophora woodsiana	(T)							+	+			
Banksia serrata	(T)							+	+	+		
Imperata cylindrica	(G)						+	+	+	+	+	+
Casuarina littoralis	(T)					+	+	+	+	+	+	
Xanthorrhoea macronema	(X)					+	+	+	+			
Casuarina torulosa	(T)					+	+	+	+			
Eustrephus latifolius	(F)					+	+	+	+			
Elaeocarpus reticulatus	(S)					+	+	+	+			
Leucopogon margoroides	(S)					+	+	+				
Eucalyptus pilularis	(T)					+	+	+				
Tristania conferta	(T)				+	+	+	+	+	+	+	
Endiandra sieberi	(T)				+	+	+	+	+			
Dianella revoluta	(F)				+	+	+	+	+			
Themeda australis	(G)			+	+	+	+	+	+			
Pteridium esculentum	(Fe)			+	+	+	+	+	+	+	+	+
Eucalyptus intermedia	(T)			+	+	+	+	+	+	+		
Acacia aulococarpa	(S)			+	+	+	+	+	+			
Banksia integrifolia	(T)			+	+	+	+	+	+	+	+	+
Astrotricha longifolia	(S)		+	+	+	+	+	+	+	+	+	
Acacia flavescens	(S)		+	+	+	+	+	+	+			
Leptospermum attenuatum	(S)		+	+	+	+						
Persoonia virgata	(S)		+	+	+	+	+	+				
Leucopogon leptospermoides	(S)	+	+	+	+			+				
Monotoca scoparia	(S)	+	+	+	+	+	+	+	+	+		

Banksia aemula	(T)	+	+	+	+	+						
Eucalyptus signata	(T)	+	+	+	+				+			
Xanthorrhoes johnsonii	(X)	+	+	+	+							
Pimelea linifolia	(S)	+	+	+	+							
Boronia rosmanifolia	(S)	+	+	+	+							
Caustis recurvata	(Se)	+	+	+	+							
Coleocarya gracilis	(Se)	+	+	+								
Acacia ulicifolia	(S)	+	+	+	+							
Phyllota phylicoides	(S)	+	+	+	+					+	+	
Aotus lanigera	(S)	+	+	+	+							
Patersonia glabrata	(F)		+		+							
Brachyloma daphnoides	(S)		+	+	+							
Angophora costata	(T)		+	+	+							
Alloteropsis semialate	(G)								+			
Petalostigma quadriloculari	(S)								+			
Eucalyptus tesselaris	(T)									+		
Petalostigma pubescens	(S)									+		
Acacia leiocalyx	(S)								+	+	+	
Alphitonia excelsa	(S)									+	+	
Phebalium woombye	(S)									+	+	+
Callitris columellaris	(T)									+	+	
Casuarina equisetifolia	(T)									+	+	
Exocarpos cupressiformis	(S)									+	+	
Pultenaea subternata	(S)									+	+	+
Dune system:		6b	6a	5	4	3	2	lb	la			
Tree species:		2	3	7	9	8	8	5	2			
Total species:		47	64	80	75	63	52	43	6			

[a]Species are coded as trees (T), shrubs (S), grasses (G), ferns (Fe), sedges (SE) and grass trees (X).

Table 9.2. Vegetation structure for each dune system.[a]

Dune system number	Tree and shrub (s) height (m)		Crown cover (%)			
	Upper layer	Lower layer	Upper layer	Lower layer	Total	General description
6b	5	1 (s)	11	50 (s)	51	Dwarf
6a	12	6	13	15	28	Low
5	20	9	36	20	56	Tall layered
4	35	15	55	25	80	Extremely tall layered
3	28	14	40	20	60	Very tall layered
2	15	0	52	0	52	Tall
1b	9	0	62	0	62	Low
1a	1 (s)	0	2 (s)	0	2	Dwarf

[a]Each value is an average for the sites comprising each group defined by the DIVINFRE classification.

Vegetation Biomass

Biomass was not measured directly because the magnitude of the task was beyond the resources available and because logging in dune systems 3, 4, and 5, and burning over all dune systems would reduce the value of the results. Instead, several indirect methods were used to estimate the relative biomass for each dune system (Table 9.3). A relative biomass index was calculated assuming that crown cover is correlated with basal area (Howard 1970), and that crown cover multiplied by height (equivalent to basal area by height) gives an estimate of biomass or timber volume (Spurr 1960), and that these relationships are relatively constant across the sand mass. The other estimates were derived from litter accumulation and its residence in the soil as organic carbon. The assumption here is that communuties with higher biomass produce more litter and that this is reflected by an increase in both the thickness of the A1 horizon and the carbon content of the top 6 m of the soil (the 6-m depth was chosen because of the large dimensions of the soil profiles). Each of the different estimates of relative biomass shown in Table 9.1 exhibits a similar trend, that is, a buildup biomass to dune system 4, then a sharp and continuing decline.

Nutrient Levels

The nutrients in a forest ecosystem can be considered in separate compartments: (a) total element content in the soil; (b) nutrients in the biomass and organic cycle; and (c) aerial nutrient input (accession). Our

Table 9.3. Estimates of biomass for each dune system.

Dune system number	Biomass index	Total organic carbon to 6 m depth (%)	Thickness of A1 soil horizon (cm)
6b	105	—	50
6a	246	0.07	80
5	900	0.09	75
4	2300	0.17	140
3	1400	0.13	60
2	780	0.11	20
1b	558	0.09	15
1a	2	—	10

study examined only limited aspects of these compartments from samples selected as representative of trailing arm or ridge sites for each dune system.

Nutrients in the Soils

Although a number of nutrients were measured, only values for Ca and P are given here (Fig. 9.8a and b) because they are the elements most likely to be in short supply in the sand dunes and the trends are similar to those for other nutrients. The concentration and distribution of nutrients per unit depth of the soil vary markedly between dune systems, and the changes in the profile pattern of calcium and phosphorus across the dune systems reflect the development of A2 and B soil horizons with increasing dune age and soil weathering. The average concentrations of calcium and phosphorus in dune systems 5 and 6 are extemely low in absolute terms and quite different from the levels in the younger systems. The overall trend is toward a reduction of nutrients in the older dune systems and a migration of nutrients down the profile as weathering proceeds.

Green Leaf Nutrient Concentrations

The concentrations of six major nutrients and two trace elements in green leaf material from tree species common to several dune systems are shown in Table 9.4. Differences in leaf nutrient concentrations across the dune systems are slight, although there is an overall tendency for lower values in dune system 6. The available data on leaf nutrient concentrations in various Australian forest communities have been summarized by Bevege (1978); the Cooloola data support the view that eucalypt forests

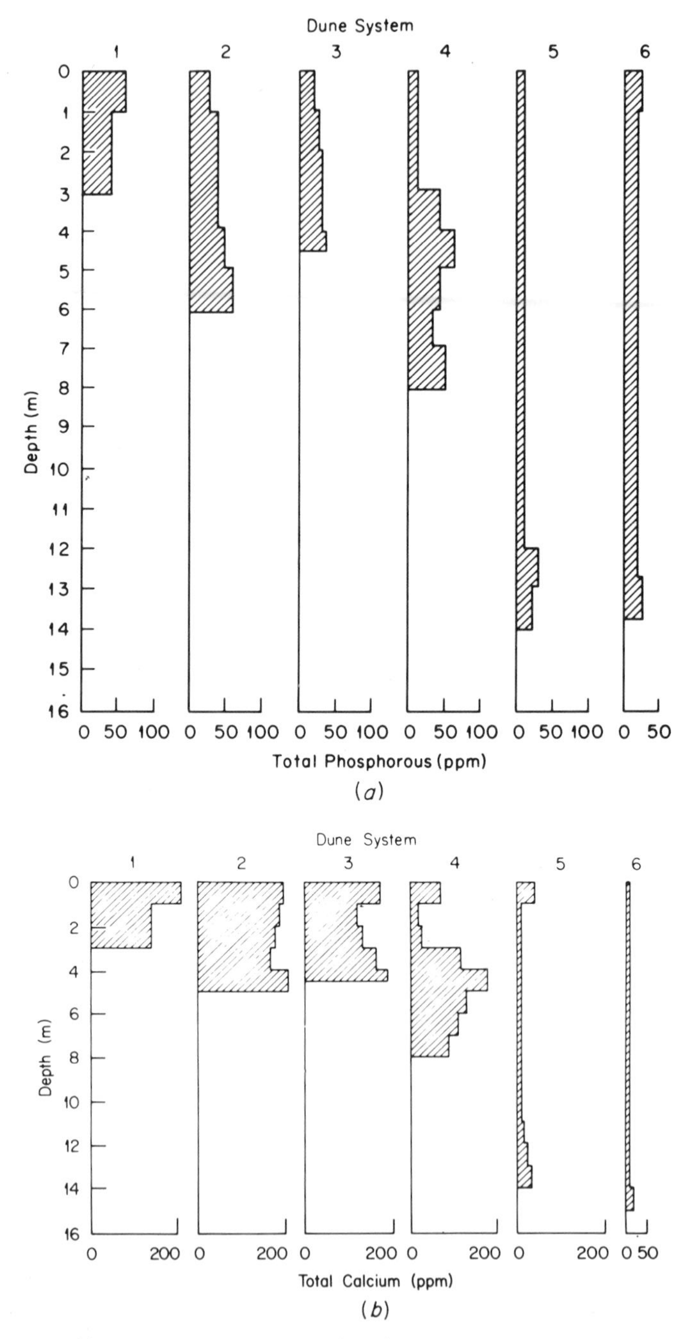

Figure 9.8. (a) Total soil phosphorus (ppm) at various depths in each dune system. (b) Total soil calcium (ppm) at various depths in each dune system (determination by x-ray fluorescence on whole soil).

Table 9.4. Concentrations of some elements in the green leaves of several species common to a number of the dune systems.

Species	Dune system number	Dry matter (%)						ppm	
		Ca	Mg	K	N	P	S	Cu	Zn
Eucalyptus signata	2	0.80	0.29	0.42	0.95	0.06	0.09	3	10
	5	0.90	0.27	0.46	1.10	0.05	0.10	6	17
	6	0.63	0.20	0.39	0.81	0.04	0.09	4	9
Eucalyptus pilularis	3	0.43	0.35	0.48	0.97	0.08	0.10	5	8
	4	0.52	0.30	0.36	0.92	0.05	0.09	5	8
Banksia serrata	2	0.65	0.15	0.32	0.60	0.04	0.09	3	5
Banksia aemula	5	0.51	0.16	0.47	0.87	0.05	0.12	5	8
Banksia aemula	6	0.64	0.12	0.15	0.55	0.02	0.09	1	4

growing on coastal sand systems have low leaf nutrient concentrations compared with most other Australian forests. Because leaf nutrient concentrations are similar across the dune systems, the amount of available nutrients stored in the sand dune systems is mainly dependent on the amount of organic matter present; thus, the available nutrients increase to system 4 and then decline in the older systems.

Atmospheric Accession of Nutrients

Atmospheric accession of most nutrients from rainfall and aerosols declines with increasing distance from the coast (Fig. 9.9), most of the decline occurring within 0.5 km of the coast. Phosphorus accession (not shown) is independent of distance inland from the coast and is consistently about 0.2–0.3 kg ha^{-1} yr^{-1} over the sand mass (Reeve and Fergus 1980). Accession has the general composition of dilute seawater, and the amounts received across the sand mass change little between dune systems 2 and 6; dune system 1 receives by far the greatest accession of ions, other than phosphorus (Reeve and Fergus 1980). The peak in biomass in dune system 4 is therefore not associated with high levels of nutrient accession.

The overall nutrient levels agree with those for a eucalypt forest site equivalent to dune system 2 on North Stradbroke Island (Westman 1978). The data presented in that article show that no single nutrient compartment of that forest ecosystem is adequate to provide nutrients indefinitely for the growth of the forest and therefore nutrient cycling is critical. Our data, obtained from a range of dune systems, support this conclusion, and they indicate that long-term soil weathering leads to the downward migration of nutrients.

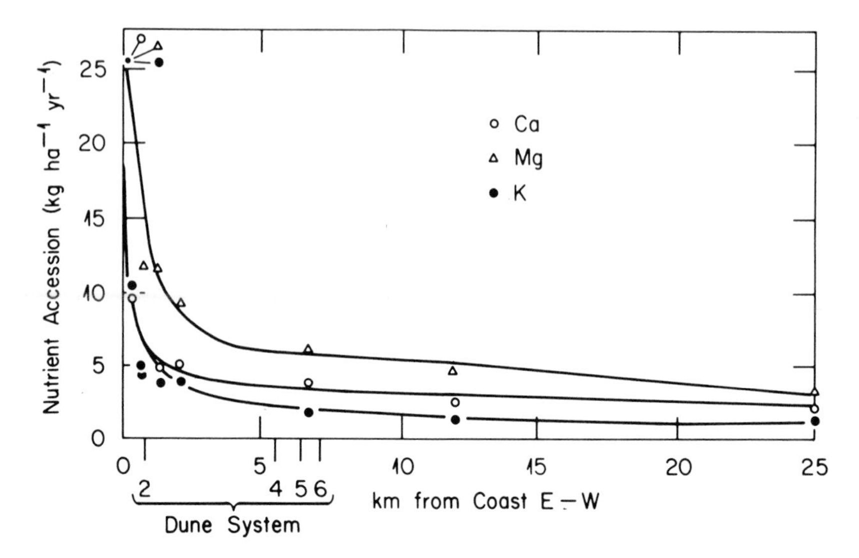

Figure 9.9. Atmospheric accession of nutrients at Cooloola (after Reeve et al. 1980).

Fire will undoubtedly influence nutrient turnover rates, particularly those in the upper soil layers, and could add to long-term nutrient loss, particularly nutrients in short supply, such as phosphorus (Raison 1980). However, fire does not explain the close relationships among the distribution of nutrients, the depth of soil weathering, and plant communities defined floristically and structurally; for these reasons it is not considered to be the primary cause of nutrient loss of of major vegetation patterns.

Long-Term Forest Succession

In the Cooloola study presented here we examined the vegetation on a series of freely drained sand dunes, which form an age sequence. The study has several features distinctly different from previous studies of long-term sand dune succession (e.g., Olson 1958; Jenny et al. 1969). While all long-term successional studies require the homology assumption, the Cooloola study has the advantage that the age sequence can be demonstrated in several independent ways. The parent material is a uniform nutrient-poor quartz sand, and the sites selected have not been affected by seasonal water tables.

The vegetation sequence described for the Mendocino coast of California, U.S.A. (Jenny et al. 1969), is superficially similar to that in the Cooloola study. Later work in this area (Westman and Whittaker 1975) extended the vegetation sequence to include "pygmy forests," which actually occur on beach terraces with seasonal water tables to the surface. The development of the pygmy forests is probably controlled by oscillating water tables and is not equivalent to the dwarf woodlands described in system 6 at Cooloola.

Trends in Long-Term Succession

The development of forest communities at Cooloola over a relatively long period has been recorded in floristic and structural terms. It cannot be assumed that each floristic plant community now present on a particular dune system was previously represented by a community with similar floristic composition on an equivalent site in the past. Each dune system supports a forest type that is a biomass maximum for present-day species assemblages, given the present-day climate and stage of soil development. In the past, the biomass and structure of the vegetation, rather than the floristics, are likely to have varied in the same relative fashion across the dune systems.

The structural differences that occur in the vegetation on the dunes at Cooloola can be illustrated in a ternary diagram (Fig. 9.10), which illustrates the relative abundance of the major components of the vegetation in the different plant communities. Directional changes from grass-shrub vegetation (system 1a) to tree-dominated vegetation (systems 4 and 5), followed by a trend toward shrubby vegetation, are evident in the ternary diagram. The succession is *progressive* to system 4 (that is, progressively more trees and an increase in biomass) and is *retrogressive* beyond system 4. This observation is in accord with the hypothetical scheme proposed by Stark (1978) to explain the development of vegetation on the tropical white sands in northern South America. Unfortunately, the lack of data in that study means that this scheme cannot be quantified and represented in a ternary diagram. Nevertheless, Stark does identify the retrogressive succession phase. The vital questions are, "Why does retrogression occur in the forest succession on the Cooloola Sand Dunes?" and "Are the ages of soils important in considerations of forest succession?"

Nutrient Run-Down and Retrogressive Succession

The nutrients associated with sesquioxide coatings on the quartz grains are mobilized at Cooloola, because they hold the main supply of elements in the unweathered sand (Jehne and Thompson in press). Electron microprobe analysis (K. Norrish 1979, personal comm.) has shown that sesquioxide coatings consist of silica, aluminium, and iron, associated with elements such as phosphorus, potassium, sulfur, and calcium. Circumstantial evidence suggests that endomycorrhizal fungi are important

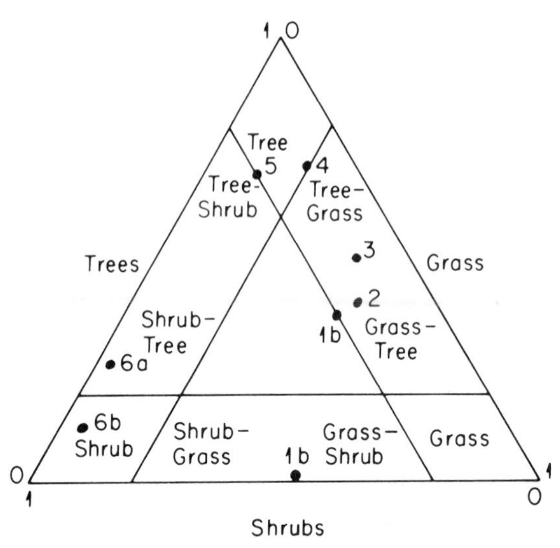

Figure 9.10. Directional changes in forest succession. The ternary diagram indicates the proportions of trees, shrubs, and grasses. The numbered points indicate proportional abundances of trees, shrubs, and grasses on the Cooloola dune systems.

in releasing nutrients from this source, at least in the early stages of forest development (Jehne and Thompson 1981) This may be the key process promoting plant colonization on the newly deposited sand. By making essential elements available from an apparently inert medium, the released elements are incorporated into the developing forest communities where their availability, as determined by the turnover of organic matter, soil weathering, and atmospheric accession, governs the process of forest succession. The major nutritional differences between the dune systems are associated with total pools in the organic cycle and mineral soil.

Because the nutrient concentrations in plant material are fairly similar for the different dune systems (Table 9.2), the vegetation nutrient store must be closely related to plant biomass. As biomass increases, there is a transfer of nutrients from the upper layers of the soils and from aerial accession to the vegetation compartment and also to deeper storage of organic matter in the soil (Fig. 9.5a and b). In dune systems 1–3 most of the nutrient store is located between depths of 1 and 4 m and in dune system 4 between 3 and 8 m. This deep nutrient store serves to buffer the system against decline, because the B horizon provides nutrients released by weathering as well as some trapped from organic cycle leakage. However, in the long term there is a loss of nutrients, as the release of nutrients from the organic pool will eventually exceed the requirements of the vegetation, and the depth to B horizon will eventually exceed the rooting depths of many species. There is a small continuous leakage of nutrients out of the system in drainage water (Reeve et al. 1980), and in time this is reflected in the reduced amounts of nutrients in both the vegetation and soil compartments of dune systems 5 and 6.

The ultimate fate of nutrients stored in the soil organic pool appears to be the key factor controlling biomass buildup and decline. As soil weathering proceeds, the loss of nutrients from the system can only be prevented through nutrient storage in the organic compartment of the rooting zone. Organic matter may be highly resistant to breakdown, but as the bulk of decomposing organic matter increases and soil weathering continues, the nutrients released into the soil solution will eventually exceed the growth requirements of the vegetation. In a situation in which factors other than nutrient availability are limiting, this causes leakage. Growth-limiting factors such as light, water availability, and temperature apparently cause a loss of nutrients from any system by limiting the ability of vegetation to take up nutrients and return them to the organic store.

While the vegetation apparently cannot prevent leakage of nutrients from the system, many of the plant species in the dune systems exhibit attributes that would appear to be beneficial for growth and survival under conditions of low nutrient availability. These attributes have been reviewed by Bowen (1980) and include proteoid roots, lignotubers, and mychorrizal associations. They relate primarily to the uptake or recycling of nutrients and may not necessarily include mechanisms for the storage of nutrients. Structures such as lignotubers or basal swellings are sometimes regarded as nutrient stores (Westman 1978), but they also occur in situations in which factors such as fire and drought appear to be operating. In these situations it is often difficult to identify adaptations that facilitate regeneration following disturbance from adaptation to low nutrient conditions.

The constraints operative in these sand dune systems are as follows:

1. An evitable slow loss of nutrients occurs through leaching and drainage.
2. Accession of nutrients from the atmosphere is less than the amount lost through leaching and drainage.
3. As podzol profile development proceeds, nutrients are removed from sand grains and the amount of nutrients stored in and released from the sand grains declines with time.
4. The nutrients can be stored in organic matter, but the nutrient content and the rate of breakdown of organic matter are essentially constant and leakage occurs during mineralization and recycling.
5. Factors other than nutrient availability can limit the production of organic matter.
6. Adaptation to low nutrient conditions are evident in some plants.

In its simplest form this situation can be modeled as a system having a limited pool of one factor, and the rate of outflow of this factor exceeds the rate of input.

Omitting interactions, the organic matter in such a nutrient-limited system will increase with time until the nutrient pool becomes sufficiently large for the nutrients available from weathering and decomposition to be adequate for the level of growth sustainable under other possible limiting factors, such as water availability. As soil weathering proceeds the condition is then reached at which the nutrients released through decomposition of organic matter alone are sufficient to maintain such a level of growth, and, in the absence of an alternate nutrient store, the nutrients released through weathering are lost from the system. Theoretically, this state can be maintained until the rate of loss of nutrients from the system exceeds the rate of release of soil nutrients through soil weathering, at which point the vegetation would again become nutrient limited and would decline in biomass and change in species composition. Such changes are called retrogressive succession.

Nutrients are lost from the soil system through weathering and leaching. The vegetation slows down but apparently cannot prevent this loss of nutrients. The review of soil chronosequences by Jenny (1980), in which the long-term decline in soil organic pools is discussed, and the analysis of the fate of phosphorus during pedogenesis by Walker and Syers (1976), indicate that this nutrient decline over long periods occurs in other vegetative systems. It appears that in any vegetative system the pattern of a buildup of vegetation and subsequent decline is inevitable where the rate of input of nutrients is less than the rate of loss.

Implications from the Cooloola Model

The concept of progressive and retrogressive forest succession associated with the degree of soil development becomes important in predicting both natural successional changes and successional changes following disturbance by humans. In young systems the vegetation may return to a facsimile of its original state following disturbance, but in old systems disturbance is likely to increase the rate of nutrient loss causing a decline in the vegetation. Such a decline can only be reversed through the input of nutrients. Given the great ages of the landscapes in Australia, most are likely to be in a state of decline.

Applying the Cooloola model to current land use practices it is suggested that, depending on the "succession type," nutrient rundown can be increased to a level that is significant in terms of management time scales. The rate of site deterioration associated with disturbances, such as burning, clear felling, or animal grazing, will depend on the nutrient buffering capacity of the system, which is related to the retrogressive or progressive successional state of the vegetation.

If a "retrogressive" type exists, these systems would not conform to the general rules established for "relative stability of mineral cycles" (Jordan et al. 1972) because nutrients could not return to predisturbance levels. It follows that plant succession could follow a course never previously observed. In the semiarid lands of Australia, for example, a "disclimax" state caused by soil loss associated with sheep grazing (Moore 1960) would not return to its preexisting condition following sheep removal if the system began as a retrogressive stage.

Retrogressive plant successions have been detected in palynological studies, and are usually attributed to climatic changes (Walker 1970). The concept of an inevitable retrogressive succession complicates the interpretation of vegetation changes resulting from climatic change, because both synergistic and antagonistic effects could be postulated, depending on the timing and direction of the climatic change. Anderson (1967) has demonstrated that pollen can be transported considerable distances before deposition. It is possibly alarming to palynologists interested in the history of sand dune vegetation to speculate that pollen currently being trapped in lakes at Cooloola could be derived from all successional stages.

This study illustrates again the ultimate dependence of terrestrial biological systems on soil fertility, which is affected by weathering and the rundown of soil nutrient reserves. In the long term and indeed in the relatively short term, given some intensive land uses, the axiomatic assumption of a self-perpetuating forest "climax" seems both biologically and geomorphologically untenable.

Chapter 10

Quaternary History and the Stability of Forest Communities

Margaret Bryan Davis

Introduction

The concept of stability of plant communities inevitably enters discussions of forest succession. Succession following disturbance often leads to restoration of the original community, which is seen as the equilibrium community for the site. In this context, succession can be viewed as a mechanism maintaining stability.

An historical perspective leads one to question the importance of stability relative to mechanisms that allow communities to change. During the Quaternary Period the forests with which we are familiar seldom maintained a constant species composition for more than 2000 or 3000 years at a time. The evidence presented in this article suggests that forest communities in temperate regions are chance combinations of species, without an evolutionary history. This view is influenced by recent geological discoveries that have shortened the time scale of Quaternary events. Biogeographers had previously thought of the Quaternary Period, during which the modern flora evolved, as a time of long interglacial intervals similar to the present, lasting hundreds of thousands of years. These were interrupted by episodes of glaciation, four in all, that disrupted the vegetation (MacArthur 1972). The time interval in which we now live, the Holocene epoch, was considered atypical because only 10,000 years had elapsed since the most recent ice sheets melted away. If evidence for disequilibrium came to light, one could argue that the normal adjustments of climate, soils and vegetation that characterize interglacial and postglacial time had not yet been attained.

Within the last 15 years new data have demonstrated that there were multiple glaciations, at least 16 during the Quaternary Period (Hays et al.

1969). Most of the two million years of Quaternary time experienced climate colder than present, with glaciers greatly expanded. Each of the 16 glaciations lasted 50,000–100,000 years, while the interglacials were brief interruptions lasting 10,000–20,000 years (Emiliani 1972; Kukla et al. 1972). The Holocene is thus a typical interglacial interval, although all interglacials are atypical for the Quaternary. The mid-Holocene temperature maximum 5000–8000 years ago was probably the maximum for the Holocene interglacial; we are now experiencing the decreasing temperatures that will lead into the next glaciation (Kukla et al. 1972; Wright 1972). Most glacial intervals have displayed highly variable climate, with oscillating but ever-increasing growth of glaciers to a maximum near the end of the interval (Broecker and Van Donk 1970).

The last glaciation, the Wisconsin, began about 100,000 years ago. During the first half of the interval the climate south of the expanding ice sheets was dryer and cooler than today (van der Hammen et al. 1971). After short interstadial periods of slightly warmer climate, glaciers expanded to reach their maximum 18,000–20,000 years ago. Then, as was the case for each preceding glaciation, the Wisconsin terminated with a sudden warming that melted the ice sheets in a few thousand years (Broecker and Van Donk 1970; Dansgaard et al. 1971). It appears now that the sudden climatic changes terminating ice sheets have been more disruptive to vegetation than the more gradual changes that caused their growth. The most recent termination, with its steep temperature rise beginning 16,000 years ago, initiated a period of rapid adjustment of species ranges to a dramatically altered climate. The vegetation instability thus initiated has continued to the present. These events have occurred many times before during the Quaternary Period. Therefore, the emphasis in studying plant communities might be placed more appropriately on changes in species composition and mechanisms that permit rapid change than on the forces that act to maintain stability.

Pollen as a Record of Geographical Distribution of Trees

Holocene pollen records characteristically show an increase through time in the number of tree genera represented. The increase in diversity was caused by the successive immigrations of trees moving northward in the wake of the retreating ice. The sequence of species arrivals is different at different sites. Species arrival times show regional patterns that become apparent when arrival times are plotted on maps.

Figure 10.1 shows pollen influx at a site in northern New Hampshire. Pollen influx (grains accumulated per square centimeter of sediment

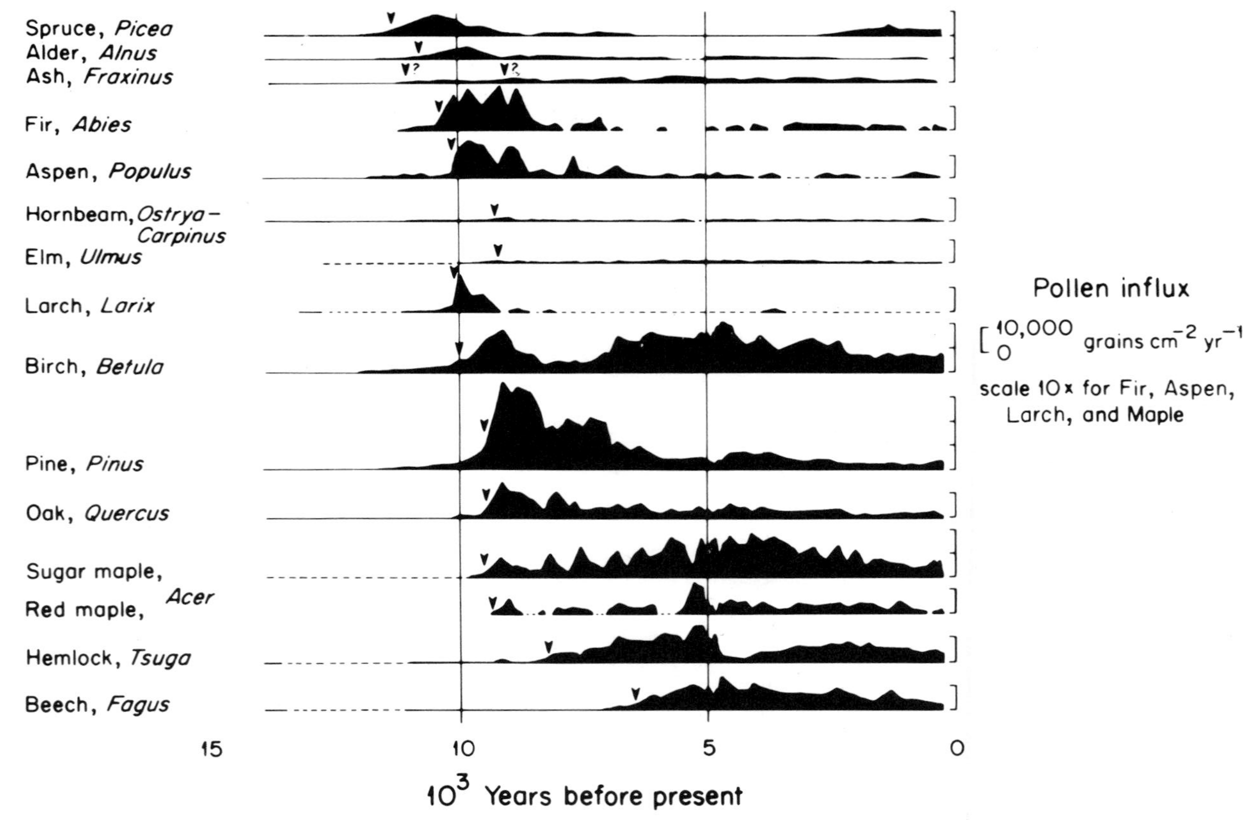

Figure 10.1. Pollen influx (grains accumulated per square centimeter per year) for the major forest trees at Mirror Lake, New Hampshire, during the last 14,000 years. A magnified scale has been used for species that produce small amounts of pollen.

surface per year) is calculated from the pollen concentration in lake sediment, corrected for the accumulation rate of the sediment matrix determined from radiocarbon dates (Davis and Deevey 1964). Pollen influx roughly approximates a population count of parent trees in the vicinity of the site. This relationship has been established by comparing modern influx rates and counts of trees on the surrounding landscape (Davis et al. 1973). There are, of course, large errors in the quantitive translation of pollen into vegetation, but the changes in pollen influx are so large when a species arrives in the vicinity of a site (often two orders of magnitude) that many of the errors are overcome. The appearance of fossil leaves or seeds in sediments also serves to corroborate pollen evidence for the immigration of a species (e.g., Davis et al. 1980). Sudden increases in pollen percentage values have also been interpreted as evidence for immigration (e.g., Watts 1973), especially when supported by comparisons of pollen percentages in modern sediment and geographic distributions of trees (R. B. Davis and Webb 1975). Although I have used pollen percentage diagrams in this paper, I find changes in percentages more ambiguous than changes in influx rates and have given them less weight in regional interpretations.

The arrival times of all the dominant tree species or genera were mapped using this method of interpreting pollen evidence. The sites where pollen data were available in the eastern half of the United States and adjacent Canada are shown in Fig. 10.2. Circled points are sites with influx diagrams. At each site I determined the radiocarbon age for the arrival of each of the important forest taxa. Examples of changes in pollen influx that were considered significant are indicated by arrows in Fig. 10.1. The resulting dates were plotted on maps, all of which showed a trend, with older dates to the south and younger dates to the north. Isopleths were drawn on the maps, connecting points of similar age. The isopleths represent the leading edge of the spreading population of each tree taxon as it moved northward. The present range for each tree is also shown; previous occurrences outside this range demonstrate that glacial-age distributions for many species were completely nonoverlapping with modern distributions. The distances between isopleths and their locations indicate the rate of spread of each species and the migration route from its glacial distribution to its modern range. It should be emphasized that the resulting figures are maps of reconstructed vegetation based on pollen data. They are not objective collations of the pollen data *per se.* The latter are illustrated by the work of Bernabo and Webb (1977), who have mapped "isopolls", or areas within which pollen percentages are similar at the same point in time. The maps presented here represent an updating and extension of more generalized migration maps, based on far fewer data, that were published several years ago (Davis 1976). The reader is referred to the earlier article for a more extensive review of the literature and discussion of European interglacial floras.

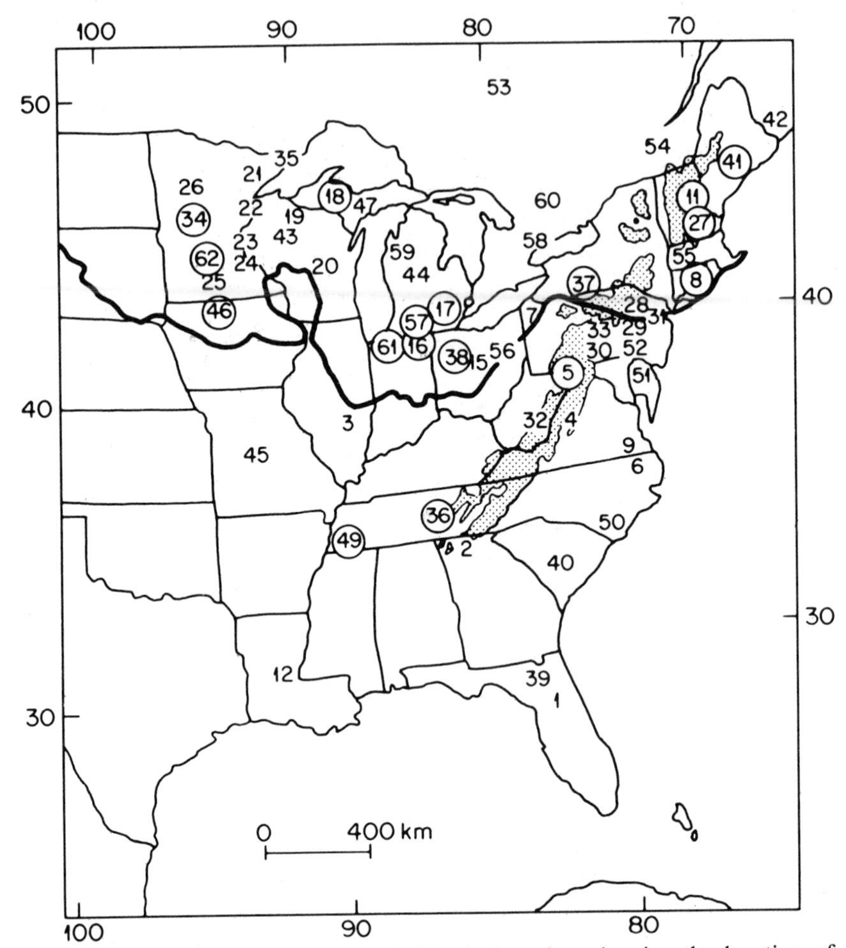

Figure 10.2. Outline map for eastern North America, showing the location of sites. The heavy black line marks the maximum extent of the Wisconsin ice sheet. Circled numbers identify sites where pollen influx has been calculated; the others are percentage diagrams. Numbers refer to authors: 1, Watts (1969); 2, Watts (1970); 3, Grüger (1972); 4, Craig (1970); 5, Maxwell and Davis (1972); 6, Whitehead (1973); 7, Walker and Hartman (1960); 8, Davis (1969b); 9, Whitehead (1972); 11, Davis (1978); 12, Delcourt and Delcourt (1977); 15, Ogden (1966); 16, Williams (1974); 17, Kerfoot (1974); 18, Brubaker (1975); 19, Webb (1974a); 20, West (1961); 21, Fries (1962); 22, Wright and Watts (1969); 23, Cushing (1967); 24, Wright et al. (1963); 25, Jelgersma (1962); 26, McAndrews (1967); 27, Davis (1981); 28, 29, 30, 31, 32, 33, Watts (1979); 34, Birks (1976); 35, Craig (1972); 36, Delcourt (1979); 37, Spear and Miller (1976); 38, Shane (1975); 39, Watts and Stuiver (1980); 40, Watts (1980); 41, Davis et al. (1975); 42, Mott (1975); 43, Peters and Webb (1979); 44, Kapp (1977); 45, King and Lindsay (1976); 46, Van Zant (1979); 47, Davis (1978); 49, Delcourt et al. (1980); 50, Frey (1953); 51, Sirkin et al. (1977); 52, Martin (1958b); 53, Terasmae and Anderson (1970); 54, Richard (1977); 55, Whitehead (1979); 56, Shane (1980); 57, Bailey (1977); 58, McAndrews (1970); 59, Lawrenz (1975); 60, McAndrews (1981); 61, Bailey (1972); 62, Waddington (1969).

Forest Communities at the Time of the Last Glacial Maximum

At its maximum the Wisconsin ice sheet covered half of the North American continent, extending from the Canadian Rockies across the northern plains and southward to central Illinois, northern Pennsylvania, and southern New England (Fig. 10.2). Before data were available documenting the nature of vegetation south of the ice sheet, paleoecologists assumed that climatic changes associated with growth of glaciers displaced existing forests, as intact communities, southward. Although there were intense arguments about the distance of displacement (Braun 1950, Deevey 1949), few scientists questioned the idea that the ice sheet was bordered by a zone of arctic tundra, with belts of boreal forest and deciduous forest arranged in sequence farther to the south.

Recent investigations have given a very different picture. Tundra did occur along the edge of the ice sheet, especially in the East, where tundra ranged several hundred kilometers to the south at high elevations in the Appalachian mountains (Maxwell and Davis 1972, Watts 1979). But boreal species were associated very differently from the modern boreal forests of Canada (Wright 1968a). Spruce and pine grew on the coastal plain, but in the Carolinas, Georgia and eastern Tennessee, jack pine was the dominant tree, with only minor admixtures of spruce and fir (Watts 1970, 1980; Delcourt 1979). In contrast, spruce and larch dominated the central plains of the continent and pine was completely absent; it died out in Illinois and adjacent regions 23,000 years ago (Wright 1968; Grüger 1972). When the ice stood at its maximum, and during its retreat, there were extensive forests of spruce and larch extending westward toward the Rockies in regions that are now prairie. Trees that now characterize the temperate deciduous forest grew in relatively small populations in the lower Mississippi valley and in northern Florida (Delcourt et al. 1980; Watts and Stuiver 1980). In some areas they grew in close proximity to boreal species such as spruce (Delcourt and Delcourt 1977). Southern Florida, traditionally assumed to have been a refuge for deciduous forest (Deevey 1949; Martin 1958a), was too dry during most of the last glacial period. Large areas of Florida that are now forest or scrubland were prairie or had active sand dunes (Watts 1969; Watts and Stuiver 1980).

Northward Migrations of Boreal Trees

As the climate warmed 16,000–10,000 years ago, causing the continental ice to retreat, vegetation changed rapidly. The changes began in the south, where boreal species were replaced by deciduous forest as early as 15,000 years ago. Replacement of boreal species took place progressively later to the north. Spruce moved into tundra, and tundra plants became established where the retreating glacial ice left newly exposed landscape. In an interesting article Watts (1979) demonstrates the differences between tundra communities outside the glacial boundary in Pennsylvania and the pioneer tundra communities that invaded newly deglaciated landscape a few kilometers away.

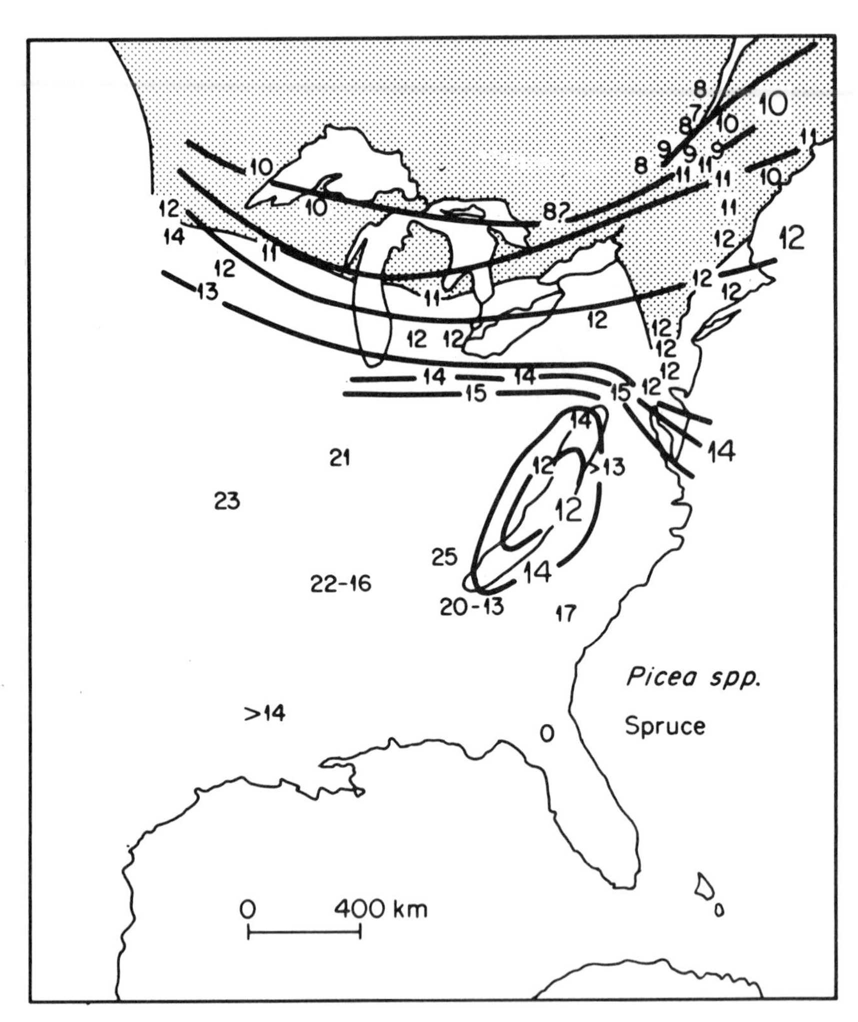

Figure 10.3. Migration map for spruce. The numbers refer to the radiocarbon age (in thousands of years) of the first appearance of spruce at the site after 15,000 years ago. The first appearance was determined from increased pollen abundance or presence of macrofossils (see text). Isopleths were drawn to connect points of similar age; they represent the leading edge of the expanding spruce population. The stippled area represents the modern range for spruce (Little 1949; Fowells 1965).

Spruce (*Picea*) was an aggressive pioneer, rapidly moving northward into the tundra. The arrival of spruce is time transgressive from south to north (Fig. 10.3). Spruce became established only a few years after glacial retreat in parts of the Great Lakes region (Watts 1967; Cushing 1967) but in New England, where deglaciation was very rapid, at least 2000 years elapsed between the time the ice left and spruce arrived (Davis 1969b; Davis et al. 1980). Expansion of spruce onto the higher elevations in the Appalachians was also somewhat delayed (Fig.10.3). Recent species identifications based on pollen morphology (Birks and Peglar 1980; Watts 1979) suggest that the early spruce populations were white spruce (*P. glauca*) followed by black spruce (*P. mariana*), and in the East, by red spruce (*P. rubens*) as well. Low rates of pollen influx suggest that the spruce forests were open and patchy in distribution, rather like the

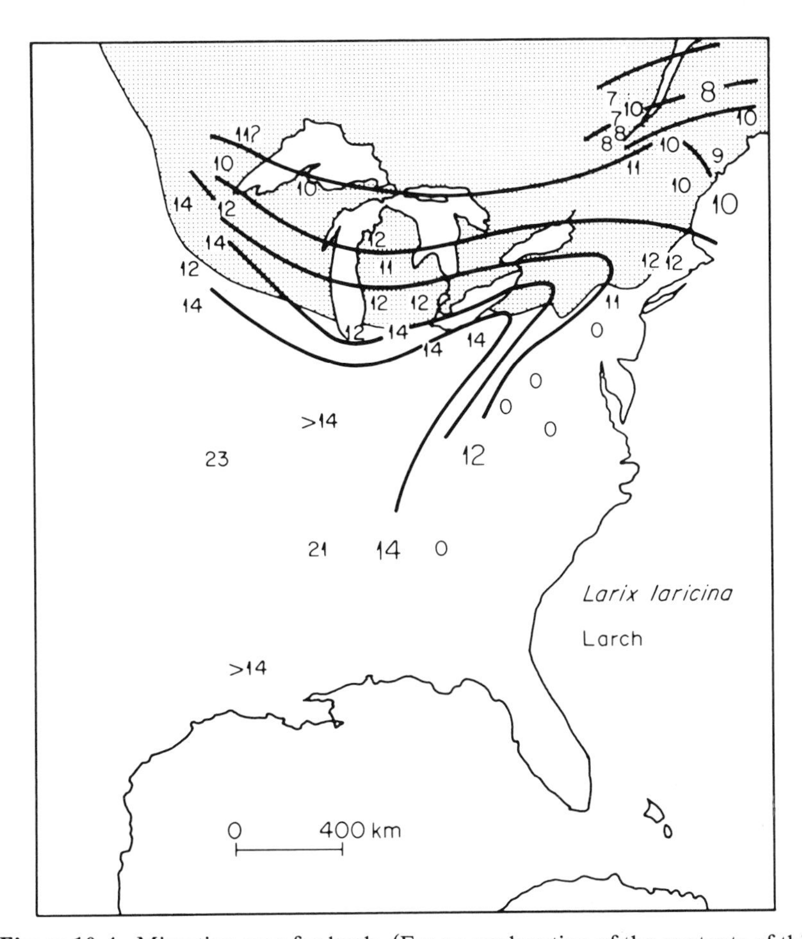

Figure 10.4. Migration map for larch. (For an explanation of the contents of this figure and Figs. 10.5–10.14, see the legend for Fig. 10.3.)

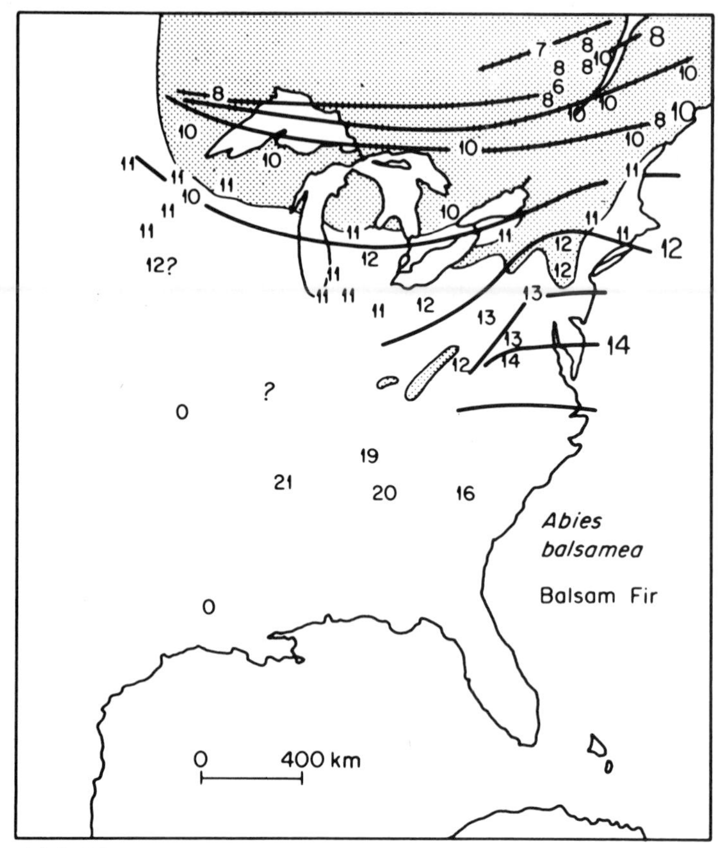

Figure 10.5. Migration map for balsam fir.

forest-tundra region north of the boreal forest in Quebec. This may also have been the case in the Middle West where values for spruce pollen influx are similarly low. The high percentage values for spruce at many mid-western sites may be a statistical artifact, caused by the absence of any prolific pollen producers in the vegetation such as pine or birch.

An interesting aspect of the migration of spruce is that everywhere spruce preceded alder (*Alnus*), the pollen of which peaked *after* spruce began to decline (Fig. 10.1). This is somewhat surprising, given the well-known primary succession sequence at Glacier Bay, where nitrogen fixation by alder is important in allowing spruce to grow (Crocker and Major 1955). Of course, it should be kept in mind that spruce was in most cases invading tundra which had persisted, presumably improving nutrient levels in the soil, for several thousand years, rather than 100 years as at Glacier Bay.

Larch (*Larix laricina*) shows an intriguing pattern of migration from its glacial-age distribution on the Great Plains to its present range (Fig.

10.4). Larch grew together with spruce on the Great Plains, but was missing from full-glacial sites east of the Appalachians (Watts 1979). It moved eastward north of the main Appalachian mountain chain, across Pennsylvania and central New York. Balsam fir (*Albies balsamea*), in contrast, seems to have expanded northward more rapidly east of the mountains than in the west (Watts 1979), reaching Ohio only 11,000 years ago (Fig. 10.5). In Connecticut, fir arrived almost simultaneously with spruce. It lagged about 1000 years after spruce in New Hampshire, forming a large population simultaneously with larch, aspen (*Populus*), and birch (*Betula*) that lasted for about a millenium after spruce had declined (Fig. 10.1).

Jack pine (*Pinus banksiana*) moved northward from the southeast beginning about 13,000 years ago and advanced onto deglaciated landscape faster than any other tree species, including spruce (Fig.10.6). Its route was north of the central plains, spreading west as far as Wisconsin before the climate south of the Great Lakes became too dry for its growth. Jack

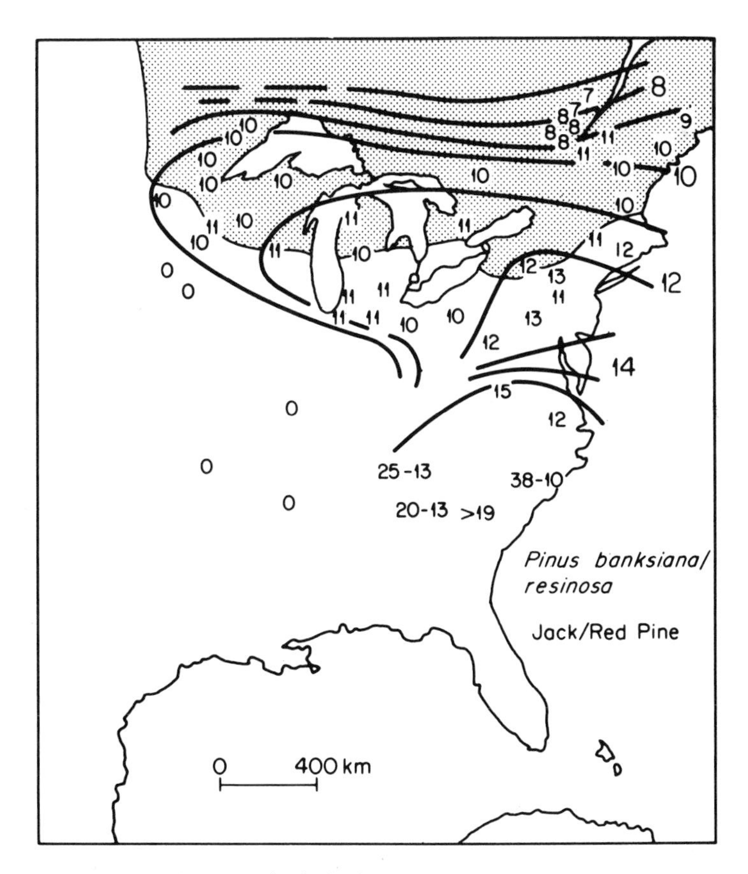

Figure 10.6. Migration map for jack pine and red pine.

pine arrived in Minnesota 10,000 years ago, but further expansion was prevented by drought; in western Minnesota spruce was replaced directly by prairie without an intervening pine stage (McAndrews 1967). It seems likely that these early populations were primarily jack pine, although from pollen morphology red pine (*P. resinosa*) might also have been present. Macrofossil evidence suggests that jack pine was the early migrant. At Canyon Lake, in upper Michigan, needles from jack pine were found in sediment deposited as early as 9500 years ago, whereas red pine needles occurred only in sediment younger than 3200 years (Davis, unpublished data).

Migrations of Species from the Temperate Deciduous Forest

Four deciduous tree genera—ash, elm, oak and hornbeam-ironwood (*Fraxinus*, *Ulmus*, *Quercus*, and *Ostrya-Carpinus*)—show very early increases in pollen representation at many sites, causing debate as to whether these trees grew together with boreal species in mixed stands. Both ash and elm now range very far to the north (Fig. 10.7), and so an early appearance together with spruce is not unexpected.

Hornbeam and ironwood (pollen of *Ostrya* and *Carpinus* is indistinguishable) are problematical because they are both relatively unimportant in modern forests and consequently their distribution and ecology is not well known. Because *Ostrya* ranges farther north, I presume that *Ostrya* is the genus represented by the early pollen increase. An early increase in hornbeam and ash pollen percentages appeared in the Cumberland Plateau 12,000 years ago (Delcourt 1979), but in New England there was both an early increase 12,000 years ago and a stronger increase 9500 years ago. I am uncertain which of these represents immigration. Hornbeam displays rather low influx rates later in the Holocene; it may have been much more important in the species-poor forests of the late-glacial and early Holocene than at any time since. Today it occurs most commonly as an old-field tree, suggesting that it competes poorly in closed forest communities. Black ash has been identified from pollen by several authors (Van Zant 1979; Birks 1976). Their work shows that black ash was the first ash species to arrive at midwestern sites. Black ash could easily have grown together with spruce, as it ranges far north into the boreal forest today.

Elm also arrived early, especially at sites in the Great Lakes region (Fig. 10.7). It makes up a high proportion of the pollen rain during the early Holocene in southern Michigan and adjacent regions; elm pollen is much less important in the east. The pollen productivity of elm, ash, and

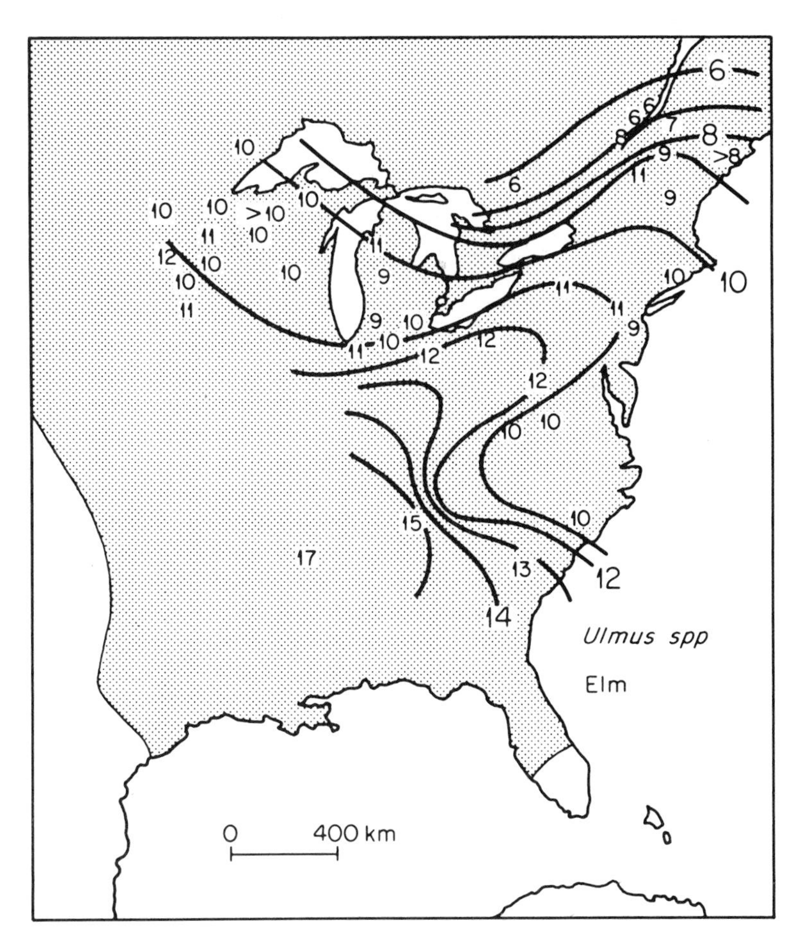

Figure 10.7. Migration map for elm.

hornbeam relative to other species has not been studied intensively. Consequently, it has been difficult to determine whether the first increase in pollen and influx represents local arrival for these species.

Oak is a different story, since the propensity of American oak species to produce large quantities of well-dispersed pollen is well known (Davis and Goodlett 1960; Webb 1974a,b). Unlike elm, hornbeam, and ash, oak is now restricted to latitudes south of 47° N (Fig. 10.8). Its northern range limit is similar to white pine, and it does not really penetrate the boreal forest. Oak generally shows two increases in pollen influx: an early increase 12,000 or more years ago and a later, much larger increase to Holocene levels. Amundson and Wright (1979) have argued that the high percentages of oak pollen (2–7% in Minnesota) 12,000 years ago indicate its presence in the spruce forest. On the other hand, I have argued that although the percentage is high, the influx of oak pollen is low relative to

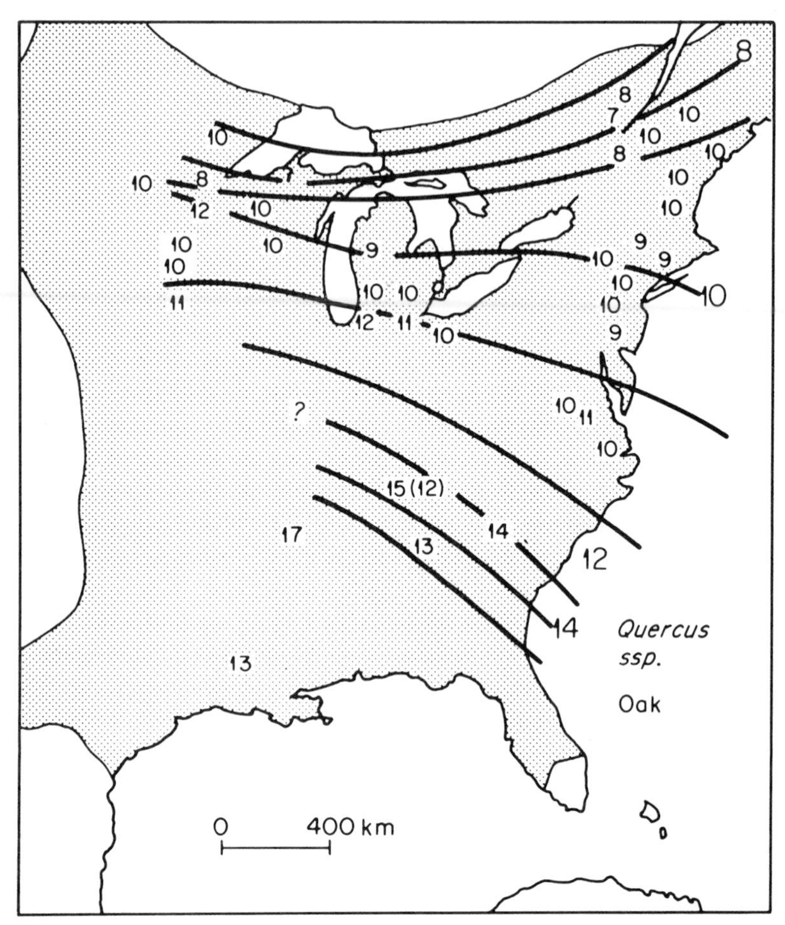

Figure 10.8. Migration map for oak.

Holocene rates, indicating long-distance transport of pollen. The map shown as Fig.10.8 is based on the later, large increase, which I believe represents the actual local arrival of oak trees. The earlier increase in New England 12,000 years ago (cf. Fig. 10.1) must be pollen transported from populations to the south or west, where oak populations were expanding at this time. Alternatively, it might represent a change in prevailing wind direction. This was a time of rapid ice retreat, and winds may have shifted from the north, off the ice sheet, to a western origin, which crossed the central plains. Figure 10.8 shows that oak spread very rapidly between 11,000 and 9000 years ago, reaching its present range limit 7000 years ago. Northern sites, such as Mirror Lake, New Hampshire (Fig. 10.1), show maximal percentages and influx 9000–7500 years ago; sites from southern NewEngland, such as Rogers Lake, where oak is still abundant today, show a rapid increase 9500 years ago and high influx rates continuing to the present.

Two species display patterns suggesting refuge areas on the east coast or in the foothills of the Appalachians. These are white pine (*Pinus strobus*) (Fig. 10.9) and hemlock (*Tsuga canadensis*) (Fig. 10.10), both of which occur today in temperate forest, often in mixed stands with hardwoods. White pine is less tolerant than hemlock and grows more abundantly on disturbed sites or on poor soils. Both species are absent from full-glacial sites in the Mississippi Valley and Florida. White pine (identified from pollen morphology and from needles at some sites) appeared first at a site in Virginia (Craig 1970) and then expanded rapidly northward and westward (Fig. 10.9). Seven thousand years ago, when white pine arrived in Minnesota, dry climate prevented its expansion westward (Wright 1968b). There has been renewed movement in recent millenia, however, in response to late-Holocene climatic change (Jacobson 1980). Meanwhile, white pine continued to expand northward, reaching sites

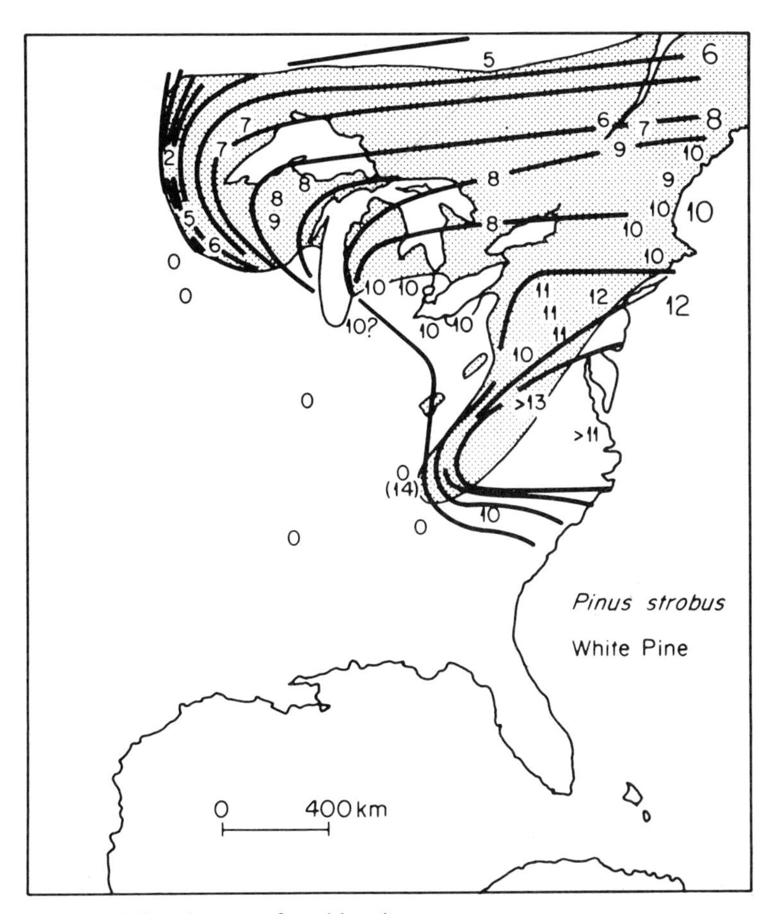

Figure 10.9. Migration map for white pine.

north of its present range limit in Canada 5000 years ago (Terasmae and Anderson 1970), apparently because the climate was warmer. At this time white pine also grew 350 m above its modern elevation limits in the mountains of New Hampshire (Davis et al. 1980). White pine was very abundant at many sites in the northern United States when it first arrived, forming large populations for a brief time—a few hundred years in Ohio (Shane 1980), 1000 years in Connecticut (Davis 1968), and 2000 years in New Hampshire (Davis et al. 1980) (Fig. 10.1). Lack of competition from deciduous trees, most of which had not yet arrived on the scene, may have been an important factor, allowing temporary population expansion in these regions, where white pine is no longer abundant.

Hemlock did not grow in Georgia or Tennessee during the full-glacial. Its pollen appeared in very low percentages 13,000 years ago at White Pond in South Carolina, but apparently it did not grow at the site (Watts 1980). The refuge area for hemlock during the glacial maximum is not known, but could have been somewhere in the Appalachians, on the coastal plain or on the continental shelf.

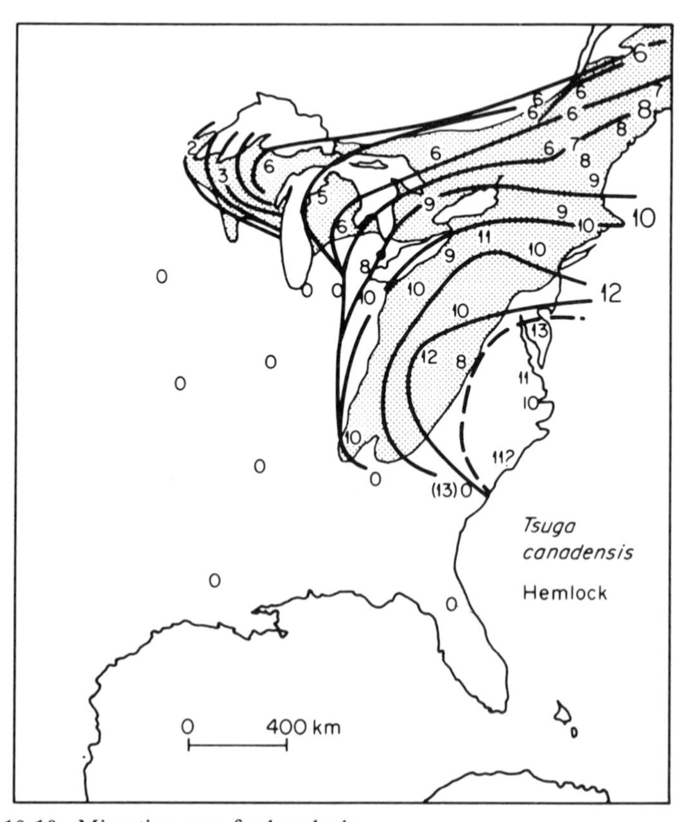

Figure 10.10. Migration map for hemlock.

Hemlock moved north and west from the central Appalachians-coastal plain area, starting about 1000 years after white pine (Fig. 10.10). Watts (1979) reports high percentages at Cranberry Glades at a level just above a stratigraphic level which was radiocarbon dated at 12,185 years. The steepness of the increase in hemlock pollen percentages, from 0 to 30% from one sample to the next, leads me to suspect a sedimentary hiatus. I have given the arrival time as 12,000, but I believe it may have been similar to other sites in the vicinity, which show arrival times of about 10,500. Hemlock expanded its range more slowly than white pine, arriving as much as 3000 years later in northern Michigan. Apparently the climate was too dry by the time hemlock arrived in the Midwest to allow it to spread south of Lake Michigan into Wisconsin; white pine may or may

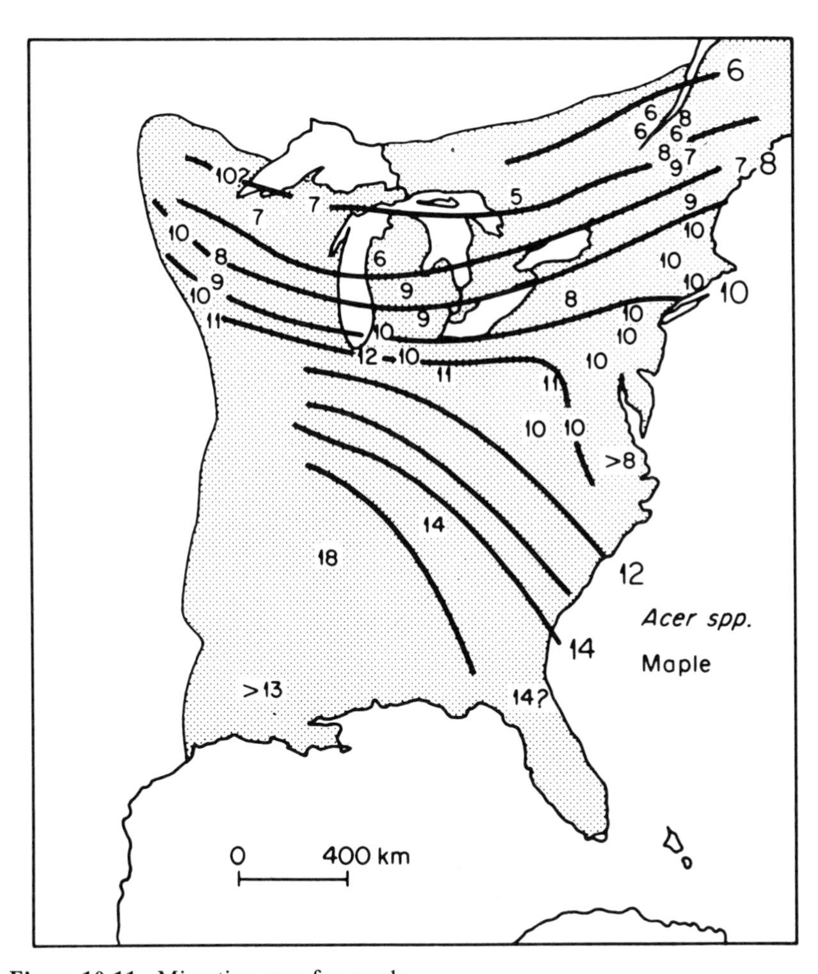

Figure 10.11. Migration map for maple.

not have taken this route. Hemlock, like white pine, moved up the mountain slopes in New England as soon as it arrived there (8000 years ago) to sites 350 m above its present elevation limit (Davis et al. 1980). It disappeared from the highest sites around 5000 years ago, when hemlock experienced a sudden drastic decline in abundance throughout its range. This decline may have been caused by an outbreak of disease, as it occurs virtually simultaneously throughout the range of the species. Hemlock recovered its previous abundance about 3000 years ago (Davis 1981). It never recolonized the high mountain sites, however, although it continued to grow a few tens of meters above its present limit until a few hundred years ago (Davis et al. 1980).

Maple, hickory, beech, and chestnut (*Acer*, *Carya*, *Fagus grandifolia*, and *Castanea dentata*) all appear to have grown in the lower Mississippi

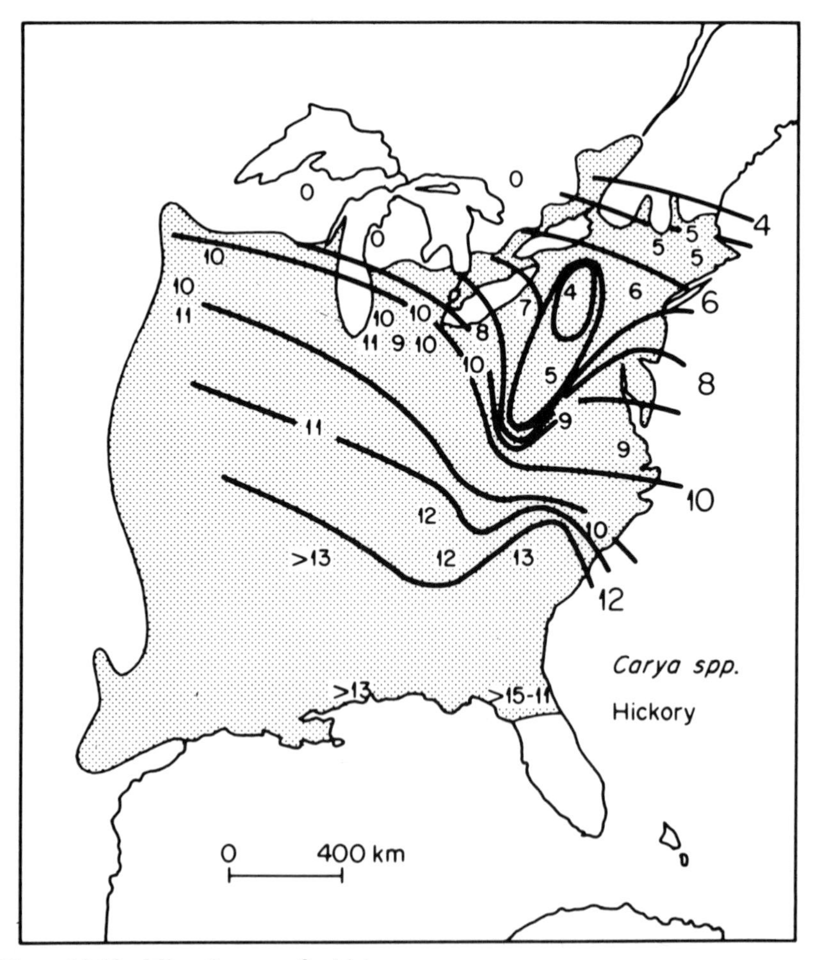

Figure 10.12. Migration map for hickory.

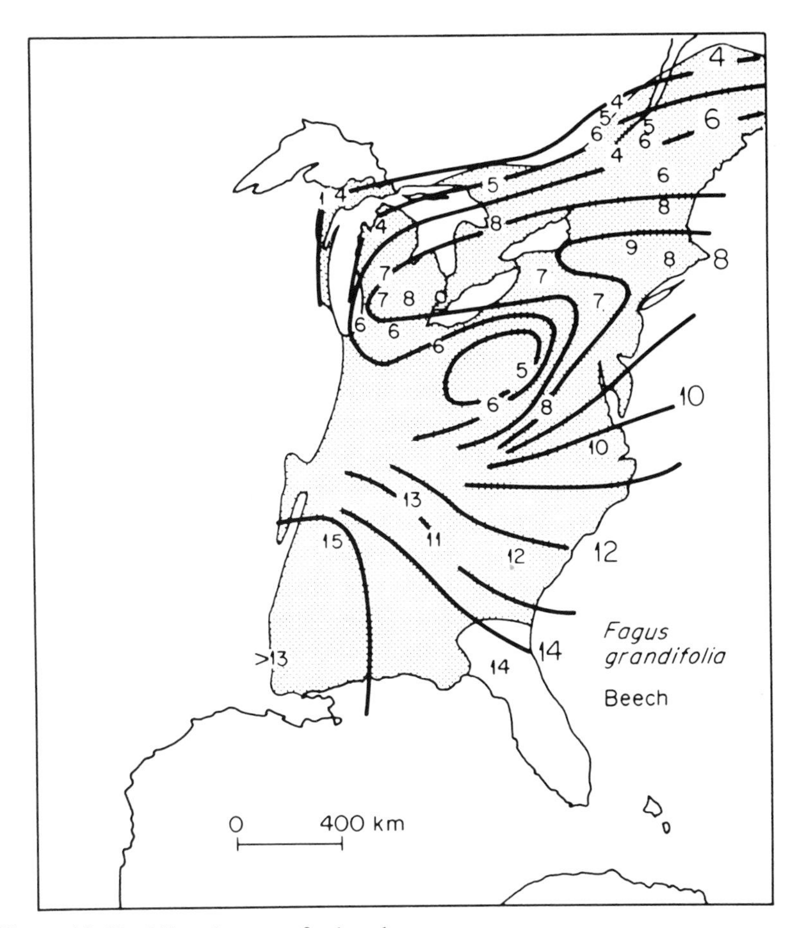

Figure 10.13. Migration map for beech.

Valley or adjacent regions during the last glacial maximum. They occur as fossils at a site near Memphis (Delcourt et al. 1980), and maple, hickory, and beech are also known from the Tunica Hills farther to the south (Delcourt and Delcourt 1977) (Figs. 10.11, 10.12, 10.13, 10.14). Hickory occured early at western sites, where it had reached localities near its present northern limit as long as 10,000 years ago. Its penetration of eastern forests occurred much more slowly; it arrived in New England only 5000 years ago (Fig. 10.12).

In contrast, the migration route of beech was east of the Appalachian mountains. It apparently crossed the mountains in New York State (Fig. 10.13), moved westward across southern Ontario, and then southward into Indiana and Ohio (Kapp 1977). Beech arrived in upper Michigan only 3500 years ago. Its boundary was stable for several thousand years but expanded 70 km westward in the last few hundred years, apparently as a response to climatic changes associated with the Little Ice Age (Davis 1978).

Chestnut was the slowest species to expand its range northward. The migration route is still somewhat obscure, as few fossil sites record its presence. It occurred at the site near Memphis 15,000 years ago (Delcourt et al. 1980), is found in the northern Appalachians 5000 years ago, and arrived in Connecticut just 2000 years ago (Fig. 10.14). Chestnut was dominant in southern New England once it arrived, but reasons for its slow migration relative to other species are not known.

Important trees that have not been included in the discussion are aspen (*Populus*), cedar (*Juniperus* and *Thuja*), and birch (*Betula*). These species were not mapped because I experienced difficulty in determining arrival times. Unlike Europe during late-glacial time or the modern North American Arctic, juniper and birch shrubs were apparently not abundant in the tundra before the invasion of spruce. Pollen from all three genera began to increase during the spruce maximum and peaked just after spruce pol-

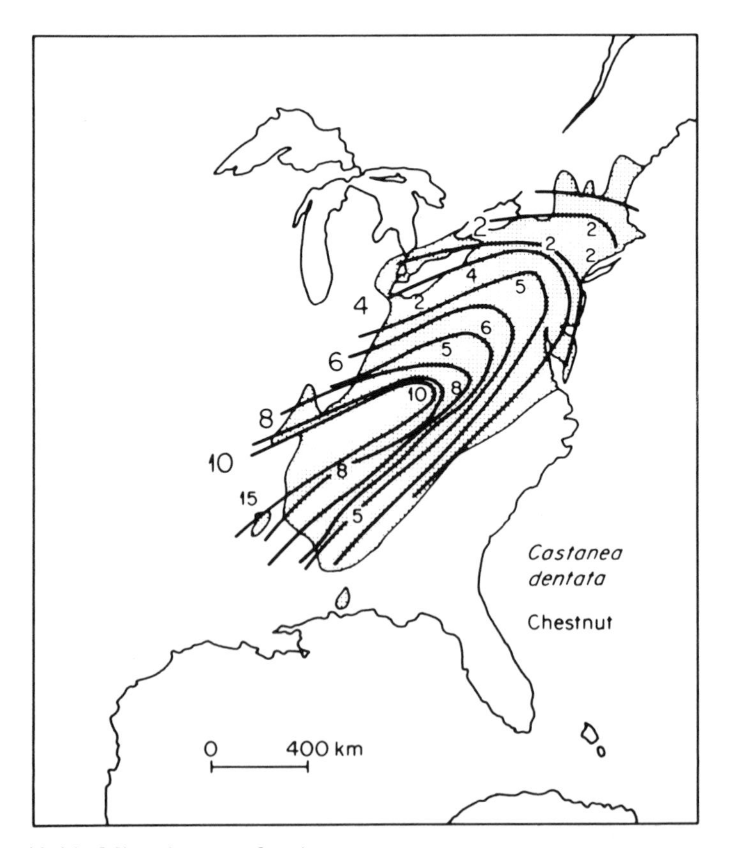

Figure 10.14. Migration map for chestnut.

len declined. Influx diagrams show the maximum after spruce pollen declined, but percentage diagrams often show an additional false maximum just before the spruce peak. Percentages after the spruce peak are sometimes suppressed by the very large amounts of pine pollen at these levels. Changes in percentages are, therefore, somewhat misleading. When additional influx diagrams are available, it will be possible to map the spread of these genera.

Discussion

The history of the spread of trees northward during the present interglacial leads inevitably to an individualistic view of plant communities. Even forests mapped as a single community (Braun 1950) have had very different histories. For example, oak-chestnut forests in the central Appalachians have included chestnut as a dominant for 5000 years or longer, while oak-chestnut forests in Connecticut have included chestnut for only 2000 years. Deciduous forests in Ohio were penetrated first by hickory and then by beech 4000 years later; in Connecticut beech arrived first, followed 3000 years later by hickory. Beech and hemlock entered northern hardwood forests in the Appalachians at about the same time, but because hemlock spread more rapidly, it arrived 2000 years before beech in New Hampshire and 2500 years before beech in upper Michigan.

The survival patterns of tree species during the long and variable glacial intervals may have differed from one glaciation to the next, resulting in different migration patterns and, therefore, different forest communities during each interglacial (Davis 1976). These differences are well illustrated by the interglacial floras of Great Britain (Table 10.1). The communities there are different from one interglacial to the next,

Table 10.1. Occurrences of forest genera in fossil deposits from six interglacial stages and from the Holocene of England (modified from West, 1970).

	Interglacial stage						
Genus	Ludhamian	Antian	Pastonian	Cromerian	Hoxnian	Ipswichian	Holocene
Quercus	+	+	+	+	+	+	+
Ulmus	+	+	+	+	+	+	+
Tilia	–	–	–	+	+	–	+
Carpinus	+	+	+	+	+	+	+
Picea	+	+	+	+	+	–	–
Abies	–	–	–	+	+	–	–
Tsuga	+	+	+	–	–	–	–

depending on the order of arrival of species entering Great Britain before rising sea levels isolated it from the continent (West 1970). Evolutionary changes, which affected the competitive success of different species and thus their migration rates, may also have been involved.

Climatic changes associated with the termination of the Wisconsin glacial phase were the ultimate cause of the expansion of tree species during the Holocene. However, much of the time the rate of spread was not controlled by climate, and the geographic distributions of many species were not in equilibrium with climate, depending instead on the availability of propagules and the ability of seedlings to survive in competition with plants already growing on the site. The idea that species ranges expanded northward in parallel with northward movements of isotherms as the climate gradually warmed is disproved because the migration routes for individual species differed greatly; some moved from east to west, others moved from west to east. Rates of movement also differed. The migration history of the four boreal species, which now have overlapping, almost coinciding ranges, are a persuasive example (Figs. 10.3–10.6). The barrier to migration presented by the Appalachian mountains (Figs. 10.4, 10.7, 10.13) also suggests that seed availability controlled the distribution of many species for several thousands of years during the Holocene.

Nevertheless, climate also affected species distributions and migration patterns. Several species, which expanded without climatic limitation on one frontier, were limited by climate on another. Both white pine and hemlock, for example, spread to the north of the central plains 10,000 years ago when this area was becoming dry and the vegetation was changing to prairie. When these species reached the southern end of Lake Michigan, western areas were too dry for expansion in that direction. After reaching Minnesota, white pine was in equilibrium with the climate at its western limit for several thousand years but continued to expand northward. Climate may also affect rates of migration. The evaluation of climatic periods favorable or less favorable for the expansion of particular species may be possible from migration maps, once isopleths can be drawn with more precision. Adjustments to changing climates have continued to the present, with the expansion of several species (e.g., white pine) westward in the last 2000–3000 years as rainfall has increased (Jacobson 1980). Beech has expanded 70 km westward in upper Michigan within the last 500 years, an apparent response to climatic changes associated with the Little Ice Age (Davis 1978).

During much of the Holocene, however, the distributions of many species were in disequilibrium with climate. Genera like hickory, which were slow to reach the northeastern United States, reached New England after the warmest part of the Holocene. The potential spread of hickory to the north was not realized because the climate had already passed its temperature maximum by the time hickory arrived. The speed with which forests can adjust to climate is thus far exceeded by the speed

Table 10.2. Average rates of Holocene range extensions in eastern North America.

Species	Rate (m/yr)
Jack/red pine	400
White pine	300-350
Oak	350
Spruce	250
Larch	250
Elm	250
Hemlock	200-250
Hickory	200-250
Balsam fir	200
Maple	200
Beech	200
Chestnut	100

of climatic change. This is true although the rate at which trees have extended their ranges is quite remarkable—about 300 m/yr on the average (Table 10.2). Some species spread more rapidly than others, but the reasons for these differences are poorly understood. Seed dispersal is clearly not the sole explanation for the differences, as some heavy-seeded, animal-dispersed genera such as oak migrated very rapidly. The survival ability of invading seedlings may be the most important factor permitting species to spread. The mechanisms whereby a species can so easily disperse and invade an established forest community should be studied more intensively.

Many studies of forest communities focus on the adaptations of species and on evolutionary changes that have allowed species to partition the environment and avoid competition. Fine adjustments of species to one another must indeed have occurred as the result of evolution. One should inspect the paleoecological record, however, to be sure that the species of interest have co-occurred long enough for evolutionary adjustments to have taken place.

Acknowledgments

This work has been supported by the National Science Foundation. I gratefully acknowledge the generosity of the following individuals who have made unpublished data available to me: Linda C. K. Shane, Robert E. Bailey, Ron Lawrenz, and John J. McAndrews.

Chapter 11

Simulating the Role of Climate Change and Species Immigration in Forest Succession

Allen M. Solomon, Darrell C. West, and Jean A. Solomon

Introduction

Forest succession in its traditional sense implies two important features that resist direct examination. First, classical definitions of forest succession generally connote directional changes in species composition and in community structure through time. Following a major disturbance, a directional succession of tree species in our temperate forests may involve 200 (Oosting 1942), 400 (McAndrews 1976), or even 1000 years (Franklin and Hemstrom Chapter 14). There is hardly the opportunity to observe the direction, let alone the species succession, that occurs within our lifetime.

The second feature of forest succession that defies quantitative description is the end point, or that phase of succession frequently described as climax. A broad definition implies that, under environmental conditions which vary within narrow limits, the initially rapid rate of succession diminishes either to a state of dynamic equilibrium without net change or to a state of constant, moderate change controlled by soil-forming processes and associated shifting availability of nutrients and water. The potential for both of these conditions is rare at present and is likely to become more so, as the earth's growing human population demands increasing consumption of natural resources and of open space.

Forest succession (or its absence!) must be observed and quantified if we are to describe it. Langford and Buell (1966) and McIntosh (1979, and Chapter 3) outline the primary basis for our problems in succession studies, some of which are aptly illustrated by Heinselman (Chapter 23), Horn (Chapter 4), Franklin and Hemstrom (Chapter 14), Cooper (Chapter 21), Christensen and Peet (Chapter 15), and Brian Walker (Chapter 25). Forest stand succession models may allow objective evaluation of the bases for our disagreements.

A careful examination of the modeling approaches, discussed by Botkin, Doyle, and Shugart et al. (Chapters 5, 6, and 7, respectively) reveals that additional validation tests are required, in both time and space, to establish models as reliable tools in studying forest succession. Yet other studies indicate that at least rudimentary examination of long-term processes in forest succession may proceed by means of simulation model output, verified with independent pollen evidence. Particularly appropriate for this task are Davis' (Davis et al. 1973; Davis 1974, 1976, Chapter 10) perceptive, far-reaching conclusions on the role of successive invasions by immigrating species in controlling the composition of the climax communities of New England during the Holocene. If these conclusions apply throughout the eastern forests of the United States, then they provide substance for the belief in individualistic theories of vegetation development. If, instead, the conclusions cannot be generalized because climate provides the primary source of stability and change in other records of forests during Holocene time, then we must consider the potential relevance of seral vegetation dynamics that attain a pervasive climax community.

The work described in this chapter evaluates the separate and combined effects of climate and of species immigration upon composition of forest vegetation, via forest stand simulation. We utilized the FORET model, which is particularly appropriate in the evaluation because the forest dynamics it simulates are based upon interactions among species as a function of individual growth characteristics and not upon the assumption of any sequential order of species assemblages, or superorganismic properties of ecosystems. We also used Delcourt's (1979) pollen record from middle Tennessee, which is also appropriate because of its location. The site is far south of glaciated terrain, and near full-glacial refugia of dominant forest species (e.g., southern Alabama, Delcourt 1980; west Tennessee, Delcourt et al. 1980), minimizing effects of the individual species migration periods that are so important in New England forests. The presence of mixed mesophytic forests and the associated high diversity of dominant species (Braun 1950) should reduce the impact upon the forest dynamics of single species invasions in comparison to the more depauperate forests of New England. Indeed, if maintenance of a climax community mediated by climate occurs, it should be clearly expressed in the Middle Tennessee region.

Parameters for the Model

Simulation Modeling

The FORET simulation model described by Shugart and West (1977) was modified to accommodate additional species, bringing to 65 the number of species simulated (Table 11.1). The model simulated species dynamics

Table 11.1. Scientific nomenclature, common names, and species parameters used in the FORET model.

Scientific names	Common names	Age max. (yr)	Ht. max. (ft)	Dbh max. (ft)	Deg max. (days)	Deg min. (days)	Tc[b]	Reproduction[c] 1	2	3	4	Sprouting[d] No.	min. age	max. age	K time[e]
Aesculus octandra	Buckeye	100	100	5.0	6640	4820	1	T	F	T	F	1	12	200	
Abies balsamea	Balsam fir	200	116	2.2	4328	2205	1	F	F	T	F				
Abies fraseri	Fraser fir	200	160	6.0	5007	4826	2	F	F	T	F				
Acer negundo	Box elder	75	75	6.0	9000	1600	1	F	T	T	F	1	6	40	
Acer rubrum	Red maple	150	136	5.0	13,390	2300	1	F	T	T	T	3	12	200	
Acer saccharinum	Silver maple	125	120	7.0	9000	1600	2	F	T	T	F	1	6	50	
Acer saccharum	Sugar maple	300	135	6.0	7500	2200	1	F	F	T	T	3	12	80	
Betula lenta	Sweet birch	265	70	4.5	5551	2556	1	F	T	T	T	3	12	100	
Carya aquatica	Water hickory	250	110	3.5	11,100	5800	2	F	F	F	T	3	6	70	
Carya cordiformis	Bitternut hickory	300	171	4.0	9170	3470	1	T	F	T	F	1	12	200	20
Carya glabra	Pignut hickory	300	145	4.0	12,560	3470	1	T	F	T	F	2	12	200	30
Carya illinoensis	Pecan	250	180	6.0	9900	4300	1	F	F	T	F	1	6	25	
Carya laciniosa	Shellbark hickory	300	130	4.0	8340	4520	1	T	F	T	F	2	12	90	
Carya ovata	Shagbark hickory	275	130	4.0	12,560	3700	1	T	F	T	F	1	12	200	
Carya texana	Black hickory	300	130	4.0	9170	4820	1	T	F	T	F	2	12	110	
Carya tomentosa	Mockernut hickory	300	112	3.0	10,820	3470	1	T	F	T	F	1	12	200	
Castanea dentata	Chestnut	300	120	10.0	8260	3470	1	T	F	F	F	3	12	200	
Celtis laevigata	Hackberry	125	110	2.0	12,560	4820	1	T	F	F	F	2	6	115	
Cornis florida	Flowering dogwood	100	40	1.5	10,820	3470	1	F	F	T	F	3	12	200	
Fagus grandifolia	Beech	350	120	5.0	10,000	2420	1	F	F	T	F	2	6	30	
Fraxinus americana	White ash	300	125	7.0	10,820	2550	1	F	T	T	T	2	6	20	
Fraxinus nigra	Black ash	300	87	4.9	4103	2205	2	F	F	F	T	2	6	20	
Fraxinus pennsylvanica	Green ash	150	105	4.0	9900	1600	2	F	T	F	F	1	6	50	
Ilex opaca	Holly	200	100	4.0	10,820	4820	1	T	F	T	F	1	6	40	

Juglans cinerea	Butternut	100	100	6.0	5913	3399	1	F	T	T	F	1	6	12	20
Juglans nigra	Black walnut	250	150	8.0	8260	3470	2	T	F	T	F	1	6	20	20
Juniperus virginiana	Juniper	300	120	4.0	10,000	3130	2	F	T	F	F				20
Liquidambar styraciflua	Sweetgum	250	150	5.0	10,820	4820	2	F	F	F	T	2	2	12	
Liriodendron tulipifera	Yellow poplar	300	198	12.0	10,820	3470	2	F	T	F	T	2	12	200	
Morus rubra	Red mulberry	100	70	3.5	13,390	3470	1	T	F	T	F				
Nyssa spp.	Blackgum	300	130	6.0	12,560	3470	2	F	T	F	F	2	25	100	
Ostrya and Carpinus	Hornbeam	150	78	3.0	10,820	2300	1	F	F	F	F	2	6	50	
Picea glauca	White spruce	200	128	3.1	3473	804	1	F	F	F	F				
Picea mariana	Black spruce	250	83	1.6	3473	477	1	F	F	T	F	1	10	20	
Picea rubens	Red spruce	400	125	4.5	4644	2278	2	F	T	T	F				
Pinus banksiana	Jack pine	230	70	2.4	4022	384	2	F	T	T	T				
Pinus echinata	Shortleaf pine	400	146	4.0	9170	4820	2	F	T	F	F	2	6	20	
Pinus resinosa	Red pine	310	141	3.2	3696	1578	2	F	T	T	F				
Pinus strobus	White pine	450	220	6.0	5730	1580	1	F	T	T	F				
Pinus taeda	Loblolly pine	350	182	5.0	10,820	573	2	F	T	F	F				
Pinus virginiana	Virginia pine	300	120	3.0	6640	4820	2	F	T	F	F				20
Planera aquatica	Planer tree	60	40	0.8	9300	4800	2	F	T	T	F	1	6	25	
Platanus occidentalis	Sycamore	500	175	12.0	9900	3500	2	F	T	T	F	2	6	50	
Populus deltoides	Cottonwood	150	175	12.0	10,000	1600	2	F	T	T	T		6	30	
Quercus alba	White oak	400	150	8.0	10,000	3130	1	T	F	F	F	2	12	40	
Quercus coccinea	Scarlet oak	400	102	5.0	8260	3700	1	T	F	F	F	2	12	80	
Quercus falcata	Southern red oak	400	122	7.0	10,820	4820	1	T	F	F	F	2	12	30	
Quercus falcata pagodaefolia	Cherry bark oak	300	140	7.0	10,000	4600	2	F	F	T	F	1	6	12	
Quercus lyrata	Overcup oak	250	114	5.0	9600	5300	2	T	T	T	T	1	6	12	
Quercus marilandica	Black jack oak	400	50	3.0	10,000	4520	2	F	F	F	F	2	12	40	
Quercus michauxii	Swamp chestnut oak	300	130	9.0	10,200	5000	2	F	F	T	F	1	6	12	
Quercus nigra	Water oak	250	125	6.0	10,700	5500	2	T	F	T	F	2	6	12	
Quercus nuttallii	Nuttall oak	250	120	4.0	9500	6100	2	T	F	T	F	1	6	12	
Quercus phellos	Willow oak	250	130	6.0	9600	4800	2	T	F	T	F	1	6	12	

Table 11.1. (*continued*)

Scientific names	Common names	Age max. (yr)	Ht. max. (ft)	Dbh max. (ft)	Deg max. (days)	Deg min. (days)	Tc[b]	Reproduction[c] 1	2	3	4	Sprouting[d] No.	min. age	max. age	K time[e]
Quercus prinus	Chestnut oak	267	100	7.0	7430	3470	1	T	F	F	F	2	12	40	
Quercus rubra	Northern red oak	400	160	8.0	8260	1580	1	T	F	F	F	2	12	40	
Quercus shumardii	Shumard oak	300	180	8.0	10,820	4520	2	F	T	F	F	2	12	160	
Quercus stellata	Post oak	400	100	4.0	10,820	4820	2	T	F	F	F	1	12	30	
Quercus velutina	Black oak	300	150	7.0	9170	3290	1	F	T	F	F	2	12	40	
Salix nigra	Willow	70	140	8.0	11,200	2900	2	F	T	T	T	3	6	30	
Tilia americana	American basswood	140	115	7.1	5679	2205	1	F	F	F	T	3	12	80	
Tilia heterophylla	White basswood	150	130	5.0	8260	4820	1	F	T	T	F	3	12	80	
Tsuga canadensis	Hemlock	1000	160	7.0	6559	2416	2	T	F	T	T				
Ulmus alata	Winged elm	125	80	3.5	10,820	4820	2	F	F	F	F	1	6	110	
Ulmus americana	American elm	300	160	11.0	12,560	2200	1	F	F	T	F	2	6	240	

[a]Scientific nomenclature and common names follow Little (1953).

[b]Tc, tolerance class of each species. Class 1 is considered shade tolerant and Eq. (6) from Shugart and West (1977) is used to calculate growth. Class 2 is considered intolerant, and Eq. (7) from Shugart and West (1977) is used to calculate growth.

[c]Reproduction switches used in the birth subroutine take values of T (true) or F (false). Switch 1 is T if the species requires leaf litter for successful reproduction. Switch 2 is T if the species requires mineral soil. Switch 3 is T of hot years reduce species reproduction. Switch 4 is T if the species if a preferred food of deer or small mammals.

[d]Sprouting No. is the number of sprouts that can occur annually between minimum (min. age) and maximum (max. age) ages of each tree species.

[e]K time is the age after which a parent tree must be present on the plot before new seedlings can be added to the plot.

during a time span of 16,000 years using the processes described by Shugart and West (1977), and by Solomon et al. (1980). Four simulation conditions were each run ten times, and the results were averaged. The output is the average of ten, 1/12-ha. plots. Output was stored from each 25th annual sample, then was smoothed with a seven-level moving average, in order to approximate the estimated 150-year duration of each pollen sample (Solomon et al. 1980).

Condition 1: Seed sources of all species were available for the entire simulated period. Annual values for growing degree-days were drawn from a normal distribution with a mean matching that at Crossville, Tennessee, about 44 km east of the pollen collection site, and a variance matching that of the more detailed meteorological records at Oak Ridge, Tennessee, about 108 km east of the site. The mean value used (5395) may be as much as 500 degree-days (1°C) cooler than the mean value at Anderson Pond, which lies 300 m lower in altitude than the Crossville Meteorological Station. This simulation condition was run to provide a base case in which neither climate nor immigration changed, and to determine the magnitude of variability which must be accounted for by changes in exogenous forces.

Condition 2: Seed sources of each species became available at the dates specified in Table 11.2. The pollen record was the basis for inferring these dates. Seed sources for all palynologically indistinguishable species in a

Table 11.2. Species immigration dates estimated from Anderson Pond pollen stratigraphy.

Species	Year	Species	Year
Fir	16,000	Butternut	16,000
Box elder	16,000	Walnut	1800
Red maple	8600	Yellow poplar	16,000
Silver maple	13,300	Sweetgum	9200
Sugar maple	13,300	Mulberry	12,600
Buckeye	15,600	Blackgum	8600
Birch	14,900	Spruce	16,000
Hornbeam	16,000	White pine	15,300
Hickory	16,000	Hard pines	16,000
Chestnut	8600	Planer tree	11,000
Hackberry	12,000	Sycamore	11,000
Dogwood	7300	Cottonwood	8600
Beech	14,900	Oak	16,000
Black ash	16,000	Willow	15,600
White ash	9200	Basswood	12,600
Green ash	9200	Hemlock	9800
Holly	9200	Elm	15,800
Red cedar	16,000		

genus became available at the same time. The annual values for growing degree days were generated as in Condition 1. This condition was run to simulate the case in which variance of the pollen record is due to competitive interactions among established and invading species, without the influence of climatic change.

Condition 3: Seed sources of all species were available for the entire simulated period. Mean growing degree-days were changed at 500-year intervals (Table 11.3). The changing degree-days were calculated from climatic data (United States Geological Survey 1965) at the geographic locations where modern pollen composition (Davis and Webb 1975; Webb and McAndrews 1976) could be matched to the fossil pollen composition from the middle Tennessee pollen record. The locations of modern pollen spectra are illustrated and discussed by Delcourt (1979, Fig.13). This condition was meant to simulate the case in which variance of the pollen record is due to modulation of competitive interactions by climate change, without the influence of sequentially invading species. This simulation is the basis for the earlier article by Solomon et al. (1980).

Condition 4: Seed sources of each species became available as in Condition 2 (Table 11.2), and growing degree-days changed as in Condition 3 (Table 11.3). This condition was meant to simulate the case in which climate shifted, allowing the immigration of new species that were not previously available to grow at the site.

Table 11.3. Annual growing degree-days (GDD) at each 500-year interval, estimated from Anderson Pond pollen stratigraphy.

Years ago	GDD	ΔGDD	Years ago	GDD	ΔGDD
0	5395	0	8500	4842	+105
500	5395	0	9000	4737	+189
1000	5395	0	9500	4548	+87
1500	5395	+136	10,000	4461	+334
2000	5259	−57	10,500	4127	+416
2500	5316	+114	11,000	3711	+300
3000	5202	+107	11,500	3411	+227
3500	5095	−79	12,000	3184	+211
4000	5174	+95	12,500	2973	+298
4500	5079	+90	13,000	2675	+73
5000	4989	−90	13,500	2602	+249
5500	5079	+84	14,000	2353	+31
6000	4995	−100	14,500	2322	+59
6500	5095	+73	15,000	2263	+55
7000	5022	0	15,500	2208	+168
7500	5022	+67	16,000	2040	—
8000	4955	+113			

Measurement of Actual Vegetation History

The pollen record used is described in detail by Delcourt (1979). The pollen samples were drawn from a 35-ha sinkhole at 300 m altitude in White County, Tennessee (36°02′N latitude, 85°30′W longitude). The immediate area consists of flat to gently rolling topography termed the eastern Highland Rim (Hardeman 1966). The Cumberland Plateau at about 600 m elevation lies to the east, as close as 1 km southeast of the pond. Prior to the chestnut blight, the forest vegetation of the area was in the transition from western meosphytic forest to the west, and mixed mesophytic forest to the east (Braun 1950). The vegetation was dominated by mixed oak and oak-chestnut forest, with mesic hardwood species such as beech, basswood, and sugar maple restricted to the slopes of deep ravines along the west margin of the Cumberland Plateau.

Pollen samples were taken at depth intervals of approximately 400–500 years, each representing approximately 150 years of sedimentation. Percentages of arboreal taxa were determined from light microscopic analysis of at least 300 arboreal pollen grains in each sample.

Comparison of Model Results and the Pollen Record

The Record of Vegetation History

The Anderson Pond pollen record consists of taxonomic pollen frequencies arrayed in time-ordered profiles (Fig. 11.1). A more authentic image of the vegetation history may be obtained if we could correct for the taxonomic differences in pollen productivity and transport. One approach is to measure the ratio of pollen frequency to plant frequency on the modern landscape, then to treat the sample pollen estimates with these "representation values" (R values, Davis 1963) before percentages are calculated.

However, there are strong theoretical and practical reasons for viewing corrected percentages with caution. Representation of each species is apparently not constant (Davis 1963; Webb et al. 1978), but instead varies systematically such that corrected pollen percentages are too small in the case of abundant tree species, and are too great in the case of sparse tree species (Janssen 1967). In addition, actual changes in tree species frequencies may differ greatly from the apparent changes in corrected species frequencies when the species mix represents widely divergent representation values (Solomon and Harrington 1979). Therefore, we utilized only the uncorrected pollen percentages (Fig. 11.1), and we mention the most important inaccuracies in representation of vegetation by pollen.

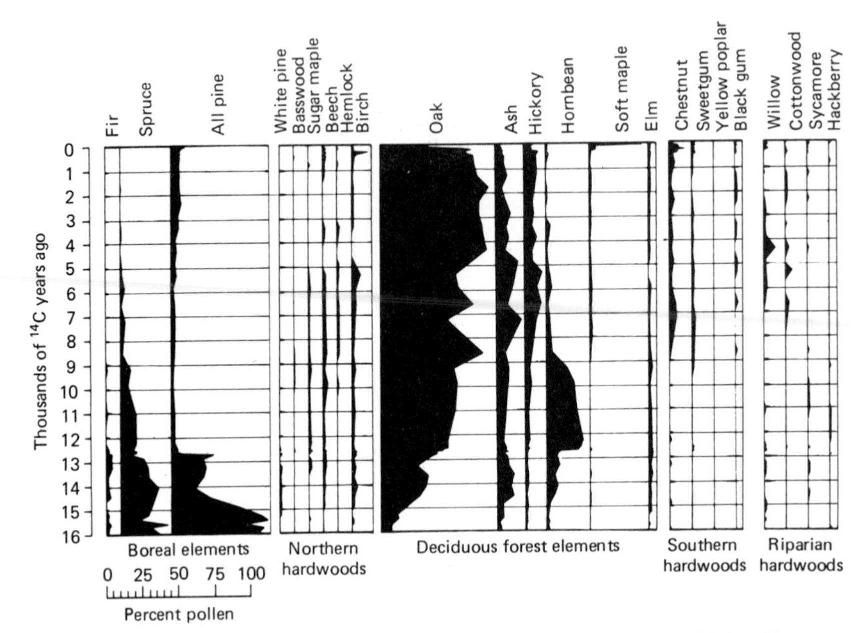

Figure 11.1. Arboreal pollen diagram from Anderson Pond, White County, Tennessee, for the past 16,000 ^{14}C years.

The earliest three millenia of the pollen diagram (Fig. 11.1) are dominated by spruce and hard pine (jack pine constitutes the only pine macrofossils identified; Fig. 12, Delcourt 1979), with significant amounts of oak and ash. Spruce probably dominated the vegetation, considering the low productivity and weak dispersal of its pollen, compared with pollen of, e.g., pine, oak and ash. Fir, white pine, hickory, hornbeam, and elm probably were present, though not particularly important, during this period. A rapid and pervasive transition from this mixed coniferous-deciduous vegetation to one dominated by deciduous elements occurred between 12,750 and 12,500 years ago. The shift was accompanied by a rapid influx of organic sediment, and by the increased presence of aquatic plant macrofossils (Delcourt 1979).

Preceded by increased abundance of sugar maple (a poor pollen producer) at about 13,000 years ago, fir disappeared, and spruce, followed by hard pine, declined drastically. Concomitantly, oak, hornbeam, hickory, and sugar maple increased in abundance. This period provides the first ecological enigma we examine with simulation modeling. Did the mixed coniferous-deciduous forest exist because its constituent species outcompeted others, or merely because the others were absent? Did the conifer populations crash because the climate changed, shifting adaptive advantages among species already present, or because new species better adapted to unchanged conditions there arrived between 13,000 and

12,500 years ago? Or, were both phenomena responsible for the drastic change in character of the vegetation?

Between about 12,500 and 9500 years ago oak, with hornbeam, hickory, and sugar maple dominated the forests. Other taxa characteristic of mixed mesophytic forest regions (Braun 1950, p. 40) were present, including birch, beech, basswood, hemlock, and walnut (Fig. 11.1, walnut not shown). Many of the mixed mesophytic forest taxa (e.g., yellow poplar, white basswood, sugar maple, sweet buckeye) are such poor producers of wind-borne pollen that each also could have been abundant.

We can model the apparent dominance during this period of taxa found today in Braun's mixed mesophytic forest region. The same questions asked above apply to this period, albeit with different taxa: "What separate or combined roles did climate change and species immigration play in the composition of forests recorded by fossil pollen?"

From about 9000 to 5000 years ago chestnut, sweetgum, black gum, and red (soft) maple were consistently present in the forests near Anderson Pond, although not apparently abundant. Chestnut is another dominant of Braun's mixed mesophytic forest, but the "gums" can be classed as southern hardwoods. Hickory reached maximum abundance between 7000 and 5000 years ago, bracketed by peak values of ash at 7250 and 4800 years ago. Yellow poplar may have been present in forests during this period and at other times, but its entomophily precludes its presence in the pollen record. It grows today in forests surrounding Anderson Pond although its pollen is absent there. About 5000 years ago, taxa characteristic of the mixed mesophytic forest declined as oak assumed primary dominance, with lesser amounts of hickory and ash.

The presumed immigration of new species into the Anderson Pond area ceased 7000–8000 years ago (Table 11.2), while the major climatic shifts ceased by 8000 years ago with a minor increase in degree days at 3000 years (Table 11.3). Thus, the simulation output should be quite similar under all four conditions after about 7000 years ago and should converge to the same stable "climax" after about 3000 years ago if indeed climate exerts the ultimate control upon the assemblage. This expectation will be treated below.

Simulated Vegetation History

Species Complement and Climate Constant

The simulated forest diagram of the "base condition" (BA case, Fig. 11.2) indicates the degree to which community composition near Anderson Pond can be explained on the basis of internal competitive dynamics alone. The diagram illustrates a forest dominated by oak. According to simulation output (not shown), about half of the oak biomass is red oak,

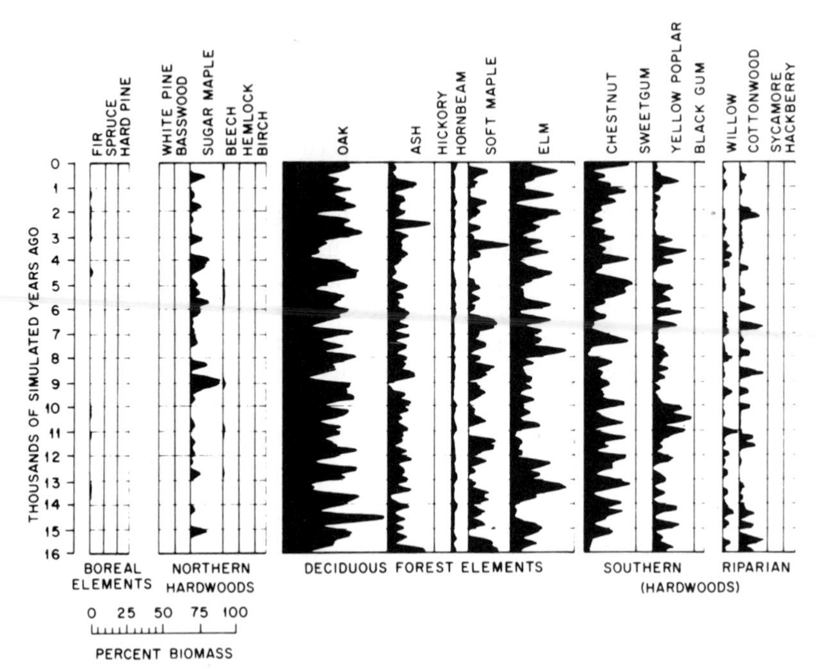

Figure 11.2. Simulated forest composition for 16,000 years when all species are available and growing degree-days are constant at the modern value.

the remainder in order of importance being black, chestnut, and white oak. The other important simulated dominants are chestnut, elm, soft maple (silver maple and box elder), ash, and yellow poplar. Sugar maple was the only one of the nine taxa (17 simulated species) classed here as boreal elements and as northern hardwoods (Fig. 11.2, Table 11.1), that achieved more than minor, sporadic occurrence. Hickory was conspicuously absent among the deciduous forest elements. Sweetgum, black gum, sycamore, and hackberry were also absent in the simulation, but none of these was particularly important in the forests near Anderson Pond.

More striking than compositional disparities is the difference between the low-frequency, low-amplitude changes in pollen profiles (Fig. 11.1), and the variable-frequency, high-amplitude oscillations displayed in the profiles of all important simulation taxa (Fig. 11.2). The model output oscillations appear to mimic the dynamic properties of species populations over long time periods (Emanuel et al. 1978a,b) rather than reflecting, e.g., erroneous silvical information or fallacious model routines. Certainly the oscillations would be reduced had we run the model more times under each condition. Yet, strong oscillations also characterize FORET output, which consists of averages of 100 runs (Shugart and West 1977, 1980, Chapter 7; Emanuel et al. 1978a,b).

In contrast, the stability visible in the pollen diagram (Fig. 11.1) is more apparent than real. Pollen data magnify uniformity while attenuating high-frequency and high-amplitude fluctuations. Population dynamics that involve less than ca. 1000 ha cannot be recorded by pollen spectra except in unusual circumstances because of the large (minimally 3000–100,000-ha.) areas pollen reflects (Tauber 1965, Fig. 6; Tauber 1977, p. 71). Temporal variations are smoothed by bioturbation (R. B. Davis 1967, 1974), by sediment redeposition (M. B. Davis 1967, 1968, 1973), and by wide intervals between collected samples.

Similar modeling work now in progress at Oak Ridge National Laboratory should reduce the disparity through the use of pollen records from small water bodies, short-interval sediment sampling, increased simulation runs, and stratification of simulation run conditions to match the habitats reflected by the pollen records. In this work, however, we assume that the high variance in the simulation output has been damped out in the pollen record, and this disparity will be treated as a source of statistical noise, which limits the precision of comparisons between model output and pollen data.

Species Complement Varies, Climate Constant

The simulated effects of invasion by new species into forest stands is illustrated in Fig. 11.3 (migration variable, or MI case). The species unable to enter the stand under unchanging environmental conditions (BA case, Fig. 11.2), are absent here as well. There is no perceptible effect on the modeled stand owing to early arrivals that successfully entered the stand (silver and sugar maple, elm, willow), perhaps because any minor changes would be masked by the high variance in model output. Effects of later arrivals (green ash, chestnut) are more evident. Yellow poplar abundance declines, coincident with the invasion by green ash (neither white ash nor black ash generated measurable biomass on the plot) 9200 simulated years (sy) ago. The chestnut entry at 8600 sy occurs at the expense first of elm, then of oak, particularly of white and black oak species (not shown). Green ash and chestnut subsequently may act in concert to constrain oak and yellow poplar abundance.

A similar oak decline between 8500 and 5500 sy ago, coincident with increases in chestnut and ash, is visible in the pollen diagram. Ash appears in sufficient abundance to have played a role in the oak decline. The minor amount of chestnut may still be erroneously great as our experience (Solomon and Kroener 1971, Fig. 7 and discussion; Solomon 1970, Fig. 15) indicates that chestnut pollen overrepresented the importance of chestnut trees in the vegetation. Even if yellow poplar grew at Anderson Pond during this time, we cannot determine its actual response to increases in green ash and chestnut.

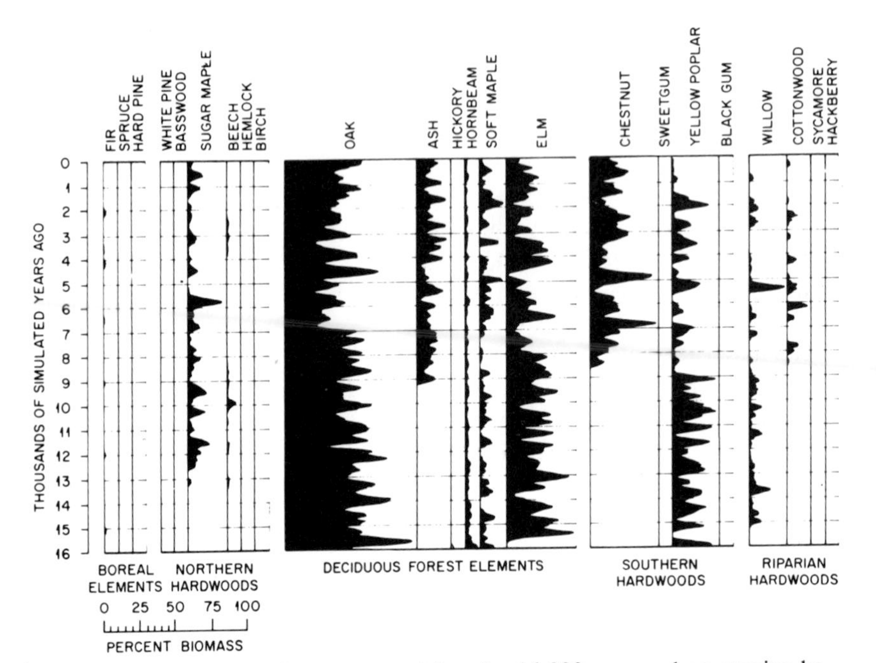

Figure 11.3. Simulated forest composition for 16,000 years when species become available as in Table 11.2, and growing degree-days are constant at the modern value.

Species Complement Constant, Climate Varies

The independent role of climate in controlling simulated community composition is illustrated in Fig. 11.4 (climate variable, or CL case). Most of the boreal elements and northern hardwoods that were missing in the first two simulations successfully grow under the climate-variable condition. Spruce (primarily white spruce with some black spruce, and no red spruce; Fig. 9, Solomon et al. 1980) and hard pine (about 90% jack pine with red pine) dominate the stand during the initial 2500–3000 sy period. Fir (primarily Fraser fir) grows in approximately the same abundance as at Anderson Pond (Fig. 11.1) though it does so at the wrong time (Fig. 11.3). Balsam fir grew on the plot between 13,500 and 11,000 sy ago, but its biomass was too slight (<0.2 metric tons/ha) to be graphed.

Among the northern hardwoods, white pine, basswood and beech are successful on the plot at about the times they were detected growing near Anderson Pond. The simulated extensive abundance of basswood may actually have occurred at Anderson Pond, as basswood is entomophilous and is recorded only sporadically in sediments near which it grows. Sugar maple both enters and declines 2–4 millenia later than it apparently did at Anderson Pond. Hemlock and birch, never abundant at Anderson Pond, were unable to reach maturity on the simulated stand.

After about 13,000 years ago, oak dominates the stand, primarily red oak at first, followed by increasing amounts of black, chestnut, and white oak (Fig. 9, Solomon et al. 1980). Simulated red oak is responsible for the maximum in the oak profile between 13,500 and 10,500 sy ago. Profiles of most other deciduous forest elements are similar to those at Anderson Pond with the notable exception of hickory. Elm appears later and more abundantly in the simulations than in the fossil materials, and hornbeam biomass does not include the hornbeam peak recorded at Anderson Pond. Modeled biomass from the soft maples is composed of silver maple (ca. 70%) and box elder without the red maple, which dominates the uppermost pollen spectra (Delcourt 1979).

Chestnut appears in the simulation diagram at about the time that it occurred in the Anderson Pond record, indicating that its entrance into the forests there can be explained by climate change alone. As in the previous two simulations, sweetgum, blackgum, sycamore, and hackberry did not become established on the plot (Fig. 11.4). Yellow poplar appears in the stand about 8800 years ago. Its possible response to immigrating ash or chestnut, discussed under the MI case, is apparently a moot point. Simulated willow and cottonwood profiles appear to be generally similar to the respective pollen profiles, given the systematic differences in variability that occur between the model output and the vegetation record.

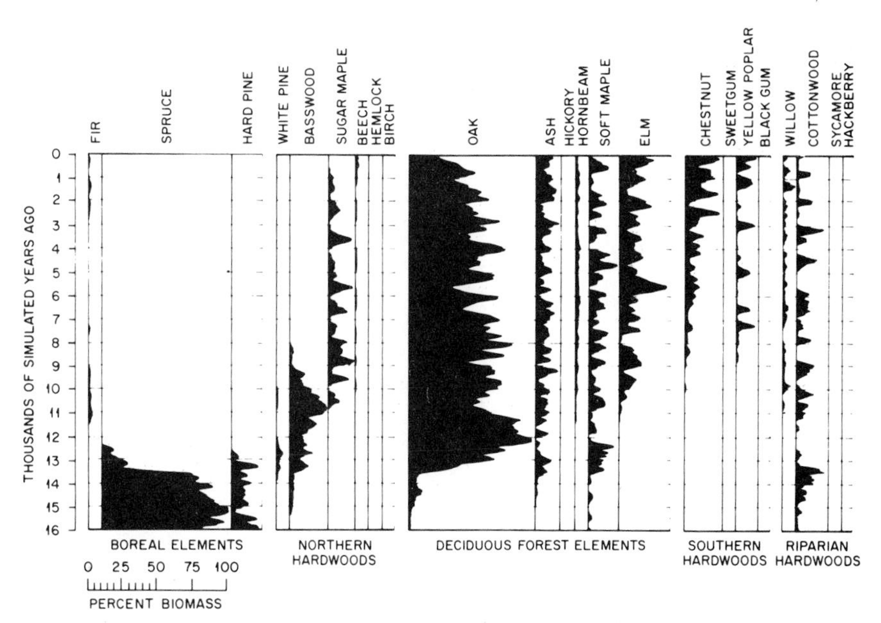

Figure 11.4. Simulated forest composition for 16,000 years when all species are available, and growing degree-days change as in Table 11.3.

Species Complement and Climate Vary

With certain minor exceptions, the simulation diagram in which immigration and climate both vary (MI-CL case, Fig. 11.5) is about the same as the diagram in which climate alone varies (Fig. 11.4). Basswood enters the plot precipitously in the former case, and gradually, at an earlier date, in the latter case, but basswood pollen is too poorly represented to determine if either profile is more accurate. Ash appears in the CL case and the pollen record shortly after the beginning of the records, but appears in the MI-CL case after 7000 years. Here, ambiguities in ash pollen identification (Delcourt 1978, Appendix) resulted in simulating the immigration of black ash at 16,000 years and of green ash at 9200 years (Table 11.2). Green ash, however, was the only successful ash simulated and therefore may have been the species that actually grew near Anderson Pond 16,000 years ago.

An overall comparison between the simulation and vegetation diagrams (Figs. 11.1–11.5) indicates a strong similarity between only the two climate-controlled model runs and the vegetation history (Figs. 11.4, or 11.5 and 11.1), particularly during the critical transition to oak dominated forest 12,500 years ago, the establishment of mixed mesophytic forest taxa in the early Holocene, and the subsequent entry of certain southern hardwoods after about 9000 years ago. The other simulations appear to explain little variation in the vegetation record. The most dramatic difference between the climate-variable simulation and the

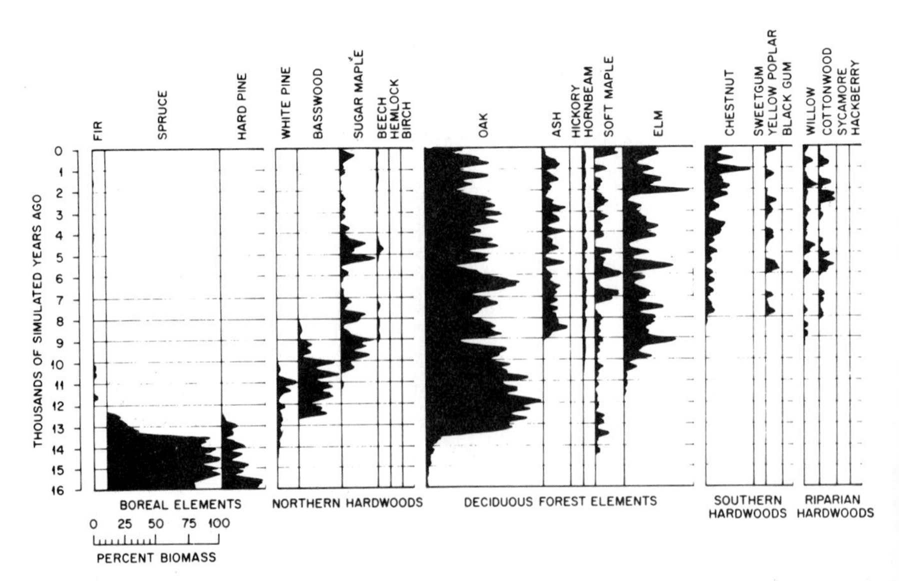

Figure 11.5. Simulated forest composition for 16,000 years when species become available as in Table 11.2, and growing degree-days change as in Table 11.3.

vegetation diagrams is the lack of hickory in simulation output, in contrast to its importance during the last 8000 years at Anderson Pond.

Statistical Analyses

The problems that we wish to solve through application of statistical techniques include (a) measuring the similarity between modeled vegetation assemblages, and the pollen ("actual vegetation") assemblages, as a function of time; (b) measuring the similarity between modeled taxon profiles and the pollen profiles; and (c) determining the amount of variance in the vegetation history that is explained by the results of vegetation modeling, under each simulation condition.

The data sets consisted of pollen percentages and simulated biomass percentages at 500-year intervals (33 levels) for each of the 22 comparable pollen and biomass taxa listed in Table 11.4. A 23rd taxon, "other trees," was not analyzed, but was used to calculate percentage values in all data sets.

We used Pearson's product-moment correlation coefficient r. Probabilistic significance tests of r, based on relative percentages, are tenuous for interspecies comparisons, and are inappropriate for intercollection (assemblage) comparisons (Sneath and Sokal 1973). In the former case, we improved the validity of the significance test by (1) the use of 500-year intervals between samples (reduced autocorrelation between levels); and (2) the use of percentages that totaled less than 100% (the 23rd taxon) within any level (reduced interdependence among percentage values within each level). In the case of the assemblages, we transformed the data via the arcsine transformation (Sokal and Rolfe 1969) to normalize percentages (producing the bivariate normal distributions required).

Statistically significant positive correlations between pollen profiles and simulated biomass profiles (Table 11.4) represent less than 1% of the pollen and of the biomass in the base case (BA case), and about 8% in the migration-variable (MI) case. In contrast, statistically significant positive correlations include 66% of the biomass and 70% of the pollen in the climate-variable (CL) case. The migration and climate variable (MI-CL) case represents 67% of the biomass and 72% of the pollen grains analyzed from the 16,000-year period of record. Oak, pine, and spruce—the significantly correlated taxa in the two climate-variable cases—also dominate the pollen and CL case biomass. The lack of significant correlations among minor taxa in all simulation cases is probably due primarily to the difference in magnitude of variance between pollen and biomass profiles (discussed above).

Correlation coefficients between pollen assemblages and biomass assemblages are uniformly high in all four cases, the majority being significant well beyond the $p = 0.01$ level. The CL case produced the best

Table 11.4. Correlation coefficients comparing pollen and biomass profiles.

Species	Biomass proportion versus pollen proportion when			
	Immigration climate constant	Immigration varies	Climate varies	Immigration climate vary
Fir	−0.25	0.15	−0.17	−0.18
Soft maple	−0.18	−0.10	−0.15	0.42[a]
Sugar maple	0.31	0.18	−0.13	0.33
Birch	0.03	−0.04	0.02	−0.19
Hickory	0.00	−0.35[a]	0.02	0.03
Chestnut	−0.02	0.59[b]	0.36[a]	0.41[a]
Hackberry	0.17	0.19	−0.18	0.12
Beech	−0.10	0.32	0.07	−0.06
Ash	−0.13	0.20	0.13	0.17
Sweetgum	0.41[a]	0.25	−0.22	0.38[a]
Mulberry	−0.02	0.58[b]	−0.29	−0.23
Blackgum	−0.22	0.44[a]	0.44[a]	0.27
Hornbeam	−0.13	0.04	−0.43[a]	−0.42
Spruce	0.05	0.49[b]	0.85[b]	0.86[b]
Hard pine	0.27	0.13	0.88[b]	0.88[b]
White pine	−0.02	−0.05	0.23	0.00
Sycamore	−0.18	0.09	−0.21	0.26
Cottonwood	−0.02	0.00	−0.08	0.29
Oak	0.01	−0.46[b]	0.62[b]	0.49[b]
Willow	−0.19	−0.01	0.09	0.19
Hemlock	−0.10	0.24	−0.26	0.08
Elm	0.10	0.08	−0.13	0.03
	r^2=0.03	r^2=0.08	r^2=0.09	r^2=0.16

[a] r significant at $\alpha \geqslant 0.95$; critical value = 0.343.
[b] r significant at $\alpha \geqslant 0.99$; critical value = 0.442.

overall match to the pollen record, accounting for the highest of the four correlations, in 16 of the 33 cases. Average r^2 values (a measure of the variance in the pollen variate, explained by the simulation variate) are about equal for the CL and MI-CL cases (about 0.45, or 45%), compared with considerably lower average r^2 values in the MI and BA cases (34 and 31%, respectively).

The correlation sequence can be subdivided on the basis of the distribution of highest correlations, and of trends in three-level moving average r^2 values (Fig. 11.6). The climate-controlled simulations (CL and MI-CL cases) compare well with the pollen record during the initial 3500 years of the sequence (16,000–13,000 years ago). Climate change (CL case, r^2 = 0.44) and to a lesser degree, migration (MI case, r^2 = 0.10) are both required to provide the best match with pollen (MI-CL case, r^2 =

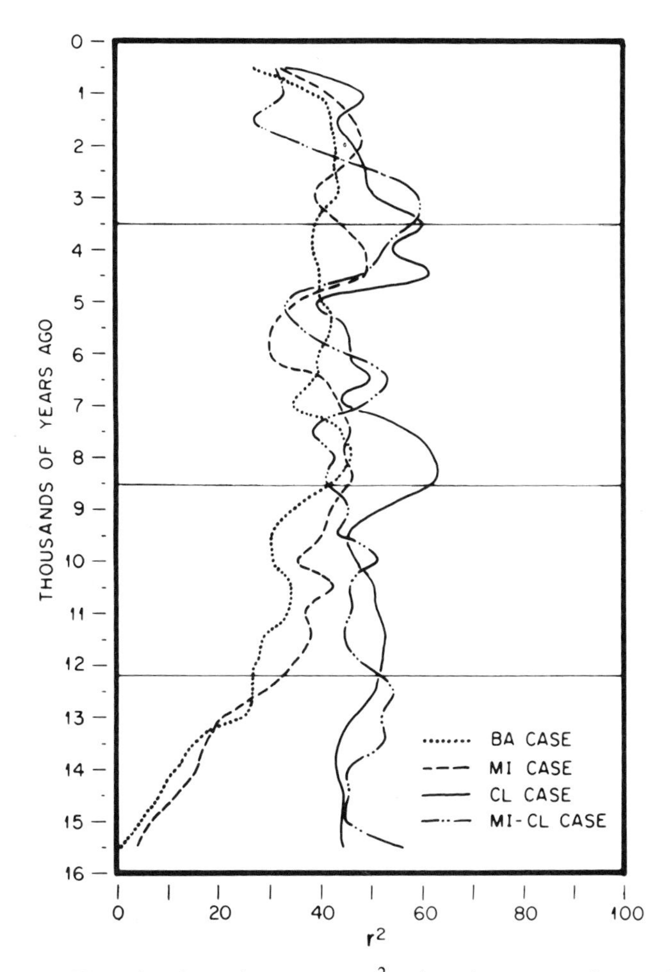

Figure 11.6. Three-level moving average r^2 values between pollen and simulated biomass under four different simulation conditions.

0.53) during this time. The CL case thereafter provides the majority of the highest correlations. During the period 12,500–9,500 years ago, climate changes most rapidly (Table 11.3; 933 growing degree-days, or GDD, before the period; 1764 during the period; 658 GDD after the period), and the climate variable simulation matches the pollen record most closely (CL case, $r^2 = 0.50$). Migration as a variable appears to reduce the match with pollen (MI-CL case, $r^2 = 0.48$; MI case, $r^2 = 0.37$) perhaps because only two new taxa (Table 11.2) migrated into the simulated plot during the period.

The steadily increasing correlation (i.e., decreasing difference) between pollen and simulated biomass in the BA and MI cases appears to stabilize after about 9000 years ago (Fig. 11.6). The CL case still provides the best fit with pollen from 9000 to 4000 years ago (average $r^2 = 0.53$; cf. MI and MI-CL case, $r^2 = 0.43$; BA case, $r^2 = 0.41$), but the wide variation from level to level in all four simulations precludes close examination of the overall differences. Correlation values generally decline after about 4000 years ago, perhaps because the climate values provided to all four runs during this period are increasingly incorrect (owing to elevation differences between Anderson Pond and the Crossville meteorological station). However, the average values of (r) among all four cases converge during the final 3500 years (BA, $r^2 = 0.37$; MI, $r^2 = 0.39$; CL, $r^2 = 0.42$; MI-CL, $r^2 = 0.41$), as we previously suggested they might.

The strikingly high correlation coefficients in all four simulation conditions and the dominance of oak (40–50% of all simulated biomass and all pollen), indicated the utility of correlating biomass with pollen assemblages in the absence of oak percentages. None of the resulting correlation coefficents is significant at the $p = 0.01$ level except during the early period under CL case and MI-CL case conditions. Dominance by spruce and pine in the climate-variable simulations and in the pollen record appears to be responsible for those significant correlations.

The lack of significant correlations in the balance of the temporal sequence probably reflects the absence of any simulated or palynological dominants. In the absence of oak, the relationships among the remaining taxon profiles are obscured by the strong differences in variance displayed by simulated biomass and pollen (Table 11.4 and discussion above). This feature does not invalidate the simulation and pollen comparisons because the pollen indicates that the Anderson Pond forest was indeed strongly dominated at first by spruce, and subsequently by oak species.

Vegetation Migration and Development

Recall our suggestion that individualistic theories of succession would probably be verified if successive migration proved to be important in the long-term vegetation dynamics of our test area. We further suggested that primary control of stability and change in the forest system by climate would indicate the possibility of truth in the climax concept.

The closest match for pollen-derived vegetation was the climate-controlled simulated vegetation. We consider remarkable the fact that the CL case output explained 40–60% of the variance in the pollen record, considering the contrast between variance patterns in the pollen and simulation chronologies, the inaccurate representation of vegetation by pollen profiles, the different areas of vegetation represented by pollen and by simulations, and the absence of important taxa in pollen (e.g., yellow poplar) and biomass (e.g., hickory) records. The climate-controlled simulation output is therefore worthy of closer examination.

Loss of the Boreal Forest

The first enigma we considered was the initial 3000 years of vegetation history, particularly the subsequent rapid diminution in spruce and pine, and the equally rapid increase in abundance of oak and other deciduous taxa. Similar abrupt shifts from dominance by spruce to that by deciduous trees or pines during the early Holocene have long been the subject of speculation (Ogden 1967; Wright 1968b, 1971; Davis, Chapter 10). Ogden's (1967) estimate of a required associated climate change, equivalent to 1000 km (600 mi) of latitude, has not been challenged. Yet our simulations required very little climate change to affect the shift from dominance by spruce to that by deciduous trees. Simulated warmth increased by about 250 degree-days at 14,000 years ago (Table 11.3), equivalent to about 60 km (37 mi) in latitude, or 0.5°C in annual temperature increase. White spruce was growing under conditions approximating 92% of its degree-day optimum at 14,000 simulated years ago. Five hundred years later simulated degree-days had shifted to 80% of its optimum, yet spruce virtually disappeared from the plot under 80% of its degree-day optimum.

The slight change in spruce growth potential also indicates that the abrupt decline in spruce populations did not necessarily require a drastic effect of climate upon spruce growth and reproduction alone as previously hypothesized (Ogden 1967; Wright 1968a, 1970, 1971; Davis 1969a). It appears likely that the simulated climate shift at 14,000 years ago had little affect on spruce, but instead augmented the competitive position of the deciduous elements, particularly that of red oak. Red oak and silver maple each were shifted from 40% to 50% of optimum growth as a function of degree-days. Potential growth of green ash and cottonwood shifted from 35 to 45%, and that of American basswood from 20 to 45% of optimum growth. Although all five are sprout hardwoods, it seems more important that red oak grows 10 m taller than white spruce, allowing it to shade out spruce on the simulated plot. The same spruce decline occurs even when American basswood, green ash, and cottonwood are excluded from the plot (Fig. 11.5).

Certainly the behavior of spruce in the simulated plots does not preclude the role of a major climate change in the spruce decline detected at more northerly latitudes. Probably the actual climate change during early Holocene time was more intense at greater latitudes. In addition, the direct replacement of spruce by oak is apparent only in the pollen records of the eastern midwest (e.g., Indiana, Otto 1938; central Ohio, Ogden 1966; western Pennsylvania, Walker and Hartman 1960; central Virginia, Craig 1969). Elsewhere, spruce is replaced by jack pine (e.g., in New England, Davis 1969b), which disappeared in our simulation, and by pine or birch (paper birch, Minnesota, Cushing 1967), which we did not simulate. The grasslands that replaced boreal forest on the great plains (Wells,

1970) could hardly have outcompeted white spruce for sunlight. Yet despite these differences, a great climate change unfavorable to spruce is not required to explain the abrupt demise of spruce in early Holocene forests of the eastern United States.

Development of a Mixed Mesophytic Forest

The appearance of mixed mesophytic forest elements at Anderson Pond during the early Holocene was the second "major forest change" we sought to model. This occurred when the CL case was clearly the best of the four simulations, accounting for 50% of the variance in pollen data.

The dominant species of the mixed mesophytic forest described by Braun (1950) are listed in Fig. 11.7. The entry of six of the nine species into the vegetation at Anderson Pond cannot be positively identified through pollen evidence. Yellow poplar and sweet buckeye produce very little atmospheric pollen. Sugar maple and the basswoods produce very

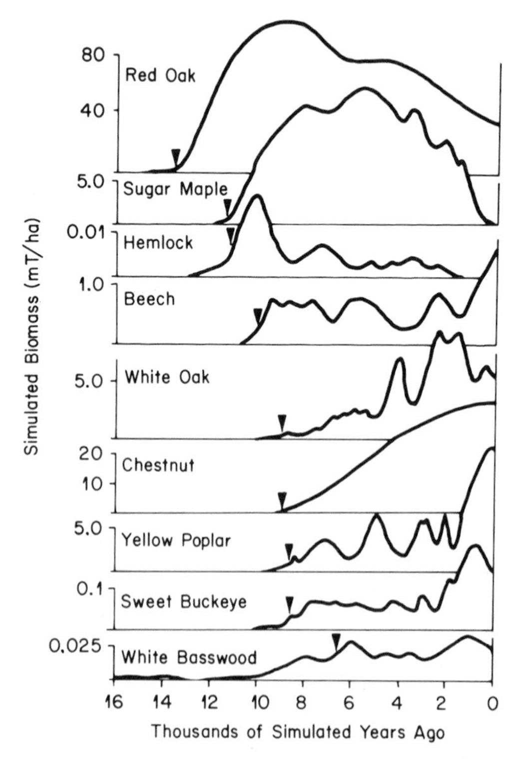

Figure 11.7. Simulated biomass (CL-Case 3 conditions) of species that characterize Braun's (1950, p. 40) "mixed mesophytic forest formation." Arrows indicate the point at which mature trees of each species grew on the simulated plot.

little pollen, and in addition, the basswoods are indistinguishable based upon pollen morphology, as are red and white oak. Thus, the pollen record can provide "indicators" that a mixed mesophytic forest assemblage was likely to have been present without giving substantive description of its composition. From these indicators, Delcourt (1979) hypothesized that the mixed mesophytic forest that dominates the present landscape developed between 12,500 and about 9500 years ago.

The CL case simulation appears to confirm this hypothesis. Only red oak occurred in simulated vegetation before 12,500 years ago (Fig.11.7). Hemlock and beech appear at about 11,000 years ago, followed by sugar maple (10,800 years ago), white oak (10,000 years ago), chestnut (9000 years ago), yellow poplar (8800 years ago), and sweet buckeye (8600 years ago). White basswood first occurs on simulated plots 6400 years ago. With the exception of the two chronological extremes, the simulated mixed mesophytic forest required about 2600 years to develop (11,200–8600 years ago).

Although the resulting association of species in central Tennessee is hardly a cohesive community of great antiquity, as surmised by Braun (1950), it certainly is a climatic climax, self-perpetuating under the relatively stable climatic conditions of the past 8000–9000 years.

Convergence of Vegetation Composition under Differing Simulation Conditions

We suggested above that if climate exerts the ultimate control upon forest assemblages, then species composition of simulation output in all four conditions should converge, following the cessation of new species immigrations about 7000 years ago, and the stabilization of climate about 3000 years ago. This indeed was the case. Between 7000 and 5000 simulated years ago, the amount of variance explained by each of the four simulation cases (Fig. 11.6) was low, rising thereafter, and converging as all four r^2 curves declined in the final 3000 years of simulation.

The low r^2 values between 7000 and 5000 years ago probably result from effects of maximum warmth and dryness upon Anderson Pond and its nearby vegetation. Hickory and ash pollen percentages were high and oak pollen percentages were low during this period. In addition, Anderson Pond may have decreased in surface area during the period (Delcourt 1979), inducing changes in pollen sedimentation patterns. The simulations, in contrast, do not change in the way they reflect vegetation.

As discussed earlier, the declining r^2 values in the final 3000 years probably reflect the erroneous degree-day values presumed for the present climate. Thus, the simulated vegetation converges toward oak-chestnut and mixed mesophytic forests that grew on the Cumberland Plateau, while the pollen reflects the oak-hickory forests of the eastern Highland Rim during this period.

Conclusions

The long-term pollen records and parallel model output provide a diagrammatic picture of climax vegetation history in the central portion of the eastern deciduous forests. The simulation and pollen records are similar enough to allow the conclusion that the simulated biomass broadly represents the same dynamic processes and vegetation that the pollen recorded at Anderson Pond. The dynamic processes explicitly built into the model are a function of species life histories, acted upon by stochastic processes of birth, growth, death, and competition for light and space. This individualistic and stochastic process of forest growth, which apparently occurred at Anderson Pond, produced features of "climax" vegetation that coincide with many traditional Clementsian concepts.

Evidence to support the superorganism concept is obviously absent, but the climatic climax concept is clearly present in these results. Self-perpetuating communities occurred as long as the climate remained stable. When minor or major climatic changes shifted competitive advantages among species, new climax communities developed that were similarly self-perpetuating. The convergence of vegetation from distinctly different origins to a singular community structure and composition under the same climatic regime, are all distinctly Clementsian.

The results presented here may appear to contradict the picture provided in the preceding chapter by Davis. In large part, this is because of the contrasts between Davis' data and those from our study location. The emphases of the two articles are simply quite different. The Anderson Pond site is near the origin, rather than at the periphery or along the path of species that migrated during postglacial time. The Anderson Pond site has supported a highly diverse species assemblage for the past 12,000 years, unlike the steadily more depauperate communities, which occur along latitudinal and temporal gradients.

The result is that our study area and techniques highlight the development of climax vegetation within a relatively stable landscape, where the gradually changing complement of species has little effect and where climate is the only significant changing variable. The data Davis presents, in contrast, emphasize individual species and their occurrence at discontinuous points in time and space. The only assemblages of species discussed are those occurring near the edge of deciduous forests, where nonclimatic disturbance (e.g., fire; Heinselman Chapter 23) and few dominant species are the rule.

This is not to say that all conflicts between the two chapters result from differing emphases of data sets. For example, we examined the pollen records on which Davis' migration diagrams are based, and found no evidence of the characteristic pattern of simulated biomass adjustment under conditions of variable migration and constant climate (Fig. 11.3 and discussion). Instead, the pollen profiles were much more similar to patterns of simulated biomass adjustments that occurred when climate changed with or without species immigrations (Figs. 11.5 and 11.4, respectively). Clearly, the climax concept is considerably less useful in areas that (1) contain few dominants; (2) occur far from full-glacial refugia; and (3) retain high-frequency environmental disturbance, all of which rarely allow climax to be attained. In areas like the central deciduous forest we studied, individual species life histories combined with landscape stability provide communities of predictable species structure and composition. The internal stability of the climax vegetation may be quite dynamic (e.g., Fig. 11.2), but it is clearly modulated by climate within definable limits. The set of species that characterizes these communities occurs repeatedly as climatic climax vegetation.

Acknowledgments

This research was supported by National Science Foundation Ecosystem Studies Program under Interagency Agreements No. DEB-77-26722 and DEB-77-25781, with the U.S. Department of Energy, under contract W-7405-eng-26 with Union Carbide Corporation. We also acknowledge with gratitude the data on Anderson Pond pollen supplied by H. R. Delcourt, and the computer programming assistance from M. L. Tharp. Publication No. 1664, Environmental Sciences Division, Oak Ridge National Laboratory.

Chapter 12
Patterns of Succession in Different Forest Ecosystems

Grant Cottam

Introduction

The next four chapters contain descriptions of the successional process in four materially different regions on the North American continent. It is by no means a description of all the forest types that occur on the continent, and there are some important omissions. All four of these studies are of forests dominated by evergreen trees throughout most of the successional sequences, and the deciduous forests of the upper Midwest—a region that was the professional birthplace of many of the early workers in plant succession—is unrepresented. This is an important omission because one thing that emerges from a study of the four articles presented here is that the ecology of the ecologists is an important determinant of their philosophy. The forest vegetation of an area as large as a continent can be remarkably different, and these differences color the thinking of the people who study the successional process in these different regions.

Boreal Forest

In the cold climate of interior Alaska, where growth and decomposition of organic matter are slow and nutrients are low, plant diversity is minimal (Van Cleve and Viereck, Chapter 13). Only four tree species are mentioned as dominants in the three examples presented in their contribution, and two of these (perhaps only one) constitute what Van Cleve and Viereck postulate as being climax forests. In a situation such as this the species involved are not the primary object of study. It is easy to postulate that succession is a fairly straightforward and constant process and that the stages are predictable. The biggest problem concerns the fact that there are practically no stands more than 200 years old. Fire is universal and relatively frequent—so much so that it seems unproductive to try to concern oneself with speculations about what the forest would be in the absence of fire. This disturbance is so all pervasive that the most reasonable way to handle it would be to consider it to be part of the natural environment and not a disturbance at all. Just as the sun rises every morning, fires burn an area every 70 years. The frequency of fire is no more variable than day length over the course of a year in these northern regions, although it is not quite as predictable. Given this constraint, climax becomes a metaphysical concept. Perhaps this is just as well for adherents to the Clementsian climax philosophy, since Van Cleve and Viereck postulate that there might be one further stage over some of this area were it completely protected from fire—a sphagnum bog or a moss-lichen stage. Succession from black spruce to sphagnum is a little hard to fit into the Clementsian mold.

Van Cleve and Viereck focus most of their attention on temperature effects and on nutrient accumulation. In an area where permafrost can exist, the depth of the organic layer on the soil becomes of major importance in determining what will grow, and there is clearly a direct relationship between the kind of vegetation, particularly the presence or absence of a moss layer, and some physical aspects of the environment, especially soil temperature and nutrient availability. Their interpretation of succession in interior Alaska is fairly close to classical Clementsian doctrine.

Coniferous Forest

In the Pacific Northwest (Franklin and Hemstrom Chapter 14) the major characteristic seems to be the magnitude of scale. Trees there appear to be bigger and older than anywhere else. Some trees live for 1000 years and take another 500 years to decompose. They get to be over 3 m in diameter and 80 m tall. The species list, while somewhat larger than that from the Alaska study, is still not overwhelming. Only a few species, all conifers, make up the dominants. The time scale is even more impressive than the size scale. Franklin and Hemstrom state that a large, long-lived Douglas fir tree influences what happens on the site for 2000 years–1000 years while it is growing, 500 years as a dead tree, part of which time is usually spent as a standing snag, and another 500 years while the area is under the influence of whatever species grew on the decaying log. Given this kind of a time scale, climax status would be difficult to attain, especially since fire frequency is something on the order of 400–500 years. As with the Alaskan forests, fire frequency is shorter than the life span of the dominant trees. Again, disturbance in the form of fire is a major determinant of the vegetation, and one has to question whether fire can be considered to be a disturbance, since it is a fairly predictable (within the life span of the dominants) part of the environment. This chapter separates shade tolerance from the successional process and points out that there are several tolerant species that can act as pioneers. Franklin and Hemstrom cite examples of "climax" forests, but "suspect that these were typically (but not exclusively) developed where the climax species was the major or sole colonizing species." Nevertheless, they state: "The climax concept is useful and climax forests do exist."

Two other major points are developed in the chapter. One of these is the fact that the presumed even aged Douglas fir forests arising after a catastrophic fire are hardly even-aged. There is an age difference of over 100 years in most of the stands examined, and in some cases age differences of up to 200 years in 600-year-old Douglas fir stands. Apparently this forest did not arise all at once after a catastrophe, but continued to colonize the site for periods of time that exceed the life span of some forest trees. The other major point is the extreme difficulty in modeling this type of forest, with its long life span and the huge size of the trees. Annual mortality is very small, plot sizes must be large, and the model should run for thousands of years.

Christensen and Peet (Chapter 15) were fortunate to be able to locate a number of experimental plots in North Carolina on which records had been kept for 50 years. Individual trees could be identified, and it was possible to arrive at excellent figures for changes in density and biomass over this 50-year time span. The availability of these data permitted a different kind of analysis than was possible to the authors of the other articles; consequently, the authors attack the problem of succession in a different way. Here, we get the trappings of multivariate statistical ecology applied to real data over a time period which, in this vegetation type, is long enough to permit the investigators to identify significant trends. The trends they found were complex. The vegetation appeared to be most significantly correlated with the environment in intermediate-aged pine stands and in old hardwood stands, and less well correlated in young and old pine stands. There appeared to be periods of convergence and divergence correlated with degree of use of the environmental resources. There were also marked differences in the species of hardwoods invading the pine stands dependent on degree of stocking and proximity to heavy-seeded hardwood trees.

Christensen and Peet have provided us with some real insights into the vegetational changes that occur within the North Carolina forests, and these changes are not simple, not uniformly convergent, and not always predictable. Many factors, including site quality, nutrient availability, initial stocking, proximity to seed sources, all effect the course of these changes. One factor that had surprisingly little effect was thinning. The remaining pines simply grew to fill the gaps in the canopy.

Tropical Forest

Succession in the tropics, as described by Gomez-Pompa and Vasquez-Yanes, is an even more complex process. Growth is rapid and species diversity is high—so high that these authors found it impossible to define vegetation change in species terms. They emphasized growth rates, seed size, seed dormancy, and tree geometry. There are so many species capable of filling a particular niche that emphasis was placed on physiognomy and tree physiology. Vegetational change can be described in terms of the life history characteristics of the succeeding species, but cannot be described in terms of the particular species. These authors found that the sun-loving, shade-intolerant species that are capable of occupying disturbed sites are small seeded, with seeds that are capable of lying dormant in the soil of a mature rain forest for periods of several years, while the primary species of the rain forest produce large seeds that germinate immediately. They emphasize the importance of the nature of disturbance that initiates the successional process, and the subsequent treatment of the area. A final conclusion was that the study of succession in the tropical rain forest must await a better understanding of the life history characteristics of the species involved.

Discussion

The purpose of these chapters of the book is to illustrate some of the differences imposed by the variations in climate and other environmental factors over a large area. Any succession concept should have a degree of universal applicability over all the conditions that occur on earth, and it is necessary to examine the different insights into the succession process provided by the study of the different regions. What are the universalities that are true wherever succession is studied, and what are the regional differences between succession in the near arctic and succession in the tropical rain forest?

One of the major variables seems to be the degree of control that the vegetation exerts on the environment and how important this control is in determining the course of vegetational change. In Alaska, it appears that vegetation is very important in determining the nature of change. The organic layer on top of the soil plays a major role in determining soil temperature, which, in turn, determines rate of decomposition and the ultimate formation of permafrost. Under these conditions it is clear that vegetation plays a dominant role in modifying the physical environment, and that particular species are especially important—the mosses that grow under the spruce trees. In this case we appear to have autogenic succession in its most classical form. Christensen and Peet, on the other hand, have a situation in which there is greater or lesser specificity to environmental conditions, depending on the age and density of the trees in the forest, and species of tree appears to be much less important than biomass. They report convergence and divergence, depending on the degree of utilization of resources, and this appears to occur, regardless of species. In the tropics, Gomez-Pompa and Vasquez-Yanes state flatly that succession can only be identified as changes in physiogmony, and that it is almost impossible to determine succession floristically.

It would appear that the degree of control exerted by the vegetation is quite variable. In some cases the vegetation is very important; in others it is not important at all. In Chapter 9, Joseph Walker describes the formation of a giant podzol with a leached layer 20 m deep. Given his description, it is unlikely that the kind of vegetation had very much to do with the formation of this giant podzol, yet its presence is very important in determining the kind of vegetation that exists. In the tropics a 10-year-old secondary forest appears to exert as much control over the environment as a mature rain forest, being just as successful in absorbing nutrients, casting shade, and producing litter as its much older successor. The autogenic nature of succession is not a universality.

Is succession predictable? Christensen and Peet question the convergent nature of vegetational change in North Carolina. Gomez-Pompa and Vasquez-Yanes state: "(It is) impossible to predict what the regeneration process will be in the next hundred years." Franklin and Hemstrom state that recent data show highly variable species composition in early stages of forest succession in the Northwest. They say that species play varying successional roles and cite two shade tolerant species that reestablish directly on disturbed sites. They note that shade-tolerant and shade-intolerant species may become established at the same time, with the shade-intolerant species becoming the dominant due to its faster growth and persisting because of its longer life. On the physiognomic scale a degree of prediction is possible. Only in the arctic does it appear that succession on the species level is predictable.

Is succession directional? Perhaps this question should be: Is succession unidirectional? Here it is necessary to consider the question of climax. If succession is directional, it must be going somewhere, and agreement is general that it is proceeding toward climax. None of the chapters in this section attempt to provide a working definition of climax, and one is left with the impression that the authors feel that everyone knows what the term climax means. One gathers that climax is stable, yet Clements warns that other communities may also be stable. He also states that the degree of stability of the climax must be reckoned in "thousands or even millions of years" (Weaver and Clements 1938). The general view seems to be that succession proceeds from shade-intolerant species to shade-tolerant species, and from small, herbaceous species to trees, if the climate will support trees. Yet Van Cleve and Viereck postulate an ultimate stage of sphagnum bog or lichens for their black spruce community, and Franklin and Hemstrom disavow the necessity for succession to proceed from intolerant to tolerant species. Daubenmire (1968) defines the climax community as the community that gains essentially permanent occupancy of the habitat and perpetuates itself there indefinitely. Apparently permanency is the most important criterion of the climax. Climax as a concept is more than a little fuzzy, and the determination of the climax community requires one to wait an interminable length of time, during which the climate does not change appreciably.

Until one knows what the climax is, it is difficult to answer the question about the directionality of succession, so perhaps it is best to state that the chapters in this section do not shed much light on the question. Gomez-Pompa and Vasquez-Yanes at least imply that they can identify a climax physiognomy in the tropical rain forest, and Van Cleve and Viereck, except for their suspicion that succession may overshoot their black spruce climax and get bogged down in sphagnum, appear to be comfortable with their concept of climax. Franklin and Hemstrom state that the climax concept is useful, but do not elaborate.

In Chapter 3 of this volume McIntosh considers the organismic-holistic versus the reductionist-individualist approaches to succession. There is no clear consensus among the following four chapters. Gomez-Pompa and Vasquez-Yanes are the most forthright, and conclude: "The individualistic concept of the plant association for the humid tropics is the only concept that can usefully help us." Van Cleve and Viereck would probably be on the other extreme. My own conclusion is that both approaches are necessary. The community functions as a whole, but it functions the way it does because of the particular physiological and structural attributes of the component species. It would be impossible to deduce the functioning of the complete community without such knowledge, at least for the major species. The interactions of the species with each other and with the physical environment create synergisms and inhibitions that cannot be deduced by simply examining a species list, no matter how familiar one is with the ecology of the individual species. In the workshop on which this volume is based very little time was wasted on arguing the relative merits of reductionism versus holism.

Conclusions

1. The vegetation of North America is remarkably diverse. The four regions considered in this section each contain forests that have little in common with the others.

2. Methods of studying succession, and to a degree the concepts within which the studies are conducted, must be adjusted to fit the particular characteristics of the region.

3. Scientists with different philosophies, and using different methods, can produce valuable insights into the process of vegetational change. Other scientists with different methods and concepts might come up with different insights, but these insights would be complimentary, not contradictory.

4. The study of succession suffers from over 50 years of conflict. The controversy discussed by McIntosh has been very real and still continues. The term succession means too many things, and much of the argument is probably one of semantics. It would appear, however, that some of the old arguments are no longer valid. Nowhere in this group of articles is there any evidence of a unidirectional regional succession leading to a sin-

gle end point. Nowhere does there appear any justification for continuing the concept of the superorganism.

5. Vegetation does change, and many of the causes of this change can be understood. The vegetation change can be modeled with a good degree of accuracy, even over periods as long as 16,000 years (Solomon et al. Chapter 11, Botkin Chapter 5, Shugart et al. Chapter 7). This indicates that good progress is being made toward understanding the underlying concepts that lead to vegetational change.

6. The straitjacket of formal and rigidly adhered to concepts does more to inhibit open inquiry into vegetation change than it helps by providing guidelines.

Chapter 13

Forest Succession in Relation to Nutrient Cycling in the Boreal Forest of Alaska

Keith Van Cleve and Leslie A. Viereck

Introduction

The successional sequences described in this chapter were located in the central part of Alaska, in the general vicinity of Fairbanks, but the general trends and conclusions derived from our studies should have broader application to most of the North American taiga.

Vegetation

In the uplands and on older river terraces on the taiga, the dominant vegetation on most sites consists of stands of open and closed black spruce, especially *Picea mariana*-feathermoss and *Picea mariana*-sphagnum types (see Viereck and Little 1972 for arboreal taxa nomenclature and authorities). These open and closed black spruce stands are especially widespread on poorly drained sites, including those underlain by permafrost, and on north-facing slopes. On the wettest sites tamarack (*Larix laricina*) is associated with black spruce, and near the altitudinal tree line, at approximately 1000 m, open stands of mixed black and white spruce occur primarily as woodlands with an understory rich in lichens (Viereck 1979). On many of these cold sites black spruce stands are replaced directly by black spruce after fire, but occasionally birch and rarely aspen intervene before mature black spruce stands are reestablished.

On the warmer, well-drained upland sites, the mature forests consist of tall, fast-growing, closed or open white spruce stands, primarily of the *Picea glauca*-feathermoss community type. Successional to white spruce on these sites are open and closed deciduous stands of birch, aspen, or birch and aspen mixed.

Bottomland spruce and balsam poplar (*Populus balsamifera*) forests are common along the major taiga rivers in Alaska, and the successional stands of shrubs and balsam poplar that lead to them are some of the most productive in the Alaska taiga. Shrublands are common along most of the small water courses and frequently form broad bands near timberline. These shrublands are composed primarily of alder (*Alnus crispa* and *A. tenuifolia*), willows (*Salix* spp.), and resin birch (*Betula glandulosa*). Other treeless types occurring commonly in the area are grasslands on south-facing bluffs, and bogs, marshes, and aquatic types in low-lying wet areas. Alpine tundra occurs on the adjacent hills.

Climate and Permafrost

Mean annual temperature in the Fairbanks area is −3.4°C, and precipitation averages 287 mm. Snow covers the ground from mid-October until mid- to late April, and maximum accumulation averages 75–100 cm. Degree days sums based on 5°C are 1090 for the 30-year average at Fairbanks, ranging from 1300 for south-facing bluffs to about 500 in a forested area near the tree line.

Permafrost is spatially discontinuous, occurring on most sites, but is lacking on south-facing slopes and in freshly deposited alluvium. In many sites permafrost is in a delicate balance with the present environment and is maintained only because of the shading and insulating effect of the vegetation. In many black spruce stands the depth of annual surface thaw, termed the "active layer," may be only 45–50 cm. Removal of the vegetation cover by natural or human disturbance usually results in an increase in the thaw depth. In much of the frozen layer water has been incorporated as wedges or lenses of pure ice, which in some soils may amount to as much as 50% of the substrate by volume. The melting of this ice-rich substrate may result in unequal subsidence; the terrain that results is aptly called thermokarst. In many areas of coarser soils, permafrost contains little or no ice, and an increase in the active layer resulting from disturbance of vegetation does not result in subsidence.

Soils

One of the unique features of taiga soils is the predominance of area occupied by soils that show relatively slight morphological development. Thus, for interior Alaska, inceptisols, entisols, and histosols occupy approximately 78, 12 and 7%, respectively, of the land area, a total of 97% of approximately 33 million ha (Rieger et al. 1979). Soils showing more extensive profile development, including mollisols and spodosols encompass only 3% or approximately 840,000 ha of the land area in interior

Alaska. These conditions primarily reflect the cold, semiarid nature of the subarctic climate in reducing the intensity of chemical weathering. For most Alaskan soils, the syllable *cry*, indicating cold-dominated soil, is added to form the names of the great groups.

Throughout the general region of the study area, upland soils have developed in micaceous, silt-textured loess, which was laid down during the last glacial maximum. The thickness of the deposits ranges from 30 cm on ridge tops to 30 m on low hills near the Tanana River. The predominant texture for these soils is silt or silt loam. Alluvial plain soils have developed in sandy or silt-textured alluvium, mostly of glacial origin but also including some loess and the influence of underlying bedrock (Rieger et al. 1963). Bedrock of the uplands is the Birch Creek schist, a Precambrian formation of quartz-mica and quartzite schist.

Summary of Succession

The taiga of interior Alaska is dominated by young stands in various stages of succession—mature stands of over 200 years in age are rare. Fire is the main cause of the young ages of the stands—in some areas a fire that kills all of the above ground vegetation can be expected every 50–100 years. In areas relatively protected from fires such as the river floodplains, the active erosion and meandering of the silt-laden, glacially fed rivers results in the active production of newly vegetated silt bars and the rapid erosion of older, mature stands. Even along the river, fire is common on the older terraces where stands that had developed originally on sandbars intermingle with stands developing after fire. However, unlike many areas of the world, successional sequences developing after human disturbances are relatively rare and recent, and result primarily from some early mining and logging and very limited areas of abandoned agricultural land. Humanity's attempts to revegetate sites date primarily from the past five years, and are limited primarily to pipelines and a few mining sites.

In this chapter we shall attempt to describe the similarities and differences in three naturally occurring successional sequences common in interior Alaska. These are (1) primary succession on the floodplain of the Tanana River developing from willow and alder through mature white spruce stands; (2) the most common successional sequence in interior Alaska, that which follows fire in black spruce stands on permafrost sites; and (3) the less common succession following fire on warm productive sites which pass through a shrub and hardwood stage to white spruce stands.

In the subsequent sections of this chapter we shall provide a description of each of the three successional series, including ecosystem production and nutrient cycling in relation to succession. The final section of the chapter summarizes physical and biogeochemical controls of system structure and function within and among the successional hierarchies.

Floodplain Succession

General Description

Early Successional Stages on the Tanana River

We define early successional stages as those that exist from the initial bare surface stage (I) through the open-shrub stage (III). These stages generally reflect dominance of physical-chemical control of successional processes, with Stage III representing a transition stage (Fig. 13.1).

Certain forms of physical control of ecosystem state and function generally are superimposed over the entire successional sequence (Van Cleve et al. 1980). For example, terrace elevation above the river stage reflects terrace age and determines frequency of flooding. However, the meandering nature of the river ultimately controls the fate of each terrace, regardless of age. Thus, older terraces at higher elevations may gradually be eroded along one reach, with sediment deposited along another section of river channel, initiating a new, low-level terrace. The higher elevation of old terraces and the advanced stage of forest development indicate that these stages of succession are being influenced by alluvial erosion and deposition to a relatively small degree compared with lower-elevation, younger terraces.

Until sufficient alluvium has accumulated to raise river terraces above the zone of frequent intrayear flooding, physical and chemical controls dominate ecosystem structure and function. These mineral soil surfaces may appear to be ideal locations for seed germination and establishment. However, frequent inundation, sediment deposition, and erosion make them highly unstable zones for plant establishment. Abundant seed of herbs, shrubs, and trees are deposited and germinate on these sites, but the shallow rooted seedlings will be washed away or completely buried during the next stage of high water. Nutrient cycling in this environment is largely controlled by a saturated flow of soil moisture and the leaching of nutrients with elemental concentrations similar to those encountered in river water.

A combination of physical and chemical controls exists in the next stage (Stage II), which is characterized by the formation of a salt crust on the bare surface. At this elevation above the river, less-frequent intraseason flooding occurs. Flooding generally is restricted to spring breakup and during mid- to late-summer high water. The physical removal or burial of

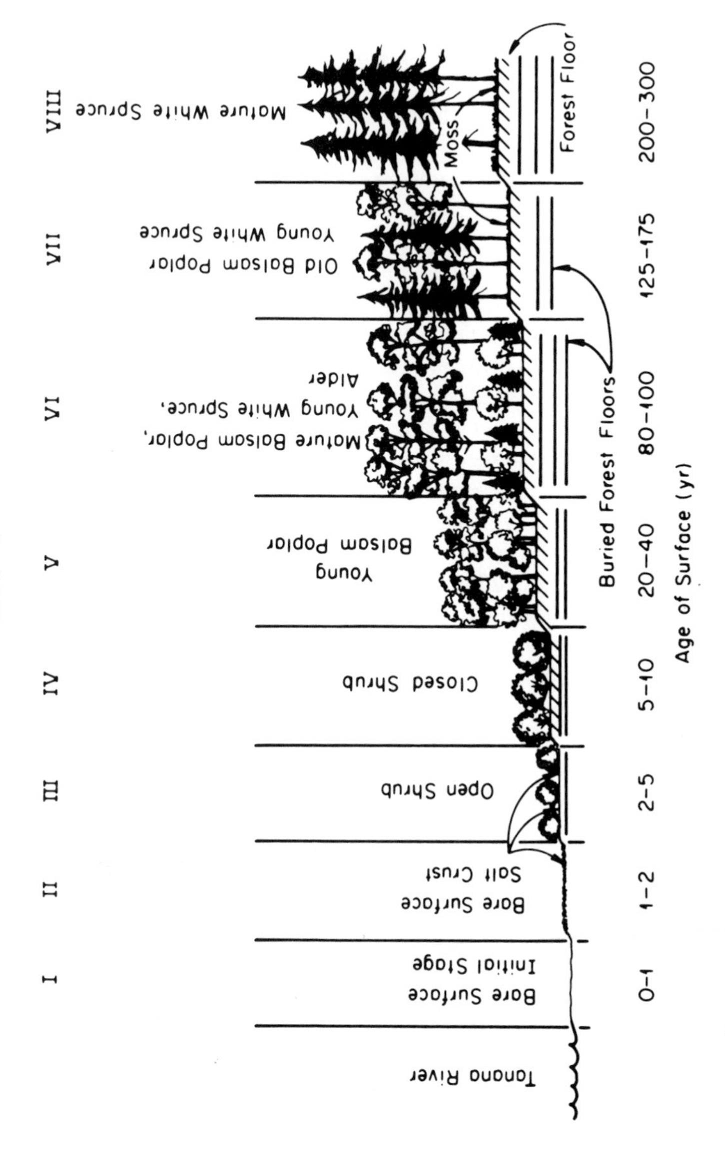

Figure 13.1. Primary succession on the Tanana River floodplain.

shallow rooted seedlings still might occur, but usually some plants do become established. The first to persist are usually herbs such as *Hedysarum alpinum*, *Equisetum pratense*, and *Calamagrostis canadensis* and the shrubs, both willow (*Salix alaxensis* and *S. interior* being the most common) and alder (*Alnus tenuifolia*). At this time plant cover is usually only 1 or 2%.

Through State II, physical-chemical controls associated with alkali soil formation dominate nutrient cycling. Groundwater rises to the surface by capillary action. Surface evaporation results in substantial concentrations of salts, especially $CaSO_4$, or gypsum, and various chloride- and carbonate-bearing salts. The salts form a crust several millimeters thick largely composed of gypsum crystals mixed with soil particles. This high salt concentration may control the germination and establishment of quaking aspen (*Populus tremuloides*). This important upland forest type is not encountered in the primary successional sequence on the floodplain, although an abundant seed source is available in the adjacent uplands. Laboratory observations of germination success on samples of the salt crust indicate that aspen seed germination may be reduced by as much as 80% on crusts collected from terraces representative of the bare surface-salt crust and open shrub stages (Stages II and III). Balsam poplar (*Populus balsamifera*), the most important hardwood species on the floodplain sites, also shows reduction in germination success in response to high salt concentrations, but to a lesser degree than aspen. Whether the effect is one of moisture stress, toxic salt concentrations, or both is yet to be determined. Aspen seedlings germinated on salt crusts show a substantially reduced rate of growth and smaller cotyledons and hypocotyls. This condition may result in the reduced ability of seedlings to obtain sufficient water to prevent dessication.

Continued sediment deposition raises terrace surfaces to levels at which greater success obtains in seedling establishment and an open shrub stage of willow and alder results. Changing of the river course results in quiescent periods during which newly established seedlings are not washed out by river erosion. The capillary rise of ground water continues. The visible manifestation of this phenomenon in the form of evaporate accumulation occurs until the terrace surfaces are higher than 1m above the river stage. Even at this point, a high river stage may result in the surface evaporation of groundwater. However, leaching by snowmelt and rainwater results in generally lower salt concentrations than those encountered in late Stage II.

Stage III represents a transition between declining physical control and the rise of biological control of succession and nutrient cycling. Once well established, willows and alder act to slow the river flow, stabilize the terrace surfaces, and provide biological control. Willow and alder (primarily *Alnus tenuifolia* and a number of *Salix* spp., but with *S. alaxensis* being the most abundant) shrub cover may reach nearly 40%, and exposed mineral soil nearly disappears because of the development of a forest floor. The presence of alder results in improved soil fertility primarily through the addition of nitrogen. This phenomenon is of substantial importance to primary production and nutrient cycling in later stages of succession. The herbaceous layer also becomes much more widespread and diversified, but is still dominated by *Equisetum pratense* and *Calamagrostis canadensis.*

Mid-Successional Stages on the Tanana River

We define mid-succession to extend from the closed shrub stage (IV) through the mature balsam poplar stage (VI), a period of approximately 75–90 years (Van Cleve et al. 1980). This period is characterized by a rapid increase in shrub cover and net productivity of balsam poplar. Within 4 or 5 years of establishment, a dense stand of willow and alder shrubs, 1–3 m in height, is well established. Stems may be 100,000/ha at the age of 5 years and the canopy so dense that only *Equisetum pratense* survives in the shade.

Seedlings of balsam poplar (*Populus balsamifera*) become established early in the shrub stage, but usually do not overtop the shrubs until 20–30 years following their establishment. During this balsam poplar stage, tree basal areas may reach 35 m^2/ha, or 75% of the maximum basal area reached during the mature stages. The net annual production of balsam poplar, standing-crop tree biomass, and tree basal area appear to reach a peak for this vegetation type in mature balsam poplar stands (Stage VI). As the balsam poplar canopy closes, the shrub layer changes drastically. A number of pioneer willows are intolerant of the shade and disappear completely. *Salix alaxensis* and *Alnus tenuifolia* persist, but begin to decline in vigor and abundance. Other, more shade-tolerant shrubs such as rose (*Rosa acicularis*) and highbush cranberry (*Viburnum edule*) become common, but do not have a high cover percentage. By the end of the young balsam poplar stage, shrub cover has markedly declined from near 100% to about 10% because of overstory shading. However, later in the balsam poplar stage during the period from about 80 to 100 years, the crown canopy has opened and shrub cover has increased to nearly 30%.

Physical and chemical controls play a declining but important role in productivity and nutrient cycling through mid-succession. For example, capillarity may supply ground water to the tree-rooting zone so that vegetation has adequate supplies of water throughout the growing season even during drought periods. In addition, even though the soil solution appears to be low in such important nutrients as nitrogen and phosphorus, the slow movement of soil solution through the rooting zone, along the gradient of the river, may result in continual replacement of these reserves and satisfy plant nutritional demands. The leaching of surface soil layers by more acidic solutions moving through the forest floor results in the decline of soil pH from 7 or above to a slightly acid range. This undoubtedly improves the availability to plants of phosphorus, which, under neutral to alkaline soil reactions and in the presence of abundant calcium, would largely be insoluble.

Flooding and sedimentation, now occurring at less frequent intervals, continue to increase terrace elevations. Buried forest floors, which are incorporated into the soil by periodic sediment deposition, are zones of low density, high water-retaining capacity, and high nutrient concentration. This environment favors abundant microbial activity and organic matter mineralization. Plant roots are markedly concentrated in these buried organic layers.

Up to this point in succession, the plant species that have been most successful in establishing themselves are broad-leaved, deciduous species (willows, alder, balsam poplar). Most of these species easily regenerate either by seed or by stump or root sprouting. They are also generally shade intolerant and have astonishingly rapid growth rates during the early stages of succession when ground surfaces are least stable and subjected either to erosion or periodic sedimentation.

Only after terrace elevation has reached the point where sediment deposition has declined to a relatively infrequent occurrence or the river course has markedly shifted to result in a quiescent period with regard to sedimentation can white spruce become successfully established. White spruce requires a mineral soil seedbed (Zasada and Gregory 1969) for germination and establishment. However, it does not appear able to survive repeated flooding within a year, or yearly sediment deposition on needle surfaces. Seedling top growth rates are not rapid enough to elevate foliage to a level at which needles should be free of siltation. In addition, root growth rates may also be slow. In combination with lack of sprouting capability, these factors may further limit the ability of spruce seedlings to withstand erosion, overcome potential oxygen limitation to root systems following sediment deposition, or seek adequate water supply. Periods of drought and/or low water and deficient moisture supply by capillarity may also decrease seedling survival.

By the time of poplar stand maturity, physical site conditions have been favorable for a sufficient time interval for spruce to become well established in the understory (Stage VI). During the late intermediate and early mature successional stages, white spruce becomes a dominant species.

Mature Successional Stages on the Tanana River

The shift from a poplar to a spruce stand is gradual and may take nearly 100 years (Van Cleve et al. 1980). Growth of the white spruce is rapid but does not equal that of the balsam poplar. The white spruce slowly become dominant in the stand as they grow larger and the balsam poplar begin to topple. We define mature successional stages to be those that have the dominant tree basal area composed of white spruce (Fig. 13.1). The young white spruce-old balsam poplar stage (VII) continues to have old balsam poplar as an overstory component, but these individuals comprise less than 10 m^2/ha basal area compared with the near 50 m^2/ha basal area contributed by white spruce. Late mature stages (Stage VIII) have an occasional or no balsam poplar remaining and a white spruce basal area ranging between 40 and 50 m^2/ha. During the mature successional stages, shrub cover declines from a peak of about 50% in Stage VII to between 10 and 20% in Stage VIII, although there is a slight increase with age as the mature spruce stands become more open. One of the most dramatic changes to take place is the development of the moss mat. As the spruce canopy closes, the feathermosses, *Hylocomium splendens* and *Pleurozium schreberi*, become dominant on the forest floor, and the moss cover reaches nearly 80%. Bare ground surface again appears (up to 5%) as mature poplar and spruce fall and the heaving of root systems exposes mineral soil. Some spruce seedling regeneration may arise on soil mounds. Rose and viburnum persist, and alders, both *Alnus tenuifolia* and *A. crispa*, are scattered in the stand. A low shrub layer of *Vaccinium vitis-idaea* and *Linnaea borealis* becomes established, and a few herbs such as *Pyrola* spp., *Galium boreale*, orchids (*Goodyera repens* and *Calypso bulbosa*), and *Cornus canadensis* are scattered in the forest floor; however, the feathermosses dominate. A few lichens, primarily *Peltigera* spp., are scattered in the moss mat.

Capillarity and sedimentation are no longer functioning as controls of primary production and nutrient cycling. Mature stage terrace levels are sufficiently high above the river level that near catastrophic flooding is required for substantial sedimentation to occur. The leaching of surface soil layers depends on the downward movement of snowmelt water and precipitation through the acidic forest floor.

In the successional sequence diagramed and described in this chapter we have ended with the mature white spruce stand. However, under some conditions the white spruce may be replaced by black spruce. Drury (1956) described this sequence from the Kuskokwim River, and Viereck (1970a, 1975) described it from the Tanana River and adjacent Chena River. In the sequence investigated on the Tanana River we have been unable to locate transitional stands connecting the mature white spruce stage with the permafrost-dominated black spruce and bog of the older terraces. We have therefore ended our description of the succession with mature white spruce stands, but can postulate that with time and no flooding these stands would eventually change to black spruce as the soils became colder and permafrost developed.

Forest Floor Thickness

One of the most important aspects of succession in the Alaska taiga is the accumulation with time of a thick organic layer (Viereck 1970b). In the floodplain succession O1 (L—fresh litter) and O21 (F—fermentation) layer develops as a result of leaf fall in the shrub and balsam poplar stages.

However, the thickness of this layer is seldom more than 4–5 cm because it is periodically buried by flood-deposited silt layers (Fig. 13.1). However, with the establishment of the spruce overstory and the rapid development of mosses, an organic layer begins to accumulate. The deposition of organic material in the mature spruce stands exceeds the decomposition rate, and the thickness of the layer gradually increases to 8–10 cm or more. Humus layers (O22) exist in the forest floor in these stands.

Productivity

A peak of 20–25 kg/m^2 soil organic matter is attained early in Stage VI, followed by a slight decline, and stands at a relatively constant level of approximately 20 kg/m^2 during the mature successional period. The forest floor biomass reaches maximum values of 5 kg/m^2 in mature white spruce stands, a 12-fold increase over early successional stages.

The standing crop of above-ground, live-tree biomass reaches 20 kg/m^2 on productive balsam poplar sites. Accumulated above-ground tree biomass in white spruce may reach 22 kg/m^2 at approximately 180 years, and undoubtedly declines to a lower level in overmature forests. From early successional to mature successional stages, the portion of total system organic matter stored in mineral soil declines from a maximum of 100% at time zero to 43% at 180 years. The remaining 57% is in above-ground, live-plant biomass and forest floor organic matter. Litter fall declines from a peak of 350 g/m^2 in Stage V to 250 g/m^2 in Stage VI and less than 100 g/m^2 from early to late mature white spruce stages.

Maximum net annual above-ground production, 950 g/m^2, is attained late in Stage V and corresponds to the peak in litter fall. Growth rate declines as poplar mortality increases. White spruce, which progressively dominate these sites, may not attain maximum productivity until later in the mature stages of succession.

Maximum net annual production in the spruce is attained by about 200 years. This growth rate undoubtedly declines at greater ages and depends on the frequency of flooding and fire. An important new component of these mature ecosystems is moss, whose net annual production may approach 100 g/m^2 or more by 200 years. The nonvascular plant cover contributes to an increasingly thick forest floor, an effective insulator for mineral soil, which promotes the reduction in soil temperature discussed earlier.

Nutrient Cycling

The soil nitrogen pool increases from about 50 (time zero) to 500 g/m^2, at about 150 years, due to nitrogen fixation by alder (Van Cleve et al. 1971). A larger phosphorus pool may reflect the higher organic matter content of soil and associated higher concentrations of these nutrients in the organic remains that are incorporated into mineral soil. Available K shows a decline, probably due to transfer of K to the vegetation (Van Cleve and Viereck 1972).

Soil nitrogen continues to increase in early mature successional stages and declines in late mature stages. We feel that this change in soil nitrogen reserves is caused by the loss of alder as a major understory species, continued removal by plant growth, and accumulation in standing live and dead biomass in the forest floor.

The most dramatic changes in nutrient storage are encountered in the forest floor. Nitrogen storage in the mature stage has increased by three times to 75 g/m^2. Storage of P increases by four times on a total basis and nearly 15 times using the available P measure. Estimates for K show approximately a five-fold increase to 5 g/m^2. The declines noted in soil nutrient storage from the mid-successional period may reflect a combination of increased standing crop of nutrients in above-ground biomass and forest floor, and increased soil weathering.

Early successional stages show annual N and P uptake by above-ground tree components is nearly three and two times, respectively, that encountered at 200 years in mature white spruce. Approximately the same amount of K is annually accumulated at both stages (2.4–2.5 $g/m^2/yr$).

Nitrogen flux undoubtedly primarily reflects the rise and decline of alder and the influence of highly productive balsam poplar later in the successional sequence. Phosphorus and potassium also reflect early maxima in productivity during the mid-successional period. Potassium shows maximum increment (5 g/m^2) early in Stage VI, a 2.5-fold increase from Stage IV. Phosphorus shows a four-fold increase and N a two- to three-fold increase during the same period of succession. Potassium declines to 60% of maximum uptake by 100 years.

In mature white spruce, annual increments of N, P, and K continue the decline from peak values realized early in Stage VI. During these stages, mosses are contributing to forest floor buildup and may be an effective competitor with trees for nutrients through physiological uptake and physical adsorption of cations at internal and exchange sites.

During the mature successional stages, dominated by white spruce, approximately a three-fold greater net above-ground tree production occurs per gram of leaf tissue produced compared with earlier successional species.

Upland White Spruce Succession

General Description

On dry upland sites, especially south-facing slopes, the mature forest vegetation is white spruce, paper birch, aspen, or some combination of these species (Fig. 13.2). Information on the succession following fire on

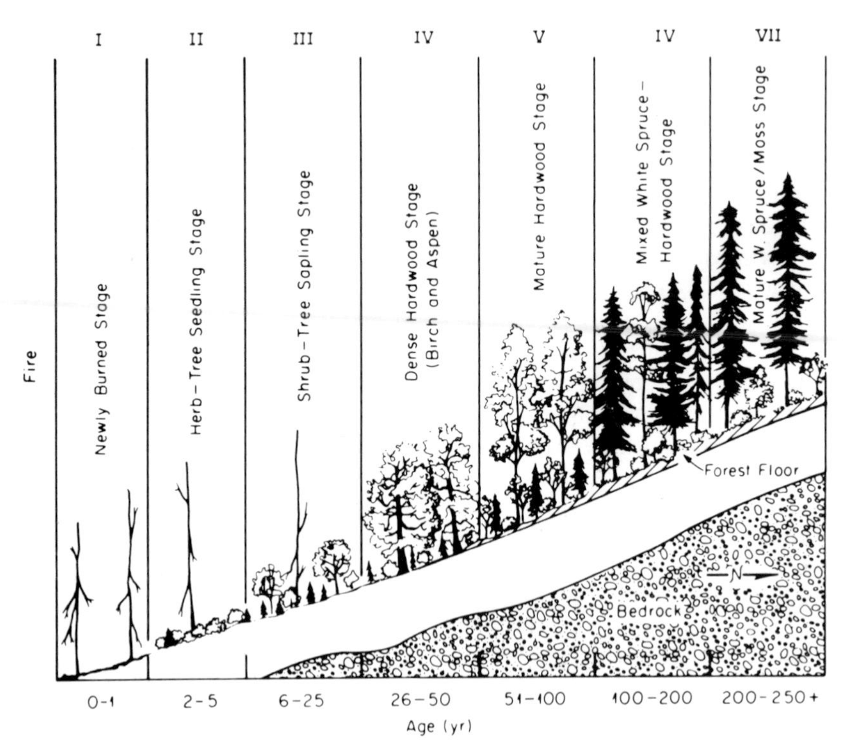

Figure 13.2. Upland white spruce succession after fire.

these sites has been compiled by Lutz (1956), Viereck (1973), and Foote (1976), but the data are not as abundant as those for black spruce because white spruce stands are not as widespread as black spruce and fire return time is greater in white spruce than in black spruce.

Succession on these sites seems to follow two separate patterns. In most cases the deciduous trees, aspen and birch, develop as a successional stage after fire but before the development of a spruce stage. However, occasionally, if a seed source is available and site conditions optimal, white spruce will invade with the hardwoods or only a few years after the fire, and even-aged white spruce stands will develop without an intervening hardwood stage. The more common situation is represented by an intermediate stage of aspen or birch, although a spruce stand may develop directly following fire.

In the typical fire sequence the following stages have been recognized and described (Fig. 13.2). Following the fire and an intervening few days to weeks when no live vegetation is present (Stage I), there develops an herb and seedling stage (Stage II). The usual first invasion is by light seeded species such as fireweed (*Epilobium angustifolium*) and willow shrubs, especially *Salix bebbiana* and *S. scouleriana*. A few species apparently have viable seeds buried in the soil that are stimulated to germinate by fire. Thus, *Geranium bicknellii* and *Corydalis sempivirens* appear quickly following fire. Mosses and liverworts, especially *Marchantia polymorpha* and *Ceratodon purpureus*, quickly invade areas where mineral soil has been exposed by fire. A number of species resprout from stumps and roots, especially shrubs such as *Viburnum edule*, *Rosa acicularis*, willows, and aspen and birch.

The next stage in the developing vegetation is dominated by shrubs (Stage III), primarily willows, and deciduous tree saplings. The heavy shrub and sapling canopy shades out most of the pioneer mosses and greatly reduces the herbaceous strata. This is a period of heavy litter fall. The density of birch and aspen may be as great as 30,000 stems per hectare and shrub cover as much as 50%. Spruce seedlings may be abundant at this time, but their slow growth makes them relatively inconspicuous in the stand.

In the dense hardwood stage (Stage IV), young aspen and birch trees form a dense canopy that tends to shade out much of the understory that has developed since the fire. The more shade-tolerant highbush cranberry and prickly rose replace the willows in abundance in the shrub layer. Herbaceous cover drops to less than 10%. Heavy leaf litter fall prevents the development of a continuous moss layer, but moss species of the mature spruce stages such as *Hylocomium splendens* and *Pleurozium schreberi* become scattered on mounded areas. This stage usually occurs during the period of 25–50 years following the fire.

For the next 50 years deciduous trees dominate the site (Stage V), although white spruce may become conspicuous in the understory. Both the aspen and birch stands become more open, and the density drops to about 700 trees per hectare for aspen and 300 for birch. Rose, highbush cranberry, and alder (*Alnus crispa*) form a loose shrub layer, with only isolated clumps of *Salix bebbiana* remaining. The low subshrubs, twinflower (*Linnaea borealis*), lingonberry (*Vaccinium vitis-idaea*), and Labrador tea (*Ledum groenlandicum*) form a scattered layer. *Equisetum arvense* and *Cornus canadensis* replace *Calamagrostis canadensis* and *Epilobium angustifolium* in the herb layer. The moss cover of *Hylocomium splendens*, *Pleurozium schreberi*, *Polytrichum* spp., and *Dicranum* spp. is only about 5%.

At about 100 years after the fire, white spruce becomes dominant (Stages VI-VII), often with a component of birch trees. Tree densities are approximately 500/ha. As the stands become older, the hardwoods are less abundant. Rose and highbush cranberry are the common shrubs; *Equisetum* spp. and bunchberry are the most common herbs. The greatest change is the development of a dense and thick layer of feathermosses, previously mentioned. The mature spruce stage, with only scattered remnant birch and with a continuous moss mat, is reached at about 200 years following the fire.

What might follow these mature white spruce stands, if there were no recurrence of fire, is not known. Few, if any, examples of upland white spruce stands of over 200 years of age have been located, and none has been studied. Lutz (1956) and Foote (1976) consider the white spruce stands to be the climax vegetation on these sites—the end point in succession following fire on well-drained upland sites. It has been suggested by others that some upland white spruce stands may be replaced by black spruce and bog. Wilde and Krause (1960, p. 272) investigating spruce stands in the uplands near Fairbanks, made the following statement:

> The poor regeneration of white spruce on these moss-covered soils casts doubt on the climax nature of these species in the subarctic environment. A wide opening in the canopy is likely to cause invasion by sphagnum and black spruce, an association which would preclude the regeneration of white spruce.

While this may be a possible future direction of the white spruce successional sequence just described, we (the authors) have not observed black spruce replacing white spruce on south-facing slopes in interior Alaska, although this does occur in the successional sequence on the floodplain.

Soil Temperature and Forest Floor Thickness

Fire on south-facing slopes often consumes all or most of the organic layer. The buildup of a new organic layer is slow and is primarily through the addition of leaf litter. However, decomposition is more rapid on these south-facing slopes, and the buildup on a forest floor organic layer is slow and not as great as that on the cold black spruce sites. The organic layer of the mature white spruce stand may be only 8–12 cm.

Soil temperatures are warmer than in the other two successional sequences described in this chapter. A newly burned stand with all or most of the organic layer removed and with a layer of dark ash may have soil degree-day sums of over 2000 degrees. This changes only slightly through the hardwood stages because of the thin organic layer that is added and the south-facing aspect. As the white spruce stand develops and the organic layer reaches its maximum of 12 cm, soil temperatures become somewhat cooler but the degree-day summation still remain above 1000. There is no permafrost in the successional sequence on south-facing slopes.

Productivity

Our data base is not complete for standing-crop biomass immediately following fire. Observations on destruction of tree and forest floor organic matter during this period and estimates obtained from black spruce ecosystems indicate that above-ground biomass may be reduced by up to one-half or more in tree and forest floor compartments. The patchy nature of burn intensity is reflected by the fact that all conditions of forest floor combustion are observed, from complete ashing to light scorching.

The effect of fire on soil organic matter content also is not clear at this time. On the basis of soil organic matter content 15–20 years following severe fire, it appears that the effect may be slight, especially for deeper soils. For ecosystems which have not experienced postfire soil erosion, it appears that soil organic matter content remains fairly constant throughout the entire successional sequence.

Recovery of these systems following fire is rapid, primarily due to sprouting from tree and shrub root systems (Van Cleve and Noonan 1975). Over the entire sequence soil organic matter storage has declined from 94 to 45%, while above-ground storage has increased from 6 to 55% of the ecosystem total. The distribution of above-ground standing crop of organic matter undoubtedly shifts in favor of the forest floor in mature white spruce ecosystems because of tree mortality. However, the soil organic matter pool appears to remain constant. Decomposition rates probably decline because of cooler soil temperatures, and therefore incorporation of organic debris into mineral soil approaches minimal levels.

Annual above-ground tree production appears to reach a maximum during Stages III and IV. Maximum production approaches 800 g/m^2 then declines to less than half this value (340 g/m^2) early in mature successional stages (VI and VII). The annual productivity of moss in the mature successional stages may approach one-third that (100 g/m^2) of white spruce. We predict that moss production will increase to approximately 200 g/m^2, while tree production will decline to the same level or less in overmature white spruce ecosystems.

Nutrient Cycling

Mineral soil nitrogen reserves may increase from 200–300 g/m^2 in Stage III to 500–600 g/m^2 in Stages V and VI. These trends would reflect the early influence of alder and other nitrogen-fixing shrubs in the buildup of N reserves following fire. Phosphorus, on the other hand, shows the highest total and available pool sizes in Stage III. Available K pools tend to follow the same trends, the highest values occurring in the 6–25 year period following fire, and declining 50% by Stage VI (160 years). Early maxima for mineral soil P and K pools reflect the "rapid decomposition" of organic matter by fire and associated release of these nutrients. A decline in pool sizes primarily indicates nutrient transfer into the new standing crop of plant biomass.

Subsequent to the fire, decomposition would reduce forest floor biomass further, but by 10–25 years (mid-successional period), pool sizes show increasing N, P, and K storage in association with new forest floor organic matter. The standing crop of nitrogen in above-ground tree biomass ranges from 10 to 50% of that stored in the forest floor. The same condition exists for P in the mid-successional stages. However, in the mature successional period, this distribution shifts to a larger portion encountered in above-ground tree biomass. The standing crop of potassium follows the same trends noted for N and P.

The progressive dominance of white spruce in mature successional stages is reflected in increasing efficiency of biomass production per unit of nutrient incorporated into leaves and on a total above-ground production basis. Ratios for total production are 60–90 g/g N, 500–800g/g P, and 150–200 g/g K through Stage V. For total production the mature stage shows 300, 1600, and 400 g production per gram of element incorporated for N, P, and K respectively.

Upland Black Spruce Succession

General Description

Revegetation following fire in the black spruce type is usually rapid (Fig. 13.3). Stage I is brief: within weeks of the fire, sprouts from underground plant parts of shrubs and herbs are abundant. Even where burning is

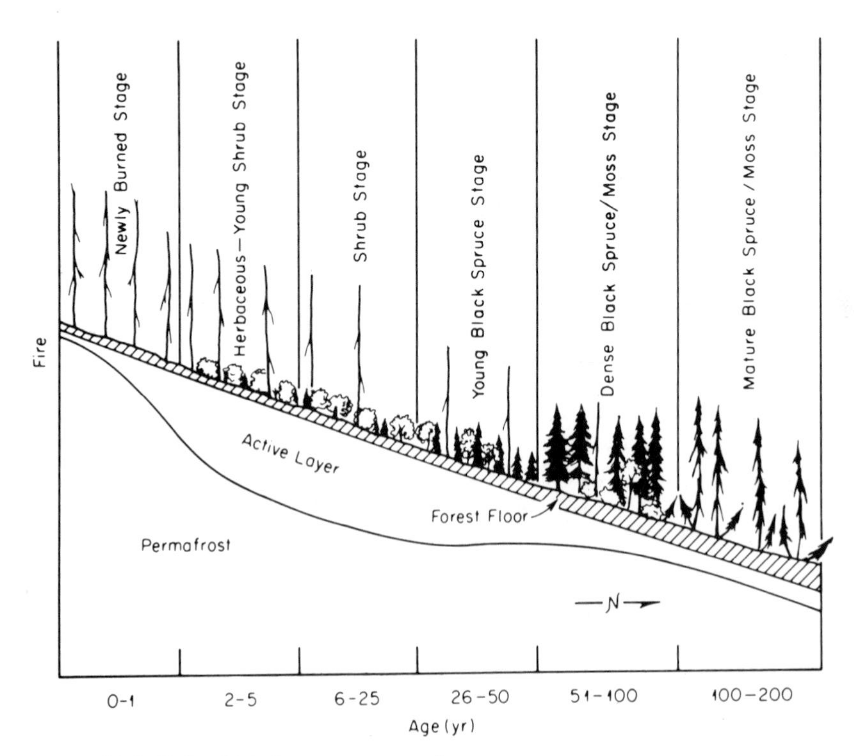

Figure 13.3. Upland black spruce succession.

severe and all of the underground plant parts are killed, invasion by mosses and liverworts and the establishment of seedlings of light-seeded species is usually accomplished by the spring of the following year.

The first stage of revegetation, the seedling-herb stage (Stage II), usually lasts from 2 to 5 years, depending on the site and fire conditions. Areas of bare mineral soil are covered with *Marchantia polymorpha*, *Ceratodon purpureus*, *Polytrichum commune*, *Epilobium angustifolium*, and other light-seeded herbaceous species. Spruce seedlings usually become established during this stage, often in the *Marchantia* and *Ceratodon* mats. In the lightly to moderately burned areas, regeneration is primarily by the continued sprouting of underground parts of the shrubs such as *Salix* spp., *Betula glandulosa*, *Rosa acicularis*, *Vaccinium uliginosum*, *Ledum groenlandicum*, and herbs, primarily *Calamagrostis canadensis*, *Rubus chamaemorus*, and *Equisetum silvaticum*. Plant cover during this stage increases quickly from 0 to as much as 40–50%. It is during this stage that most of the vascular species that will continue on through the vegetation sequence are established. Few of the mosses and lichens that are abundant in the mature stages become established.

In the shrub stage (Stage III), the shrubs that have originated from sprouts and shoots continue their rapid growth and dominate the vegetation; the herb layer also continues to increase, however, primarily through the expansion of rhizomes of *Epilobium angustifolium* and *Calamagrostis canadensis*. The mosses, again primarily *Ceratodon purpureus* and *Polytrichum* spp., also continue to increase in the early part of the shrub stage, although they may begin to decrease toward the end of this stage if the shrub canopy closes and leaf litter becomes abundant. The first lichens, usually the foliose lichens *Peltigera canina* and *P. aphthosa*, become established during this stage. This stage usually occurs from 6 to 25 years following the fire.

The tree canopy begins to dominate from 25 to 30 years following the fire. The young stands of 40–60 years are dense (Stage IV), with as many as 4000–6000 trees and saplings per hectare; and seedlings may be as dense as as 12,000/ha. The tall shrub layer of willows and alders begins to thin out as the spruce become taller, but the low shrub layer of *Vaccinium uliginosum, V. vitis-idaea, Ledum groenlandicum, Betula glandulosa,* and *Rosa acicularis* continues to expand and increase in cover. The most significant event that occurs during this stage is the invasion and rapid development of the feathermosses, *Hylocomium splendens* and *Pleurozium schreberi,* which, along with some *Sphagnum* species, may develop as much as 50% cover. With the establishment of these mosses, there begins the development of a thick organic layer, which accumulates the available nutrients and creates colder soil temperatures and the freezing back of the permafrost layer on many sites. Another significant invasion is by fruticose lichens, the *Cladonia* and *Cladina* species, and additional foliose lichens such as *Nephroma arcticum*, which together may account for at least 20% cover.

Once the tree canopy is well established, the changes in the vegetation sequence are slower and more subtle. During the older stages, the tree canopy is mostly closed, although the density is less, averaging 2200 stems per hectare. The moss layer remains about the same as in the younger tree stage except for an increase in *Hylocomium splendens* and fruticose lichens and a decline in foliose lichens.

In Alaska, the development of the black spruce type into a mature stand of over 100 years follows without any major changes. Tree densities are about the same, 1700/ha for black spruce, and a few paper birch may persist into the mature stage. The spruce tend to be in clumps produced by layered branches, and there are more openings in the canopy. Because of this, the shrub layers, especially the low shrubs, are better developed than they are during the 60–90-year period when the canopy is more closed. The moss cover in mature stands is dense and covers nearly 75% of the forest floor, but the lichen cover continues to decline on these mesic sites so that the total lichen cover of both foliose and fruticose lichens averages only 2%.

The question of whether or not one can describe a climax community type in black spruce following fire relates primarily to the high frequency of fire. In many areas that have been studied, it was difficult or impossible to find an old stand that had not been burned. Fire is a recurring phenomenon, and the species and communities have evolved around this disturbance. In many northern areas, fire rotation ages may not be more than 50–100 years, but it may be 300 or 400 years before a stable situation is reached. In many of the older stands, there is no obvious self-perpetuation mechanism. This has also led to the questioning of the climax concept.

Drury (1956) and Viereck (1970b) suggest that on the floodplains in Alaska, cycles of black spruce and bog may result if fire is absent for a long time. In the lower Mackenzie River Valley, Strang (1973) suggested that if fire were eliminated from the black spruce stands, the climax vegetation would be treeless moss-lichen association. He based this conclusion on the study of several older black spruce-lichen communities, where he found little evidence of sufficient regeneration to continue the existing density of black spruce. In the same area, however, Black and Bliss (1978) found that understory species tended to become "more tundra-like" with prolonged absence of fire, but that the *Picea mariana* was maintaining itself through layering, and in fact, the density of the tree species tended to increase with age. In the open lichen woodland type, Maikawa and Kershaw (1976) also found an increase in tree density with age and suggested that in the absence of fire the open lichen woodland type was eventually replaced by closed stands dominated by feathermoss. Thus, it is suggested that the climax vegetation on both the dry and moist sites is the black spruce-feathermoss type. On other sites, however, it was concluded that open lichen woodlands would persist as a climax type (Johnson and Rowe 1977, Ahti 1959).

Soil Temperature and Forest Floor Thickness

Fire usually removes a portion of the accumulated organic layer in a black spruce stand, but seldom does it burn away all of the organic material. This removal of the live and dead organic layer plus the change in surface albedo results in warmer soil temperatures following the fire, which, in turn, results in a deepening of the active layer. The results of this increased warming of the soil may last well into the successional sequence. It is really not until the spruce stand is reestablished and the feathermoss cover nearly 100% of the forest floor that the active layer again reaches its prefire thickness.

Following fire the organic layer depth may be only 4–8 cm, depending on the amount removed by the fire. This gradually increases to a thickness of 16–20 cm in the older black spruce stands. Soil temperatures, as measured by soil degree days, is high immediately following the fire, but declines steadily during the successional sequence. We recorded soil heat sums of 1250 the first year following the Wickersham fire, but only of about 1000 5 years following the fire. Soil degree days in mature black spruce stands ranged from 500 to 800 degree-days.

The active layer, or annual thaw depth, in a mature black spruce stand is usually from 40 to 60 cm. Following fire this increases for several years and may reach a thickness of over a meter. The annual thaw remains at about 1 m or greater until the spruce and feathermosses are reestablished and then, between 25 and 50 years following the fire, may return to its original depth. These changes in soil temperature and depth to permafrost are important aspects of the increased productivity of these black spruce sites following fire.

Productivity

Available data on postfire standing crops of biomass indicate generally substantial impact on above-ground standing crop of tree biomass, possibly with up to an order or magnitude or more (10–0.7 kg/m^2) reduction. This impact is associated with effects on the forest floor, which are substantially more variable than those noted for the overstory. The impact of burning ranges from complete ashing of the organic layers to surface scorching of the green moss. The maximum accumulation of forest floor organic debris amounts to nearly 10 kg/m^2 and is approximately equivalent to the standing crop of above-ground tree biomass in late mature successional stages (Barney and Van Cleve 1973). The amount of surface organic debris accumulated in late mid- to early mature stages may be between four and five times that found in live, above-ground tree biomass. The forest floor generally plays a more significant role in biomass and nutrient storage and in maintaining soil temperature control in these ecosystems than in the other successional sequences. Substantial, but currently unknown, reduction undoubtedly occurs in tree root-system biomass following fire.

The organic matter content of mineral soil attains values in excess of those measured for other vegetation types. These trends reflect the buildup of organic matter because of restricted microbial activity in the cold soils. Direct effect of fire on loss of mineral soil organic matter generally will be minimal primarily because of cold, moist conditions. However, loss of the insulating forest floor and change in surface albedo may lead to the thawing of permafrost with associated mineral soil erosion on steeper slopes. In addition, warmer soil will stimulate microbial activity. We predict under most conditions slight decreases in mineral soil biomass primarily due to improved soil temperature and the associated increased microbial respiration during early succession (Stages I–II).

Mineral soil in late mid-successional systems may store up to 80% of system organic matter, while 17% remains in the forest floor and 3% in live above-ground standing biomass. Later in mature stages the mineral soil may store 70% of system biomass, while 15% each remains in the forest floor and above-ground standing crop.

The fire cycle controls attainment of these levels of biomass accumulation. In the absence of fire we hypothesize that vascular plant production will decline and forest floor and mineral soil biomass remain constant or slowly increase to a higher stable level as bryophyte cover becomes the dominant plant growth form in systems older than 200 years.

On the basis of current information it appears that maximum tree production is attained early in the mature successional period (Stage V). In general, litter fall and above-ground tree production are up to an order of magnitude less than that encountered in floodplain and white spruce successional sequences. Moss production (110–120 g/m^2) may nearly equal above-ground tree production in mid- to late successional stages (V and VI). While tree production is substantially less than that encountered in other successional sequences, the combined moss plus above-ground tree production of around 200–230 g/m^2 approaches the lower limits noted for above-ground tree production in the more productive sequences. We have no information on root production at this time. We predict that vascular plant production will slowly decline prior to the next fire and that moss production may actually exceed that of slowly senescing black spruce.

Nutrient Cycling

Reserves of mineral soil N, P, and K do not show obvious trends with succession. This condition may reflect the relatively limited data base including immediate postfire soil analysis. Following fire, P and K from partially or completely ashed forest floor may leach into surface mineral soil layers, increasing the available supplies of these nutrients. We have measured up to 40-fold increases in available P in ashed forest floor following intense burning. In the most extreme cases, complete ashing of the forest floor could result in loss of 60–90 g/m^2 of N and, under these hot burning conditions, loss of varying portions of the N reserve from surface layers of the mineral soil. Late successional trends in N, P, and K storage reflect increasing cold soil conditions, with reduced rates of organic matter mineralization and replenishment of available nutrient supplies (Heilman 1966). Larger portions of soil nutrient reserves become "frozen assets."

Because of the accumulation of organic matter on the forest floor, this compartment plays a larger role in nutrient storage in the black spruce than in other systems. From 15 to 30% of the N, 2–11% of the P, and 8–20% of the K may be stored in the forest floor. The distribution of these elements when root nutrient reserves are included shows little change for P, but 8 and 33% for N and K, respectively.

Storage of an unknown mass of nutrients in standing snags and slowly decaying root systems may persist for at least 60 years. Nutrient storage in the forest floor may continue to decline during the first decade following fire as a result of increased organic matter decomposition rates. During this period shrub and tree species will conserve nutrients in new biomass, including an accumulating forest floor. Advanced successional stages (200 + years), free from intervention of fire, would show declining nutrient storage in tree tissue as nonvascular plants assume production dominance. At this stage the forest floor would comprise the most important nutrient reservoir.

The magnitude of nutrient flux in litter fall remains substantially less than that encountered in the upland white spruce or floodplain successional sequences. In addition, it appears that peak litter fall values occur somewhat later (100–200 years) in late succession compared with 50–100 years in the more productive sequence. This trend reflects the importance of earlier aspen and birch phases in the white spruce series. In addition, it appears that the period of maximum litterfall nutrient flux succeeds the period of maximum productivity and nutrient uptake, possibly indicating the senescence of older age-class black spruce ecosystems (Stage VI). As pointed out earlier, moss production becomes an increasingly important component of mid- and late successional black spruce. Moss may be an equally effective competitor with spruce for N and P. As shown by Heilman (1966), spruce exist on increasingly nutrient-deficient substrates, and the principal plant competitor of the limited nutrient reserves are bryophytes. Production-to-nutrient incorporation ratios indicate that black spruce is approximately as efficient as white spruce in use of N, P, and K reserves. Foliage production to nutrient incorporation ranges from 80 to 120 g/g for N, 60–1200 g/g for P, and 130–190 g/g for K. Total production to nutrient incorporation ranges from 140 to 290 g/g for N, 1400–2200 g/g for P, and 330–610 g/g for K. No trends exist with respect to production to nutrient content ratios and successional stage.

Nutrient Cycling and Succession

From the previous discussion it is obvious that marked similarities and differences exist in processes controlling variables among the three successional sequences. Among the successional sequences a picture emerges of the most dynamic systems with regard to productivity and nutrient cycling occupying the floodplains and south-aspect sites, while the least dynamic systems with regard to productivity and nutrient cycling occupy north aspects. Within successional sequences the classical situation exists in which more productive, nutrient extravagant species appear early in succession, while less productive, nutrient-conservative species appear late in succession. A number of important biogeochemical controls of nutrient cycling among and within the successional sequences are summarized in Fig. 13.4.

One unifying physical control across all sequences is the general cold-dominated, semiarid nature of taiga regional climate. Temperatures favorable for measurable biogeochemical activity may occur only during a 140-day period from mid-April to mid-September. Tree growth may occur for the 90-day period mid-May through mid-August. Biogeochemical processes may further be restricted by insufficient water, especially during periods of most favorable temperatures and light regimes (mid-June to mid-July).

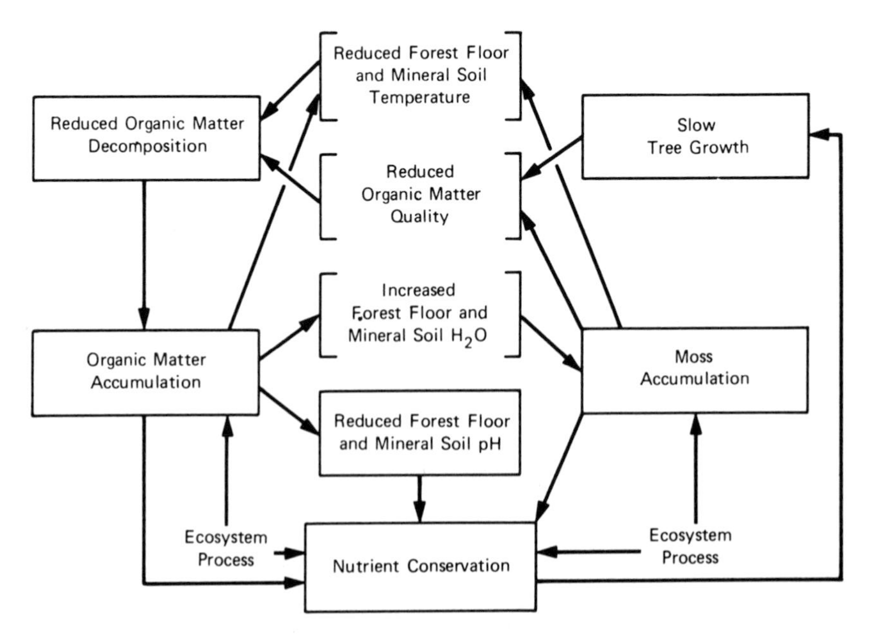

Figure 13.4. Hypothesized processes that control the distribution of biomass and the distribution and cycling of nutrients between ecosystem compartments.

However, topography (slope and especially aspect) directly modifies the climatic regime so that the north-aspect black spruce sequences generally are coolest, south-aspect white spruce sequences warmest, and floodplain sequences intermediate in temperature regime. Previously we discussed the general reduction in soil temperature as forest floor biomass accumulates during succession. Across successional sequences the effect is equally dramatic (Fig. 13.5). Black spruce systems occupy the sites displaying the coldest soil temperature; white spruce, birch, and aspen occupy sites that display increasingly warmer temperatures. The wettest soils are also encountered in the black spruce sequences with progressively drier substrates encountered in the other major vegetation types (Fig. 13.6). In summary, black spruce ecosystems occur in the coldest, wettest soils; white spruce, birch, poplar, and aspen ecosystems occur in warmer, drier soil. Aspen occurs at the warmest, driest end of the temperature-moisture spectrum.

Topography alters soil moisture regimes in another manner. Black spruce ecosystems will tend to have a higher soil moisture content due to seasonal frost melt water and to the drainage restriction imposed by permafrost. The south-aspect white spruce successional sequence displays deep, well-drained soil, or well-drained shallow soil over bedrock. The

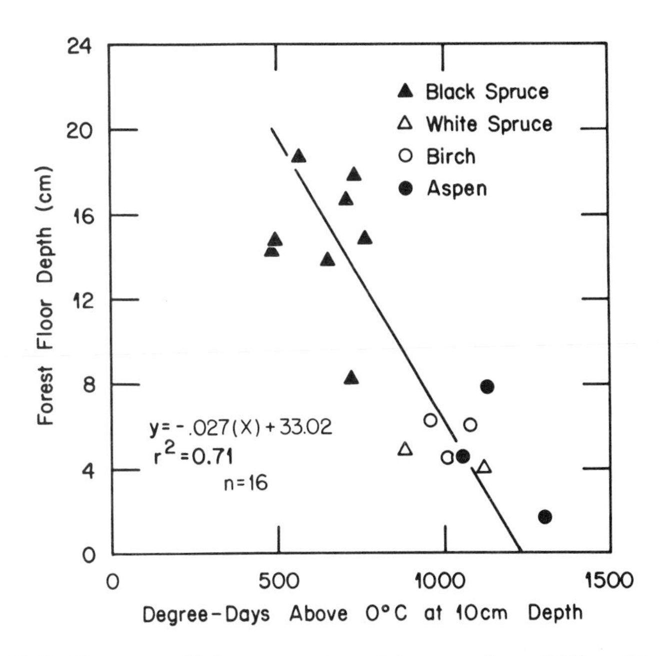

Figure 13.5. Heat sum (daily summation of degrees above 0°C) at 10 cm depth across stands in a successional sequence and with different forest floor depths. The regression of depth (y) on the degree-day heat sum (x) based on 16 (n) observations has a coefficient of determination (r^2) of 0.71.

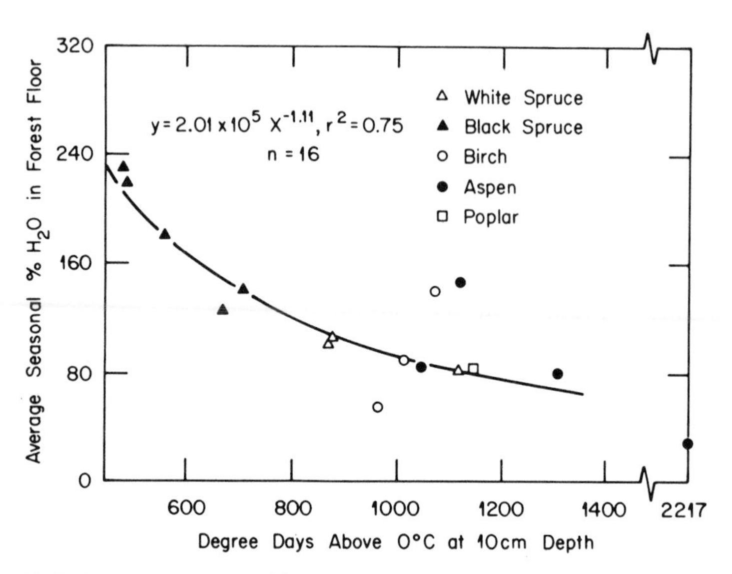

Figure 13.6. Moisture content (y) in relation to heat sum (x) for the forest floor across successional sequences. Based on 16(n) observations, the square of the correlation coefficient (r^2) was 0.75.

floodplain successional sequence soils generally are well drained. However, here the high water table and the phenomenon of capillary rise of water may result in more sustained supplies of water to plant root systems than in upland south aspect ecosystems. During the growing season periods of low precipitation, floodplain and north-aspect ecosystems may have sufficient supplies of water to sustain soil-plant biological processes in contrast to the moisture-limiting conditions on south aspects. In fact, observations of the seasonal course of soil moisture, using the neutron probe, and measurements of water and nutrient flux, using tension plate lysimeters, indicate little or no moisture or nutrient flux out of south-aspect soil profiles, which are up to 1 m deep over bedrock. With the possible exception of spring runoff of snowmelt water over the frozen soil, and during and for an undetermined but probably short period (up to 5 years) following fire, the south-aspect successional sequence may largely be closed to nutrient loss.

Of course, floodplain ecosystems experience a variable flux of nutrients, depending on periodicity and severity of flooding. Loss of nutrients may also be experienced along the permafrost table in north-slope black spruce systems. This loss may be of substantial importance to these nutrient-poor ecosystems.

Among the vegetation types that comprise the successional sequences substantial differences exist in the chemical nature of the forest floor. For example, the percentage of exchangeable base element saturation of the forest floor [(Ca + Mg + K)/cation exchange capacity] generally appears higher for earlier successional stages in the upland south aspect sequences and declines with the increasing age of system. Black spruce forest floors (north aspect sequence) display the lowest base element saturation. This condition is in turn reflected in k, the fractional annual turnover rate of the forest floor. Cause and effect with regard to decomposition is difficult to establish in this case, for the black spruce occupy sites having the coolest, wettest soil, while aspen occupy the warmest, driest sites, where leaching of base elements probably is minimal and where moisture deficiency may restrict biological activity. Carbon-to-nitrogen ratios of the forest floor provide a further indication of potential N mineralization rates. Among the successional sequences the upland south aspect sequence generally has lower forest floor C/N ratios (23 for earlier stage aspen and birch, 28 for mature stage white spruce) than north-aspect black spruce (44). Mineralized N supplies should be smaller in the latter case.

These indices of organic matter chemistry (quality) provide insight into the control of nutrient cycling and productivity, both within and among successional sequences (Fig. 13.7). Lower N, higher lignin content, and wider C/N ratio organic materials tend to accumulate in black spruce, which display the lowest plant productivity. There is some indication, especially in the case of base saturation, percentage of N, and the C/N ratio, that a similar trend exists with advancing succession in the white spruce sequence. Both vegetation types occupy the coolest, wettest positions with regard to temperature and moisture regimes. These types also produce organic matter, which has lower inorganic nutrient concen-

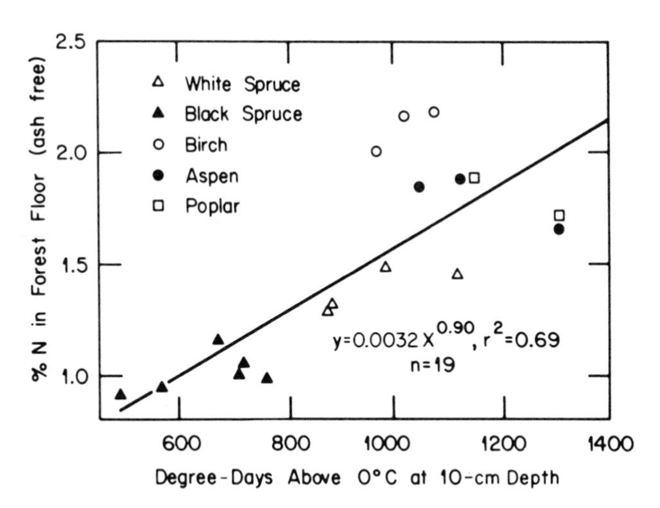

Figure 13.7. Nitrogen content (y) in relation to heat sum (x) for the forest floor. Based on 19 (n) observations, the square of the correlation coefficient (r^2) was 0.69.

trations, reinforcing the reduction in forest floor nutrient concentration. Competition for nutrients among the developing nonvascular plants as well as microbes undoubtedly limits the supplies of nutrients available to vascular plants. Primary production is reduced under these conditions.

Conclusions

We view the interaction of system processes and process controlling variables in the manner summarized in Fig. 13.4. Many of the relationships encompassed in this figure have been evaluated on a theoretical and practical basis in the literature (Odum 1969; Vitousek and Reiners 1975). However, a feature of the taiga ecosystems is the unique synergy among soil temperature, primary production, and decomposition. Regardless of successional sequence, nutrient cycling reflects a dramatic change with regard to this important control (soil temperature) and these associated processes (production and decomposition).

As primary or secondary succession progresses, organic matter accumulates both in the standing crop of trees and shrubs and in forest-floor organic debris. The latter "pool" acts as an effective insulating blanket, cooling the mineral soil as forest floor biomass accumulates. Reduced soil (and forest floor) temperatures slow the decomposition rate, promoting organic matter accumulation in this compartment. On the average, for each centimeter increase in forest floor thickness, we have found approximately a 37°C decline in heat sum accumulated above 0°C at the 10-cm depth between May 20 and September 10. Concomitant with forest floor organic matter buildup is increased forest floor and mineral soil water content and reduced forest floor and mineral soil pH. With advancing succession, a higher forest floor moisture content is reflected in the establishment and increased productivity of bryophytes. The appearance of moss is associated with a decline in chemical quality (lower nitrogen content, higher lignin content) of forest-floor organic matter, further reduction in forest-floor and mineral soil temperature, and reduced organic matter decomposition. These processes and process-controlling variables are manifested in increased ecosystem nutrient conservation, generally slowed nutrient cycling, and slowed tree growth. A condition develops of increased deficiency with regard to nutrients important to tree growth (Heilman 1966). Nonvascular plants and microbes are increasingly effective competitors with vascular plants for reduced nutrient supplies as more system nutrient capital is accumulated in slowly decaying organic remains. Under many conditions, permafrost appears in these later stages of succession, further reducing the volume of soil that acts as a nutrient reservoir for vascular plants.

At any point in time, and depending on the severity of the disturbance, fire or flooding may act to set the system to an earlier stage in succession. Coming late in succession, fire acts as a rapid decomposer, replenishing reservoirs of some nutrients (P, base elements) and depleting supplies of others (N), and the change in surface albedo is associated with a marked warming of mineral soil and increase in thickness of the active layer on permafrost-dominated sites. The nutrient conservation-reduced primary production scenario temporarily is reversed until shrub and tree species capable of rhizome or root sprouting or developing from seed begin to take advantage of improved temperature and nutrient regimes by accumulating nutrients in new plant biomass.

Acknowledgments

This work is part of a continuing, cooperative research effort between the Institute of Northern Forestry, USDA Forest Service and the Forest Soils Laboratory, University of Alaska, Fairbanks. Research support was obtained through grants to the University of Alaska from the McIntire-Stennis Cooperative Forestry Research Program and the National Science Foundation (Grant Numbers DEB 76-12350 A01 and A02, DEB 78-11594).

Chapter 14

Aspects of Succession in the Coniferous Forests of the Pacific Northwest

Jerry F. Franklin and Miles A. Hemstrom

Introduction

Many forest phenomena achieve ultimate expression in the coniferous forests of coastal northwestern America. Coniferous species dominance is, in itself, unusual in a moist temperate region (Waring and Franklin 1979). Nearly every arboreal genus represented attains a size and life span at or near the generic maximum. Except for some *Eucalyptus* forests in Australia, biomass accumulations and leaf areas routinely exceed levels encountered in any other forest type. Accumulations of coarse woody debris are similarly large. Time scales are expanded by species longevity, infrequent disturbances, and relatively slow vegetative regrowth.

In this chapter we shall explore aspects of succession in the northwestern forests found from the crest of the Cascade Range to the Pacific Ocean. Emphasis is on newer findings since general descriptions of successional sequences are already available (e.g., Franklin and Dyrness 1973). These include the persistent effects of the pioneer species, the variable nature of seres as revealed in age structure analyses, and importance of coarse woody debris. Our objectives are to: (1) provide contrasts with succession in temperate deciduous forests as well as with conifer forests in boreal and arid regions; (2) contrast recent understanding about succession in the Pacific Northwest with previous concepts; (3) examine the role of coarse woody debris in forest succession; and (4) suggest some problems northwestern forests present in successional modeling.

Regional Setting

The coastal Pacific Northwest is a mountainous region dominated by the Cascade Range, Olympic Mountains, Coast Ranges, and Siskiyou Mountains of Washington and Oregon (Heilman et al. 1979). Topography is typically mountainous and rugged. Gentle intermountain regions, such as the Puget Trough and Willamette Valley, are not extensive. Forests extend from essentially sea level to 1600 m or more. Soils are typically youthful but moderately fertile and commonly surprisingly deep due to depositional processes.

The climate of the region favors forest development (Waring and Franklin 1979). Winters have mild temperatures and heavy precipitation. Summers are warm and relatively dry. This regime is believed to be a major factor in the dominance of conifers over deciduous hardwoods. Climate does vary substantially with geography and elevation. The mildest climates are in the coastal *Picea sitchensis* zone. The coolest environments are in the subalpine zones above 800–1200 m elevation, where winter snowpacks of up to 3 or 4 m may be developed.

Synecological studies in the region have shown that moisture and temperature are the primary environmental controls on plant community composition (Zobel et al. 1976). Both vary sharply over short distances in the broken mountain topography. Consequently, similar contrasts occur in the composition and structure of forest communities.

General Features of the Forests

Forests in the Pacific Northwest differ markedly in composition, structure, and function from the deciduous forests that dominate most mesic segments of the world's temperate zones. There are relatively few major tree species, and most are conifers (Table 14.1). Productivity can be very high as a record of 36.2 metric tons ha^{-1} yr^{-1} in a 26-year-old *Tsuga heterophylla* stand indicates (Fujimori 1972). Nevertheless, forest recovery following disturbance (as measured by leaf area, for example) typically requires several decades, and is substantially slower than deciduous forests (Bormann and Likens 1979a). Live above-ground biomass accumulations in old-growth forests (>250-year-old) forests average 700–900 metric tons/ha, and stands with over 1000 metric tons/ha are not uncommon (Franklin and Waring 1980). Leaf areas (one side only) range from 10 to 13 m^2/m^2 in one sample of stands (Franklin and Waring 1980). Gholz (1979) reports a maximum value of 20.2 m^2/m^2 in a *Picea sitchensis-Tsuga heterophylla* stand.

Several general successional sequences have been described in coastal Pacific Northwest forests (Franklin and Dyrness 1973). A typical sere in the *Tsuga heterophylla* zone would be

Pseudotsuga menziesii → *Tsuga heterophylla*, *Thuja plicata* → *Tsuga heterophylla*

In the *Picea sitchensis* zone a normal sere would involve

Picea sitchensis *Tsuga heterophylla*	→	*Tsuga heterophylla*

.ft 1 Alnus rubra stands could conceivably occur as the first stage in either of these seres although its successional replacement by conifer stands is still unproven. A sere for the lower subalpine in the Cascade Range is

Pseudotsuga menziesii and/or *Abies procera*	→	*Tsuga heterophylla* *Abies amabilis*	→	*Abies amabilis*

Throughout most of the region under consideration *Tsuga heterophylla* and *Abies amabilis* are viewed as the major potential climax species based upon a variety of evidence, including size class analyses. Special habitats provide exceptions, such as the dry sites where *Pseudotsuga menziesii* or *Libocedrus decurrens* appear to be self-perpetuating. Some species of moderate shade tolerance, such as *Thuja plicata* and *Chamaecyparis nootkatensis*, typically reproduce poorly in old-growth stands. This may indicate subclimax status, or they may behave similarly to *Sequoia sempervirens*, which manages a stable age class distribution despite an apparent sparsity of reproduction at any one time (Stephen Viers 1979, pers. comm.).

Most tree species in the Pacific Northwest play multiple successional roles. In this discussion it is important to distinguish ecological roles from species' shade tolerances. Species colonizing a site early in a sere are playing a pioneer role. Species apparently capable of perpetuating themselves on a site in the absence of disturbance are playing a climax role. Although shade tolerance is often equated with a climax role and intolerance with a pioneer role, this need not be the case. Several intolerant species can form a stable type of climax if environmental conditions exclude their more tolerant associates. A prime northwestern example is the development of self-perpetuating stands of *Pseudotsuga menziesi* on habitats that are too dry for *Tsuga heterophylla* or *Abies grandis.* In other cases, environmental conditions need only favor the less tolerant associate, such as snowpack favoring reproduction of *Abies amabilis* over that of *Tsuga heterophylla* (Thornburgh 1969).

Similarly, essentially any of the shade-tolerant species can and do play pioneer roles on most sites where they are also climax. *Tsuga heterophylla* is conspicuous early in seres on cut-over forest lands in the *Picea sitchensis* zone. So is *Abies amabilis* on many high-elevation burns. This can be due to an absence of seed source for faster-growing intolerant species. In other cases, the growth rate of the tolerant species may be adequate to stay with or even exceed that of the intolerant species.

Table 14.1. Typical and maximum ages and dimensions attained by forest trees on better sites in the Pacific Northwest.[a]

Species	Typical			Maximum	
	Age (yr)	Diameter (cm)	Height (m)	Age (yr)	Diameter (cm)
Abies amabilis	400+	90-110	45-55	590	206
Abies grandis	300+	75-125	40-60	—	217
Abies procera	400+	100-150	45-70	>600	270
Acer macrophyllum	300+	50	15	—	250
Alnus rubra	100	55-75	30-40	—	145
Chamaecyparis lawsoniana	500+	120-180	60	—	359
Chamaecyparis nootkatensis	1000+	100-150	30-40	3500	297
Libocedrus decurrens	500+	90-120	45	>542	368
Picea sitchensis	500	180-230	70-75	>750	525
Pinus lambertiana	400	100-125	45-55	—	306
Pinus monticola	400+	110	60	615	197
Pinus ponderosa	600+	75-125	30-50	726	267
Populus trichocarpa	200+	75-90	25-35	—	293
Pseudotsuga menziesii	750+	150-220	70-80	1200	434
Quercus garryana	500	60-90	15-25	—	220
Sequoia sempervirens	1250+	150-380	75-100	2,200	501
Thuja plicata	1000+	150-300	60+	>1200	631
Tsuga heterophylla	400+	90-120	50-65	>500	260
Tsuga mertensiana	400+	75-100	35+	>800	221

[a]Based on Franklin and Waring (1980).

Scale in the Pacific Northwest

Many of the unique aspects of coastal Pacific Northwest forests result from an increase in the size and longevity of the tree species in comparison with other temperate and boreal forest regions. Pacific Northwestern tree species are well known for achieving large sizes and long life spans (Table 14.1). Consequently, compositional effects of a disturbance can persist for centuries or even millenia. The long-term persistence of seral species and infrequent catastrophic disturbances often combine to make achievement of climax stands hypothetical. Classical old-growth stands in the Pacific Northwest are actually not climax but, rather, draw much of their character from large seral *Pseudotsuga* (Franklin et al. 1981). This is apparent in stands of *Pseudotsuga* undisturbed for over 1000 years in Mount Rainier National Park (Table 14.2). Climax forests, on the other hand, are almost exclusively dominated by species of smaller stature and shorter life span, such as *Tsuga heterophylla* and *Abies amabilis*, and are not as impressive as those containing the larger, seral species. In forests with such extended seres the potential for direct observation of successional processes is obviously very limited.

The impact of the large, long-lived pioneer species extends to structural as well as compositional aspects of succession. These trees become sources of coarse woody debris—snags and down logs—after dying. These structures are important in many aspects of ecosystem function and composition (Franklin et al. 1980). *Pseudotsuga* produces relatively decay-resistant logs of large size. Between 480 and 580 years are believed necessary to eliminate 90% of an 80-cm diameter *Pseudotsuga* log (Joseph E. Means 1980, pers. comm.). In forests containing these species the accumulations of woody debris are greater and individual pieces are larger and more persistent than would be the case in a climax forest of *Tsuga heterophylla* or *Abies amabilis*. In this way the influence of a large seral tree is extended several centuries beyond the death of the last specimen. While climax composition might be achieved soon after the death of the last *Pseudotsuga* (say, in 1000 years), a stable forest structure based on the climax species could be delayed another 500 years. If one views *Thuja plicata* as a seral species (albeit one that extends the duration of its persistence on a site by some modest reproduction within the stand), many millenia would be necessary to achieve the idealized climax forest. Live trees would persist for several generations, and the large *Thuja* logs are even more resistant to decay than those of *Pseudotsuga*.

In spite of these problems, the climax concept is useful, and climax forests do exist. Forests that compositionally and structurally approximate hypothetical climax conditions are present in both temperate and subalpine regions of the Pacific Northwest. We suspect that these were typically (but not exclusively) developed where the climax species (e.g., *Tsuga heterophylla* or *Abies amabilis*) was the major or sole colonizing

Table 14.2. Pseudotsuga menziesii presence (percentage of plots where it occurs) and basal area and total basal area in three old-growth age classes at Mount Rainier National Park, Washington, on a series of 500- and 1000-m^2 plots located below 1350 m.

Age class (yr)	Number of plots	*Pseudotsuga* presence (%)	Basal area	
			Pseudotsuga	All species
350-600	114	67	26.9	98.4
750-1000	55	69	20.1	91.8
1000	9	33	11.9	73.9

species following the last disturbance. Forests dominated by climax species, with only a minor component of typical seral species, are even more common.

Disturbance frequency is a final area of contrast in scale between the Pacific Northwest and most other temperate and boreal forests. We hesitate to suggest that the geographic scale of the disturbances is necessarily any greater in the Pacific Northwest, despite the evidence for fire episodes that covered hundreds of thousands of hectares. Hurricanes and typhoons destroy vast reaches of temperate forests and wildfires destroy similar areas of boreal forests at a time in other parts of the world. The disturbance pattern in the Pacific Northwest is one of infrequent, holocaustic forest fires, however. A study of fire history over the last 1000 years indicates a natural fire rotation averaging 434 years within Mount Rainier National Park (Hemstrom 1979). Limited data from elsewhere in the region suggest that natural fire rotations of several centuries are typical. There are, of course, variations associated with habitat, topographic settings, or geographical locale (e.g., rain shadow of a mountain range). Natural fire rotation may also decline with latitude. The pattern of infrequent catastrophic, as opposed to frequent, light, burns is quite consistent, however.

Several other aspects of catastrophic fire in the Pacific Northwest are relevant to successional studies. Major fire episodes appear to correlate with climatic conditions. This is suggested by correspondance of fire episodes at Mount Rainier with reconstructed dry climatic periods (Hemstrom 1979). It is also suggested by the occurrence of similar forest age classes over large, but geographically isolated, segments of the region. In addition, fuel loadings are almost always heavy in these forests. This is not an environment in which fire suppression leads to unnaturally heavy fuel accumulations. Related to this is the fact that wildfires in these forests rarely consume much of the wood. Trees die and become snags and down logs, but several subsequent fires are necessary to consume a majority of this woody debris. Even the tree foliage often escapes burning. Numerous examples of reburn, at least during historic times, suggests that young stands (e.g., 25–75 years) are more susceptible to burning than later forest stages.

Revised Concepts about Succession in *Pseudotsuga* Forests

General descriptions of succession in northwestern *Pseudotsuga* forests have been extant for many decades (e.g., Isaac 1943; Munger 1930, 1940). *Pseudotsuga* is described as rapidly occupying burned sites, followed by reinvasion of the more shade-tolerant and fire-sensitive *Tsuga heterophylla* and *Thuja plicata*. Establishment of the *Tsuga* and *Thuja* continue indefinitely under the canopy of the *Pseudotsuga*. Ultimately these species replace *Pseudotsuga*, which are lost to wind throw, pathogens, or other agents of mortality.

Such descriptions were largely anecdotal accounts rather than quantitative analyses, however. They were based on observations of forest restablishment following extensive wildfires in the mid-1800s and the Yacholt (1902) and Tillamook (1933) burns. Later stages of succession were inferred from species size-class patterns in mature and old-growth stands. Modern successional studies have also, by necessity, relied heavily on inferences drawn from size-class analyses. The great sizes and ages achieved by trees limited the usefulness of tree coring and direct observation on sample plots.

Age structure analyses and historical reconstructions of *Pseudotsuga* forests are now underway, however, utilizing sites scheduled for timber cutting. Quantitative data on succession are accumulating rapidly and providing new insights, many of which run counter to previous dogma.

One surprising discovery from age structure analyses is the wide range of ages in dominant *Pseudotsuga* in at least one age class of old-growth (400–500-year-old) forests. This phenomenon was observed on Experimental Watershed No. 10 at the H. J. Andrews Experimental Forest. This 10-ha watershed is located at 400-600 m elevation on the western slopes of the central Oregon Cascade Range. In its pristine state the watershed was occupied primarily by old-growth forests of *Pseudotsuga, Tsuga heterophylla, Thuja plicata, Pinus lambertiana,* and *Castanopsis chrysophylla* (Hawk 1979). Prior to logging, all trees greater than 5 cm diameter at breast height were mapped, measured, and tagged. Followinging logging in 1975, approximately 600 of 2800 trees, including a large proportion of the dominant *Pseudotsuga*, were relocated and aged from ring counts on stumps.

Watershed No. 10 appears to have been burned sometime around

1800, based on a younger generation of *Pseudotsuga* (Fig. 14.1) and sparsity of *Tsuga* over 200 years old. The most notable feature of the age structure is the broad range of ages in the old-growth *Pseudotsuga*, however (Fig. 14.2). This age structure indicates a very slow reestablishment of the new forest, following the catastrophic disturbance(s) that destroyed the forest present on the site over 500 years ago.

These results were so unexpected that confirmation was needed from age structure analyses in other stands before they were accepted. Much of Watershed No. 10 was relatively droughty and covered by a mosaic of

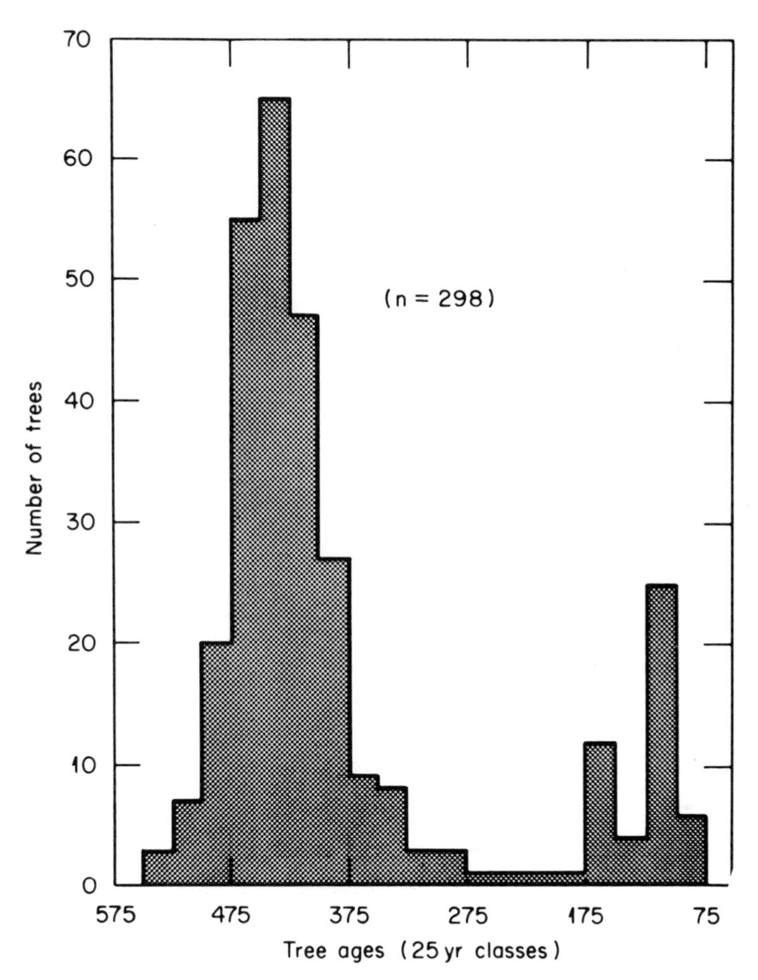

Figure 14.1. Age-class distribution of all *Pseudotsuga menziesii* by 25-year age classes on Watershed No. 10 at the H. J. Andrews Experiment Forest, western Oregon Cascade Range.

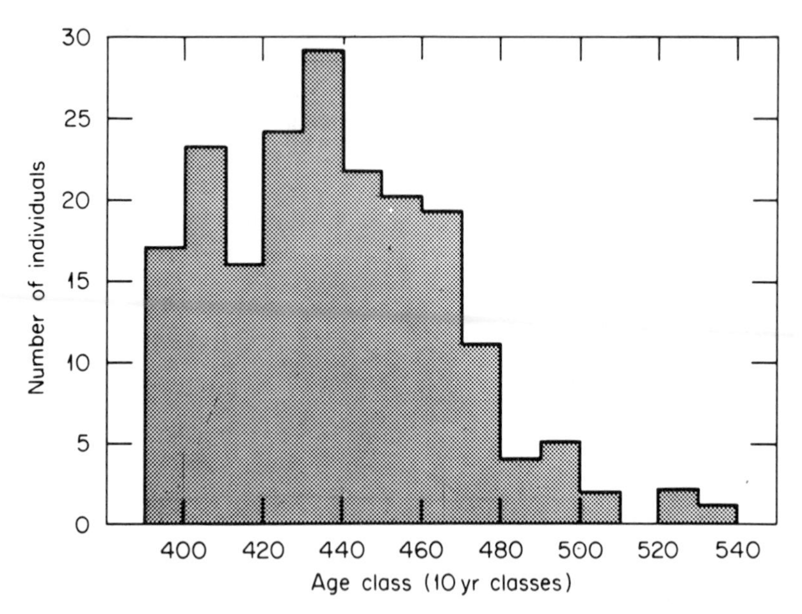

Figure 14.2. Age-class distributions of the old-growth *Pseudotsuga menzesii* on Watershed No. 10 at the H. J. Andrews Experimental Forest, western Cascade Range.

older and younger forests because of the wildfire around 1800; perhaps the age range in forests of comparable age would be less variable elsewhere. Subsequent age structure analyses were conducted to test this hypothesis. One was in a heavily stocked old-growth forest at about 900 m (Joseph E. Means 1980, pers. comm.), also on the H. J. Andrews Experimental Forest, and in several clear cuts on the Wind River Experimental Forest in the southern Washington Cascade Range. The results confirm a wide age range in dominant *Pseudotsuga* in forests established 400–500

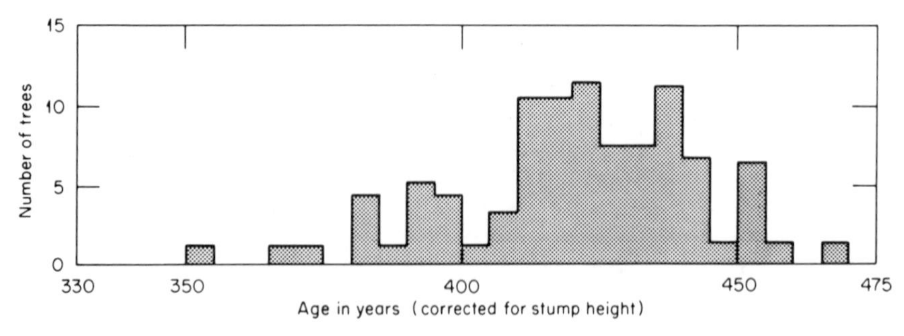

Figure 14.3. Age-class distribution of *Pseudotsuga menziesii* in an old-growth stand located at about 900 m on the H. J. Andrews Experimental Forest, western Oregon Cascade Range (courtesy Joseph E. Means).

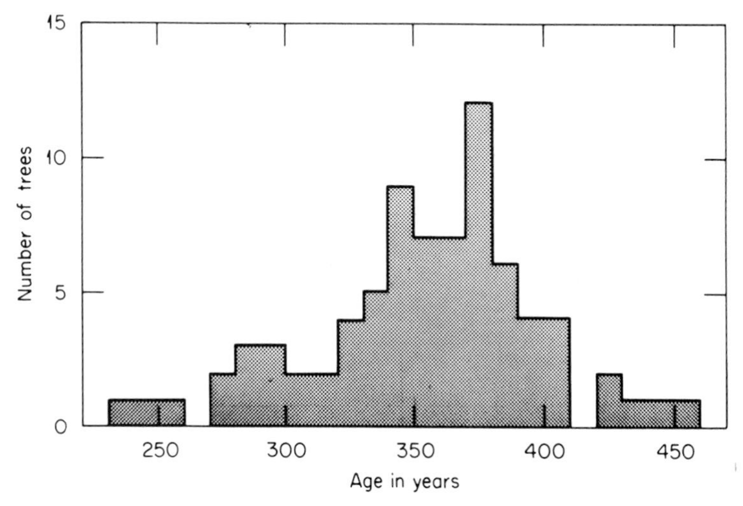

Figure 14.4. Age-class distribution of *Pseudotsuga menziesii* in old-growth stands located around the periphery of the T. T. Munger Research Natural Area in the Wind River Valley, southern Washington Cascade Range.

years ago. *Pseudotsuga* occurred only as dominants in this stand and covered an age span of about 140 years (Fig. 14.3). Stands at Wind River had the widest range of *Pseudotsuga* ages yet encountered—230–460 years (Fig. 14.4).

A number of hypotheses have been proposed to explain these results. Slow recolonization is an obvious possibility. The disturbance episode that occurred around 500 years ago was certainly extensive, covering hundreds of thousands of hectares in the Cascade Range. The episode may also have been exceptionally severe involving repeated wildfires. Such circumstances could result in a lack of seed source. The age distributions resemble those proposed for such a scenario by Harper (1977a). Multiple disturbances are a second possibility. Young *Pseudotsuga* forests are particularly susceptible to wildfire during their first 75–100 years. Partial reburns during this time can provide opportunities for establishment of additional *Pseudotsuga* cohorts. This has been documented in young *Pseudotsuga* stands (Mark Klopsch 1980, pers. comm.) but is very hard to verify in old-growth stands, where there are fewer trees and greater errors in aging. Surface erosion on steep sites or climatic damage early in stand history could also contribute to the wide age range in dominant, old-growth *Pseudotsuga.* Another possibility is competition from shrub or hardwood trees, such as *Ceanothus* and *Alnus.* They could delay conifer establishment or suppress early growth of conifer seedlings and saplings. There is very little evidence for the suppression phenomenon in early growth patterns of the dominant conifers, however.

We currently suspect that lack of seed source and multiple disturbances are both major contributors to the wide age range in *Pseudotsuga* dominants in 400–500-year-old stands. This has certainly been the case in some young stands at Mount Rainier National Park, Washington (Hemstrom 1979). These stands are sufficiently open that recruitment of intolerant *Pseudotsuga* and *Abies procera* is still occurring over 100 years after the last recorded wildfire (Fig. 14.5).

Other age classes of *Pseudotsuga* forests appear to have had different developmental histories. Old-growth forests of around 250 years in age

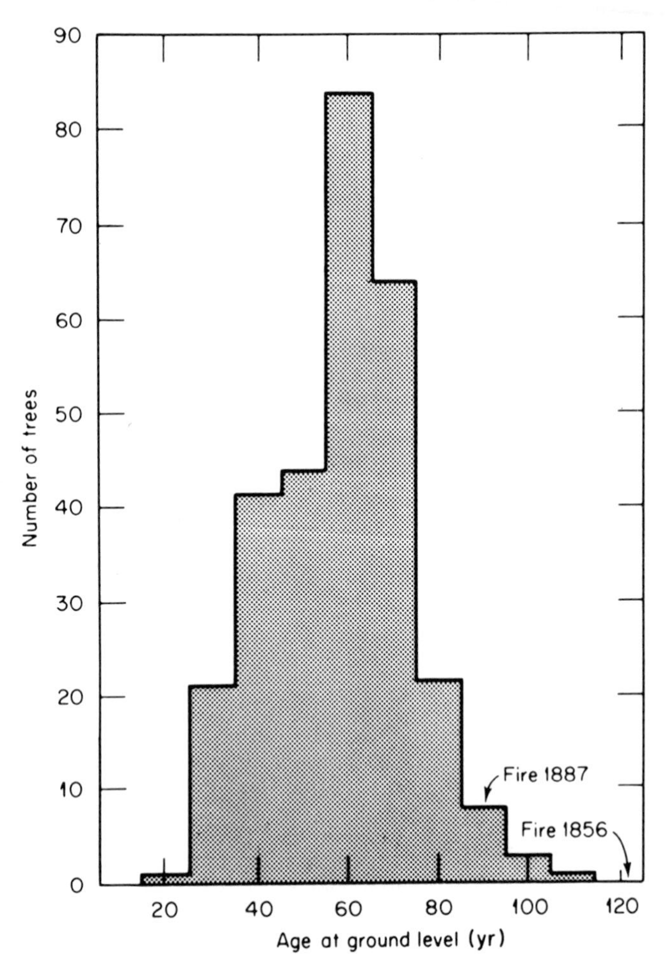

Figure 14.5. Age-class distribution of seral tree species (*Pseudotsuga, Pinus monticola*, and *Abies procera*) in stands developed following wildfires in the late 1800s in the Cowlitz River valley, Mount Rainier National Park, Washington. (From Hemstrom 1979.)

occur at scattered locations in the Cascade Range. These forests have been noted for their relatively high densities of rather uniformly sized *Pseudotsuga* dominants. Tree ages in several stands, were much more tightly clustered than in the older forests (Fig. 14.6). More abundant seed supplies and an absence of reburns would both contribute to the more rapid complete restocking of the site. Age structure analyses of stands on sites of small, single-burn wildfires may reveal even tighter cohorts—comparable to these proposed by Munger (1930, 1940) and Isaac (1943).

Recent data show species composition to be as variable as stocking levels in early stages of forest succession in the Pacific Northwest. Many species play varying successional roles, depending on site and seed source. Shade-tolerant species, such as *Tsuga heterophylla* or *Abies amabilis*, have the potential for direct reestablishment on many disturbed sites. Even more commonly species mixtures, such as *Pseudotsuga* and *Tsuga*, become established on disturbed sites, followed by structural differentiation due to contrasting species growth rates. Dr. Chadwick Oliver and his students at the University of Washington have studied several young forest stands in the Pacific Northwest that follow this pattern, including mixtures of *Pseudotsuga* and *Tsuga heterophylla* (Wierman et al 1979) and *Alnus rubra* and associated conifers (Stubblefield et al 1978).

Analyses of age structure and developmental history are just beginning in the Pacific Northwest, but current work has been revealing. The seres

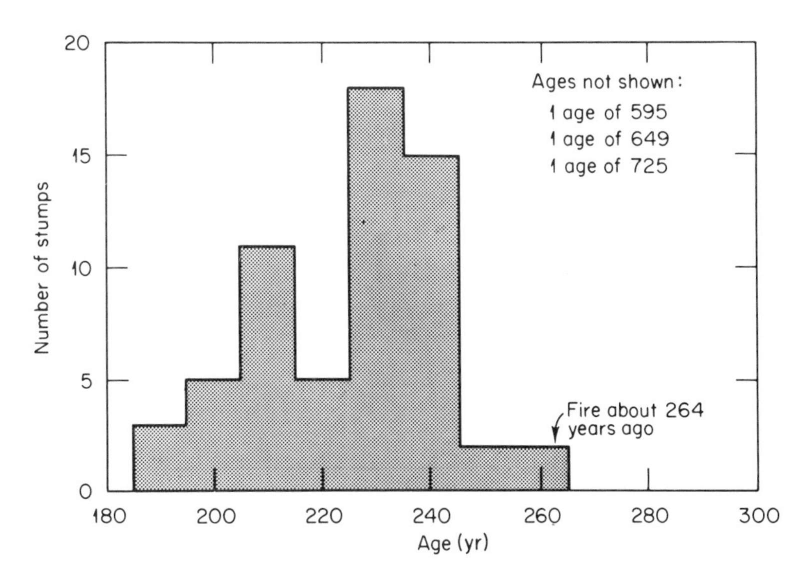

Figure 14.6. Age-class distribution of *Pseudotsuga* in a mature forest near Bagby Research Natural Area, Clackamas River drainage, western Oregon Cascade Range.

following each disturbance episode appear very individualistic. Initial forests have been widely divergent in composition, stocking levels, and times to canopy closure. Forest chronosequences must be interpreted conservatively. For example, a 450–year-old *Pseudotsuga* forest has not necessarily evolved through stages comparable to existing 135–250-year-old stands.

Role of Woody Debris in Forest Succession

Coarse woody debris in the form of snags and down logs is a major structural feature of northwestern forests (Franklin et al. 1981). An average of 194 metric tons/ha was measured in a series of 10 natural stands ranging from 100 to 1000 years in age (Franklin and Waring 1980). Down logs averaged 106.7 metric tons/ha with a range of 60–156 in this data set. The maximum value recorded for down logs was 418.5 metric tons/ha in an intact stand in the Carbon River, Mount Rainier National Park.

Trends in coarse woody debris were weak across the chronosequence of stands due partially to surprisingly high levels in three young (100–150-year-old) stands. Large amounts of dead wood are apparently carried over from previous stand, since massive boles of the fire-killed trees are typically not consumed. This material persists until the replacement stand begins to provide woody debris of comparable size.

Table 14.3. Densities and total numbers of *Picea sitchensis* and *Tsuga heterophylla* reproduction (stems ⩽ 8 m tall) by substrate type and log decay class in a mature forest on a high terrace along the South Fork Hoh River, Olympic National Park, Washington.[a]

	Density (number/m2)		Absolute (number/ha)	
	Picea	*Tsuga*	*Picea*	*Tsuga*
Substrate type				
Picea logs	36.0	15.3	14,500	6170
Tsuga logs	29.9	9.6	15,300	4910
Unknown logs	30.0	7.5	4920	1230
Stumps and root wads	5.1	1.6	550	170
Ground humus	0.08	0.01	700	90
Log decay class				
Early (II)	24.8	15.5	7030	4340
Middle (III)	38.5	10.7	19,000	5250
Late (IV)	28.6	9.2	7800	2500
Very late (V)	28.2	8.2	820	170

[a]From McKee et al. (in press).

Rotten wood seedbeds are very important throughout the coastal *Picea sitchensis* zone (Minore 1972), but nurse logs attain maximum importance in the forests of *Picea sitchensis* and *Tsuga heterophylla* found in the valleys of the western Olympic Mountains. Reproduction in these alluvial forests is essentially confined to log substrates (Table 14.3) (McKee et al. in press). The absence of seedlings on the forest floor contrasts sharply with the dense reproduction crowded onto logs, which occupy only 11% of the area. Logs also vary markedly in their suitability as a seedbed with stage of decay (moderate stages of decay have highest seedling densities) and log species (*Tsuga* logs are poorer seedbeds than those of *Picea sitchensis* or *Pseudotsuga*).

Past patterns in amount and distribution of woody debris may also have strongly influenced the structure of existing mature *Picea-Tsuga* forests in the western Olympic Mountains. Tree densities and stand basal areas and biomass values are less than expected in this climatically favorable forest environment, resulting in relatively open and well lighted stands. The lower densities could result from the fact that suitable seedbeds (logs) were limited at the time of stand establishment and have still not been produced in sufficient numbers to allow development of full stocking. In any case, *Picea sitchensis* reproduces successfully and attains climax status in these relatively open alluvial forests in contrast to its seral role elsewhere in the dense forests typical of the *Picea sitchensis* zone (Franklin and Dyrness 1973).

Long-Lived Seral Species in Succession Modeling

We are interested in developing and utilizing successional models for Pacific Northwestern conifer forests for use in both basic and applied research. The philosophical construct that underlies the JABOWA-FORET family of models appears to us as a concise general statement summarizing our knowledge of succession: vegetational or ecosystem change is a function of life histories of available species interacting with environmental conditions (including disturbances) and with stochastic elements to produce a probabilistic array of alternative outcomes. This construct encompasses the diversity of ecosystems and processes outlined in this book and provides a basis for developing and testing hypotheses. Some of the distinctive features of Pacific Northwest forests require special consideration in a FORET-type model (Shugart and West 1977). Rates and scales rather than types of processes distinguish succession in these forests and require model modification. Specific modeling concerns in the northwestern forests include the following: (1) mortality must be modeled on an annual time step; (2) plot size is critical; and (3) previous stand history is important.

Long-lived trees complicate the task of developing mortality probability functions. Two to four generations of shorter-lived species may germinate, mature, and die during the life span of one *Pseudotsuga, Thuja plicata*, or *Chamaecyparis nootkatensis.* Annual mortality of these short- and long-lived species is not constant between species nor within species by diameter. There seem to be five patterns of suppression-related mortality (Fig. 14.7):

1. Long-lived early seral species, which undergo high mortality rates in small, suppressed size classes but are able to endure long periods of slow growth in large size classes (e.g. *Pseudotsuga, Pinus ponderosa*) (Fig. 14.7a).
2. Short-lived early seral species, which experience constant annual mortality (e.g., *Alnus rubra*) (Fig. 14.7b).
3. Long-lived early to late seral species, which are relatively shade tolerant and experience little suppression mortality in small or large sizes (e.g., *Chamaecyparis nootkatensis* and *Thuja plicata*) (Fig.14.7c).
4. Long-lived early to mid-seral species, which experience moderate suppression mortality in small sizes, reach minimum suppression mortality at about half their maximum diameter, then experience increasing suppression mortality in large sizes (e.g., *Abies procera*) (Fig. 14.7d).
5. Relatively shorter-lived, mid- to late seral, shade-tolerant species, which experience the least suppression mortality at small sizes and slowly increasing suppression mortality in larger sizes (e.g., *Abies amabilis* and *Tsuga heterophylla*) (Fig. 14.7e).

The current model requires that trees die any year based on their estimated annual mortality probability. Because all species are relatively long-lived and at least a few usually survive to old age in a given stand, annual mortality probabilities are very small (Table 14.4). In addition, fires or wind storms, which occur at intervals of 100–1000 years, cause episodes of mortality. Episodic disturbances must be included to accurately simulate patterns of regeneration and growth in understory trees and inputs of dead wood to the forest floor.

Large trees also place some constraints on plot size for simulated stands. Internal computer storage limits require that number of trees modeled as individuals be relatively small. A plot size for simulated stands large enough to prevent a single large *Pseudotsuga* from completely dominating the stand could contain large numbers of small trees. Raising the minimum recruitment size to compensate results in loss of information about regeneration patterns. In addition, the simulated death of an old-growth *Pseudotsuga* in a small plot causes a sharp increase in available light; gap-phase reproduction becomes possible. In reality, a small open area in a matrix of trees 80 m tall does not receive enough light for suc-

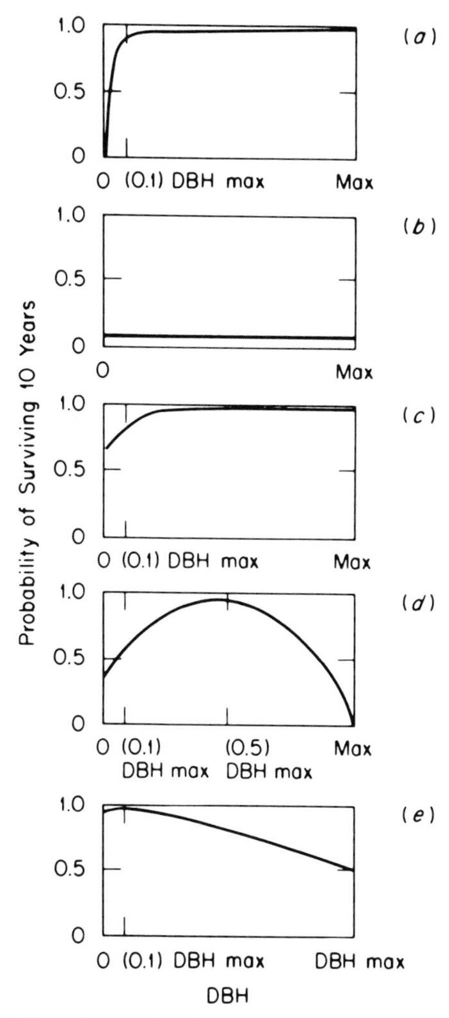

Figure 14.7. Probability of surviving 10 years of suppressed growth, diameter increase <0.1 cm/yr, for five groups of important tree species west of the Cascade Crest, Washington and Oregon. (a) long-lived early seral species; (b) short-lived early seral species; (c) long-lived early to mid-seral species, low suppression mortality; (d) long-lived early to mid-seral species, increasing suppression mortality; (e) long-lived late seral species (dbh, diameter at breast height). Curve a is compiled from McArdle (1949) and a chronosequence of stands in western Washington and Oregon (data on file at Forestry Sciences Laboratory, Corvallis, Oregon).

cessful establishment of intolerant species. Unrealistically high levels of intolerant species regeneration may occur in closed forest situations. Eventually, stand models should be altered to accept larger plot sizes,

Table 14.4. Example of annual mortality in *Pseudotsuga menziesii*.[a]

Age (yr)	p_1[b]	p_{10}[c]
20-50	0.940	0.586
50-100	0.981	0.825
100-150	0.996	0.961
150-250	0.999	0.990
250-450	0.996	0.961
450-750	0.995	0.951
750-1000	0.994	0.942

[a]Compiled from McArdle (1949) and unpublished data on file at the Forestry Sciences Laboratory, Corvallis, Oregon.
[b]Probability of surviving 1 year of diameter growth less than 0.1 cm.
[c]Probability of surviving 10 years of diameter growth less than 0.1 cm/yr.

perhaps by modeling cohorts of small trees and then graduating them into the stand as individuals at a particular diameter.

Finally, stands and disturbance over the previous centuries have an important bearing on the development of future stands by determining the availability of log seedbeds and seed sources. Currently, the model assumes time zero to be bare ground with all species available for establishment. As emphasized by Franklin and Waring (1980), abundant dead wood carries over from one stand to another even though sequences of severe wildfires or other disturbances. Thus, adjustments in the model must be made to account for variable initial conditions that follow natural disturbances. Disturbance intensity and frequency are important determinants of those conditions.

Conclusions

1. The forests of coastal Pacific Northwest differ from forests in other temperate and boreal regions in several respects. Trees can grow very large and old. Leaf areas can reach 20 m^2/m^2, and above-ground, live biomass values can exceed 1500 metric tons/ha.
2. Northwestern forests recover slowly after disturbance. Canopy closure may not occur for 40 years or more. Early seral intolerant trees may have an age range greater than 100 years in one cohort.
3. Due to infrequent, catastrophic disturbance and long-lived seral species, climax conditions rarely occur when intolerant, early seral trees are important in the initial stand.
4. Late seral or "climax" species are often important pioneers.

5. Stand history, as it determines the availability of log seedbeds and seed sources, plays an important role in determining the composition and structure of future stands.

6. The longevity and size of the early seral species pose interesting problems in developing FORET-type successional models, particularly in modeling mortality, determining plot size, and verifying models.

7. The northwestern forests provide the maximum expression of many forest features as well as trends observable in other temperate forest regions. These include size and longevity of species, biomass accumulations, and overall importance of coarse woody debris. General successional theories must take these northwestern phenomena into account and may find their ultimate tests in this region.

Chapter 15

Secondary Forest Succession on the North Carolina Piedmont

Norman L. Christensen and Robert K. Peet

Introduction

In 1916 F. E. Clements outlined a set of general principles that he felt described not only the events but also the mechanisms of succession. In Clements' view, succession was a directional, predictable process of community change driven by the biotic reactions of dominant species converging onto some stable community composition and structure that was ultimately determined by the climate. The concept of the monoclimax or climatic climax was judged to be less than useful by a few ecologists soon after its exposition (DuRietz 1930; Gleason 1917; Tansley 1920) and by the majority of ecologists by the 1950s and 1960s. However, many of the other tenets of this model were, until recently, widely accepted. Indeed, Odum's (1969) "strategy of ecosystem development" presupposes most of Clements' doctrines.

It is clear from other chapters in this book as well as recent reviews (e.g., Connell and Slatyer 1977; Drury and Nisbet 1973; Egler 1977; Horn 1974; Pickett 1976) that all aspects of the classical successional paradigm are in need of reexamination. In particular, several workers have questioned the notion that succession is necessarily a replacement sequence driven by autogenic environmental modification. Alternatively, Drury and Nisbet (1973), Egler (1954), and Niering and Egler (1955) suggest that eventual dominants invade early in succession and simply outlive early succession species. If this is the case, initial conditions and historical features play a large role in vegetational development. Furthermore, successional convergence under these conditions might be too slow to be of practical significance. We should point out, however, that most of Clements' generalizations were based on observations of primary succession. Indeed, intensive study of forest regeneration following disturbance (i.e., secondary succession) began for the most part after the publication of Clements' 1916 exegesis on succession.

It is perhaps worth asking why a theory that enjoyed a long period of support should so suddenly come under critical scrutiny. The answer probably lies in the fact that appropriate data to rigorously test questions such as the importance of competition or the extent of successional convergence were, until recently, nonexistent. The traditional approach to the study of succession has been to compare representative sites of different ages. Often, a single stand is chosen as a sort of type specimen for a particular age class. Given the slowness of forest succession in particular, it is hard to fault this approach as a means of describing general successional trends. However, it is now widely recognized that this approach provides little information concerning the nature of successional change and virtually no information concerning the extent and nature of variability in the process. For example, the presence of oak seedlings in young pine stands might be cited as evidence for Egler's differential longevity hypothesis, even though the evidence is purely circumstantial without a clear knowledge of establishment and mortality rates.

In Clements' view, biotic reactions of seral species result in environmental alterations (i.e., soil development, microclimatic change) and in convergence to an association of species determined by regional climatic patterns. Thus, vegetational variability should initially be high owing to variations in initial site conditions, and should decrease as the stands mature. Margalef (1968), in contrast, suggested that the question of convergence might better be asked with respect of the predictability of community properties as an ecosystem develops. In his view, early successional composition would be highly stochastic, converging to a predictable steady state.

It has proved useful to view climax vegetation as a continuum changing in relation to a complex set of environmental variables. However, this concept has not entered into considerations of successional convergence. With respect to Clements' view of convergence, vegetational diversity along environmental gradients (beta diversity) should diminish as succession proceeds. Using Margalef's definition, we might ask to what extent vegetation-environment relations of successional stands converge towards the vegetation-environment relations characteristic of mature vegetation. Put another way, how predictable is community composition given a knowledge of initial environmental conditions?

In this chapter we examine the nature of successional change as revealed by the study of permanent sample plots over a 50-year period and an analysis of data from an additional 176 stands of varying site characteristics and successional age. After a brief review of previous work in this region, we consider the population biology of the early dominant pines and present a general model of pine stand development and decay. We then examine the relationship between pine population changes and patterns of hardwood establishment and specifically address the mechanisms of succession proposed by Connell and Slatyer (1977). Last, we measure successional convergence, using a comparative study of vegetational gradients at different successional ages. We shall show that vegetation in this region neither converges nor diverges in a monotonic fashion. Rather, the degree to which vegetational patterns of successional communities resemble those of mature stands is related to changes in resource availability and disturbance during succession. Resource availability may, in turn, vary as a consequence of changes in populations of the early dominants.

Past Studies on the North Carolina Piedmont

By the mid-nineteenth century, most of the arable land on the Piedmont of North Carolina was under cultivation. The remainder was exploited for wood or grazed by domestic stock. Poor farming practices and economic hard times resulted in sporadic land abandonment over the subsequent 130 years. Thus, much of the southeastern Piedmont is today a mosaic of fields and forests in different stages of secondary succession.

The early herb-dominated, old-field communities of this region were first described by Wells (1932), Crafton and Wells (1934), and later, in a more quantitative fashion by Billings (1938) and Oosting (1942). Because vegetation change proceeds quickly during this stage, old fields have been very useful systems for the study of many aspects of vegetation change. Indeed, much of the conventional wisdom of succession has had its primary tests in such systems. Regarding mechanisms of succession in old fields, Keever's (1950) work is important in that it verified the importance of dispersal and life-history characteristics in the observed species sequence.

Pinchot and Ashe (1897) described the structure of successional forest stands in the Piedmont, but they did not appreciate their dynamic nature. Clements (1916) and Wells (1928, 1932) agreed that the climatic climax for this region was oak-hickory forest and that the more common pine forests were anthropogenic and would eventually be replaced by hardwoods.

The first quantitative study of forest development on the southeastern piedmont was that of Billings (1938), which was followed closely by the studies of Coile (1940) and Oosting (1942). Confining his analyses to shortleaf pine (*Pinus echinata*) stands of a single soil series, Billings showed that hardwood invasion was correlated with changes in soil organic matter content, porosity and water-holding capacity, and the thickness of the humus layer. Coile (1940) concluded that changes in soil characteristics were not related causally to forest development in stands of loblolly pine (*Pinus taeda*), which is in general agreement with the work of Hosner and Graney (1970), who concluded that soil changes associated with succession are likely the "result of and not the cause of successional patterns."

Oosting's 1942 analysis of the plant communities of the North Carolina Piedmont remains the basic reference on forest vegetation of this area. Oosting divided successional vegetation into upland and lowland categories and analyzed stands of varying age in each category. Although species composition was variable, the same general trends emerged as those observed by Billings (1938): rapidly diminishing pine reproduction and gradual invasion of broad-leaved hardwoods. Oosting did, however, describe distinct variations in composition between bottomland and upland stands. Barrett and Downs (1943) also studied numerous shortleaf and loblolly pine stands throughout the Piedmont and made similar observations.

Much research on the autecology of forest succession has been done on the southeastern Piedmont. Kramer and Decker (1944) demonstrated that broadleaved hardwoods have lower light compensation points than loblolly pine and can better tolerate the lower light intensities beneath the pine canopy. Prior to this work, Korstian and Coile (1938) had shown root competition to be important for establishment and survival at all stages of succession. They observed decreased mortality and increased vigor of seedlings of a variety of species in trenched plots beneath forest stands ranging from young pine to mature oak-hickory forest. Subsequent research revealed that higher rates of photosynthesis under low light gave hardwoods an advantage over pines in the competition for water and nutrients (Oosting and Kramer 1946; Kozlowski 1949; Ferrell 1953). The importance of seed dissemination and rapid establishment of pines in the early stages of succession was clearly demonstrated by Bormann (1953). He further noted that pines, because of their ability to compete for water under high light conditions, were better adapted to old-field conditions than most hardwoods.

Species that characterize late successional stages appear early in succession, sometimes in old fields (Oosting 1942; Barrett and Downs 1943). However, relatively little information is available on the dynamics of these populations. Such information, which requires permanent plot studies, is necessary in order to accurately characterize the nature of successional change. We have reported preliminary results of such studies (Peet and Christensen 1980a) and will discuss results pertinent to this article in a later section.

Little attention has been given in any forest ecosystem to the variability in secondary succession resulting from variation in site quality and history, or the composition of adjacent vegetation. Oosting (1942) was aware of this variation and noted that: "Deciduous species coming in under pine are consistently the same, but their time of appearance, numbers, and proportions are radically different." Variation in piedmont forest vegetation related to soil, topography, and hydrologic conditions has been described by several workers (Applequist 1941; Oosting 1942; Bourdeau 1954; Della-Bianca and Olson 1961; Dayton 1966; Nehmeth 1968; Peet and Christensen 1980b). Duncan (1941) and Bourdeau (1954) found that properties of seedling growth and root development were important determinants of which species of oaks invade a given site. Diftler (1947) found relationships between variation in hardwood invasion and soil texture, although too few sites were studied for firm conclusions. Holman (1949) and Coile (1949) found more convincing correlations between stand hardwood invasion, and both site index and overstory pine density. Density of *Cornus florida* and *Liquidambar styraciflua* were positively correlated, and that of *Juniperus virginiana* was negatively correlated with site index. *Quercus alba*, *Q. rubra*, and *Q. stellata* were more prevalent in the understory of less densely stocked pine stands.

Much of our work in the initial phases of our study was directed toward a thorough documentation of vegetational variation in all stages of secondary forest succession. Both herbaceous and woody plants were sampled, and the techniques of sampling and analysis, as well as a description of the major vegetational patterns among 105 hardwood stands, are described in Peet and Christensen (1980b). Our data indicate that both herbs and trees are segregated in mature forests along gradients of moisture availability and soil fertility. As might be expected, herbaceous plants appear to show greater environmental fidelty than trees whose distributions reflected disturbance history.

Population Biology of Pines

Except for sites with limited seed input, establishment of pine seedlings during old-field succession is compressed into a rather short time span. Thereafter population change is largely a consequence of growth and mortality. Because pines account for well over 90% of the total biomass and production during roughly the first 100 years of forest succession, changes in their populations almost by definition influence stand development in general. We have reported elsewhere on the specific patterns of pine populations (Peet and Christensen 1980a; Peet Chapter 20) and on the relationships between those changes and stand production. Here we discuss a general model for such changes. This model is based on data from 12 permanent sample plots in pine stands of differing ages and initial densities monitored at 5–8-year intervals over a 50-year period (see Peet and Christensen 1980a for details).

Changes in pine density and total pine biomass through time are illustrated in Fig. 15.1 for six permanent plots of differing initial densities. Notice that the initiation of mortality in these stands is density dependent, beginning first in the most densely stocked stands. The linearity of the semilogarithmic mortality curves indicates that a constant proportion of the population is dying each year once mortality is initiated. There is no

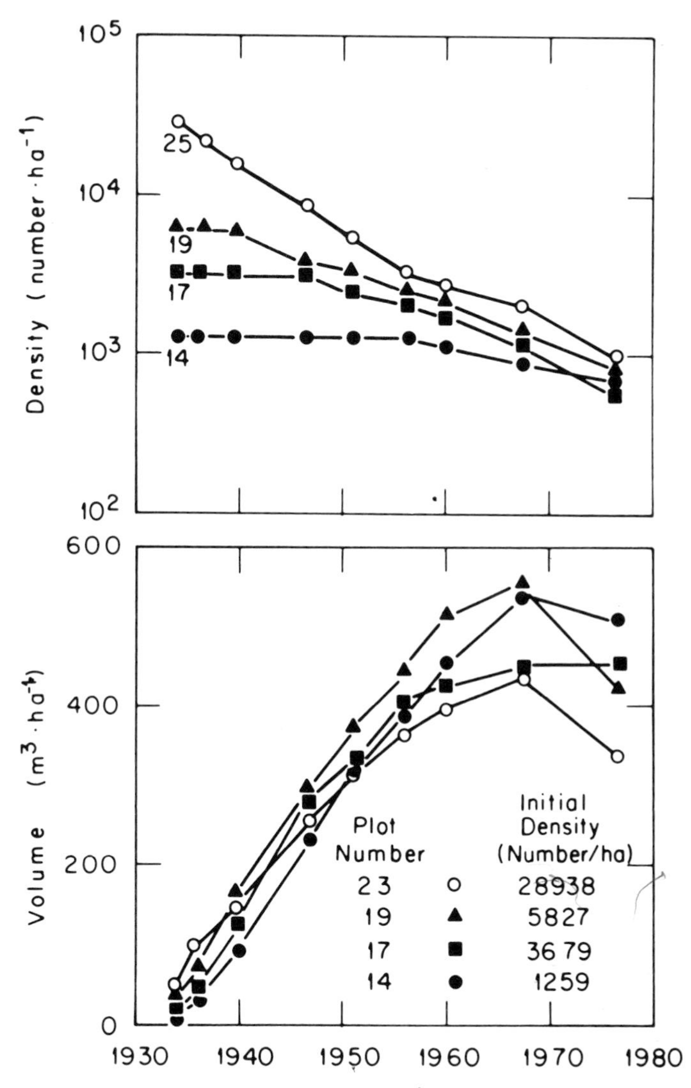

Figure 15.1. Changes in total tree volume (m^3/ha) (a) and log density (stems > 1.0 cm dbh/ha) (b) in four pine stands with initial densities as indicated. At the initiation of this study in 1934 these stands were each 8 years old.

evidence of an eventual release from this constant death, such as has been reported by Schlesinger and Gill (1978) for *Ceanothus.* To the contrary, plots that were experimentally thinned showed no change in the slope of the mortality curves (Peet and Christensen 1980a). Pine biomass has a logistic response through time, reaching an apparent maximum as trees mature and reach limiting dimensions (particularly height). The rate of volume increase was initially highest at the highest densities, probably reflecting greater photosynthesizing area. However, the rate of increase subsequently drops in these high-density stands below rates for stands of lower initial densities. The actual volume attained at the asymptote was proportional to the soil site index (Coile 1949) but may also be affected by initial density (Peet Chapter 20). We are continuing to investigate the relative importance of site and density on stand yield.

Stand development can perhaps be better visualized by looking at changes in individual tree biomass with changes in tree density on a double log scale (Fig. 15.2). These curves show only tangental conformity to the classical $-3/2$ thinning law of Yoda et al. (1963). Rather, the slope of the density-mean biomass relation changes as the stand matures, reflecting, we believe, changes in biotic and abiotic environment and the capacity of the pine population to respond to these changes.

Initially, individual tree growth is rapid and mortality is low, undoubtedly reflecting initially high levels of resources and low levels of competition. During this phase some individuals of hardwood trees such as *Liquidambar styraciflua* and *Liriodendron tulipifera* also become established. As density-induced mortality begins, the slope of the thinning curve approaches -1.5 and may tract along that slope for a few decades. This situation is thought to reflect increased competition for resources in a limited space (Harper 1977a). As the stand approaches the total biomass asymptote, the slope increases to -1.0. During this phase mortality is matched by growth of the survivors. Eventually, the curve becomes less steep than the -1.0 slope, a period during which each tree death releases considerably more resources than can be compensated for by increased production of surviving pines.

Thus, we can characterize development of the pine forests in terms of an initial phase of rapid recruitment and growth owing to available resources and space, a second phase of competition coupled with net production ($-3/2$ thinning), a third stage of reciprocal growth in relation to thinning, and a fourth phase of population degeneration. Resources should be most available during the first and fourth phases. The timing of these various phases is dependent, in part, on initial pine density. Phase 1 is brief in densely stocked stands and long in poorly stocked stands. Furthermore, Phase 4 arrives much sooner in stands with initially high densities. We do not yet have abundant data from sites of differing quality, but, based on observations in other tree populations (Ilvessalo 1937; Yoda et al. 1963), high site quality should lead to increased production, which is expected to accelerate the thinning process.

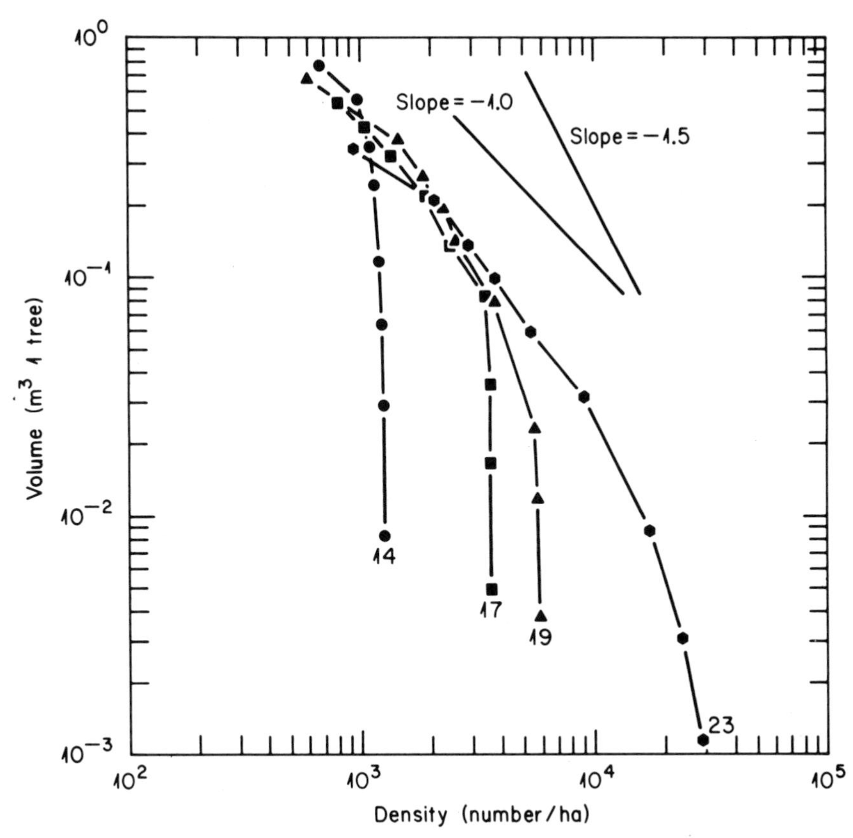

Figure 15.2. Changes in log mean tree volume relative to log density in four stands of differing initial density. These plots are based on data presented in Fig. 15.1.

Patterns of Hardwood Establishment

A rigorous evaluation of Egler's differential longevity hypothesis or Connell and Slatyer's alternative successional mechanisms requires permanent mapped seedling plots established in stands of varying successional age coupled with appropriate experimental manipulations. We have such studies in progess now, but several years will be required for their completion. In the interim, observations from our comprehensive vegetation sampling and long-term permanent sample plots (trees $\geqslant 1$ cm dbh) can provide us with some insight into these questions.

Table 15.1 lists the changes in constancy (percent age of occurrence in 0.1-ha plots) for the ten most common hardwood species in different age stands. Large-seeded, animal-dispersed tree species, such as oaks and hickories, are relatively infrequent in young pine stands and increase in constancy through succession. White oak, while relatively uncommon in

Table 15.1. The fraction of stands (0.1 ha) within which a species occurs (constancy) at different successional ages.[a]

Species	20-40 years N=18	40-60 years N=22	60-80 years N=38	80+ years N=20	Hardwoods N=74
Acer rubrum	1.0	1.0	1.0	1.0	1.0
Carya glabra	0.33	0.55	0.74	0.86	0.92
Carya tomentosa	0.35	0.55	0.79	0.93	0.94
Cornus florida	0.88	0.90	0.98	1.0	0.96
Liquidambar styraciflua	1.0	1.0	0.97	0.93	0.46
Liriodendron tulipifera	0.94	0.86	0.74	0.86	0.68
Oxydendrum arborea	0.64	0.64	0.74	0.93	0.80
Quercus alba	0.23	0.59	0.89	0.93	1.0
Quercus rubra	0.55	0.67	0.86	0.93	0.94
Quercus velutina	0.71	0.77	0.86	0.93	.72

[a]N refers to the number of stands sampled.

young pine stands, is ubiquitous in old pine and hardwood stands. Those young pine stands with populations of large-seeded trees were those with a nearby source of seeds. Species that produce abundant wind-dispersed seeds, such as *Liquidambar styraciflua*, are ubiquitous in immature stands and have more restricted distributions, often favoring mesic or moist sites, in mature vegetation. *Liquidambar* and *Liriodendron*, in particular, are considerably less shade tolerant than oaks and hickories and do not successfully establish from seed beneath closed canopies. The absence of oaks and hickories from many young pine stands and their eventual high frequencies in old stands suggest that early invasion is not a necessary prerequisite for eventual colonization. Similarly, we might conclude from *Liquidambar* that early establishment is no guarantee of later success. *Acer rubrum* also supports this view as it is the most abundant seedling species in most pine types, but is virtually never encountered as a canopy dominant. While these data say little about the potential effects of early pine dominance on hardwood establishment, they do emphasize the role of seed dispersal and the large-scale vegetational mosaic (that is, the proximity of seed sources) in determining the composition of successional stands.

The response of hardwoods in pine stands of varying initial pine densities is indicated in Table 15.2. No patterns were evident with respect to hardwood stem densities: however, hardwood basal area was greatest in the stands with the highest (Plot 23) and lowest (Plot 12) initial pine stockings. The hardwood communities of these stands differed markedly in composition and structure. The broad-leaved trees of Plot 12 were mostly wind-dispersed species (*Liquidambar*, *Acer*, *Liriodendron*), which

Table 15.2. Hardwood densities and basal areas in pine stands of different initial densities.[a]

	Density (no./ha)			Basal area (m^2/ha)		
	Pines		Hardwoods	Pines		Hardwoods
Plot number	1934	1978	1978	1934	1978	1978
12	617	493	914	2.6	48.2	3.5
14	1259	667	1038	4.6	39.6	1.8
15	1950	642	1704	7.9	39.5	1.9
17	3654	815	395	8.5	34.5	0.9
19	5802	592	815	9.9	27.8	1.7
21	10617	1111	445	14.4	39.4	0.6
23	28173	938	1248	17.6	29.0	5.3

[a]Pines were 8 years old in 1934.

presumably established early in stand development. The large size of many of these trees (many > 30 cm dbh) is also testimony to their early establishment. Plot 23 is populated by many larger-seeded species, in particular *Cornus florida*, with few individuals larger than 12 cm dbh. It seems possible that these differences may be due in part to differences in the timing of resource availability, which arise as a consequence of these initial differences in pine density. In the poorly stocked stands, there appears to be a longer period of resource availability early in stand development that might allow increased establishment and growth of small seeded hardwoods. Hardwood density in the heavily stocked stands was initially very low, and most hardwoods appear to have established more recently, perhaps as thinning intensified and the pine population entered the final phase of the thinning curve. These conclusions are, of course, tentative, and alternative hypotheses are possible. However, the conclusions do point to the potential importance of population changes in the dominant pine population in determining the development of the hardwood communities.

Connell and Slatyer (1977) have suggested that succession can be viewed as the result of one of three mechanisms: facilitation, tolerance, and inhibition. Our data suggest these hypotheses to be overly simplistic. Certainly all three processes can be viewed as operating at one time, and the relative importance of each mechanism is not as easily tested, as Connell and Slatyer claim. If, for example, pines were prevented from becoming established or were removed early in succession, a hardwood community would form dominated by small-seeded, "weedy" trees such as *Liquidambar* and *Liriodendron*, rather than the late successional oaks and hickories (Bormann 1953). In a number of our permanent plots, pines were experimentally thinned in various ways to a basal area of 23m^2/ha.

The result in each case was accelerated growth among the already established trees with no increased rate of recruitment. Indeed, removal of the largest pines actually resulted in the diminished establishment of new trees, particularly species more characteristic of later successional stands (Peet and Christensen 1980a). In contrast, extreme canopy removal stimulated regeneration. The pines appear not only to interact with the understory tree species directly, but also seem to greatly affect competitive relations among these species. Large-scale removal of pine trees favors shade-intolerant competitors. The more typical process of gap replacement would logically favor tolerant species. Although pines are, in a general sense, inhibiting invasion of species such as oak, their inhibitory effects on species such as *Liquidambar* may in the long term facilitate oak establishment. Furthermore, establishment of large-seeded species may depend heavily on establishment of substantial animal populations, which, in turn, may depend on the development of the pine forest.

Based on these data, invasion and establishment during succession can clearly be seen to be episodic, the episodes correlating with changes in resource availability. Early invaders affect subsequent replacement probabilities in complex ways precluding description of successional processes in simple terms. The best we are likely to be able to do is to recognize the changing importance of different processes during succession. In the old-field stage, before biotic interactions are as important, the tolerance model would seem appropriate. In young to intermediate age pine stands inhibition predominates owing to high pine density. This is followed by a facilitation-dominated period as the pine canopy breaks up. As a mature hardwood forest develops, all three mechanisms are probably important.

Vegetation-Environment Relations

Because vegetation varies in a continuous fashion, the question of successional convergence might best be asked with respect to the extent of convergence of the successional stages with the vegetational mosaic or gradients that characterize mature vegetation. We have completed an analysis of over 230 forest stands representing all successional stages in as many environmental situations as possible. These data are based on 0.1-ha plots, within which all trees were tallied and herbs sampled. The details of sampling are given in Peet and Christensen (1980b). In addition, measurements were made of soil physical and chemical factors in each plot. These included exchangeable calcium, magnesium and potassium, pH, phosphate, organic matter content, and water retention at pressures of 1 and 15 bars. Ordinations were performed on the herbs (all plants $<$ 1 m tall) and trees (woody plants $>$ 1 m tall) separately, using detrended correspondence analysis (DECORANA, see Hill 1979). In addition to being statistically more robust than other ordination techniques (Gauch et al. 1977), correspondence analysis has the virtue of providing simultaneous species and stand coordinates in the same statistical space. Stand scores thus derived were compared to variations in environmental factors, using Spearman's rank correlation. In addition, canonical correlation was performed between the matrix of environmental data and the first and second ordination axes.

Our analysis of the hardwood stands (a total of 105 plots) revealed that site moisture conditions and cation availability (calcium, magnesium, and pH) correlated highly with vegetational variation (Peet and Christensen 1980b). To assure comparability we removed from consideration extreme sites, unrepresented in our successional stands (bottomland, riparian, bluff, and monadnock hardwood stands), leaving us with 74 hardwood stands. The first axes of an ordination of herbs and of trees from these remaining stands correlated very strongly with pH (also calcium and magnesium availability; Tables 15.3 and 15.4).

Pine stands were stratified into four age classes (based on ring counts): 20–40-year-old, 40–60-year-old, 60–80-year-old, and 80 or greater-year-old stands. Similar ordinations and correlation analyses were performed within each age class. As in the hardwoods, the first axis scores in each case correlated most strongly with pH.

We shall use this relationship to pH as a tool to compare species environment relations. However, it should be noted that pH was highly positively correlated with most other soil cations, and reflects a complex soil fertility gradient (Peet and Christensen 1980b).

Table 15.3 summarizes the results of the ordinations of tree data for each successional age class and the relationship between first-axis scores and pH. Eigenvalues for the second and third axes were successively lower than the first, indicating that they accounted for less variation in the data set. The number of half-changes in species composition along the first axis (Hill 1979), a measure of beta diversity (Whittaker 1972), is also indicated. We shall use this measure as an index of vegetational change along the first axis. The correlation between first-axis scores and pH is clearly highest in intermediate-age pine stands (40–60 years) and diminishes to insignificance in old pine stands (80+ years). In these old stands, there was a weak significant correlation between second-axis scores and pH. Beta diversity (as measured by half-changes in species composition) along the first axis remained relatively constant from age to age, with a peak in 40–60-year-old pine stands.

The results of similar analyses of herb data are shown in Table 15.4. As with the trees, the relationships between first axis scores and pH are most intense in intermediate-age pine stands and hardwood stands, with canonical correlations coefficients greater than 0.90 in each case. With respect to the herbs, there is a general increase in beta diversity with successional age. This implies that young pine stands are relatively similar to each

Table 15.3. Relationships between first axis ordination scores for trees and variations in pH (see text for explanation).[a]

Age-class years	N	1st-axis eigenvalue	1st-axis half-change	Spearman rank r	Canonical correlation coefficient	Range of pH
20-40	18	0.237	2.1	0.61	0.65	4.1-6.2
40-60	25	0.407	2.9	0.74	0.83	3.8-5.9
60-80	38	0.245	2.3	0.41	0.75	3.7-6.5
80+	20	0.200	1.9	0.27 (ns)	0.18	4.1-5.5
Upland hardwood	74	0.290	2.3	0.59	0.85	3.7-6.2

[a]Half-change refers to the number of half-changes in species composition along the first ordination axis. Canonical correlation coefficient is the r value for pH and the first canonical variable-based axis 1 and 2 of the ordination. "Ns" refers to nonsignificant correlations ($p < 0.05$).

Table 15.4. Relationships between first axis ordination scores for herbs and variations in pH (see text for explanation).[a]

Age-class years	N	1st-axis eigenvalue	1st-axis half-change	Spearman rank r	Canonical correlation coefficient
20-40	18	0.406	1.85	0.71	0.78
40-60	25	0.358	2.72	0.90	0.97
60-80	38	0.272	2.32	0.86	0.88
80+	20	0.273	2.61	0.28 (ns)	0.38
Upland hardwoods	74	0.350	3.26	0.74	0.94

[a]Half-change refers to the number of half changes in species composition along the first ordination axis. The canonical correlation coefficient is the r value for pH and the first canonical variable based on axis 1 and 2 of the ordination. "Ns" refers to nonsignificant correlations ($p<0.05$).

other with respect to the herbaceous species as compared with older pine stands and hardwoods. In this respect succession might be better viewed as a divergent process. Immature stands are composed of somewhat more weedy, widely dispersed species. These may be eventually replaced by species with greater site specificity, thus increasing beta diversity.

Both the herb and tree data suggest that the fidelity of the vegetation-environment relationship increases from young to intermediate-age stands, decreases markedly in old pine stands, and increases again in hardwood stands. In this respect, vegetation can be seen to go through periods of convergence and divergence (*sensu* Margalef 1968) in the course of succession. In a general sense, these observed changes in the segregation of species with respect to environment parallel the observations of Werner and Platt (1976) regarding the segregation of *Solidago* species along a soil moisture gradient in old fields and mature prairie. They found predictable patterns only in the mature prairie. In the case of the forests discussed here, we should note that predictability does not increase monotonically with succession but rather oscillates during succession, presumably as resources become more or less available.

Because fewer 20–40- and 80+-year stands were sampled than in other age classes, it is possible that some of the observed differences are a consequence of sample size. Also a narrower range of site conditions relative to pH was sampled in the 80+-year class. However, an analysis of each age class using only stands with pH values within the range of those of the 80+-year class did not markedly alter the results.

Canonical correlation analysis using the first and second ordination axis resulted in higher correlation coefficients in a few cases than those obtained by univariate comparison of the first axis and pH.

The data presented in Tables 15.2 and 15.3 provide a measure of total vegetational response to environment, but how do the distributions of individual species change through succession? Because correspondence analysis generates species and stand scores in the same ordination space, we can compare species scores along a particular axis. In this analysis species scores on the first axis for each age category of pine stands were compared with similar scores for the hardwood ordinations using Spearman's rank correlation (Table 15.5). Correlation with hardwood scores is highest in the intermediate-age stands and lowest (nonsignificant among trees) in old-age (80+ years) stands, a result consistent with

Table 15.5. Spearman rank correlation coefficients for species ranking along the first ordination axis in successional pine stands compared to upland hardwood forests.[a]

Age-class years	N	Trees	Herbs
20-40	18	0.02 (ns)	0.32
40-60	25	0.43	0.47
60,80	38	0.31	0.60
80+	20	0.24 (ns)	0.25 (ns)
Upland hardwoods	74	1.00	1.00

[a]"Ns" refers to nonsignificant correlations ($p<0.05$).

the patterns of convergence illustrated in Tables 15.2 and 15.3. Thus, species distributions along the major axis of variation for an age class most closely resemble those of the mature forest, not in old-age stands, but in intermediate-age stands.

Conclusions

Based on changes in the pine populations, we suggest that availability of resources (light, soil nutrients, and water) decreases in the early stages of succession and is perhaps lowest in intermediate-age stands (40–60 years) as stand biomass approaches its peak. Degeneration of the pine population appears to release resources, although this point will remain speculative until more data are available.

Among the factors affecting the timing of these changes is initial pine density. In densely stocked stands, the initial period of resource availability is brief, perhaps also limiting the period for successful early establishment of hardwoods. However, thinning in such stands is more rapid, and degeneration of the pine population occurs at an earlier stage. Conversely, the initial period of resource availability is long in poorly stocked pine stands, but the dominance of the pine population is prolonged. This appears to allow more establishment, particularly of wind-dispersed hardwoods, early in stand development. Thus, the status of the initial pine population may greatly affect the rate and course of secondary forest succession in this region.

Although more data are still needed on the demography of hardwood trees at various stages of succession, it is clear that early establishment is not a necessary prerequisite nor is it a guarantee of later success.

If successional convergence is measured in terms of the extent of vegetational variation along a gradient, variation is not seen to decrease with age. Among the established trees, such variation remains more or less constant along the sere from young pine to hardwood stands. In the herbaceous layer, greater successional age is characterized by increased site specialization and a general increase in beta diversity.

With respect to the specifics of the relationships between vegetational variation and environmental variation, the question of convergence is considerably more complex. Among pine stands, vegetation-environment relations are most similar to mature hardwood forests in intermediate-age stands and least similar in very young and also old-age stands. Species environment fidelity appears to be highest when resources are most limiting and competition is most intense. As pine stands deteriorate, resources are not only more available, but disturbance sites, such as wind-throw tipups and light gaps, are more abundant. These suggestions remain hypothetical and require further testing, but they do conform to the view that a mechanistic model of succession that predicts changes in community and system level attributes of piedmont forests can be constructed, based on the population biology of the major constituent species.

Acknowledgments

This work was supported by National Science Foundation grants DEB 77-08743, DEB 78-08043, DEB 77-07532, DEB 78-04041, and a grant from the North Carolina Energy Institute. The collaboration of Orin Pete Council is gratefully acknowledged.

Chapter 16

Successional Studies of a Rain Forest in Mexico

Arturo Gómez-Pompa and Carlos Vázquez-Yanes

Introduction

Secondary succession, although examined in various climatic regions, has been most intensively studied in the north temperate zone. The motivation for these studies has been scientific interest in the investigation of vegetational changes through time and space, and not because secondary succession in these areas represents the best expression of changes during the regeneration of terrestrial ecosystems.

In comparison, succession in the hot, humid tropics is more pronounced, occurs in a shorter period of time, and apparently lacks noticeable changes during the year, primarily as a result of the favorable climate. Although the stages of tropical succession are well defined, the number of successional studies in the tropics are few. Representative examples include: Budowski (1961); Kellman (1969, 1970); Symington (1933); Wyatt-Smith (1955); Webb et al. (1972); Opler et al. (1977); Snedaker (1970); Harcombe (1972); Ewel (1971); Blum (1968); and the studies in Mexico, which will be cited in this chapter. Due to the lack of tropical studies, a large part of successional theory is not based on information from the ecological region most favorable for the expression of the successional process.

This simple, but important fact has caused considerable confusion in our understanding of the temporal changes in terrestrial ecosystems following disturbance. In the temperate zone, the length of winter drastically affects the behavior of plant and animal species, resulting in a successional development composed of a series of annually interrupted events. While the process of temperate succession is fascinating, this annual interruption has obscured our understanding of the continuous population changes that occur in a given locality through time after disturbance.

Our studies of rain forest regeneration in the hot, humid tropics of Mexico (Gomez-Pompa et al. 1976) have resulted in the construction of a simple, graphic model, which we feel explains more completely the general process of plant succession in a site. The model is based primarily on the life cycles and survival of individual species in time and space. Figure 16.1 illustrates this model, and presents plant life cycle patterns along a successional time gradient.

This model accounts for all the possible ways that a species could exist in an active form (excluding seeds and other propagules) in a given moment. Also included are the presence and behavior of certain species that are dominants at different successional times. For each case and place, the number and frequency of distinct types of species that are dominants at different successional times. For each case and place, the number and frequency of distinct types of species may vary, but species with well-defined life cycles will always be present in each successional stage. Thus, for example, the first stages of succession will always contain some short-lived species, and a successional stage of 4–5 years will have some long-lived trees present. According to the scheme presented here, some areas may lack specific life cycles or stages. Such is the case in some hot, humid savannas, where small trees or shrubs dominate through long periods of time, even though long-lived trees are predicted through succession. Succession in these areas is arrested by the lack of arboreal germ plasm (Gomez-Pompa et al. 1972). The model allows us to study any site. Use of the floristic composition and life cycles of the species present permits evaluation and prediction of future changes in the vegetation, assuming that environmental conditions do not vary greatly.

This model is incomplete, since it does not include animals or microorganisms, which play important roles in succession. However, these two groups are subordinate to the primary producers. Thus, though incomplete, the model provides a clear idea of vegetational changes through succession. On the other hand, it should be noted that our knowledge of animal population changes and life cycles through secondary succession is rather rudimentary. The same could be said about successional changes in the microbial flora and saprophytes. Clearly, lack of information about these two components of the changing ecosystem restricts our complete understanding of plant life cycles. The presence or absence of a plant species may depend on the presence or absence of certain animals that are dispersers, disseminators and predators, or on microorganisms that may inhibit or stimulate plant growth.

The complexity and nature of succession varies according to the type of ecosystem examined. For this reason it has been impossible, so far, to develop a general model of regeneration that can be applied to all ecosys-

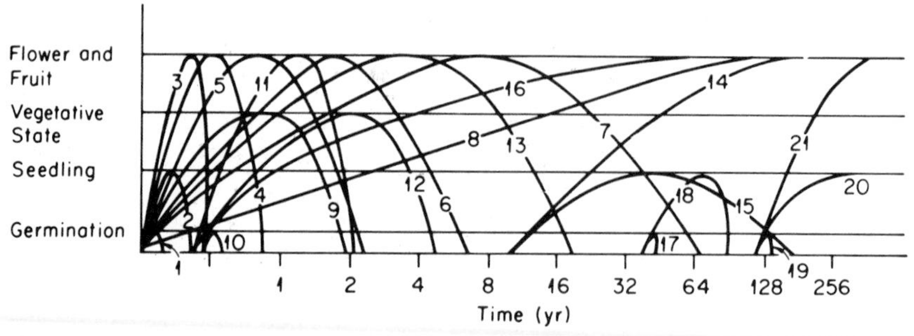

Figure 16.1. A model of life cycle patterns along a time gradient in the successional process. Each one represents a complete or an incomplete life cycle. (1) Species that germinate and die. (2) Species that germinate, produce a few leaves, and die. (3) Species that complete their life cycle in a few months. (4) Annual species. (5) Biennial species. (6) Species with a life cycle of only a few years (less than 10). (7) Species with a long life span, several decades, but eventually dying (species from old secondary forests). (8) Primary species with life spans of hundreds of years that have lived since the beginning of the succession. (9) Species that never reach the stage of sexual reproduction. (10) Species that germinate a few months after the succession begins and soon die. (11) Annual species that germinate after the succession begins. (12) Species that germinate after the succession begins but do not reach the stage of sexual reproduction. (13) Species with a short life cycles (less than 30 years). (14) Species with long life cycle that germinate when succession is well advanced. (15) Species that germinate when the succession is stage for a few years. (16) Species that germinate a few months after succession begins and then have a life cycle hundreds of years long (primary species). (17) Species that germinate and die in old successional stages. (18) Species that germinate and live at the seedling or young plant stage in old successional stages, and then die. (19) Species that germinate and die in the primary forest. (20) Species that germinate and grow to a seedling or "young" plant stage inside the primary forest and remain there, waiting for suitable conditions for continuing growth. (21) Species that germinate and grow in the primary rain forest and may reach the reproductive stage after having long life cycles (primary species). (Taken from Gomez-Pompa and Vazquez-Yanes 1974.)

tems. For example, in extreme environmental conditions with low species diversity, such as in arid regions, regeneration is a simplified process, which consists of the replacement of the disturbed community by another composed of the original species. On the other hand, in the humid tropics, with its diversity of species, ecosystems, and structures, regeneration is complicated. Through time, many diverse, intermediate successional stages occur before the original ecosystem is reestablished.

Regeneration in a Tropical Forest

Rain forests can be disturbed for numerous reasons. Sometimes the changes caused by the disturbance are severe, such as the massive death of the dominant tree strata when the rain forest is cleared for agriculture.

Less severe changes might include the destruction of the herbaceous strata by herbivory. Between these extremes exists a range of activities that disrupt the ecosystem and cause a series of distinct species responses.

The types of disturbance experienced by the primary tropical forest are: lightning, fires, vulcanism, tree falls caused by strong winds or animals (elephants), mortality or loss of vigor of trees resulting from disease, chemical pollutants, herbicides, arborcides, logging to extract forest products, conversion to agriculture or pasture, slash-burn agriculture, floods, and earthquakes.

The effects of forest disturbance vary widely, depending on the severity of the disturbance, the history and size of the area disturbed, and the environmental conditions during the disturbance. The sum of these initial factors, coupled with the available flora of the region, can determine the course of the first stages of regeneration. The series of species that appear through succession depends not only on external factors, but also on intrinsic species characteristics such as the length of the growing cycle, photosynthetic efficiency, and other physiological properties. From this point of view, regeneration in the hot, humid tropics can be thought of as a series of environmental changes through time. In the graphic model shown in Fig. 16.1, time represents an environment. Based on this model, it is possible to predict the existence of patterns of species establishment during a specific ecological succession.

There exist marked differences in the morphology and physiology of early successional species versus those of later seral stages of the primary forest. These differences have been observed and documented by various authors (Richards 1952; Budowski 1965). By studying the life cycles of primary and secondary species, within the context of a species distribution model through time, succession can be viewed as an autecological process and, at the same time, regeneration understood in ways outside the traditional realm of synecology. In addition, this approach permits an analysis of the direct relationships between particular species and their environments. These relations can be examined in two different ways: the first would be to study species adaptations to the environmental conditions of the particular successional stage in which they establish themselves, while the second focus would be to study environmental evolution during succession and its effect on the species of a community. Species interactions during succession play an important role in understanding this process, but their investigation has been extremely challenging, since it is very difficult to achieve a holistic approach to the problem. Nevertheless, this is a very promising area of study from the scientific viewpoint.

As Gleason and Cook (1927, p. 27–33) mentioned in their study of Puerto Rico:

> It may be repeated that plant associations are the basic units of vegetation, that they are the result of immigration and environmental selection, and their duration is short or long, depending on the rate of environmental change . . . .

Each area of ground in which a particular complex of environmental conditions is repeated tends to be occupied by the same groups of species.

Each type of disturbance produces characteristic changes in the biotic and abiotic environments. These changes influence the floristic composition at the onset of recovery, and thereby affect later changes. Thus, for example, a very severe disturbance that eliminates practically all plant propagules in the soil will result in a long and less predictable succession. On the other hand, a small clearing in the rain forest will result in rapid, predictable changes.

Rain forests throughout the world, when disturbed, whether by a tree falling, a small fire, or the opening of a clearing for cultivation, experience a series of rapid changes. These include the "arrival" of a group of species that are adapted to the ecological conditions caused by the disturbance. With time, these species, which Corner (1932) has termed "weed trees," promote the regeneration of the original ecosystem. Generally, this is the process of regeneration or cicatrization (so named by some authors) of the rain forest. This phenomenon has been described, in general terms, on several continents. Pioneer cicatricial species, especially the trees, that characterize the regeneration of disturbed rain forest are similar in physiognomy, physiology, and behavior. Notable examples are the genera *Cecropia* (Blum 1968; Vazquez-Yanes 1979; Schulz 1960) in tropical America, *Musanga* (Lebrum and Gilbert 1954; Aubreville 1947) in Africa, and *Macaranga* (Kartawinata 1977; Kochummen 1966; Whitmore 1975) in Asia.

Cicatrization of the Rain Forest

Floristic dominance changes with time. The first strong change occurs between the first months (characterized by the dominance of short lived herbaceous species) and first two years (according to the climatic zone). The species with short life cycles (Fig. 16.1) eventually disappear, and the vegetation (floristic dominance) changes, resulting in a temporarily stable shrub stage. The next stage is dominated by small trees. Later these species are replaced by larger trees, which dominate the upper story previously occupied by the small trees. This process (Fig. 16.1) has been documented for many tropical areas (Whitmore 1975; Halle et al. 1978). The similarity of the biological forms that dominate different stages, as well as the similarity of many species from related geographical or biogeographical zones, is remarkable.

Five stages in the regeneration of high evergreen rain forests in Mexico have been identified. The first stage is dominated by short ephemeral-lived (weeks or months) species, such as herbs. This stage can last for months and may also contain shrub or woody pioneer seedlings.

However, if the area is used as a pasture, it can remain in an arrested successional stage. The second stage is dominated by secondary shrubs that eliminate the herbs via shading. During this stage, shade species, which require lower temperatures and light levels for germination, may appear. This second stage, depending on the area, may last from 6 to 18 months. It is a period of rapid growth, dominated by short-lived shrubs (*Piper*, *Myriocarpa*, *Urera*, *Solanum*) from 1.5 to 3 m in height. In the shade produced by these shrubs some herbs from the previous stage persist, as well as new species with different ecological requirements. The third stage, which can last from 3 to 10 years, is dominated by secondary trees of low height stature (*Heliocarpus*, *Trema*, *Hampea*, *Miconia*), yet also contains taller secondary trees. These latter species, with heights of 10 m or more (*Cecropia*, *Didymopanax*, *Ochroma*, *Robinsonella*) characterize the next successional stage, which lasts from 10 to 40 years or more. Some primary trees, which ultimately reach heights of 25 m or more, also occur in this stage. These trees comprise the dominant vegetation in the fifth successional stage.

In reality, the process described above is a complex, continuous process, which is difficult to subdivide precisely. Actually, as presented in Fig. 16.1, the process is seen as a continuum through time, which can be characterized at any discrete moment by the type of life cycle of the dominant species.

Vegetation changes through time have traditionally been studied by measuring changes in one site through time and also by sampling neighboring sites with different aged vegetation. Different studies indicate that both these approaches have their limitations: the amount of time required (in the case of Method 1), and spatial variation and uncertainty concerning the original causes of disturbance in the various sites (in the case of Method 2).

Interestingly, the first studies in Mexico on secondary succession were oriented toward obtaining information on a species of enormous economic interest, *Dioscorea composita*, which is a raw material used to produce steroids. It was precisely to learn more about the behavior of this species, which acts typically like both a secondary and primary species, that led us to explore the process of ecosystem regeneration after disturbance in greater depth (Gomez-Pompa and Vasquez-Yanes 1974).

The enormous quantity of information on secondary succession in the tropics is difficult to interpret. The complexity of the process is overwhelming. The occurrence of this phenomenon in nature is readily observed, but difficult to describe with any desirable level of precision. The only thing that seems certain is that the species that appear in a particular stage possess common characteristics. This leads to two totally opposing conclusions. On the one hand, it is practically impossibile to make predictions about an ecofloristic process and difficult to identify indicator species for the different stages (Sousa 1964). On the other hand, there are certain repetitive characteristics, not only for the samples from Mexico, but from other parts of the world, both in the floristic composition and the physiognomy of the species present through succession.

Our first studies generated questions that required another type of investigation. Consequently, the strategy of our subsequent studies included investigations that could be conducted in a relatively short time, analyzing the regeneration process from several aspects (Gomez-Pompa and Vazquez-Yanes 1974). In planning these studies, we started with a very simple, graphic model (Fig. 16.2) for the process of ecosystem regeneration and regeneration triggering after a severe disturbance to the original rain forest. Using this model, we can ordinate our data to answer some of the important questions that arose when analyzing samples from different secondary successional areas. It was decided to study only one region in order to generate a large amount of information on one zone. Eventually, the study could be repeated and broadened to include other ecological zones in the same hot, humid tropical region. In the devised model, the process of regeneration through time culminates with the cicatrization or regeneration of an altered ecosystem.

The graphic model identifies various levels of investigation. One of the first objectives was to understand the causes of disturbance, and the floral and faunal composition of different successional stages in distinct ecological zones. This first level could be classified as the description of regeneration. However, an investigation at this level is insufficient for making generalizations to describe or predict regeneration patterns. We therefore decided to proceed to another level of investigation, which would help us understand the biological phenomenon that affect the presence of a species in time and space and studied patterns of individual behavior. Our investigations were oriented to regeneration of the high rain forest (Gomez-Pompa et al. 1976).

Studies on ecosystem recovery in the tropics have concentrated on areas abandoned after cultivation (old-field succession). Few studies exist on regeneration inside the rain forest. Below we present some data on both types of succession.

The development of vegetation, whether in natural clearings, cut areas, or abandoned fields, can have diverse origins. Vegetative growth from tree stumps, rhizomes, and other surviving vegetative structures can play an important role in vegetation regeneration. In certain cases, particularly in small clearings, seedlings of species removed by disturbance can be quite important.

The rate at which vacant land is covered by sun-loving, rapidly growing species, which are generally scarce or absent from the primary vegetation, suggests that plant development from seeds is a principal mechanism of vegetation regeneration, particularly in abandoned fields. Much has been said about the origin of the seeds of early colonizers. Supposedly, because of the intensity of seed predation in the tropical rain forest, few

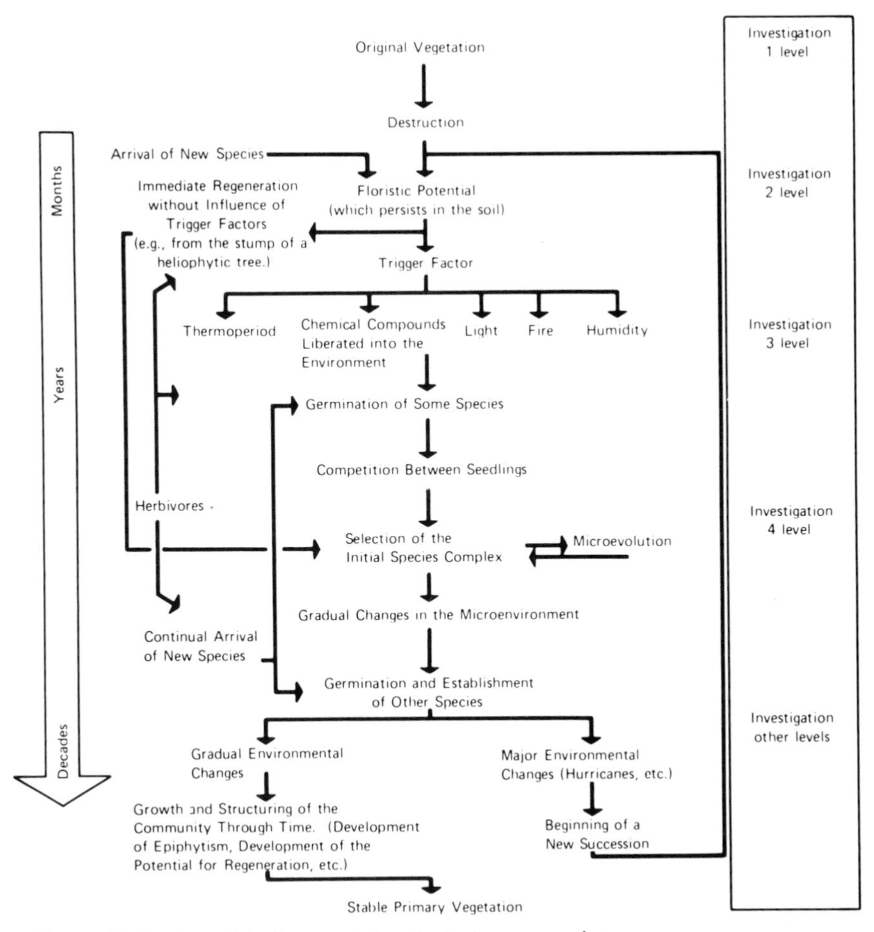

Figure 16.2. A model of research levels during succession.

seeds can survive in the soil. Therefore, colonization must occur by seeds that arrive in the disturbed area from outside. However, the presence of a great number of seeds in the rain forest soil and dormancy studies on seeds of pioneer species appear to refute the above hypothesis. This suggests that a large number of the individuals that colonize denuded soil come from seeds already present in that soil.

Few studies have examined the viable seed content of tropical rain forest soils; among the most important are Symington (1933), Liew (1973), and Ashton (1978) in Malaysia, Keay (1960) in Nigeria, Guevara and Gomez-Pompa (1972) in Mexico, and Blum (1968) in Panama. All these studies arrived at a similar conclusion: the most abundant seeds in the soil are from herbs and sun-loving, rapidly growing tree species, characteristic of large rain forest clearings or early successional stages on devegetated land. Seeds of the majority of mature rain forest species are part of the seed bank only for brief periods of time, during which they tend to germinate or are very rapidly eaten (Table 16.1).

Available data indicate that animal-dispersed pioneer species have long fruiting seasons, frequently lasting all year, and can be dispersed by very diverse species of birds and mammals. Trejo (1976) presented a list of bird species that had in their digestive tracts seeds of pioneer species common in the Los Tuxtlas region. He examined 167 birds of 37 different species and found seeds of *Cecropia obtusifolia and Trema mircrantha.* The digestive tracts of 27 of 97 birds (19 different species) contained seeds of *Cecropia*, and 49 of 122 birds (24 different species) contained seeds of *Trema.* Six bird species had seeds of both species of trees in their digestive tracts.

Wind-dispersed species, on the other hand, normally fruit during the dry season. They produce large volumes of seeds, which are simultaneously dispersed on days with low relative humidities. For the same re-

Table 16.1. Seed species found by germination from soil samples taken in primary and secondary rain forest vegetation in Veracruz.[a]

Primary vegetation soil only	Both primary and secondary soils	Secondary vegetation soil only
Primary species	Secondary species	Primary species
Enterolobium cyclocarpum	*Ageratum conyzoides*	*Brosimum alicastrum*
Sapium lateriflorum	*Amaranthus hybridus*	*Lauraceae* spp.
Turpinia occidentalis	*Bidens pilosa*	
	Clibadium arboreum	Secondary species
Secondary species	*Eupatorium* spp.	*Clibadium grandifolium*
Axonopus compressus	*Eupatorium macrophyllum*	*Dioscorea* spp.
Belotia campbelli	*Heliocarpus* aff. *donnell-smithii*	*Emilia sonchifolia*
Cecropia obtusifolia	*Iresine celosia*	*Eupatorium pensamalense*
Costus spicatus	*Nuerolaena lobata*	*Euphorbia heterophylla*
Croton draco	*Panicum trichoides*	*Heliocarpus appendiculatus*
Desmodium adscendens	*Piper umbellatum*	*Jacobinia* spp.
Erechtites hieracifolia	*Phytolacca rivinoides*	*Lasiacis papillosa*
Eupatorium pycnocephalum	*Robinsonella mirandae*	*Mikania micrantha*
Mirabilis jalapa	*Solanum cervantesii*	*Paspalum* spp.
Pleuranthodendron mexicana	*Vernonia* aff. *deppeana*	*Physalis pubscens*
Sida acuta		*Solanum nigrum*
Trema micrantha		*Solanum torvum*
		Spigelia palmeri
		Urera caracasana
		Verbesina greenmani

[a]Data from Guevara and Gomez-Pompa (1972).

gion, this phenomenon has been observed for *Heliocarpus* spp. (Vazquez-Yanes 1976a) and for *Ochroma lagopus* (Garcia-Gutierrez 1976). This strategy leads such individuals to form a wide and uniform seed shadow. The pioneers most efficient in colonizing clearings and recently devegetated land undoubtedly produce uniform seed shadows over extensive areas; otherwise, it would be difficult to explain the presence of their seeds in soil of the mature rain forest after disturbance. The seed shadow (seed rain) of animal-dispersed species can be produced by efficient dispersal agents, for example, birds and bats, with a wide activity radius, which indiscriminately visit both disturbed and undisturbed areas (Trejo 1976; Vazquez-Yanes et al. 1975).

The most constant and abundant colonizing species of clearings and devegetated land combine wide dispersability, prolonged survival of viable seeds in the soil, and specialized mechanisms for triggering germination. One of the best-known species of this type is *Cecropia obtusifolia*, which produces seeds all year that are dispersed by numerous vectors, are abundant in the rain forest soil, and exhibit a prolonged photoblastic dormancy (Vazquez-Yanes 1979, 1980).

In comparison to the above, the majority of mature rain forest trees normally have large seeds with a high water content that tend to germinate rapidly after dissemination. This might be explained by the fact that seedlings of these species experience shady conditions during the initial moments of life. Therefore, they possess abundant food reserves, which allow them to achieve a certain size and form a large photosynthetic area in order to achieve a rate of net photosynthesis sufficient to maintain life. Rapid germination, moreover, creates seedlings, which due to their active metabolism and lower nutrient content, may be less susceptible to predation than the large and very conspicuous seeds.

We conclude that those seeds that tend to establish themselves in the continuous, stable, and shady rain forest habitat are large and have a short viability and brief dormancy. Species found in a discontinuous, unstable, sunny habitat, represented by clearings, have small seeds with long viability and specialized dormancy. In temperate and dry tropical forests, seeds of the mature vegetation normally have longer viability and specialized dormancy. In these cases, species are adapted to a climate with a long season, unfavorable for plant establishment, which the seeds must survive before they can germinate. Meanwhile, in the humid tropical rain forest, this unfavorable season is usually brief, or nonexistent in certain equatorial regions. Moreno-Casasola (1976) compared seed viability of rain forest and temperate forest species. In the first case, the majority of seeds are viable for only a few weeks, while in the latter, viability can last on the average of one or more years.

With respect to the mechanisms that trigger germination of pioneer species, these show, in general, exogenous dormancy or stress by unfavorable external factors. When these factors are removed, the seed population germinates simultaneously, thereby promoting rapid colonization. The most frequent type of dormancy is photoblastic, caused by an inadequate intensity or quality of light. This type occurs in important species, such as *Cecropia obtusifolia*, *Trema micrantha*, *T. guineensis*, and *Piper* spp. (Vazquez-Yanes 1976a,b, 1978, 1979). Integument dormancy occurs in *Ochroma lagopus*, *Heliocarpus appendiculatus*, and *H. donnell-smithii* (Vazquez-Yanes 1974, 1976a).

The quality of diffuse light within the rain forest inhibits germination due to its richness in far-red versus red light. Opening the canopy allows direct sunlight to enter, which changes the red-to-far-red ratio and triggers germination (Vazquez-Yanes 1976b).

With respect to the longevity of viable seeds in the soil, little information exists for the tropics. Juliano (1940) in the Philippines found that buried seeds of certain tropical weeds can remain viable for 7 years. Castro and Guevara (1976) observed that 21 species germinated from samples of soil stored in the laboratory for 1 year. Among the species that appeared were *Cecropia obtusifolia*, *Piper auritum*, and *Trema micrantha*. Lebron (1979) in Puerto Rico reported that seeds of *Palicourea riparia*, a rubiaceous shrub present in the mature rain forest, but more abundant in cut areas, can remain viable in the soil for at least 3 years.

The seed pool in the soil will generate mixtures of species with different floristic compositions, depending on the treatment received by the soil. One clear example of this is the work of Vazquez-Yanes (1974) on *Ochroma lagopus*. When soil samples were treated with heat similar to that of a forest fire, the number of species that subsequently germinated was less than in unheated soils. *Ochroma lagopus* dominated the seedlings in the treated soils, since the seeds of this species are resistant to high temperatures and are actually stimulated to germinate by heat (Table 16.2). At the same time, if seeds of rapidly growing weed invaders are abundant in a soil, their growth can affect the composition of the vegetation. In an experiment by Lopez-Quiles and Vazquez-Yanes (1976) examining soil samples with known quantities of weed species seeds, *Bidens pilosa* was so aggressive in its growth that it inhibited pioneer species such as *Ochroma lagopus* and *Cecropia obtusifolia*. Germination can generate interspecific competition among the components of the seed pool. This competition, modified by external factors, will ultimately determine the composition of the initial vegetation of a cleared site.

The Role of Pioneer Species in Regeneration

According to Whitmore's (1975) classification, rain forest species can be divided into four groups based on their responses to openings in the canopy. These are: (1) trees whose seedlings are established and grow

Table 16.2. Effect of heat on the composition and number of seedlings that appear in soil samples from a secondary forest of *Ochroma lagopus*.[a]

		Unheated soil				Dry heated soil (80°C)				Soil heated with boiling water			
Seedlings	Replicates	10	20	30	Total	10	20	30	Total	10	20	30	total
Ochroma	1	23	2	0	25	60	4	1	65	66	3	0	69
lagopus	2	3	3	0	6	23	1	0	24	30	0	0	30
Total:		26	5	0	31	83	5	1	89	96	3	0	99
Other	1	81	260	98	439	8	5	4	17	31	38	18	87
species	2	31	117	72	220	2	0	1	3	25	20	11	56
Total:		112	377	170	659	10	5	5	20	56	58	29	143
Number of	1	8	9	2	19	2	0	1	3	3	3	0	6
different species	2	5	4	1	10	2	0	0	2	2	0	1	3

under the intact canopy of the rain forest; (2) species that germinate and grow principally under the rain forest canopy, but appear to benefit from the existence of openings; (3) trees that establish themselves under the rain forest canopy but definitely need openings to grow; and (4) pioneer species that establish themselves principally in clearings and grow only in these openings, this group being divided into species with long or short life cycles. Pioneer, short-lived trees and shrubs, or "weed trees," which do not form part of the mature vegetation, grow rapidly, reaching sexual maturity in a few years, and have a short life span, in comparison to species of later seral stages. These species generate peculiar environmental conditions after becoming established, which inhibit their own replacement and allow the growth of species characteristic of later successional stages (Vazquez-Yanes 1980).

The abundance of pioneer trees and shrubs in a rain forest region depends on the frequency of openings in the canopy that allow direct sunlight to strike the forest floor for a large part of the day. In undisturbed primary rain forest areas, these species develop only in large clearings caused by tree falls. They can also be very abundant along river banks, on rocky soils, or on slopes, which experience more frequent tree falls because of overland water flow and winds. The vegetative cover of the rain forest is subject to relatively frequent alterations independent of human activity. Whitmore (1978) and Hartshorn (1978) calculated that tree falls that uncover the soil of the rain forest occur on the average of every 80–90 years. Small clearings are filled by species that existed in the site as seedlings before the disturbance, while large openings are colonized by rapidly growing pioneer nomads (Schulz, 1960). Kramer (1933) in Indonesia observed that clearings of 0.1 ha or less were promptly colonized by the surviving seedlings of the rain forest. However, in clearings of 0.2–0.3 ha or more, the growth of pioneer species prevented the growth of rain forest seedlings. Knight (1975), as part of a phytosociological study of Barro Colorado, Panama, noted that *Cecropia obtusifolia* was more abundant in the oldest and highest rain forest on the island than in the late secondary rain forest. This may be due to the fact that clearings, produced by tree falls, are much larger and more frequent in the former than in the latter. This suggests that a group of species with nomadic behavior, which forms typical cicatricial vegetation, evolved and diversified in rain forest clearings. The abundance of these species has grown enormously due to human activity, which probably also has favored genotypic and phenotypic variation in these species, as with other taxa characteristic of anthropogenic vegetation (Gomez-Pompa 1971). The most common pioneer trees of the world's tropics belong to the genera *Adinandra* spp. (Ternstroemiaceae), *Anthocephalus* spp. (Rubiaceae), *Cecropia* spp. (Moraceae), *Didymopanax* spp. (Araliaceae), *Harungana* spp. (Hypericaceae), *Heliocarpus* spp. (Tiliaceae), *Macaranga* spp., *Mallotus* spp. (Euphorbiaceae), *Musanga* spp. (Moraceae), *Ochroma* spp. (Bombacaceae), *Piper* spp. (Piperaceae), *Trema* spp. (Ulmaceae), etc.

The relationship between clearing size and establishment of pioneer species can be affected by nutrient availability in the soil. Plants of the mature rain forest have mycorrhizae (Singh 1966; Went and Evans 1968), while apparently many pioneer species lack mycorrhizae or do not require them to achieve rapid growth. This suggests that for a pioneer species to establish itself, the clearing must be large enough to allow the plant to exist as well as contain sufficient space for root colonization and proper conditions for mineralization of soil nutrients. In small clearings, the dense mat of roots and superficial mycorrhizae of the mature community is so closed that only seedlings with roots integrated into the mat can survive. In large clearings, the death and fall of some trees reduces root competition, thereby allowing the establishment of plants that obtain nutrients, at least initially, in the free ionic form.

Pioneer tree establishment is impeded or retarded in soils impoverished by cultivation or pasturing. In these soils, perennial grasses such as *Bambusa* spp., *Imperata* spp., and slow-growing sclerophyllic trees and shrubs such as *Guazuma* and *Curatella* become established. When this occurs, the successional process is retarded or interrupted, resulting in the persistence of anthropogenic vegetation for considerable periods of time, depending on the recurrence of disturbance. In some cases succession is arrested indefinitely if species of the mature vegetation from the area (Gomez-Pompa et al. 1972) are absent. This process was described according to the following diagram by Whitmore (1975) for Southeast Asia:

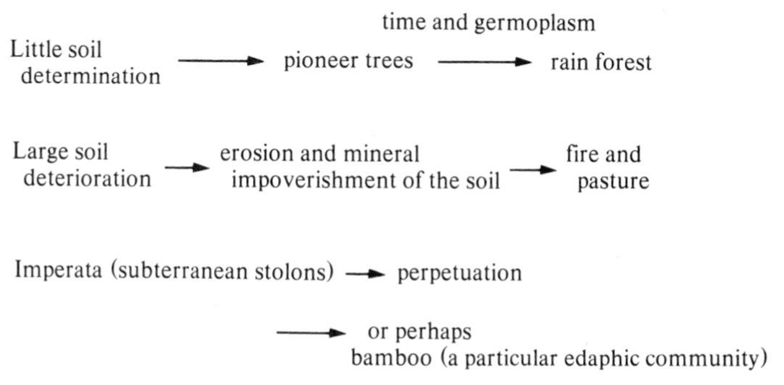

We have observed something similar to the above in southeastern Mexico, although the process has not been analyzed quantitatively.

The rapid growth of pioneer species is due primarily to an energetic investment, directed at forming new tissue for increasing the photosynthetic surface of the plant. This process has been described in some detail for *Trema* and *Musanga* by Coombe (1960) and by Coombe and Hadfield (1962); these authors conclude that the rapid growth of these species is not due merely to an increase in dry weight, which is characteristic of more slow-growing tree species, but to the continual and efficient development of new foliar surface in a climate constantly favorable for photosynthesis.

According to the studies of Lugo (1970), the photosynthetic saturation and compensation points of certain pioneer trees allow high net photosynthetic values under intense direct sunlight. Apparently, their photosynthetic apparatus is not adapted to low intensity and diffuse light. Meanwhile, mature rain forest species, though less efficient photosynthetically at high light intensities, are adapted to low light conditions. Therefore, the latter can take advantage of increased light intensity due to small openings in the canopy. Consequently, the same individual can have leaves functional in the sun or shade.

The rapid growth of pioneer trees is due overall to their peculiar architecture and the ways they utilize photosynthate in their structure. Their energetic investment in woody tissue is comparatively small. Consequently, these tissues are light and rich in cellulose, but little lignified. The wood is usually fragile, and susceptible to fungal and parasite attack when its growth is interrupted by competition with other trees for nutrients and light at the end of the life cycle.

Growth rate, measured by Aubreville (1947), Lebrum and Gilbert (1954), and Blum (1968), for various pioneer tree species can be as high as 2–3 m annually (Fig. 16.3). From studies of pioneer tree architecture, Ashton (1978) deduced that their growth form leads to the formation of parasol type canopies in trees occupying the highest strata in early secondary succession (growth type Rauh, according to Halle et al. 1978). However, those species that occupy lower height strata exhibit a different architecture. Thus, utilization of available light is optimized at each level.

The secondary rain forest composed of pioneer species affects the environment in three different ways, which are extremely important for the development of vegetation through succession. These are: (1) transference of free nutrients from the soil to the biotic community, thereby decreasing the likelihood of their loss from the ecosystem; (2) edaphic structure improvement through the production of a large quantity of organic matter; and (3) microclimate modification, which reduces thermal fluctuations and increases atmospheric relative humidity. These effects promote the establishment of late successional species that subsequently will replace the pioneer trees in the community.

Immobilization of nutrients occurs rapidly. Communities at 10 months old contained the same quantity of nutrients as a mature grass field (Tergas and Popenoe 1971). By 6 and 14 years, secondary rain forests dominated by *Cecropia obtusifolia* and *Musanga cecropioides*, respectively, had immobilized almost as much phosphorus as a 50-year-old rain forest (Bartholomew et al. 1953; Greenland and Kowal 1960; Golley et al. 1975) (Table 16.3). In addition, the levels of certain soluble ions in the soil may also increase. Kellman (1969) reported that soil in sites dominated by *Trema orientalis* in the Philippines contained 110 milliequivalents of assimilable phosphorus per 100 g of soil, or almost double the highest values obtained under the other types of vegetation studied.

In reference to litter production, Golley et al. (1975) in Panama and Ewel (1976) in Guatemala found that secondary rain forests, between 6 and 14 years old, dominated by *Cecropia*, produced as much litter as mature rain forests in the region. This is due to the rapid replacement of pioneer species leaves, which enrich the organic matter content of the

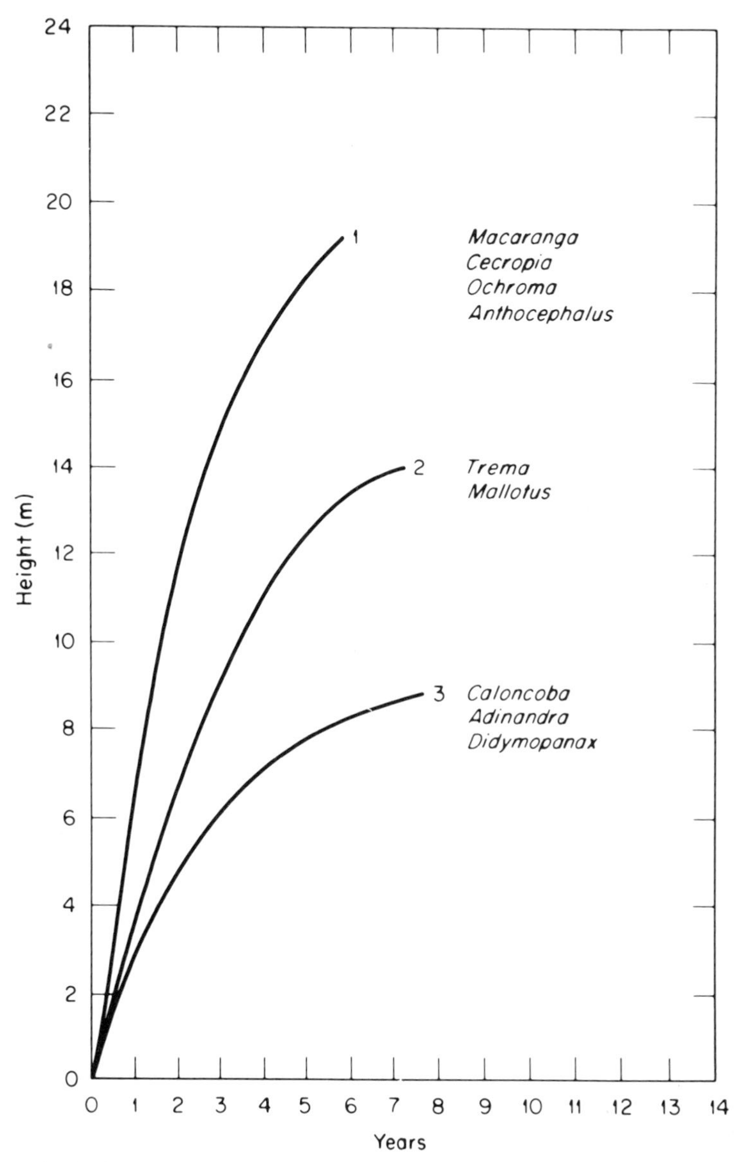

Figure 16.3. Growth rate among certain pioneer tree genera from the humid tropics.

Table 16.3. Nutrient content in primary and secondary vegetation from different places.[a]

Nutrient[a]	A	B	C	D	E	F	G	H
N	579	701	2061	—	—	—	—	—
P	35	108	137	85	241	72	154	163
K	839	601	906	1606	4598	524	1256	903

[a]A, B = 8- and 18-year-old secondary vegetation with Musanga in Zaire (Bartholomew et al. 1953); C = 80-year-old forest in Ghana (Greenland and Kowal 1960); D, E = two primary forests in Panama (Golley et al. 1975); F, G, H, = 2-, 4-, and 6-year-old secondary forest with Cecropia in Panama (Golley et al. 1975).

[b]Nutrient content in kg/ha.

soil. Ultimately, this enrichment facilitates the establishment and growth of seedlings with mycorrhizal roots from later seral stages, which obtain their nutrients, in part, from the decomposition of organic matter.

According to Golley et al. (1975) the structural characteristics of the vegetation change rapidly as the vegetation ages. The leaf area index of a 6-year-old rain forest, dominated by *Cecropia*, is almost the same as that of a mature rain forest, even though the average tree height and dbh of the 6-year-old forest were three times less. These data suggest that the productivity of the rain forest is rapidly reestablished, while the supporting structures, composed of stems and branches, develop slowly during succession. From 4 to 6 years, root biomass increases much faster than above-ground biomass, which explains in part the rate of nutrient immobilization (Table 16.4).

The characteristic microclimate of cleared land is modified in a short time by the establishment of trees. Ross (1954) and Snedaker (1970) measured temperature, relative humidity, and light in secondary rain

Table 16.4. Volume and biomass values for 2-, 4-, and 6-year-old tropical forests.[a]

	Age (yr)			
	2	4	6	Mature
Foliar surface (m^2/m^2)	8	12	17	11-22
Foliar volume (m^3/ha)	4	6	7	8-12
Total volume (m^3/ha)	16	42	57	276-338
Overground biomass (kg/ha)	13,020	38,040	42,550	
Underground biomass (kg/ha)	2600	4500	14,200	
Total biomass (kg/ha)	15,620	42,540	56,750	

[a]Foliar surface and volume and total above-ground volume is also given for mature forest.

forests of 5 and 14 years. Their results indicate that by year 14 microclimatic conditions are very similar to those of the mature rain forest. The environmental conditions caused by the growth of the pioneer tree community prevents their continued existence, and leads to their replacement by other species. Pioneer species, therefore, play a decisive role in regeneration and are a fundamental key to understanding this process. Their study can, undoubtedly, resolve many uncertainties that exist with respect to succession.

The chemical interactions between plants during succession undoubtedly can have some effect on vegetation development, since species with high allelopathic potential can affect the growth of other plants. In this respect, the work of Gliessman and Muller (1972) showed that phytotoxins of "bracken" fern (*Pteridium aquilinum*) can affect the establishment of other species. In our region, the work of Anaya and Anaya (1976) and Anaya and Rovalo (1976) indicates that certain typical secondary successional species of the genera *Piper* and *Croton* contain terpenes. When liberated into the environment, these compounds markedly affect the germination and growth of other species. However, these results were obtained under experimental laboratory conditions (Fig. 16.4). The demonstration of allelopathy in the tropical rain forest is extremely difficult under natural conditions. Nevertheless, the study of allelopathy is indispensible to understanding population dynamics, especially during the initial stages of regeneration.

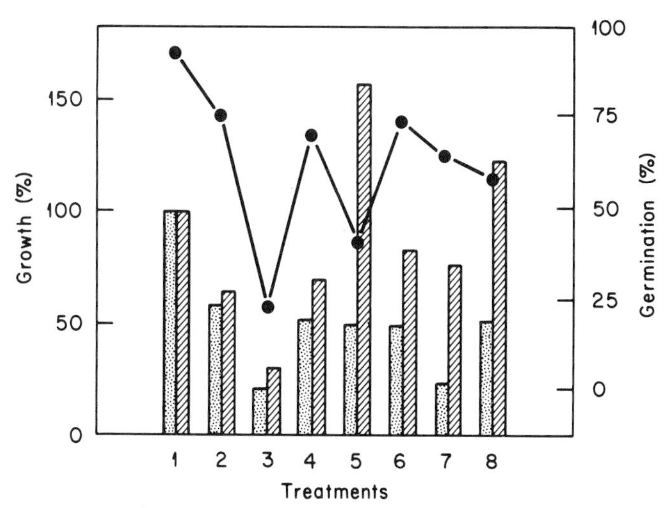

Figure 16.4. Effect of water extracts from secondary vegetation plants on the germination and seedling development of *Achyranthes aspera*. The treatments were: (1) pure water; (2) *Piper auritum*; (3) *P. hispidum*; (4) *Croton pyramidalis*; (5) *Cecropia obtusifolia*; (6) *Siparuna nicaraguensis*; (7) *Myriocarpa longipes*; (8) *Urera caracasana*. The hatched bars represent the seedlings shoots and the dotted bars the roots (taken from Anaya 1976).

Regeneration within the Rain Forest

One important fact to remember is that the primary rain forest is not a static community, but is in a process of continual change. Individuals in the herb and shrub layer die, other individuals appear, the epiphyte community is constantly changing, and some animals and microorganisms change seasonally. The rain forest trees, both juveniles as well as adults, grow in height and diameter. Some change their leaves seasonally; other replace foliage without apparent periodicity. The overall sensation of stability is due to the large trees, whose presence dominates in many senses, physiognomically as well as physiologically. Nevertheless, rain forests are in a dynamic process of change, which includes the slow substitution of the dominant tree species. In relation to this, Schulz (1960) says: "They are regenerating rather continuously and maintaining the existing proportions in the population."

If we examine a profile of the rain forest (Richards 1952), we will find dominant tree species, represented by individuals of different ages, living in lower height strata. This fact has been interpreted to mean that the primary community is stable and is regenerating. However, this is not totally correct. Only some species exhibit this pattern. Other species require another process for their regeneration, such as following obligate, sun-loving species (Whitmore 1975).

As part of the process of rain forest regeneration, we should also consider regeneration in old secondary or stable primary rain forests. We are therefore referring to the process of regeneration within a rain forest that is not prompted by a noticeable ecosystem disturbance. This process is actually more difficult to observe and perhaps of lesser importance than the process of tropical forest regeneration via secondary succession, which is probably more widespread in the tropics. However, if we assume that the process of internal rain forest regeneration under original conditions, especially before the appearance of humanity, was the precursor of the process of secondary succession, then understanding it can be essential for rain forest management.

If we visit a mature rain forest in a hot-humid region, we find it is relatively easy to walk on the forest floor, which is covered with litter and different-sized plants that are juveniles of many species found in the upper canopy. In some seasons, especially after fruiting, seedlings are extraordinarily abundant, covering the rain forest soil. However, in a few weeks or months the majority will die, and only a few will continue growing. These juvenile rain forest trees are so characteristic that we suppose they are enormously important to the process of internal rain forest regeneration. Studies in Los Tuxtlas (del Amo 1978) provided very interesting information on the behavior of primary tree seedlings and saplings with respect to the length of their growth (Table 16.5) and their enormous capacity to regenerate foliar tissue. These results suggest that

Table 16.5. Growth and leaf formation in seedlings of tropical rain forest trees under different natural light conditions.[a]

Species	Growth (cm/yr)			New leaves per year		
	S	A	CPS	S	A	CPS
Poulsenia armata Standi	2.83	2.55	17.04	2.55	—	—
Rheedia edulis (Seem) Triana & Planchon	1.35	4.14	11.75	4.81	4.74	12.15
Nectandra ambigens	2.73	3.24	33.94	N.S.	5.03	10.69
Licaria alata Miranda	3.32	ns	5.11	—	—	—
Chamaedora tepejilote Liebm.	1.05	2.44	11.16	2.55	1.752	2.92

[a]S, primary forest; A, 15-year,old secondary forest, CPS, cleared primary forest.

in these species there exists a very delicate physiological equilibrium between photosynthesis, growth of the individual, and the relation of growth to the root system. These individuals will continue growing, through time, if a severe disturbance, such as the fall of a close tree or large tree branch, does not occur. They become the regenerative potential of the species, living as shade-tolerant individuals in the lower strata of the rain forest. Seedlings of the same species in different environmental conditions, such as those under the shade of secondary species or in the direct sun, behave differently. This suggests that these species are preadapted to disturbance, since they can behave like sun-loving species before the opening of the canopy and as shade-tolerant species under secondary cicatricial trees. This interesting phenomenon is currently being studied and offers enormous opportunities for understanding the original regeneration process through species evolution. Natural selection acts on different stages of the life cycle of the species, selecting those characteristics that will be most competitive within the regeneration process under different environmental conditions. According to the calculations of del Amo (1978), *Poulsenia armata*, a primary rain forest species, reaches its height maximum in nondisturbed conditions at an age of 180 years. However, if a light gap appears, growth time to maximum height may be cut in more than half, to 70 years, if other tree species do not interrupt its growth through premature shading. In reality, the latter happens in a majority of large rain forest clearings, where the juveniles of the primary trees are overtopped by pioneer cicatricial species before they can increase their growth into the light gap. Investigators who have studied light gap regeneration (Hartshorn 1978; Whitmore 1978) observed that the rate of regeneration can vary, depending on many factors that occur in a site and/or distinct events that occur through time. Many of the secondary species are preadapted to severe changes in the environment, while the primary species that have a more predictable behavior and are more vulnerable, are not.

Conclusions

The most obvious conclusion to be derived from the present status of our knowledge of succession in tropical ecosystems is the enormous importance of understanding the life cycles of the species involved in the process of regeneration of the forest. These cycles can provide ways to investigate regenerative stages in time as well as space.

There exists an enormous diversity of tropical ecosystems in the world. Of these, only a few sites that are probably not comparable climatically or edaphically have been studied. Of the millions of species of plants and animals in the tropics, we know with some detail (although superficially) the life cycles of only a few. Until our knowledge increases, it will not be possible to arrive at a general concept of succession in the tropics.

Our only course is to continue along the same lines, to investigate more species in greater depth, and to broaden our investigations to include more aspects of the life cycles of the species. Only after the life cycles of the most important species are clearly understood, and, through this knowledge, an understanding of the interrelationships of the species is understood, will it be possible to investigate the succession of communities as a whole. Perhaps then we can ultimately discover the true nature of succession in the tropics.

Acknowledgments

This work has been funded by a grant of the National Ecological Program at the National Council of Science and Technology (CONACYT) of Mexico. We want to express our gratitude to Joan Roskowski from the National Institute of Biotic Resources of Mexico for the translation of the manuscript into English.

Chapter 17

Process Studies in Succession

Peter M. Vitousek and Peter S. White

Introduction

As the remainder of the chapters demonstrate, process studies of disturbance and succession are carried out on a number of different levels of ecological organization. Despite this diversity of approaches, however, none of these studies systematically addresses the importance of nutrient cycling processes in succession, and none directly discusses the possible contributions that ecosystem-level process studies can make to the understanding of plant succession (McIntosh 1980a, Chapter 3). Further, while all of the studies demonstrate the important of disturbance, only one provides a systematic framework by which disturbance may be characterized. Accordingly, we shall begin this chapter with a brief discussion of some of the interactions of ecosystem-level nutrient cycling with successional vegetation change. We shall then examine some of the general properties of disturbance regimes.

In this discussion, we shall make a somewhat arbitrary distinction between whole ecosystem and holistic ecosystem approaches to the study of succession. Both approaches involve similar measurements of fluxes of energy and material. In holistic studies, the regulation of these fluxes is considered to be an emergent ecosystem property, not entirely explicable by a reductionist examination of ecosystem components (Odum 1969; Patten 1975a). A high degree of integration is assumed, and it is considered productive to discuss the costs, benefits, and even evolution of ecosystem-level processes (Richardson 1980). Whole ecosystem studies consider the same fluxes, but they are viewed as collective properties of ecosystem components rather than emergent properties of ecosystems (Salt 1979). An explicitly reductionist approach is used: whole ecosystem studies seek explanations in geochemistry and the population and physiological ecology of the species present. It is considered valuable to study system-level fluxes even when ecosystem components are inadequately known, however, because study at the ecosystem level can help to elucidate which components and processes require further study.

The question of which of these approaches is more correct in describing ecosystems is presently one of world view, and a fuller discussion of it is properly beyond the scope of this book. The question of which approach now has more to contribute to the understanding of plant succession is more easily examined, however. If it is granted that patterns of nutrient availability and flux generally strongly affect plant succession (as is illustrated in Christensen and Peet Chapter 15, Cromack Chapter 22, Van Cleve and Viereck Chapter 13, B. Walker Chapter 25, J. Walker et al. Chapter 9, and later in this chapter), either approach can provide valuable input to a study of succession. The explicit population and physiological orientation of whole ecosystem studies, however, makes them more directly relevant to studies of vegetation change. The ecosystem-level integration sought by such studies can only be developed from an understanding of important properties of the species present. Conversely, the information on nutrient fluxes obtained in whole ecosystem studies tends to be at a level of resolution appropriate for examining how organisms and populations are constrained by ecosystem-level processes.

Nutrient Cycling in Succession

We shall very briefly review the importance of nutrient cycling in succession largely as a way to draw the readers' attention to the voluminous and expanding literature on the subject. Many investigators have either explicitly or implicitly studied changes in nutrient cycling and ecosystem nutrient budgets during forest succession. Outstanding examples include Turner (1975), Bormann and Likens (1979a), Swank and Waide (1980), and Van Cleve and Viereck (Chapter 13). The processes controlling changes in nutrient inputs and outputs in primary and secondary succession were recently reviewed by Gorham et al. (1979).

Studies such as these, which document a relationship between successional stage and nutrient cycling or nutrient budgets (especially in terms of controlling processes), have considerable theoretical and practical value in their own right. Of more interest in this volume, though, is an examination of the reciprocal interactions of ecosystem-level nutrient fluxes and vegetation change.

The clearest examples of the probable influence of such fluxes on vegetation derive from studies of primary succession and soil development. T. W. Walker and his co-workers (superbly summarized in Walker

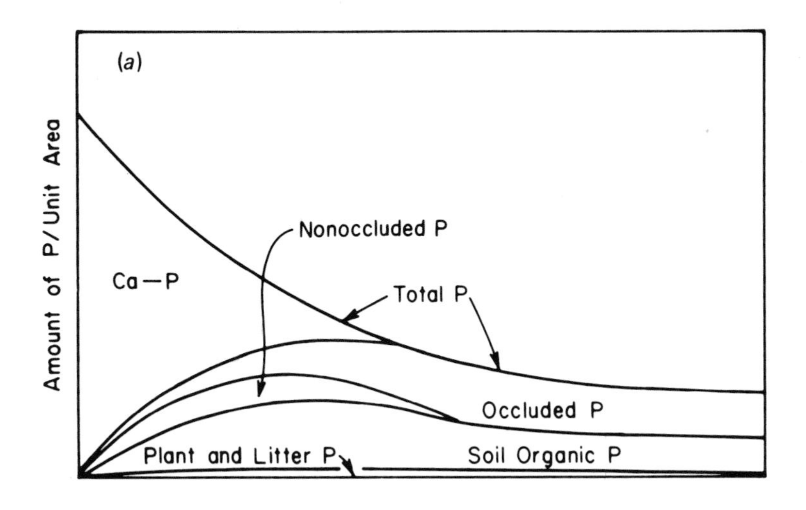

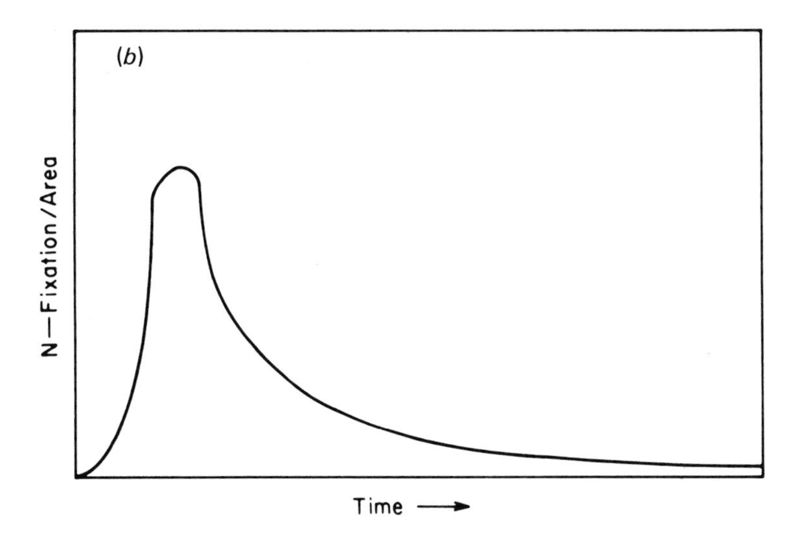

Figure 17.1. (a) Changes in total phosphorus and phosphorus fractions in the course of soil development. Ca-P is mostly primary mineral phosphorus, nonoccluded P is closely related to that fraction which is in equilibrium with soil solution P (plant-available P), and occluded P is either inaccessible primary mineral P or highly recalcitrant secondary mineral P. The soil organic P becomes progressively less available in succession as the organic carbon:organic phosphorus ratio widens. (Redrawn from Walker and Syers 1976, except for plant and litter P, which is taken from Cole and Rapp 1980.) (b) Hypothesized changes in nitrogen fixation in the course of soil development. Nitrogen fixation requires relatively high levels of available P; it can also be suppressed when the N:P ratio is high enough to allow the growth of large populations of nonnitrogen fixers. (Redrawn from Gorham et al. 1979.)

and Syers 1976) have demonstrated that both the total amount and the chemical forms of phosphorus change irreversibly during soil development (Fig. 17.1a). Phosphorus is present in weatherable minerals at the beginning of soil development, in a range of forms including presumably plant-available forms in early and mid-development, and bound in organic matter with a wide C:P ratio or in highly recalcitrant (usually inaccessible) mineral forms. The total amount of phosphorus present declines through soil development as a consequence of phosphorus leaching losses. In extreme cases, plant biomass and production can drop markedly late in soil development, probably in response to the low phosphorus levels (Walker et al. Chapter 9).

The time scale of the processes shown in Fig. 17.1a is variable from tens of thousands to millions of years, depending on the climate (especially leaching intensity), the initial phosphorus levels, and the phosphorus adsorption capacity of the weathered parent material of a site. The time scale of this soil development is probably always very long relative to secondary succession, and the fractions and forms of phosphorus can be regarded as reasonably constant for any single secondary successional event. The overall pattern shown in Fig. 17.1 must be regarded as well established, however; given time since geological disturbance (such as glaciation), the pattern of phosphorus amounts and availability will be as summarized in Fig. 17.1a.

Walker and Syers (1976) further suggest that rates of nitrogen fixation are controlled in part by phosphorus availability and in part by available N:P ratios. Nitrogen fixation increases in primary succession to a relatively early peak, and then declines as the N:P ratio in the system increases (allowing the growth of nonnitrogen fixers) and as available P decreases (Fig. 17.1b). This suggestion is consistent with our knowledge of vegetation changes in primary succession, patterns of nitrogen accumulation in succession, and the physiology of symbiotic nitrogen fixers.

The net result of these processes is a system relatively deficient in nitrogen early in soil development, one with N and P in near equilibrium, maximum production, and the largest absolute amount of nutrient cycling in the middle, and with substantially decreased nutrient availability late (Walker et al. Chapter 9). Some of the consequences of this pattern for secondary succession are drawn by Walker et al., who rightly stress the irreversibility of this developmental pattern once most of the phosphorus has been lost or is in unavailable forms.

Further, rather more speculative consequences for secondary successional mechanisms follow from Fig. 17.1. A secondary succession-initiating disturbance relatively early in soil development (before substantial N has accumulated) should favor recolonization by species with access to atmospheric nitrogen (legumes and other symbiotic associations), since nitrogen is likely to be limiting to plant growth. At least in the long term, nitrogen fixers should have the effect of increasing N availability in the system, thus facilitating the growth of other species on the site (Cromack Chapter 22).

Later in soil development, when N:P ratios more closely match plant requirements and production and mineral cycling are at their greatest (Walker et al. Chapter 9), disturbance can lead to the release of large amounts of the circulating nutrients (as discussed below), and thus to highly favorable conditions for plant regrowth. Dispersal ability, size and longevity, and the relative ability of species to compete for available resources (or, in other words, inhibition and life-history properties—Botkin Chapter 5) should determine the patterns of plant growth and species replacement following disturbance. Most of the glaciated regions of the temperate zone are currently in this portion of Fig. 17.1a.

Once nutrient availability drops to low levels (whether by this mechanism or by others), plants may internally recycle nutrients more effectively (Lamb 1975; Turner 1977), and little or no flush of available nutrients occurs upon disturbance (Vitousek et al. 1979). Only the restricted set of species capable of tolerating extremely low nutrient levels will then be able to occupy a site at any point in secondary succession. Below-ground investment (in fine roots and/or mycorrhizae) must be substantial among such species.

We are not suggesting that facilitation, inhibition, and toleration (*sensu* Connell and Slatyer 1977) can only be observed at the stages of soil development described—resources other than nutrients, and indeed nutrients other than nitrogen or phosphorus, could be more important. We do suggest that the ecosystem-level patterns of nutrient availability and flux described in Fig. 17.1 can systematically affect the species composition, functional properties, and mechanisms of species replacement in secondary succession. Moreover, the processes summarized in Fig. 17.1 provide a means of examining differences in secondary succession resulting from different disturbances. For example, fire (which mineralizes phosphorus as it volatilizes nitrogen) should yield a sere relatively rich in nitrogen fixers like those observed at the left of Fig. 17.1a (Gorham et al. 1979), while secondary succession on fertilized abandoned agricultural fields of any geological age should be more like that expected in the middle of Fig. 17 (Reiners 1981).

Ecosystem-level studies of secondary succession also have important implications for the control of vegetation change. As has long been known (cf. Hesselman 1917), disturbance, including commercial clear cutting, often causes some increase in ecosystem-level nutrient availability and loss. Changes in the availability and loss of nitrate nitrogen are generally particularly marked (Vitousek and Melillo 1979). Often this pulse of increase in nutrient availability is small or long delayed, but in some sites (notably the well-studied Hubbard Brook Experimental Forest) it can be rapid and substantial. This pulse is followed by a prolonged period of lower nutrient losses (and presumably lower availability) during which nutrients are retained by aggrading biomass in the forest and forest floor (Vitousek and Reiners 1975; Bormann and Likens 1979a). Competition for nutrients is presumably most important at this stage of secondary succession. Finally, measurements of nutrients in streams draining some old-growth versus aggrading forests (Leak and Martin 1975; Vitousek and Reiners 1975) and an extension of the JABOWA model to old-growth northern hardwood forests (Bormann and Likens 1979a) suggest that nutrient losses increase as biomass growth becomes more or less in balance with decomposition and respiration. The increase in nutrient losses probably reflects increased availability of nutrients in patches resulting from natural turnover within an old-growth forest (Vitousek and Reiners 1975; Bormann and Likens 1979a). As sites develop from an even-aged aggrading forest to an uneven-aged, patchy, steady-state forest, there may be a peak in nutrient losses (Bormann and Likens 1979a). Such a peak during a transition phase would occur in either the shifting mosaic model or the time-lag model of production in succession (Peet Chapter 20).

The pulse in nutrient availability that occurs early in many (though not all) secondary seres has received considerable attention, probably owing to its potential effects on downstream ecosystems. The reasons in the occurrence of such a pulse are reviewed in Vitousek and Melillo (1979), and examinations of why elevated losses do not always occur following disturbance appear in Stone et al. (1978) and Vitousek et al. (1979). Based on the mechanisms they propose, elevated nutrient availability in the soil probably occurs more commonly than elevated nutrient losses.

A number of studies (most notably Marks 1974) have investigated the interactions between nutrient availability, the growth of early successional species, and the retention of nutrients within early successional forests. Certain early successional species (including the pin cherry studied by Marks) can achieve extremely rapid growth rates on the high levels of resources (including nutrients) available in disturbed sites. Increased resource availability may not be the only cause of high growth rates in early succession (see Peet Chapter 20). Nonetheless, the understanding of how this increased nutrient availability interacts with the available species pool to yield vegetation change requires a fuller understanding of the physiological tradeoffs inherent in plant growth under different levels of nutrient availability. The reciprocal understanding of how the species affect ecosystem-level nutrient losses requires an understanding of plant population dynamics and life history characteristics.

On the basis of current evidence a number of reasonable interactions between ecosystem, population, and physiological processes in early secondary succession can be suggested. Where nitrogen availability is high, nitrate is generally produced rapidly following disturbance (Vitousek et al. 1979). A number of early successional species (including pin cherry) can use nitrate as a germination cue (Peterson and Bazzaz 1978;

Auchmoody 1979). This provides a mechanism whereby such species can enter suitable sites (Covington and Aber 1980) and further suggests an association in evolutionary time between elevated soil nitrate and suitable sites for the growth of such species. Nitrate is far more mobile than ammonium in the soil solution (Nye and Tinker 1977), so a lesser investment in roots and mycorrhizae should be required to yield the same nitrogen uptake. Consequently, greater leaf and shoot production could be maintained where nitrate is abundant. Additionally, photosynthetic capacities of individual leaves are strongly affected by leaf nitrogen levels (Mooney and Gulmon 1979), as most leaf nitrogen is in phytosynthetic systems. This suggests another important way that early successional species could maintain high growth rates at high nutrient levels.

Once the pulse of readily available nutrients is utilized, however, investment in roots and mycorrhizae probably increases. The amount of carbon fixed per unit of nutrient (Small 1972) may become more important than maximizing the rate of photosynthesis, and the occupancy of a site may become more important than a rapid rate of growth.

Many of the mechanisms described above derive directly from whole ecosystem studies of nutrient cycling. Others derive from explicitly physiological studies, and, together with the context provided by ecosystem-level observations, they are very useful to any process-oriented examination of plant succession.

Disturbance Regimes

The interaction of disturbance with successional processes is a general theme in the chapters that follow. Disturbance sets the stage for succession and determines the time during which succession proceeds without interruption. A growing concern with disturbance is evident in this section and in the accumulating literature on this subject (see White 1979; Vogl 1980; Pickett 1980).

The ability to model succession in regions subject to large-scale disturbances (and hence a generally useful understanding of succession in such areas) is dependent on our ability to model disturbance regimes. We believe that that necessary descriptors of disturbance regime include size of disturbance, frequency, recurrence interval, predictability, rotation period, and magnitude. *Recurrence interval* is the mean time between disturbance events and the inverse of frequency. *Predictability* is a measure of variance in return interval. *Rotation period* is the mean time interval to disturb an area equivalent to the study area at hand (see Heinselman Chapter 23). We divide *magnitude* into *intensity* (a physical measure of the force of the event, e.g., windspeed) and *severity* (a measure of the impact on the community, e.g., biomass of trees blown down). For most disturbances, severity and intensity are inversely related to frequency. Indeed, flood magnitude is often described by recurrence interval—as in a 20-year or 100-year event.

The occurrence and effects of specific disturbances, and hence disturbance regimes, are further influenced by community state. Susceptibility to disturbance may increase with successional age (time since last disturbance) in many systems—as when fire fuels build up during postfire succession (see Heinselman Chapter 23). Patches of large senescent trees are more susceptible to blowdown in windstorms of a given intensity than are patches of younger trees. The influence of community state means that disturbance severity is not always a direct function of disturbance intensity. Figure 17.2 depicts a hypothetical relationship of wind disturbance to topographic gradients of wind exposure. The frequency of windstorms is probably directly correlated with exposure but the severity of wind damage to the community probably peaks at mid-gradient. Above this point, the entire community may become structurally resistant to wind (e.g., krummholz on New England mountains—Reiners and Lang 1979). Systematic comparative studies of disturbance regimes will be necessary to establish this point and to validate a general framework for the modeling of disturbance regimes.

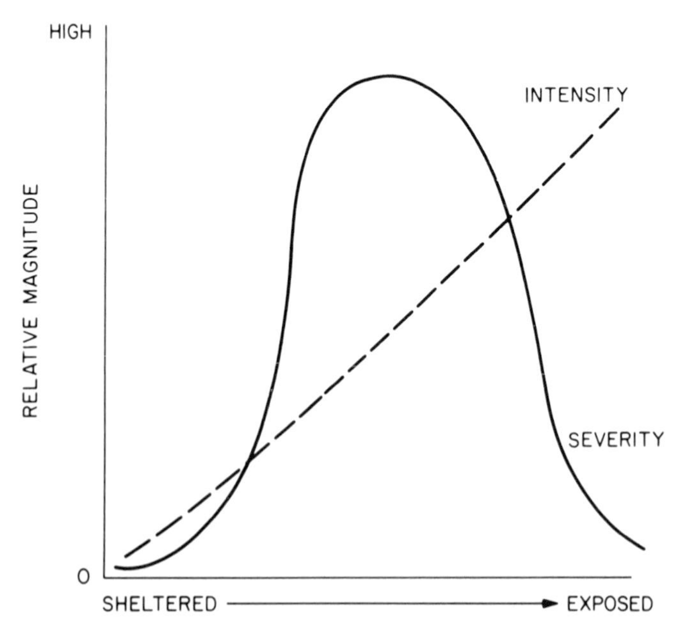

Figure 17.2. The relationship of wind disturbance to topographic exposure. High winds are most frequent on the exposed sites, but the destructive impact of wind on the plant community is more substantial in more sheltered sites.

Disturbance and Process Studies

The chapters that follow employ a number of approaches to the study of successional processes. MacMahon (Chapter 18) demonstrates that the processes originally described by Clements remain appropriate to process-level discussions of succession today, and further illustrates reciprocal influences between vegetation and animals in succession. Woods and Whittaker (Chapter 19) carefully document the processes of species replacement in old-growth hemlock-northern hardwood forests, and convincingly demonstrate that replacement is nonrandom and potentially cyclical. They suggest physiological and population-based mechanisms that could account for the patterns observed.

Peet (Chapter 20) provides a general description of variations in the patterns of plant production with succession, and offers and effectively supports a mechanism that can explain many of the disparate patterns which have been observed. Cooper (Chapter 21) summarizes a detailed, long-term study of production on a series of permanent plots. Cooper's approach and results demonstrate the assumptions and practical problems that are always encountered in any study of successional processes. His results illustrate Peet's (Chapter 20) time-lag paradigm very clearly. Both Peet and Cooper examine production as a collective property of forest ecosystems, and both identify population properties and physiological mechanisms that could yield the patterns they observe and discuss.

Cromack (Chapter 22) discusses the too little appreciated importance of below-ground processes in forest succession. The energy and nutrient resources that plants allocate below ground are truly impressive; they make any discussion of succession that is based solely on competition for light (or even light and water) questionable. Further, some below-ground biotic interactions in succession (especially the interdependent dispersal and establishment of mycorrhizae and plants) are illustrated. Despite their difficulty, studies of below-ground processes promise to yield exciting insights in the next few years.

Heinselman, Zedler, and B. Walker (Chapters 23, 24, and 25, respectively) all examine the patterns of vegetation change in ecosystems that are less familiar to temperate forest ecologists. All survey various types of vegetation change, and evaluate the comparative importance and significance of successional vegetation changes in these systems. Heinselman evaluates the large-scale, relatively frequent changes brought by fire in northern conifer forests, while Zedler demonstrates the importance of postdisturbance establishment in controlling the composition of desert and chaparral systems. B. Walker discusses vegetation change in South African savannas on several time scales, ranging from plant responses to frequent fires to long-term geological changes. Interactions between species and between life forms are emphasized, and an insightful preliminary model for the maintenance of savanna vegetation is developed. In all of these chapters new physiological measurements are strongly suggested (implicitly or explicitly); in no chapter can the physiological bases of the processes discussed be considered thoroughly worked out (although B. Walker does go a considerable distance toward a fundamental physiological understanding).

The JABOWA-FORET family of models (Botkin Chapter 5, Shugart et al. Chapter 7) can be described similarly. They are based on the physiological characteristics of individual species and their effects on population size and age structure. They succeed in describing a number of collective properties of interest in particular systems, and they further illustrate very clearly which physiological and population characteristics of species require further study.

Whether the successional process under study is species replacement, production, or nutrient cycling, a major challenge facing ecologists interested in integrating process studies with vegetation change is the need to put our interpretations on a fundamental, comparable physiological basis. A number of physiological ecologists are already meeting this challenge (Bazzaz 1979; Grime 1979; Bazzaz and Pickett 1980; Chapin 1980) with examinations of the physiological properties and tradeoffs inherent to early and late successional species. Which physiological mechanisms are important and how they fit together is not likely to be obvious to a physiologist working with a plant in isolation, however. It may be true that ecological interactions are reducible to physiology, but both species replacement and whole ecosystem studies can show which physiological mechanisms need study and evaluation in a way that studies which lack an ecosystem context cannot.

Acknowledgments

We thank K. O'Connor for pointing out the implications of T. W. Walker's work, and F. S. Chapin, R. K. Peet, and T. Wood for critical comments on an earlier draft of this chapter.

Chapter 18

Successional Processes: Comparisons among Biomes with Special Reference to Probable Roles of and Influences on Animals

James A. MacMahon

Introduction

This chapter is an attempt to present a series of ideas, based on several conceptual models, which are related to variations in the importance of successional processes in different terrestrial biomes of the world. The approach used is to explain the reasons for choosing the rows and columns of a matrix concerning successional processes and to apply this matrix first to plants and then to animals, to estimate their impact on succession. Finally, factors regulating animal species mixes during succession will be addressed.

Throughout this discussion the assumption is made that what is generally termed succession is best described as the change in the physiognomy, species composition, or proportion of species on a plot of ground, over a moderate time interval (decades to a few centuries) following a disturbance to that site. Thus, the emphasis is on secondary succession, though I have argued (MacMahon 1980a) that the primary-secondary succession dichotomy is often unclear, and does not represent qualitatively different processes.

At the outset, I implore the reader to approach the discussion with the same perspective I have used, i.e., looking for and emphasizing the presence of worldwide, generalized patterns of succession. This approach is different from the one that many of us instinctively use when confronted with generalities, i.e., we attempt to think of a few highly specific exceptions. My claim is not that this discussion applies equally to every species of organism for every community type on every continent, but rather that there is a surprising degree of pattern to successional processes in various parts of the world. The information concerning such patterns can thus form the basis for the erection of numerous, testable hypotheses—the work of science—which may help to elucidate the nature of the phenomenon termed succession.

Justification of the Matrix Approach

In the following two sections the bases for choosing the variables on the two matrix axes are developed. First, the order of biome types is justified on the basis of the conspicuousness of the successional phenomena within each biome. Second, a model is presented to show the conceptual adequacy of Clements' (1916) description of the successional processes.

Arrangement of the Biomes

The acceptance of succession as a phenomenon is intricately tied to the obviousness of its occurrence. Few workers doubt that succession, by some definition, occurs following the leveling of most forest ecosystems in the world. In contrast, strong arguments have been advanced suggesting that succession does not occur in tundra or deserts (see discussions in Muller 1940, 1952, and especially Whittaker 1974). Recently it has been argued that the processes of succession occur in most biomes. In extreme environments, however, fewer life history strategies are viable, the environment is much less buffered by the resident biota, and hence the only species complex that can occupy a site after a disturbance is one composed of those species already in the area and adapted to the relatively harsh environment. Because these species are predominantly the climax admixture, self-replacement occurs, i.e., autosuccession (Muller 1940). This interpretation is not novel, but has been resurrected and reemphasized (MacMahon 1980a), especially for desert situations (MacMahon 1980b).

Climate is one important feature (another is physiognomy) that defines a biome. Mean annual temperature and precipitation separate biomes reasonably clearly (Fig. 18.1). Tundra and deserts, lacking conspicuous succession, are both on the low end of the precipitation axis. This is unlikely to be the common factor, though, since all grasslands fall within the desert-tundra precipitation range but exhibit reasonably conspicuous succession. Also, the effect of low precipitation in the tundra is partly offset by slow water percolation (frozen subsoil) and low vapor pressure deficits (cool air).

Attempting to use the precipitation factor and to separate mean annual rainfall from the unpredictability of rainfall, MacMahon (1980a,b) took

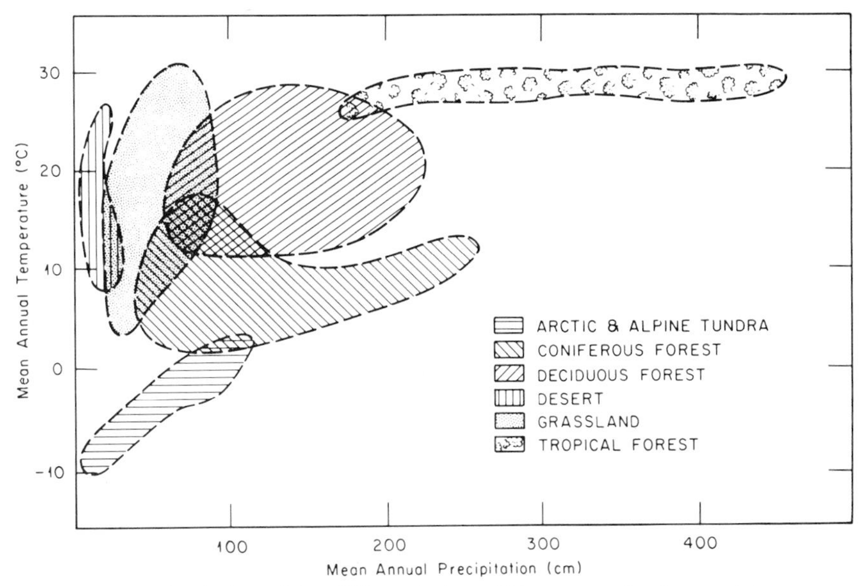

Figure 18.1. Depiction of the differentiation of biomes on the bases of temperature and precipitation. (Redrawn by permission from Hammond 1972.)

weather station data from various world localities, plotted these data on an overlay of Whittaker's (1975b) biome limits, and thus assigned each station to a biome type. Mean annual precipitation data were then plotted against the variance of the log of precipitation (Fig. 18.2). The reason for including a measure of rainfall variability is that while some organisms, in this case plants, can adapt to extremes of environmental factors (Levitt 1972), many fewer species seem capable of adapting to the combination of extreme and unpredictable environments (Grime 1979). The diminished pool of appropriately adapted species might foster autosuccession. In Fig. 18.2 tundra and desert form the ascending arm of the curve, and all forest types (biomes with conspicuous succession) fall on the horizontal arm. For any biome type there is a broad span along the curve. We might expect that if the conspicuousness of succession is related, in significant part, to the unpredictability of rainfall, then our assessment of succession would also vary. The order of biome types on the curve (Fig. 18.2), i.e., desert, tundra, grassland, coniferous forest, deciduous forest, and rain forest, is now the fixed sequence to be used for subsequent analyses, i.e., their order on graph and matrix axes.

It could be argued that assignment of sites to a few biome types based on two climate variables is too simplistic and that a more complex set of factors need be considered, e.g., Holdridge's (1967) "life zone" scheme. For our purposes this was not deemed necessary nor feasible because of data requirements (MacMahon and Wieboldt 1978).

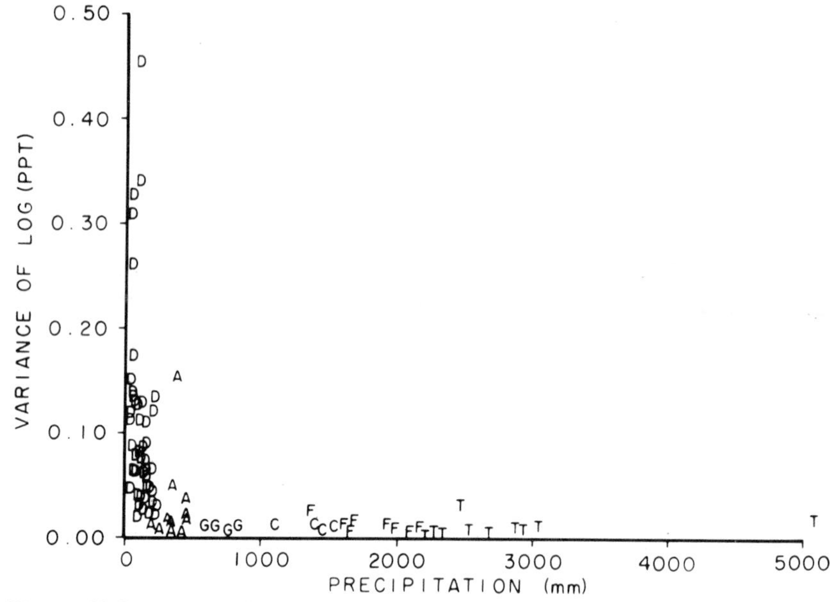

Figure 18.2. A plot of the variance of the log of mean annual precipitation versus mean annual precipitation for various world weather stations (Clayton 1944; Clayton and Clayton 1947): D, desert, F, deciduous forest; T, tropical rain forest; G, grassland; C, coniferous forest; A, tundra. (Taken, by permission, from MacMahon 1980a.)

Choice of Successional Processes

The term "process" has several meanings to ecologists. I use the term here in a generic sense to indicate general phenomena involving continuous action or operation taking place in a specified manner. Thus, what I term a successional process is, simultaneously, a category of processes and each of the processes within that category.

Clements (1916) produced a succinct categorization of successional processes. To Clements, the development of a "climax formation" involved "initiation" of a site. For secondary succession this depended on the kind of disturbance involved (nudation), what propagules remained in the soil after disturbance (here categorically termed residuals), what migrants got to the site (migration), how successful both residuals and migrants were at establishment and growth (ecesis), and how these individuals altered their abiotic environment (reaction). Additionally, as these species grow, they may compete with one another (competition), again altering their environment. These processes continue until the species present on a site come to an equilibrium among themselves and with the environment (stabilization)—essentially the "climax" state and in one sense the outcome of all the other processes. Other than doubts about the nature or existence of the stable point (climax), my only conceptual problem with Clements' processes is his identification of competition as a primary species-sorting process following establishment. "Competition" should be expanded to include any biotic interaction, as shown in Fig. 18.3, though I only discuss competition, *per se*, with regard to Figs. 18.4 and 18.5.

That Clements' processes are sufficient (see Botkin Chapter 5) to characterize succession is suggested by Fig. 18.3, a graphical model,

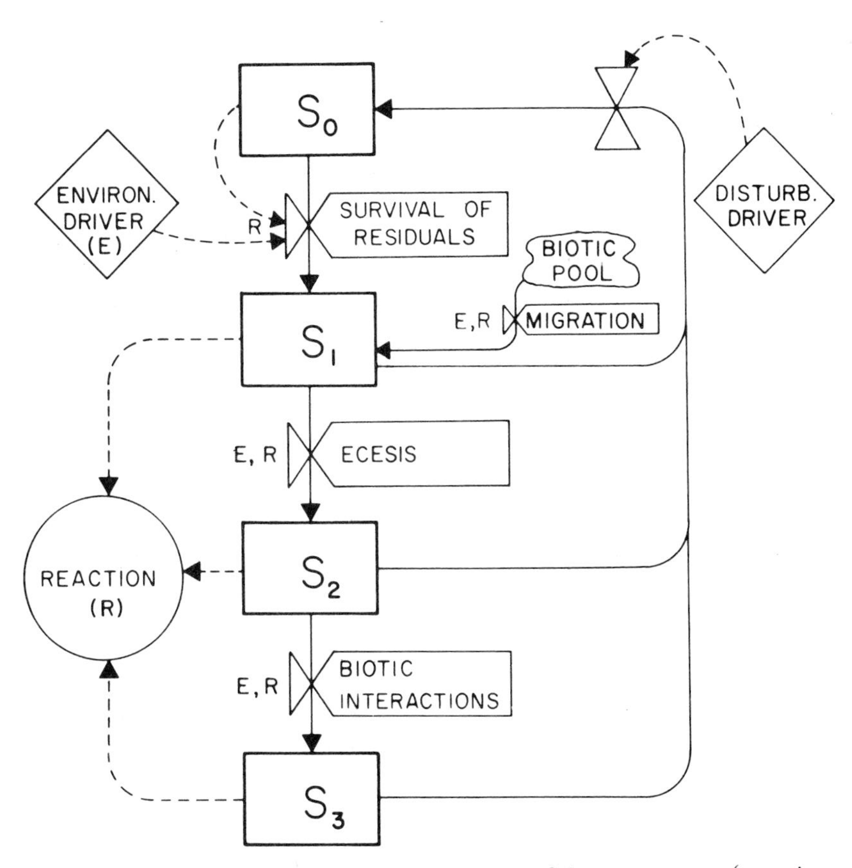

Figure 18.3. A model of the change in the status of the components (organisms and chemicals and physical conditions) of a plot of ground over time. The boxes are states of the plot at any instant. The diamonds are system drivers. The circle is an intermediate variable. Dashed arrows show information flows. Letters next to control gates replace dotted lines from that point to the control for the sake of graphic simplicity. (Taken, by permission, from MacMahon 1980a.)

which, using contemporary ecosystem approaches, adequately represents successional processes. In fact, the model permits progressive or retrogressive succession (see J. Walker et al. Chapter 9, also B. Walker Chapter 25), primary or secondary succession, or for that matter any ecosystem change over time—including year to year fluctuations, as well as the longer-term changes (see Davis Chapter 10) termed community evolution (MacMahon 1980a).

	DESERT	TUNDRA	GRASSLAND	CONIFEROUS FOREST	DECIDUOUS FOREST	RAIN FOREST
NUDATION	drought	cryo-planation	fire	fire/wind	wind/senescence	senescence/wind
MIGRATION	seed reserves	non-seeds	seeds and non-seeds	seeds	seeds	seeds
ECESIS	periodic	slow-periodic	moderately fast	mod. fast (variable)	fast	very fast
COMPETITION	water	water	water/light	light/water	light	light
REACTION	low	moderately low	moderate	moderately high	high	very high
STABILIZATION	fast	fast	moderately fast	slow	slow	moderately slow
MISCELLANY	no physiog. no species	no physiog. no species	mod. physiog. high species	high physiog. high species	high physiog. high species	high physiog. high species

Figure 18.4. A comparison of the differences in successional processes involving plants among various biomes. The sequence of biome types is derived from the sequencing of biomes in Fig. 18.2. The specific comparisons summarized are: nudation, factors most important in creating a disturbance; migration, the major type of plant migrules or propagules initiating site recovery, other than those residual, following disturbance; ecesis, the constancy of and rate at which plant establishment occurs when propagules are available on a site; competition, relative importance of water and light as limiting resources; reaction, the degree to which the seral biota (plants) alters the chemical and physical environment of a site; stabilization, rate at which the biota (plants) on the site stabilize (physiognomically and compositionally), i.e., "climax" occurs; miscellany, degree of physiognomic and species turnover during succession. (Taken from MacMahon 1981.)

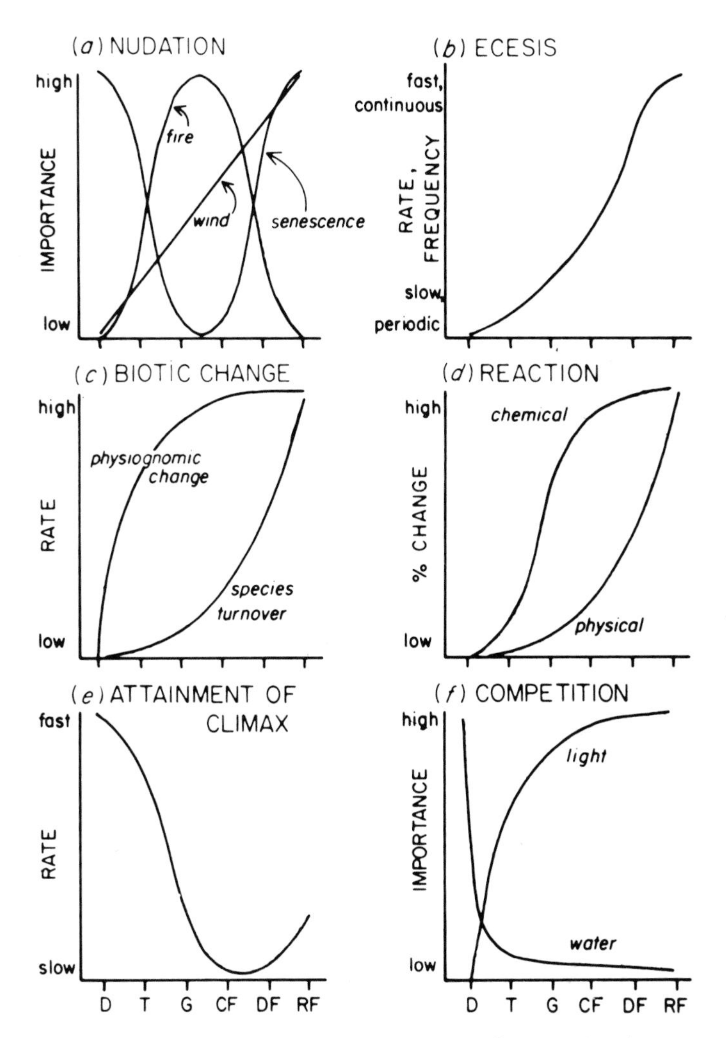

Figure 18.5. Hypothetical trends for some aspects of successional processes as they vary over a diversity of biomes based on Fig. 18.4. Biome sequence is as in Fig. 18.2. D = desert, T = tundra, G = grassland, CF = coniferous forest, DF=deciduous forest, RF = rain forest. (Taken, by permission, from MacMahon 1981.)

Comparison of Successional Processes in Various Biomes: Plants

Comparisons herein are made by assigning a qualitative term to describe the importance of the various processes in each biome type. An estimate of low importance for a process in a biome type does not mean that the process does not occur, only that the process is not as important in determining the vector of succession in that biome as it is in the biomes where its importance is indicated to be high. By vector of succession I imply both the rate and direction of changes in the biotic and abiotic components of a plot of ground.

After depicting the qualitative judgments in tabular form (Fig. 18.4), hypothetical curves are drawn (Fig. 18.5) to illustrate the possible forms of the relationships. References that suggest the adequacy of curve shapes are given in MacMahon (1981). Several of the curves are undoubtedly in partial error, and other processes could be depicted. Nonetheless, they may inspire more critical thought concerning successional processes and their variations from ecosystem to ecosystem, and they seem to fit available data, including those presented in other chapters of this book.

A final caveat: the information used to develop the form of the curves is somewhat biased. For example, coniferous forest is more explicitly slanted toward boreal and subalpine coniferous forests, deciduous forests are mainly those of the north temperate zone, and grasslands are, more specifically, temperate grasslands, which may not be representative of tropical savannahs. These biases are due to my subjective familiarity with certain communities and to the availability of literature. Perhaps, as I have heard from two authors in this book (J. Walker and B. Walker), the southern Gondwanian continents are essentially different. I do not think so, but only more data than are currently available can resolve this.

Biome Comparisons

Deserts

Secondary succession in deserts is seldom initiated by fire except during a period following above-average rainfall and its resultant increase in plant fuels. Plants are adapted to strong wind and are unaffected by it. Once established, desert perennials are long-lived [up to 9400 years for Larrea (Vasek 19801)]. Periods of below-freezing temperatures, animal damage, pathogens, or intricate combinations of these may act as denuding forces. More commonly, however, plants simply become senescent, die, and are replaced (Figs. 18.4 and 18.5).

The critical period for biota regeneration after disturbance is that involving establishment (ecesis). The variable rainfall may make germination and subsequent establishment an episodic event (see MacMahon and Schimpf 1981 for a review). Once established, however, the "climax" species mix is usually in place, and thus attainment of "climax" is rapid—40 years or so (Vasek et al. 1975) (Figs. 18.4 and 18.5b,e).

Because of the harsh environment, desert plants do not change the physical or chemical characteristics (reaction) of the site as much as is observed in other vegetation types. An exception for deserts would be the production of allelochemicals by plants, though the exact role of these is not clear. Where changes do occur, they are generally directly under the plant canopy. While this is important within the desert system, it is minor compared to other ecosystems (Figs. 18.4 and 18.5).

Plant competition varies. If a site is an extreme desert, so little water may be available that plants never attain densities sufficient to initiate competition (e.g., Gulmon et al. 1979). On the driest sites only shrubs or low-form cacti occur either before or following disturbance (e.g., the Atacama Desert of South America); therefore, there is no physiognomic change—autosuccession. In the Sonoran Desert of North America shrubs reestablish rapidly following major disturbances such as flood (see Zedler Chapter 24), but slower-growing species, e.g., tall columnar cacti (*Cereus*), do not establish their physiognomic, visual dominance until 20 or so years later—despite their early postdisturbance presence under the cover of "nurse plants."

Tundra

Tundra plants are long lived and are seldom "killed" by agents other than soil movements (Churchill and Hanson 1958). Even extensive animal activity seldom denudes an area or changes the species present (Wielgolaski 1975). Thus, fire and wind are unimportant (Figs. 18.4 and 18.5) while cryopedomorphic changes, not shown in the graphs, are the important agents of disturbance.

Establishment may be episodic, as in deserts, and is dependent on vegetative propagules rather than sexual ones (Bliss 1971). Once establishment occurs, the "climax" mix is generally there (autosuccession, as discussed previously) (Bliss et al. 1973) (Figs. 18.4, 18.5b and e), and therefore biotic and physiognomic change are inconspicuous (Figs. 18.4 and 18.5).

Although tundra plants alter soil composition, this reaction process is overridden by the harsh temperature effects on plants. This statement does not overlook the role of plant cover as an insulator over permafrost (see Van Cleve and Viereck Chapter 13). Competition, rather than being related to light (Figs. 18.4 and 18.5) or water, may, more importantly, involve nutrient availability.

Grassland

Grasslands may be fire dependent for their existence. Periodic drought is also important. Postdisturbance establishment may be from residuals, or sexual propagules carried by wind or animals to the site. Establishment is rapid, depending on the scale of the disturbance [a qualification necessary for any nudation process (Vitousek and White Chapter 17)]. Once established, significant chemical and physical changes in soil and microclimate ur (Hinds and Van Dyne 1980); moderate species turnover and gnomic changes occur (Weaver 1954) (Figs. 18.4, 18.5c and d), attainment of final "equilibrium" may take several decades. Comtition for water is less than in deserts, though it can be severe during periods of drought (Figs. 18.4 and 18.5). Some self-shading may cause competition for light. Again, nutrient availability may be the system attribute most intimately linked to competition. The sensitive competitive milieu created by water and nutrient availability, and its effects on the successional vector are clearly shown by B. Walker (Chapter 25).

Coniferous Forest

Coniferous forests are often denuded by fire (Heinselman Chapter 23) or wind (Sprugel 1976). Disturbed areas are regenerated by both sexual and vegetative propagules, often transported by animals, but residuals are important (Archibold 1979; Van Cleve and Viereck Chapter 13). Herbs and shrubs establish early, but are often rapidly topped by fast-growing deciduous tree species (e.g., Populus spp.) (Schimpf et al. 1980; Heinselman Chapter 23; also Van Cleve and Viereck Chapter 13) (Figs. 18.4 and 18.5b).

The slower growing conifers develop later and provide strong physiognomic contrast between early and late succession (Figs. 18.4 and 18.5c). The slow rate of conifer development and long persistence of some early seral species mixes causes this succession to take centuries or even 1000 years. This is the case for boreal (Van Cleve and Viereck Chapter 13; also Heinselman Chapter 23), temperate (Franklin and Hemstrom, Chapter 14), and subalpine (Schimpf et al. 1980) coniferous forests (Figs. 18.4 and 18.5e). Warm temperate conifer forests may have much shorter development times (Christensen and Peet Chapter 15).

Deciduous Forest

Deciduous forests are denuded by many agents, but especially human beings. For all biomes, however, I ignore anthropogenic disturbance. As the moisture increases across the deciduous forest-coniferous forest transition, fire becomes less important as an agent of disturbance. Trees live to old age (become senescent) and are subject to blowdown by wind (tip-ups) (Figs. 18.4 and 18.5a). If trees fall one at a time, a regeneration mosaic, hardly discernible to the casual eye, is established. If larger areas are denuded, more obvious physiognomic and species turnovers take place—the disturbance scale phenomenon.

Ecesis on a previously forested site is rapid in the relatively benign environment. Plants with many different life history characteristics establish simultaneously, including the slow-growing climax dominants (Drury and Nisbet 1973; Christensen and Peet Chapter 15).

The establishment of climax dominants may be in proportion to a species' importance before disturbance, or may actually be its complement (see Woods and Whittaker Chapter 19). At any rate, replacement or regeneration series are sufficiently predictable that they can be modeled (Botkin Chapter 5; Shugart et al. Chapter 7), perhaps even over geological time (Davis Chapter 10; Solomon et al. Chapter 11).

Significant reaction, physical and chemical, occurs (e.g., Bormann and Likens 1979a), changing species composition and the competitive milieu (Figs. 18.4 and 18.5f). The attainment of climax is faster than in conifer forests (200–500 years—Figs. 18.4 and 18.5e) probably because of favorable climatic conditions for growth.

Rain Forest

Rain forest trees may persist for long periods—100–1000 years (Budowski 1970). The accumulation of large epiphyte loads may increase the probability of a tree of blowing down (Strong 1977). Wind is a prime denudation force since well-developed forests will not carry fire, and drought and ice storms do not occur (Figs. 18.4 and 18.5a). For some areas geomorphic mass wasting can be important.

Forest openings often form a mosaic of small patches. Because these patches have very different moisture, temperature, and light regimes than the surrounding forests, germination and establishment, while extremely rapid, are most often by species other than those represented in the mature canopy (see review in Farnworth and Golley 1974; also Gomez-Pompa and Vazquez-Yanes Chapter 16). The significant altering of the physical and chemical environment (Figs. 18.4 and 18.5d) by the vegetation is well known and has even been used as part of an argument concerning the cause of high tropical species diversity (Ricklefs 1977).

Since marked changes, environmental and floristic, occur, and because the mature canopy species grow so slowly (Richards 1952), the time to "climax" is long (250 years or more) (Figs. 18.4 and 18.5e) considering the rapidity of many biological processes and the extended growing seasons.

Competition is seldom related to water availability and is undoubtedly more often related to light (Figs. 18.4 and 18.5f) or nutrient availability. All the rain forest patterns suggested in Figs. 18.4 and 18.5 are consistent with data in Tomlinson and Zimmerman (1978) and with the detailed plant life history review and ancillary data of Gomez-Pompa and Vazquez-Yanes (Chapter 16).

Animals and Succession: General

It seems reasonable that, to infer the connections between succession and animals, one might ask at least two questions. First, how do animals affect other ecosystem components in a way that might have successional consequences? Second, how might the results of succession, by other ecosystem components, affect animals?

As an aid to organizing the discussion of the multifarious possible relations between organisms, I shall use a conceptual model of some components of an ecosystem. I shall not examine every relationship nor implication of this model. Rather, I attempt to highlight possible ecosystem relations involving animal effects on successional processes that seem to be more frequent than is suggested by the mass of extant literature. Thus, I am specifically posing some ideas, in their most polar phrasing, to stimulate critical thought and discussion.

The Ecosystem Model

The interaction of organisms in ecosystems are of two general types. One of these involves the relations based on the exchange of energy and matter; the other set of relations is one that is based on the effects of organisms on one another in ways not related to, or at least out of proportion to, matter-energy exchanges (MacMahon et al. 1978). Such effects are currently thought to be the most important influences of animals on ecosystem processes. Analysis of ecosystem models generally substantiates this (Grant and French 1980; Lee and Inman 1975).

Since succession involves ecosystems over time, any effect of an animal on ecosystem components or processes can have potential successional consequences—that is my specific inference and reason for presenting the model here.

The following discussion refers to the ecosystem relationships depicted in Fig. 18.6. This conceptualization of an ecosystem is parallel to that of Chew (1974) and Wiegert and Owen (1971) but emphasizes the nonmaterial or nonenergy interactions of ecosystem components. There are no "effect" arrows between primary producers and decomposers. While there are undoubtedly some non-energy-flow interactions, these are consciously omitted. In Fig. 18.6 animals show up in two places—as consumers (biophages) and as decomposers (saprophages). My personal knowledge of decomposers is limited and is reflected in the brevity of comments about them. Also, I do not address arrows in the model that do not involve either animal decomposers or consumers.

Flows of Matter and Energy Involving Consumers

This classic figure of energy flow in ecosystems (solid arrows) (Fig. 18.6) overemphasizes the consumers as direct handlers of primary production and underrepresents the direct handling of primary producers by the decomposers (Wiegert and Owen 1971). As Chew (1974) notes in his valuable synthesis, less than 20% of the net primary production for ecosystems is handled directly by herbivores.

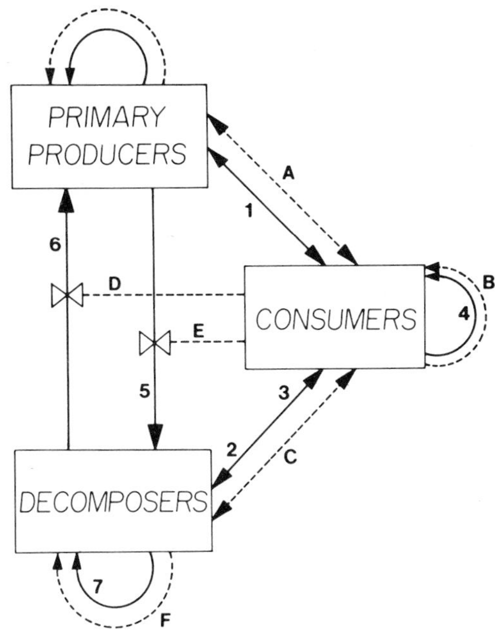

Figure 18.6. The interrelations of ecosystem components based on material flow (solid lines and numbers) and interactive affects not mediated by material or energy flow (dashed lines and letters) where: 1 = flow of energy or matter between animals and plants. The flow, from animals to plants, is considered insignificant. 2 = the flow of matter or energy from consumers to decomposers, while, 3=is the reverse flow which may have significant ecosystem consequences. 4 = predation. 5 = plants going directly to decomposers without the intermediate activity of consumers, and 6 = the matter or energy recycling to the plant from the soil via the decomposer activity. All lettered arrows are the two-way effect of the connected boxes on one another, or in the case of D and E, represent the potential influence of consumers on the rates of plants going to the decomposers (for example wastage) or of the recycling of energy or nutrients back to the plant from the decomposers. (After Anderson et al. 1980.)

It is difficult to separate the ecosystem effects of energy consumption *per se* from other indirect—but potentially important—effects. For example, is the consumption of decomposers (arrow 3) an important ecosystem relation or is it the dispersal of hypogeous fungal spores after passage, unscathed, through the consumer's gut tract (Maser et al. 1978b) that is the important function related to the consumption process?

Some cases in which the direct effects of consumption can have significant impacts are well known. Through defoliation of certain tree species, insect larvae can act as a major force destroying patches of forest and initiating secondary succession. An example is the effect of spruce budworm (*Choristoneura fumiferana*) on conifer species; it can locally eliminate balsam fir (*Abies balsamea*) (Morris 1963). Similarly, the direct consumption of decomposers can prevent the overgrowth of fungal mycelia and thus maintain the fungi in a state of exponential growth—and by implication a higher rate of nutrient cycling (MacBrayer 1973). Despite these examples, the major role of most consumers is probably related to their ecosystem effects other than those directly related to energy consumption.

Flows of Matter and Energy Involving Decomposers

The worldwide pattern of an increase in soil organic matter (SOM) from the equator to the poles, and an inverse relationship for net primary production, suggests that the direct influences of decomposers may vary latitudinally (Swift et al. 1979). The SOM accumulation pattern may be due to temperature limitation on microbial decomposition toward the poles. There is an interesting equator-to-pole shift in the size of animal decomposers. Figure 18.7 depicts the percentage of change in the biomass of the macrodecomposer fauna, compared to the summed faunal biomass of the microdecomposers and mesodecomposers (Swift et al. 1979). Since the macrofauna are implicated as important comminuters of litter and the smaller forms as chemical transformers of litter, the global percentage change in decomposer faunal composition may represent an important

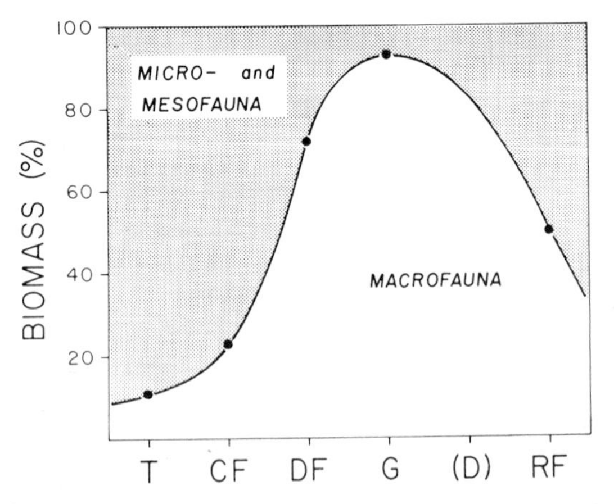

Figure 18.7. The curve of the percentage of the total biomass of animal decomposers that are macrofaunae (body width >2 mm; body length >10 mm) and therefore comminuters. Obviously, the sum of the microfaunae (body width < 100 μm; body length <200 μm) plus the mesofaunae (body width 100 μm-2 mm; body length 200 μm-10 mm) is the remaining portion of 100%. Data were estimated from figures in Swift et al. (1979, pp. 306–317). Note that the arrangement of biomes is based on increasing mean annual temperature (left to right). T = tundra, CF = coniferous forest, DF = deciduous forest, D = desert (not available), RF = rain forest.

direct effect of the rate of energy handling by decomposers (Fig. 18.6, arrows 2 and 5) and the subsequent cycling of materials back to primary producers (arrow 6). The possibility of this pattern is suggested because rates of decomposition are supposed to be related to the size of the pellets produced by the animals (see discussion and Fig. 4 in Kitchell et al. 1979).

Consumer Effects Other Than Energy Flow

The effects discussed here are distinguished from those discussed previously in that if energy flow is involved, the ecosystem effect is out of proportion to the quantity of energy consumed. In that sense arrow A (Fig. 18.6), in the direction of animal to plant, can include important processes such as: pollination, movement to safe sites, germination, altering growth form (perhaps by terminal bud clipping), regulation of species diversity, production, evolution via genetic selection, dispersal, transmission of disease, etc. Any part of this incomplete list can have significant successional consequences (e.g., Ross et al. 1970).

Similarly, any effect of a consumer on the consumer portion of the ecosystem (arrow B) can have effects that flow back to other ecosystem components and consequently have significance in the context of one or another successional process. Some such effects are reviewed by Chew (1974) for a number of organisms and Chew (1978) specifically for mammals.

Consumer impacts on decomposers (arrow C) are probably numerous, but are only now being documented, e.g., small mammals may inoculate logs with fungal spores and thus increase decomposition rates (e.g., Cromack Chapter 22; Maser et al. 1978a). This could be especially important in tropical areas where the fungal endobionts are physiologically unspecialized and thus any tree can use nearly any species with which it comes in contact (Malloch et al. 1980).

Decomposer Effects Other Than Energy Flow

The effects of decomposers on consumers (arrow C) not involving significant flows of energy are difficult to document and are probably minor in effect, except for cases in which decomposer modification of a substrate, e.g., litter, provides suitable homes or reproduction sites for consumers, or perhaps where decomposers produce toxic substances (Janzen 1977).

The effects of decomposers on other decomposers are legion. Through their effects on the physical and chemical properties of the substrate decomposers actually induce "*reactions*" that permit a separate set of decomposers to take up occupancy in what is essentially a new seral stage. Such "relay faunistics" are known for numerous cases, including the succession of arthropods on carrion (M. D. Johnson 1975), in polypore fungi during different developmental stages (Graves 1960) and in dung (Valiela 1974), to name a few.

It is important to draw a distinction between the effects of decomposers causing changes in the decomposer fauna because of their own activities in contradistinction to other successional changes in the decomposer fauna. As Usher and Parr (1977) rightly suggest, there are potentially two types of succession in decomposer communities:

1. "The succession imposed on the decomposer community due to the overall change in habitat." This is due to the pronounced successional vector of the whole ecosystem, including plant species turnover, which changes the chemical substrates available to decomposers.

2. "The succession imposed on the decomposer community by the breakdown of their own resources." This is one form of the relationship suggested by arrow F (Fig. 18.6) to which I alluded above. Microseral changes in maturing shelf fungi may simultaneously involve both types of decomposer succession.

Comparison of Successional Processes in Different Biomes: Animal Results

Most studies of succession have focused on vegetation; however, the dominant species of animals also change with succession and animals affect succession. Animals have several major effects on succession through nudation, migration, ecesis, biotic interactions, reaction, and stabilization. These will be defined and discussed in turn (Fig.18.8). Since much of the successional terminology has been developed for plants (MacMahon 1980a, 1981), application of such a lexicon to animals will require some redefinition.

All of the discussion will be directed to two summary figures (Figs. 18.8 and 18.9) which, as in the plant discussion, are a matrix of qualitative statements concerning the relative impact of animals on the process listed for the array of biomes. These are then depicted as a series of curves to show what I propose to be the general form of the relationships. A "low" importance does not mean that there is no example of an animal controlling that process for that biome type—merely that when comparing all biomes it is judged to be an insignificant overall effect.

Nudation (Figs. 18.8 and 18.9a)

Nudation is the process of baring a patch of ground. The result for initiation of secondary succession depends on the scale of the disturbance and the type of ecosystem involved. Consumption of a single flowering annual plant by an herbivore is nudation, but at an uninteresting and perhaps unmeasurable scale.

	DESERT	TUNDRA	GRASSLAND	CONIFEROUS FOREST	DECIDUOUS FOREST	RAIN FOREST
NUDATION	low	moderate	mod. high	high	mod. low	low
MIGRATION	low	low	moderate	mod. high	high	very high
ECESIS	low	low	moderate	moderate	mod. high	high
BIOTIC INTERACTION	low	low	low	mod low	moderate	high
REACTION	low	mod. low	high	high	moderate	low
STABILIZATION (SP. COMP.)	low	low	moderate	moderate	high	very high
STABILIZATION (NUTRIENTS)	mod. high	low	high	mod. low	mod. high	moderate

Figure 18.8. A comparison of the differences in successional processes involving animals among various biomes. The sequence of biomes is derived from the sequencing of biomes in Fig. 18.2. The specific comparisons summarized are: nudation, the role of animals as agents of site disturbance; migration, the role of animals in dispersing plant propagules, only positive effects are summarized; ecesis, the role of animals in enhancing plant germination and establishment; biotic interaction, the importance of species-species interactions involving animals in determining the vector of succession; reaction, the degree to which animals are agents in altering the physical and chemical environment of a site; stabilization, the role of animals in determining and maintaining a given, quasi-equilibrium, plant species mix and in determining the cycling of nutrients.

In deserts, animal-induced nudation is unimportant; none of the animals completely destroys the flora, though some plants, e.g., saguaro (*Cereus giganteus*), may be destroyed by pathogens that gain entrance via an animal agent (Niering et al. 1963). Burrowing activities of animals are not of sufficient extent to have significant effects. However, grazing in ecosystems transitional between grassland and desert shifts the balance toward desert vegetation (MacMahon and Wagner 1981).

Tundra, like the desert, seems to be resistant to animal-caused nudation (Bliss 1975). Even intensive grazing by sheep does not seem to denude the system nor to change plant species composition (Wielgolaski 1975). My estimate of moderate denudation effects of tundra animals is

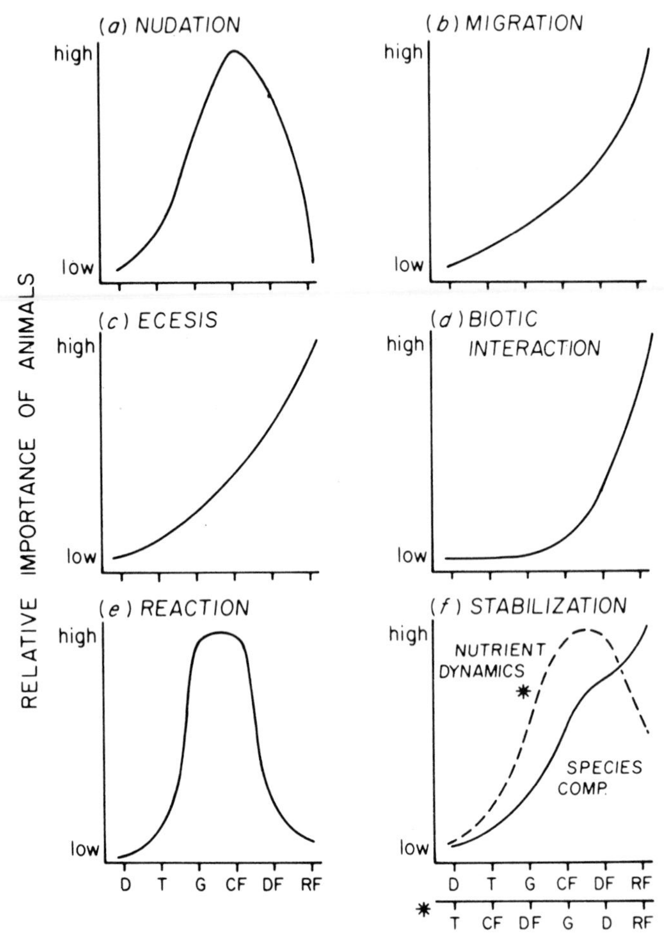

Figure 18.9. Hypothetical forms of the functions relating the roles of animals in successional processes to various biomes: (a) nudation; (b) migration; (c) ecesis; (d) biotic interaction; (e) reaction; (f) stabilization. Curves are based on qualitative assessments contained in Fig. 18.8. Note that two horizontal axes are labeled. The normal axis labels refer to species composition (solid line); the asterisked axis label refers to nutrient dynamics, the asterisked and dashed curve. D = desert, T = tundra, G = grassland, CF = coniferous forest, DF = deciduous forest, RF = rain forest.

based on activities of small mammals, especially microtine rodents, which is variable but often significant in tundra areas (Batzli 1975). During population peaks these rodents may remove 90% of the ground cover and initiate "succession." This effect, known mainly from coastal tundras, is periodic and does not usually simultaneously involve vast acreages (Dahl 1975).

Grasslands are continually disturbed by the burrowing activities of mammals (Platt 1975; Chew 1978; Grant and French 1980) and ants (Petal 1978), and are periodically overgrazed and opened to colonization by the consumptive activities of large mammals and even insects. Migratory habits of many mammals make estimates of their effect difficult (Bourliere and Hadley 1970), though the effects are clear for some species like elephants (Buechner and Dawkins 1961; Laws 1970; Wing and Buss 1970; Laws et al. 1975), and for domestic animals (Ellison 1960; Penfound 1964). In some cases grazing may actually stimulate plant growth, and plants not so adapted may not survive (McNaughton 1979a).

Coniferous forests, especially those of boreal North America, are frequently infested by defoliating insects, which can open large acreages to successional processes (Swaine 1933; Blais 1954; Lejeune 1955; Ghent et al. 1957). Even early successional species like birch can be defoliated and killed by insects such as *Oporinia autumnata* (Geometridae) (Haukioja 1975). Denudations by insects may be more likely in tree stands under additional environmental stresses (Mattson and Addy 1975).

Deciduous biome forests can be defoliated by insects (Macaloney 1966), but this is seldom as common or as widespread as in coniferous biome forests. Certainly cases such as the introduction of the chestnut blight fungus (*Endothia parasitica*) to the United States in 1904 and its spread by insects, with the subsequent decimation of American chestnut (*Castanea dentata*) (Mackey and Sivec 1973), are examples of animal-mediated denudation. However, these are uncommon in temperate deciduous forests. Animals other than insects seem not to be able to destroy deciduous forests by means other than transmittal of pathogens.

Patterns of tree dispersion in rain forests and the high plant diversity are often thought to space individuals of plant species so that no one animal can have denuding effects of major proportions. A number of local exceptions to this, i.e., species that are killed, particularly by insect pests, and cause canopy gaps are reviewed by Whitmore (1975). Nonetheless, animal-mediated denudation seems to be less important in rain forests than in grasslands, coniferous forests, or deciduous forests.

Migration (Figs. 18.8 and 18.9b)

Transport of plant propagules includes some aspects of migration and ecesis. Migration is merely moving the plant or nonanimal decomposer propagules to a disturbed site; ecesis is taken to be a plant's subsequent success.

Desert mammals, birds and insects, and especially ants (Davidson 1977a,b; Reichman 1979) disperse large quantities of seeds. By far, the majority of this movement involves seeds of annuals. However, seeds of perennials and asexual propagules (e.g., cactus "joints") are also transported. Generally, all the propagules in a desert would normally be moved by wind or the nearly laminar flow of surface runoff during storms (MacMahon and Schimpf 1981). Therefore, despite the quantity of seeds handled by animals, the absence of animals would probably not affect successful migration of plants to denuded sites.

Similarly, tundra vegetation does not depend solely on animal activities for migration. Many tundra species, even though they produce seeds, are actually propagated vegetatively (Bliss 1971).

Grasslands have significant complements of ants, birds and mammals (large and small) that can disperse seeds or other plant migrules—although few data are available to indicate how important this is. Fires in grassland cause nudations of great areal extent, and despite a fair degree of residual asexual propagules, wind-blown and animal-carried propagules are certainly more important than in the tundra (regeneration of vegetation in savannahs is discussed in Walker Chapter 25).

Conifer forests are, at least in the far north, often dominated by reproduction from propagules other than seeds (E. A. Johnson 1975). In many areas, however, small mammals and birds may provide seed sources for denuded areas, particularly as a result of caching activities. Of 19 seed-consuming vertebrates in piñion-juniper forests of Arizona, Salomonson (1978) identified 10 species as seed dispersers. Lanner and Vander Wall (1980) present strong circumstantial evidence that Clark's nutcracker (*Nucifraga columbiana*) is important as a disperser of several montane conifers. Turcek and Kelso (1968) review an extensive literature showing corvid-conifer interactions in Europe and Asia. Mammals such as red squirrels (*Tamiasciurus* spp.) may have similar roles (Finley 1969; Smith 1970), as may other small mammals (e.g., Abbott and Quink 1970).

Many plant species of deciduous forests are dispersed by animals. Many canopy tree species are nut bearers, clearly adapted to animal dispersers, e.g., *Quercus* and *Juglans* (Chettleburgh 1952; Smith and Follmer 1972). Additionally, deciduous forests contain numerous fleshy fruited herbs, shrubs, and small trees, many of which seem to be early successional species (e.g., *Prunus sertotina*) and are adapted to disturbances (Thompson and Willson 1978). Even small-seeded herbs are carried by arthropods such as ants (Beattie and Lyons 1975; Beattie et al. 1979).

Finally, rain forest species, whether they are in the New World (Janzen 1975) or the Old World (Whitmore 1975), rely heavily on animal dispersal of seeds, commonly contained in fleshy fruits (Snow 1971). In fact, wind-dispersed seeds in the tropics have been found to be "inefficient" with regard to establishment (e.g., Burgess 1970). The role of vertebrates and invertebrates in vegetation migration is so pervasive that even groups like warblers, vireos, and flycatchers, generally thought to be insectivores in North America, are frequently found to contain undigested seeds in the neotropics (Gomez-Pompa and Yanes-Vasquez Chapter 16); even fish may be important dispersers of terrestrial rain forest species in seasonally inundated areas (Gottsberger 1978).

In addition to the positive aspects of animals as effectors of plant migration, many animal species have a negative influence on dispersal; for example, of 33 plant species handled by Amazonian fish, 16 species were dispersed and 17 species were destroyed (Gottsberger 1978). For further substantiation of the patterns mentioned above, see Ridley (1930), van der Pijl (1972), C. C. Smith (1975), and McKey (1975).

Ecesis (Figs. 18.8 and 18.9c)

The roles of animals in the establishment and growth of plants parallel and are often related to their roles in seed movement. Interestingly, in deserts, tundra and even grasslands (Figs. 18.8 and 18.9c) the propagules that animals handle are individual seeds or nonfleshy fruits. As rainfall increases, the propagules become increasingly fleshy fruits, culminating in fleshy fruit-dominated rain forests. This pattern can affect the impact of animals (Halevy 1975). Nonfleshy seeds are cached by seed-predatory rodents and insects. A few may escape seed predators, but cache sites are not always favorable germination sites. In contrast, fleshy fruits are ingested immediately and the indigestible seeds deposited with the feces. The presence of the fecal matter may provide nutrients and moisture that enhance germination. On this basis alone, the shape of the curve in Fig. 18.9c is justified.

Desert animals handle many seeds, and recent studies suggest that distantly related taxa vie for these in a competitive manner (Brown et al. 1979), thus reducing soil seed reserves. That these heavy seed predations may ultimately depress the abundance of annuals is suggested by increased densities of annuals when rodents, ants, and birds are excluded. Such animals probably do little to promote plant establishment, but do much to reduce it. Such relationships have a strong historical component and may vary between similar ecosystems within the same biome on different continents (e.g., Mares and Rosenzweig 1978).

Since tundra plants frequently reproduce asexually, animals are of little importance to establishment. Exceptions may be the interesting ornithocoprophilous lichens and algae of Antarctica and northern Europe, which are primarily distributed by birds via their feces (Lamb 1970).

Grassland animals may create sites favorable to plant establishment (see Platt 1975 and references therein). Such effects can be due to invertebrates or vertebrates (Brahmachary 1980). In some cases the effects, particularly the negative ones, can be quite pronounced, often depending on the specific kind of disturbance, as exemplified by the postfire role of grasshoppers in Australian parklands (Whelan and Main 1979).

In coniferous forests, mammals and insects have significant positive and negative impacts on plant establishment (e.g., West 1968; Radvanyi 1970). These range from short-term effects (e.g., moose, Bedard et al. 1978, and references therein) to the long-term effects of gophers (Tevis 1956), or of insects.

In deciduous forest, and especially rain forests, animals may play an important role in plant establishment (e.g., Livingston 1972; A. J. Smith 1974). Again, both positive and negative effects are noted (Janzen 1971; Smith and Follmer 1972). In rain forests the possibility of seeds in animals' feces being associated with generalist mycobionts' spores (Malloch et al. 1980) may be an important ecesis phenomenon. By moving seeds away from parent plants, animals may also prevent some seed predation (e.g., Janzen 1978).

Biotic Interactions (Figs. 18.8 and 18.9d)

Obviously, animal-mediated migration and ecesis are just discrete cases of biotic interactions. In general, the positive animal effects on reaction processes involving the soil can be divided into at least two important types (Kitchell et al. 1979). The first is translocation, including those processes in which the animals move material into the upper layers of soil where plants can use them. Such movements can be from above, as in the case of carbohydrate enrichment of the soil from aphid "honeydew" deposition (Owen and Wiegert 1976). Additionally, the defoliation of trees has the effect of increasing soil nitrogen (Bocock 1963a,b). Translocation of material can also come from below in cases where animals, like earthworms, can increase nitrate by 370% and exchangeable Ca by 40% (Satchell 1967; Lofty 1974) by moving mineral soil to the surface layers. Similar effects, for different chemicals, are known for a variety of small mammals (Chew 1978).

A second important soil reaction process is transformation (Kitchell et al. 1979), wherein animals alter materials in the soil and, in turn, affect decomposition rates. Frequently, changes in surface area-to-volume ratios of litter particles are thought to be caused by animals (Webb 1977). Such changes can alter decomposition rates and nutrient residence times (Swift et al. 1979; Kitchell et al. 1979). The role of animals seems especially important for these processes in tundra (Bliss 1975) and grasslands (McNaughton 1979b).

In evaluating animal impacts on succession, the curve for biotic interactions (Fig. 18.9d) was drawn to emphasize the dominance of the abiotic environments of deserts and tundra, a slight relaxing of this in grasslands but still a strong independence of animals, and finally a sharply increasing species-species coupling of plants and animals in forested systems with maximal development in the tropics. The literature cited up to this point suggests this pattern. One only has to think about examples of obligatory species-species relationships; more tropical examples exist than for any other biome, and fewest are available for deserts or tundra. Data on important biotic interactions, such as pollination, which may have successional consequences, show these patterns (Moldenke 1976).

The discussion of animal impacts up to this point, because of the nature of the processes involved, is really limited to the roles of insects and vertebrates working above the ground. The addition of below-ground decomposer dynamics is appropriate in this section and follows the proposed curve. Where there are data (e.g., tundra), decomposer animals do not seem to be tied to a particular plant community or litter type (McLean et al. 1977), and this may be true across most biomes for decomposer animals (Anderson 1978) and even animals that eat decomposers such as fungi (Dash and Cragg 1972). The rapid increase of the curve portraying the importance of animals in biotic interactions, therefore, depends in large measure, on above-ground consumers and seems a reasonable representation of available data.

Reaction (Figs. 18.8 and 18.9e)

The effect of animals on the abiotic environment, including soil characteristics (e.g., McColloch and Hayes 1922), varies greatly from biome to biome. Figures 18.8 and 18.9e depict animal impacts as low in desert, tundra and rain forests, and higher in all the other biomes.

The low animal impact in deserts and tundra is mainly because temperature (high or low) and rainfall patterns dominate soil processes. Low temperature causes soil organic matter buildup in tundra, despite animal activities. Low rainfall causes soil activities in deserts to be pulsed, in tune to the sporadic rainfall events (MacMahon 1979; MacMahon and Schimpf 1981; MacMahon and Wagner 1981). Even though desert animals contribute to soil processes, the effects of plants and microbes dominate most soil activities. Exceptions might be termites in certain warm deserts with moderate rainfall (Lee and Wood 1971; Brian 1978). Effects of termites on soil properties are often so pronounced that it has even been suggested that termites be manipulated as an aid to revegetation of mine spoils in arid areas (Ettershank et al. 1978).

In grasslands and forests other than rain forests, activities by burrowing mammals, e.g., gophers (Mielke 1977; Andersen et al. 1980), and invertebrates may have significant effects on the physical and chemical characteristics of the soil (see reviews by Chew 1978; Grant and French 1980).

Rain forests have few fossorial mammals or insects (Longman and Jenik 1974; Whitmore 1975), so the animal impact on physical soil turnover is less than that in other biomes. Additionally, the dominance of the decomposer process by leaching and microbial activity lessens the impacts of animals on chemical soil changes. Obviously, animals play some, often significant, role in regulating decomposition rates; Madge (1965) found that litter decomposed more slowly when arthropods were excluded.

Stabilization

By stabilization, I mean the persistence of species composition and physiognomy of a plot of ground for a given time interval. What role do animals have in maintaining "equilibrium" communities? If all animals were removed, could we measure a difference in some defined attribute of the plant community? In this respect, animals have two effects: one affecting nutrient dynamics and one in altering the species composition of the plant community (Figs. 18.8 and 18.9f).

The species composition curve shows an increase in animal influences along the biome sequence, as a consequence of the increase in the number of obligate plant-animal relations as one proceeds from vegetations of low form physiognomy into forested areas [see, for example, the important role of the mountain pine beetle (*Dendroctonus ponderosae*) in maintaining self-perpetuating stands of lodgepole pine (*Pinus contorta*) (Amman 1971)], and is most highly developed in rain forests. The increase of species diversity toward the equator may affect and be affected by these obligate relationships in some sort of a self-enhancing biological loop. See Whittaker (1977a) for a brief review.

The effects of animals on nutrient dynamics does not yield a sensible pattern across the biome sequence. This may be because, as long as water is available, nutrient dynamics are tightly coupled to temperature (Fig. 18.9f). Animal effects appear minimal in harsh environments (too hot or too cold) or where decomposition is largely a chemical process not requiring animals (rain forest) (Fig. 18.9). In between, animals are more important in transforming and transporting material during decomposition (Kitchell et al. 1979). Not unexpectedly, this curve parallels the world macrodecomposers biomass curve (Fig. 18.7), which was also a temperature-arrayed biome series. These semiindependent results suggest some reality to some of my assertions.

Effects of Other Ecosystem Components on Animals during Succession

It is obvious that the species mix of animals changes over time following nudation of a plot of ground. In the early part of this century Victor Shelford and his students studied such changes in animal communities along a number of Midwestern seres (Shelford 1963). The common interpretation of such changes paralleled the interpretation for plants: as the plant community reestablished and sequentially altered the environment of a disturbed site, the animals tracked the plant species changes as well as changes in the abiotic environment (Shelford and Olson 1935).

Specifically for above-ground consumers, I posit that the vegetational architecture is predominantly responsible for the animal species mix. Decomposer animals are not as closely tied to a particular substrate; that is, soil invertebrates, as individual species, feed on a number of foods derived from different plant or animal species. Data suggesting "decoupling" come from arctic tundra enchytraeid worms (McLean et al. 1977), desert microarthropods (Santos et al. 1978), and a variety of other taxa in numerous ecosystems (Wallwork 1976; Swift et al. 1979). The driving variable is probably the general nature of the material to be decomposed. This is also especially obvious in the microseral studies (e.g., Valiela 1974).

In fact, it is becoming more obvious that consumers as diverse as insects (Martin 1966; Murdoch et al. 1972; Futuyma and Gould 1979; Joern 1979; Southwood et al. 1979; Strong 1979; Strong and Levin 1979), spiders (Duffey 1978; Stratton et al. 1979; Hatley and MacMahon 1980), birds (Wiens 1974; Smith and MacMahon 1981), and mammals (M'Closkey 1975; MacMahon 1976; Fox 1978; Andersen et al. 1980; Bond et al. 1980) are more tightly correlated in their distributions to vegetation architecture, including dispersion pattern, than to particular plant species. Clearly the presence of certain plants can provide particular architectural configurations and thus form plant-animal correspondences (e.g., forest birds studied by Holmes et al. 1979). Obviously, this purely architectural relationship is less pronounced in the tropics with more obligate interspecies relationships, or in closely coevolved groups of species—the component communities of Root (1973). However, even in monotypic stands of collards (*Brassica oleracea*), plant dispersion pattern (a form of architecture) altered the herbivorous insect community composition (Cromartie 1975). Some interesting food-plant and consumer correlations, during succession, exist. For example, there are rough correlations between successional stages and plant palatability. The postulated increased palatability of early successional species compared to those of late succession (Cates and Orians 1975) does not, however, negate the relative catholicity of food choices that I allude to above. Feinsinger's (1978) hummingbird data suggest similar relationships in tropical successions.

There is the possibility that the generalist-specialist dichotomy may change with successional status of the ecosystem and/or the relative predictability of the resource availability (Odum 1969; Valiela 1971). Valiela's data are somewhat limited, however, because they pertain mainly to birds, which show significant correlations between the percentage of insectivores and the specific seral stage.

A few sample data may help to develop my point. A subalpine sere in northern Utah grades from meadow through aspen, subalpine fir, and ultimately is dominated by the long-lived Engelmann spruce. Such a site is described in detail by Schimpf et al. (1980). There is a strong physiognomic shift from meadow to aspen and aspen to fir; however, the two conifers, fir and spruce, are much more similar. Parallel to the physiognomic change, the mammals and birds are more similar in spruce and fir plots than any other pairwise comparison (Fig. 18.10). This relationship holds in years of normal precipitation (1976) or drought (1977). The change in species with community physiognomic change (architecture) is less pronounced for mammals, probably because the mammals in this study (except for squirrels) respond to the vertical structure of their ecosystems less than birds (MacMahon 1976). For this same site, similar trends, i.e., good correlations between animal species and community architecture and poor correlations between animal species and plant species *per se*, can be shown for spiders (Waagen 1980) and insects (unpublished data).

The cause-effect nature of such correlations is strongly suggested by a series of studies from my laboratory. Abraham (1980) showed that spider guild changes correlated to vegetation physiognomy over a series of plots containing different plant species. Hatley and MacMahon (1980) altered the physiognomy of a single shrub species (*Artemisia tridentata*) and found a spider guild assortment that matched the physiognomy of different plant species. Finally, Robinson (1981) made "artificial" shrubs confirming the correlation.

Thus, if succession involves conspicuous architectural change in communities, I would predict that there would be good correlation between

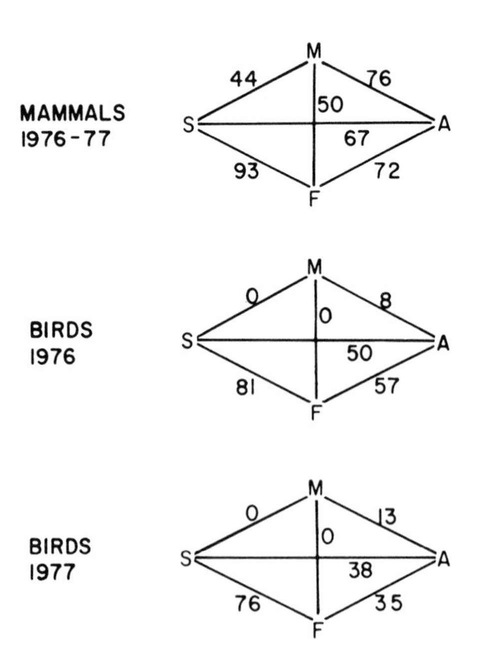

Figure 18.10. Diagram showing the degree of affinity among vertebrate assemblages in four seral stages of a subalpine sere in northern Utah: M, meadow; A, aspen; F, fir; S, spruce. The numbers between stages are the Jaccard coefficient of community (percentage of overlap). (From Schimpf et al. 1980.)

animal species turnover and physiognomic turnover along the sere, regardless of the plant species composition turnover. This is precisely the result of Southwood et al. (1979).

This architectural emphasis does not deny that the presence of a particular species may influence the presence of another species through some species-species interaction such as interference or exploitation (e.g., see Slagsvold 1980). Nor does it suggest that all animals respond equally to architectural changes within all biomes. For example, MacMahon (1976) suggests little response of desert mammals to vertical architecture changes.

In a broader context, there is an element of chance in all of the animal-succession relationships I treat, and this can alter the outcome of various interactions. For example, the time of colonization of plants by animals can alter the plant component in ways that feed back differently on the animal component of the community (e.g., Bach 1980). There are also differences in what appear to be similar relationships in different geographic areas. For example, insect effects on bacteria and fungi in cattle dung vary between Michigan and Wyoming (Lussenhop et al. 1980).

Similarly, a species in one ecosystem type may have very different relationships with other species in a different ecosystem type. Howe and Vande Kerckhove's (1979) example of the dispersal of *Casearia corymbosa* mediated by several bird species in secondary dry forest and by essentially one species (*Tityra semifasciata*) in wet forest emphasizes such a possibility.

Despite a number of necessary qualifications, such as the above and others not mentioned, the successional changes in plant architecture (physiognomy), characteristic of "conspicuous" succession, may be a very important factor in determining the nature of the animal community composition, regardless of the exact plant species contributing to the ecosystem architecture. Additionally, while there are variations in the importance of animals in the various successional processes from place to place, nonetheless, generalities supported by numerous studies can be offered.

Summary

The material presented in this chapter suggests that:

1. The conspicuousness of succession depends on the degree of harshness of the environment.
2. The unpredictability of precipitation in low precipitation ecosystems may be, in a cause and effect manner, related to autosuccession.
3. The processes identified by Frederic Clements provide sufficient explanation for what is generally termed succession.

4. The processes related to the phenomenon of succession vary along biome types in predictable ways, whether one addresses animals or plants.
5. Ecosystem processes in biomes characterized by extreme environments are dominated by abiotic processes, which in turn dominate the succesional processes.
6. In biomes of equable environments, biological factors dominate the successional processes and even alter the abiotic environment in ways that control seral changes.
7. Animals, while minor participants in ecosystem energy flow, can have important effects on other ecosystem components and processes, including succession.
8. Animal species are, to a surprising degree, decoupled from specific plant species, but are coupled to the architecture of the plants.

Finally, I hope that my speculations will stimulate the erection of explicit hypotheses about the nature of successional processes and the relative roles of plants and animals in various kinds of ecosystems. An understanding of the successional phenomenon has scientific as well as practical implications because humanity has altered the environment in ways that make most points on the earth a seral stage at some point in a sere other than at an equilibrium.

Chapter 19

Canopy-Understory Interaction and the Internal Dynamics of Mature Hardwood and Hemlock-Hardwood Forests

Kerry D. Woods and Robert H. Whittaker

Introduction: Stability and Perceptual Scale

In this chapter some of the population functions and interactions that are present in stable, self-maintaining forests are examined. We shall show by examples that these population functions or strategies are varied and suggest how they might maintain stable communities. Finally, we shall point out that these processes are not qualitatively different from those involved in successional change but may be thought of as extensions of them on various spatial and temporal scales.

Though definitions of and criteria for community stability are abundant and varied (Holling 1973; Orians 1975; Whittaker 1975c), we shall use a simple, functional definition: a community in which several species coexist, with relative importances (and population structures) remaining fairly constant for some significant amount of time, is stable. A suitably long time period might be at least three or four generations of the longest-lived species.

Consideration of scale is important in this definition. No community is stable by our definition if the time considered is great enough; secular change is universal. Similarly, the spatial scale of consideration may control the perceived stability of a community. No forest community would be stable if areas containing only one or a few trees were considered as units. At a large enough scale, almost any forest might satisfy our criteria for stability. These problems of scale receive some attention in Connell and Slatyer (1977).

One extreme might be seen in forests that suffer massive disturbance at intervals about equal to the generation time of dominant trees (e.g., in Heinselman 1973). No single stand could, under such a regime, attain stability in composition. Such a system may be perceived of as stable only in landscape units—areas of several hundreds or thousands of hectares, which could encompass stands of all ages in a dynamic equilibrium. Individual stands in such forest types are usually or perpetually undergoing successional change.

There exist, though, forest types in which we might expect individual stands to achieve and maintain stability. In forests with closed canopies, gaps caused by the death of canopy trees are generally required for the recruitment of new canopy individuals. Such forests will, like those mentioned in the last paragraph, exist as mosaics of patches in various stages of regeneration. If, however, disturbances or gaps are generally on the scale of one or a few trees rather than of whole stands, the difference in scale will favor different life history strategies. Furthermore, it is possible in such cases that a steady state might be maintained within a single stand—that is, over an area of one or a few hectares. We shall examine the possibility of stability being maintained through the interaction of canopy and understory in forests under such a regime of small-scale disturbance.

Canopy-Understory Interaction and the Microsite Mosaic

A mechanism for such a dynamic interaction may be developed on three propositions. First, we suggest that dominance by shade-tolerant species indicates that disturbances are normally single or few-tree gaps, and replacement is usually by advance reproduction in the form of suppressed saplings of the dominant species (Jones 1945). Such a regime could permit stands to reach a demographic equilibrium. Though certainly not common in some regions, we believe that such relatively stable forests do exist, are ecologically interesting, and can be interpreted in terms of population behavior and within-community pattern. Several studies support our suggestion that in stable forests in the American Northeast disturbances are small and infrequent and replacement of canopy trees is usually by saplings of the shade-tolerant canopy dominants (Fox 1977; Brewer and Merritt 1978; Canham 1978; Runkle 1979).

Second, we propose that important differences in the requirements of the dominant species of these forests might be expressed most strongly in early parts of the life cycle. The important niche differences (or limiting factors in the sense of Levin 1970) that permit coexistence of these species may apply primarily to seedlings or saplings. It seems reasonable that seedlings or saplings, which view the environment on a finer scale than mature trees, would experience a greater variability in microsite conditions. Recent related formulations include the microsite mosaic affecting seedlings (Whittaker 1975c), and the concepts of the regeneration niche (Grubb 1977) and the safe site (Harper 1977a).

Finally, we suggest that an important factor in generating diversity in seedling or sapling microsites is the influence of nearby canopy trees. Canopy trees might influence subjacent microsites differently by their needs for water and nutrients, by differences in shade or litter quality, or by differences in root competition. The spatial patterns of advance reproduction and replacement of trees would depend, then, on interactions between variables of canopy influence and requirements of seedlings and saplings that could be effective in maintaining diversity and stability in even relatively complex forests. This might occur through canopy trees favoring conspecific reproduction beneath their own canopies, or by canopy trees favoring other species in such a way as to cause a cyclic succession of species on a microsite, or by a combination of such interactions. In a theoretical discussion of the microsite mosaic Whittaker (1975c) and Whittaker and Levin (1977) have further discussed how such phenomena might occur.

Though we propose that canopy influence is a particularly important factor in generating variability in reproductive microsites for forest trees and permitting their coexistence, there are certainly other important factors (dimensions of the regeneration niche in the sense of Grubb 1977), and some of these may play roles in the dynamics of the forests we will discuss.

The gap size required for establishment and growth is known to vary among tree species. Tolerance of canopy cover ranges from that of shade-intolerant, early successional species, through gap-phase species, which may reproduce in small gaps, to very shade-tolerant species, which may germinate and survive beneath an intact canopy (though gaps are necessary for their entry into the canopy). Other dimensions of the regeneration niche involve physical microsite factors. Soil depth (Bratton 1976a), nearness to trees, logs, and rocks (Bratton 1976b; Hicks 1980), tree-fall pits and mounds (S. Beatty 1980, pers. comm.), and "nurse logs" are all known to permit differentiation in response by herbaceous species and seedlings of some tree species.

These sources of variability in the environment experienced by understory individuals of tree species might well be sufficient to permit substantially different patterns of exploitation among species, which, in turn, could allow stable coexistence. A niche space or microsite space in which seedlings of various species could be dispersed could be defined by axes of (1) biological factors such as canopy and root effects; (2) substrate factors such as soil type and microtopography; and (3) gap factors such as size and age of gap. The variables defining this space are not independent. Gap size may be influenced by species of gap-making canopy tree. Some species usually die standing; others tend to break or uproot (Runkle 1979), frequently knocking down other trees. Specific differences in types of tree fall will influence distribution of pits and mounds. Differences among canopy species in decay rates and wood characteristics will cause further interaction of biological and microrelief factors.

These niche variables are almost certainly all significant in affecting life histories of some species. We suggest that the influence of canopy species on reproduction is one factor in particular that should be important in forests with low substrate variability and infrequent disturbance, and that this influence can lead to dynamically stable mixed forest stands.

Beech-Maple Reciprocal Replacement

Aside from single-species forests with self-replacement, the simplest basis for stability through canopy-understory interaction would be reciprocal replacement (Woods 1979) in forests dominated by two species. If each species showed superior survival beneath canopy trees of the other species, the result would be a tendency toward replacement of each species by the other. These interactions could easily lead to maintenance of codominance. Such patterns are conceptually related to Watt's (1947) phasic equilibrium and the cyclic succession discussed by Connell and Slatyer (1977).

Such a pattern suggests some sort of avoidance by seedlings or saplings of conspecific canopy trees, as has been suggested in the literature. Grubb (1977) quotes Aubreville's observation that, in some forests, "Any one tree species never regenerated beneath itself." Jones (1945) mentions "a very strong belief amongst continental foresters in the tendency of one species to regenerate beneath canopy of another." Instances of such interactions are listed by Schaeffer and Moreau (1958).

In the beech-maple (*Fagus grandifolia-Acer saccharum*) forest type of eastern North America (Braun 1950), strong codominance by two shade-tolerant species is frequent. Observations in an old-growth beech-maple stand showed that beech and maple saplings seemed to be most often subjacent to canopy tree of the other species (F. E. Smith 1980, pers. comm.; Whittaker and Levin 1977). Horn (1971) makes a similar suggestion. Fox (1977), using observations of saplings in monospecific canopy patches and in canopy gaps, proposes a reciprocal replacement scheme in several two-species forests including a beech-maple stand (Table 19.1). None of these studies, however, described quantitatively the distribution of tree species in the understory with respect to distribution of species in the canopy.

In June of 1976 Woods (1979) examined such distributions in Warren Woods, a stand of old-growth beech-maple forest in southern Michigan described by Cain (1935). For each sapling (individuals >2 m tall and <15 cm in dbh) in each of six 0.1-ha plots, distance to, diameter, and

Table 19.1. Association of sapling species with canopy of cooccurring species in virgin forests.[a,b]

Sapling species	Canopy species No. 1	Canopy species No. 2	
Warren Woods, Mich.[c]			
Sugar maple	24	81	$\chi^2 = 60.0$
Beech	69	17	$p < 0.001$
Warren Woods, Mich.[d]			
Sugar maple	80	90	$\chi^2 = 4.95$
Beech	47	27	$0.025 < p < 0.05$
Great Smoky Mountains[e]			
Fraser fir	8	82	$\chi^2 = 12.1$
Red spruce	10	15	$p < 0.001$
Univ. of Florida Cons. Res.[f]			
Live oak	5	13	$\chi^2 = 0.30$
Water oak	22	35	NS
Tionesta, Pa.[g]			
Beech	46	162	$\chi^2 = 87.3$
Hemlock	67	23	$p < 0.001$
Tionesta, Pa.[h]			
Beech	21	39	$\chi^2 = 7.51$
Hemlock	37	23	$0.005 < p < 0.01$
Medicine Bow Mountains[i]			
Subalpine fir	29	77	$\chi^2 = 8.58$
Engelmann spruce	22	18	$0.001 < p < 0.005$

[a]From Fox (1977).

[b]Table gives numbers of saplings between 5.1 cm and 33 cm dbh of each species. In gaps, saplings may exceed 33 cm but are smaller than the dead trees formerly dominating them.

[c]Under monospecific patches.

[d]In gaps, 22 of maple, 25 of beech.

[e]Under monospecific patches. Yellow birch was relatively abundant in the stand. The numbers of fir, spruce, and birch saplings under canopy birch were 33, 7, and 2, respectively. There were 11 and 7 birch saplings under fir and spruce canopies. The spruce x birch association is not significant. For fir x birch, $p < 0.005$.

[f]Under monospecific patches. Upper size limit for saplings here was 25.4 cm instead of 30.5 cm.

[g]Under nearly monospecific patches.

[h]Under individual trees, 70 of each species, where mixed in canopy.

[i]In gaps, 24 of fir, 39 of spruce. Lower size limit 7.6 cm instead of 5.1 cm.

species of all canopy trees (>12.5 cm in diameter) within 10 m were recorded. The ratio of diameter of a canopy tree to its distance from a sapling was used as an index of its influence on that sapling. This index, summed for all canopy trees near a sapling or for all trees of a particular species, yields a measure of canopy influence that, though arbitrary, takes into account (1) all canopy trees within a set distance from a sapling; (2) size of these various canopy trees; and (3) distance from a sapling to canopy trees influencing it. Canopy influence as measured by this index could be through any of several mechanisms—root competition, shading, allelopathy, and so on—but one would expect it to show positive correlation with size of canopy tree and negative correlation with sapling-to-tree distance. An index that yields a comparative measure of beech and maple influence can be contrived thus:

$$I = \sum^{b} \frac{\text{dbh}}{\text{dist}} - \sum^{m} \frac{\text{dbh}}{\text{dist}}$$

where b is the number of beech trees within 10 m of a sapling, m is the number of maple trees, dbh is the diameter at breast height of a canopy tree, and dist is the sapling-to-tree distance. I is, then, our original index of beech canopy influence minus a similar index for maple canopy. When the two canopy species have nearly equal influence on a sapling, I should be near 0. The index becomes positive with increasing beech influence and negative with increasing maple influence.

This index was calculated for each sapling in the sample quadrats, and the distributions of beech and maple saplings with respect to it were compared. In the histogram showing these distributions (Fig. 19.1) maple and beech saplings show different distributions over the range of values of the canopy influence index. In particular, beech saplings show greater relative frequency at low values of the index (higher maple canopy influence). In fact, if these saplings are ranked by value of I, and a rank sum test is performed categorizing saplings by species, the difference between distributions of beech and maple saplings is highly significant ($\alpha < 0.005$). A null hypothesis of similar distribution of beech and maple saplings with respect to beech and maple canopy influence is readily rejected. The alternative strongly suggested by these data is reciprocal replacement. It appears that canopy trees exert a strong influence on distribution of species in the understory in this beech-maple forest, and this influence encourages coexistence.

These results are obtained with exclusion of beech saplings of obvious root-sprout origin. Root sprouts, of course, are usually located near the parent tree, heavily influencing a distance-weighted index. When sprouts are included in the analysis just described, the null hypothesis may still be rejected but only at $\alpha = 0.05$.

The role of root sprouts in beech reproduction is unclear. Ward (1961) has described variation over the range of beech in the relative abundance

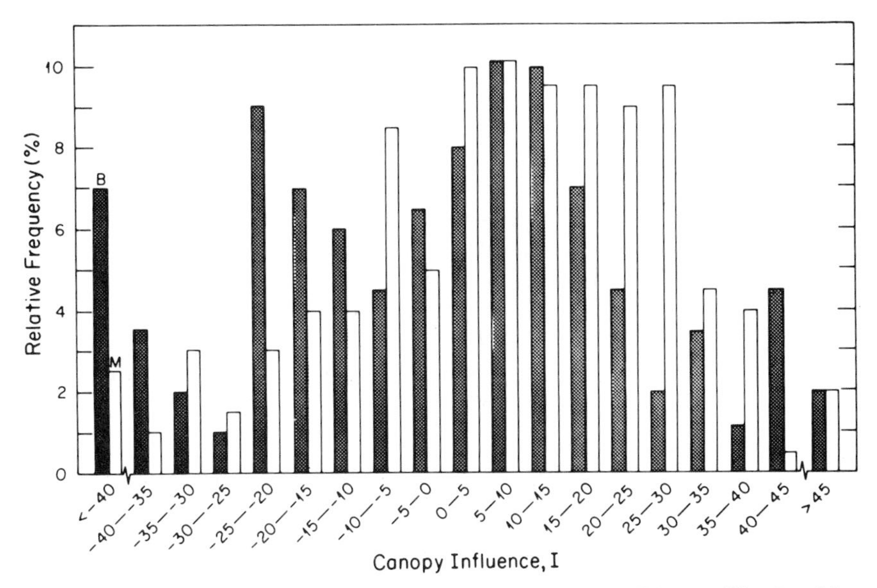

Figure 19.1. Distribution of beech and maple saplings in Warren Woods with respect to composite index of canopy influence (I). Relative frequency is the proportion of saplings of a species within a given range of the index and sums to 100 for either species. Shaded bars represent beech saplings, unshaded maple. The relative influence of beech canopy is highest at the right side of the graph, of maple canopy at the left (from Woods 1979).

of reproduction by sprouts and seedlings. In western stands, such as Warren Woods, seedlings are generally much more abundant than sprouts. There is a long-standing debate concerning the ability of root sprouts to develop into canopy trees (Fowells 1965). Some workers contend that beech sprouts, particularly those near the parent tree, rarely survive beyond sapling stage due, perhaps, to transmission of pathogens from the parent tree (Campbell 1938) or lack of development of an independent root system. Poulson (1980, pers. comm.) finds beech sprouts in Warren Woods to be a large proportion of the sick and dead category of beech saplings and not generally able to survive the death of the parent tree.

Runkle (1979) observed canopy gap dynamics, including species of gap makers and of saplings in gaps, in several old-growth forests of mixed dominance and used these data to calculate the probability of a tree of any given species being replaced by a tree of each available species (transition probabilities). One of his stands, Hueston Woods, in western Ohio, is similar to Warren Woods. Transition probabilities in Hueston Woods (Table 19.2) indicate a strong tendency toward reciprocal replacement. Assuming replacement to be a Markov process, Runkle used these transition probabilities to generate a predicted steady-state composition. His estimated transition probabilities are sufficient to maintain coexistence of

Table 19.2. Transition probabilities for Hueston Woods.[a]

Sapling species	Species of gap maker (%)[b]	
	Acer	*Fagus*
Acer	45– (*)	78 + (*)
Fagus	45 + (*)	19– (*)
Other species	10	3
Number of canopy individuals	8	26

[a]From Runkle (1979).
[b]Plus indicates positive association of sapling species with gap-maker species; minus indicates negative association of sapling species with gap-maker species; asterisk indicates significance at $\alpha = 0.005$.

beech and maple, though they suggest an increase in dominance by maple (51% maple and 49% beech compared to current values of 38% maple and 62% beech). It is likely that small changes in environment could affect transition probabilities so that canopy composition need not reflect the steady-state composition corresponding to current environment. The forest can, nevertheless, be regarded as stable, with the strong two-species codominance being maintained.

We have not suggested causal mechanisms for the phenomenon of reciprocal replacement. Horn (1971) proposed a mechanism involving differences in shade tolerance and canopy density: while maple is more shade tolerant in the understory, beech canopy casts denser shade, permitting each to prosper, relatively, beneath canopy of the other. Woods (1979) detected no difference in the response of seedlings of beech and maple to beech and maple canopy, leading to the suggestion that, rather than shading effects, which would be expected to act on all size classes, root competition or some other factor, which would act more strongly on saplings than seedlings, might be important. Ranking of shade tolerance of beech and maple is certainly not universally agreed upon (Fowells 1965), and, in fact, Poulson (1980, pers. comm.) has suggested, from detailed work on reproduction in Warren Woods, a mechanism essentially the reverse of that proposed by Horn. Using data on annual twig elongation of seedlings and saplings, Poulson suggests that beech is more shade tolerant (grows more rapidly beneath intact canopy), while maple, though barely able to survive as suppressed seedlings under full canopy, can grow much more rapidly than beech at higher light levels. Maple canopy, presumably through denser shading, inhibits both species more than beech canopy, but is especially inhibitory of maple growth. Tolerant beech saplings may persist beneath intact maple canopy, but maple is able to survive there only as seedlings, giving the slower-growing beech an edge in filling gaps left by maple trees. Beech canopy shows less difference in its effects on the two species in the understory, and maple, owing to more rapid growth, can outdistance beech saplings in gaps formed by beech trees.

Data published in Woods (1979) may be seen as supporting this conjecture. Using the separate terms of the formula for I to represent beech and maple canopy influence separately and graphing sapling response to the two resulting indices, it was noted that: (1) both sapling species are seen to decrease in frequency more rapidly with increasing maple influence than with increasing beech influence; and (2) the difference in response of the two sapling species to maple canopy influence is more marked than that in response to beech canopy influence. The first observation probably reflects the greater importance of beech in the canopy, but the second may be a result of maple more significantly affecting understory than beech does, as suggested by Poulson. Runkle's (1979) data show maple and beech saplings to be equally important in gaps left by maple canopy, while maple saplings are strongly dominant in beech gaps. This may seem contradictory to Poulson's proposed pattern. Runkle's data, though, are for saplings in gaps where growth of suppressed individuals has already begun, and the faster-growing maple saplings may have gained in relative importance already, as Poulson predicts.

Interactions among Several Species: The Hemlock-Northern Hardwood Forests

Patterns of canopy-understory interaction become difficult to detect and clarify with more than two species. We shall consider, though, whether such interactions underlie the maintenance of diversity in more complex stable forest communities.

In the upper Great Lakes region, in the hemlock-white pine-northern hardwood region of Braun (1950) can be found forest stands dominated by shade-tolerant species. Extensive wind throws, apparently the only significant large-scale disturbance in these forests, are very infrequent with a return time in excess of a millenium (Canham 1978). Individual stands of these forests are usually dominated by various combinations of two, three, or four of five species: American beech (*Fagus grandifolia*), sugar maple (*Acer saccharum*), eastern hemlock (*Tsuga canadensis*), yellow birch (*Betula lutea*), and basswood (*Tilia americana*).

In the last four years the first author has taken samples in a number of stands: nine in the northern hardwood region of northern Wisconsin and the upper peninsula of Michigan, two in similar forest types in Pennsylvania and New York, and one each in a beech-maple forest in southern Michigan and a maple-basswood forest in southern Minnesota. Sampling was by 0.1-ha (20 × 50 m) quadrats. For each sample, saplings and canopy trees were recorded by species and diameter and mapped so that.

for each sapling in a quadrat, distances to all canopy trees within 8 m could be calculated (8 m is an arbitrarily chosen distance beyond which canopy influence was considered insignificant). Community composition varied considerably from stand to stand. Several stands were west of the range boundary of beech; one was west of the range of hemlock in the maple-basswood forest region (Braun 1950). Reciprocal averaging ordination and geographical location were used to group similar samples into composite samples, which in the following treatment will be referred to as samples.

In the discussion of reciprocal replacement canopy influence was assumed to vary as the inverse of the distance between a canopy tree and the sapling (or point on the forest floor) being considered. There is no theoretical or empirical justification for assuming this form. Data by which a true relationship between canopy influence, diameter of tree, and distance to tree could be derived are almost totally unknown in the literature. For this study several such relationships were contrived, all fitting the intuitive requirement that influence of a canopy tree decrease with decreasing size of and increasing distance from it. These included, in addition to the ratio index described earlier, indices assuming influence to decrease linearly as distance from canopy tree increases, and others for which influence decreases in a form resembling a normal curve. In fact, results obtained using these various indices do not differ in any significant way and all results discussed in the remainder of this paper are for the first index:

$$I = KD/R$$

where D is the diameter of canopy tree, R is the tree to sapling distance, and K is a constant. Indices can be calculated for all canopy trees near (within 8 m) a particular sapling and added for all trees or for trees of particular species to yield measurements of total canopy influence and influence of individual canopy species. All appropriate indices were calculated for all saplings in all samples.

It is possible to calculate, for a particular sample, an expected value for any of these indices at a randomly chosen point on the forest floor. Holding D constant in any of the above formulas, the index may be considered a function of R, and the expectation for I at a randomly chosen point would be

$$e(I) = \int_0^8 \rho_i 2\pi R f(R) dR$$

where ρ_i = density of canopy trees. Canopy trees may be divided into diameter classes within which D may be taken as constant. e(I) may then be calculated for each size class and results summed for a true expectation for I:

$$E(I) = \sum_{i=1}^{n} \rho_i 2\pi \int_0^8 R f(R) dR$$

where n is the number of size classes and ρ_i the density of canopy trees in size class i; ρ_i's may be estimated from quadrat data. This expectation was calculated for influence of each canopy species and for total canopy influence in each sample. It is now a simple matter to compare, by sample, observed values of canopy influence on saplings with the calculated expectation. Such comparisons were made in all samples for all dominant canopy species and all sapling species represented by more than 20 saplings.

Displayed in Fig. 19.2 are the results for such comparisons involving the influence of hemlock canopy trees on various species as saplings. For each sample (row), average influence on particular sapling species is indicated on a relative scale by a leaf of the sapling species. The vertical line represents expected hemlock canopy influence; sapling symbols to the

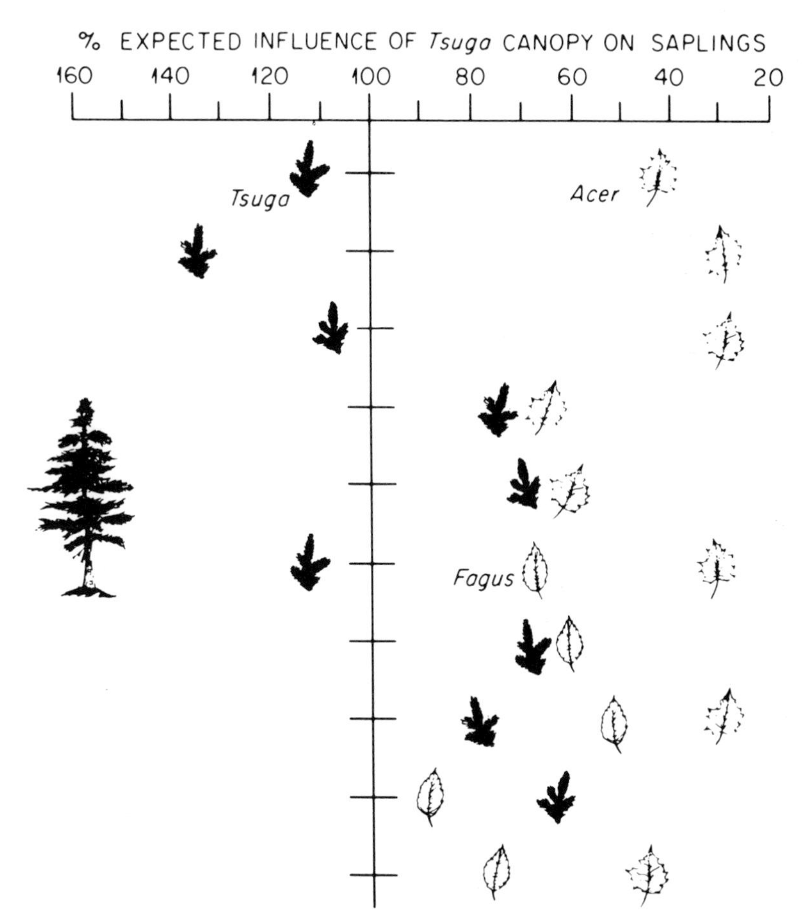

Figure 19.2. Distribution of species of saplings with respect to *Tsuga* canopy influence. Leaf symbols represent sapling species. Each row is a sample. Horizontal axis is *Tsuga* canopy influence as percentage of its expected value and is independently scaled for each sample. Each leaf is located at the mean value for all saplings of that species in that sample.

right of it suggest avoidance of hemlock canopy; to the left of the figure distributional affinity for hemlock canopy is indicated. Three observations can be made from these results: (1) only hemlock saplings show a mean value of hemlock canopy influence as great as the expected value, and in eight of the nine samples with hemlock saplings, they are found at higher hemlock canopy influence than any other species; (2) maple saplings are consistently found at low hemlock canopy influence—the lowest of all species in all eight samples where they occur; and (3) beech saplings are in four of four cases found at higher hemlock influence than maple, though not as high as the expectation nor, usually, as high as hemlock saplings. Thus, it appears that saplings of maple show strong avoidance of hemlock canopy as do, to a lesser extent, beech saplings, while hemlock saplings show no clear response.

Influence of beech canopy, on the other hand, elicits no clear response from any species as saplings (Fig. 19.3). Mean influence values for all species are scattered around the expected values with no apparent differences among species.

Maple canopy influence (Fig. 19.4) appears to affect sapling distributions, though not so strongly as hemlock. Maple saplings, on the average, are found near, but perhaps a little below, the expected value of maple influence. Beech saplings are, perhaps, found at slightly higher maple canopy influence; in three of four samples with both species as saplings, beech

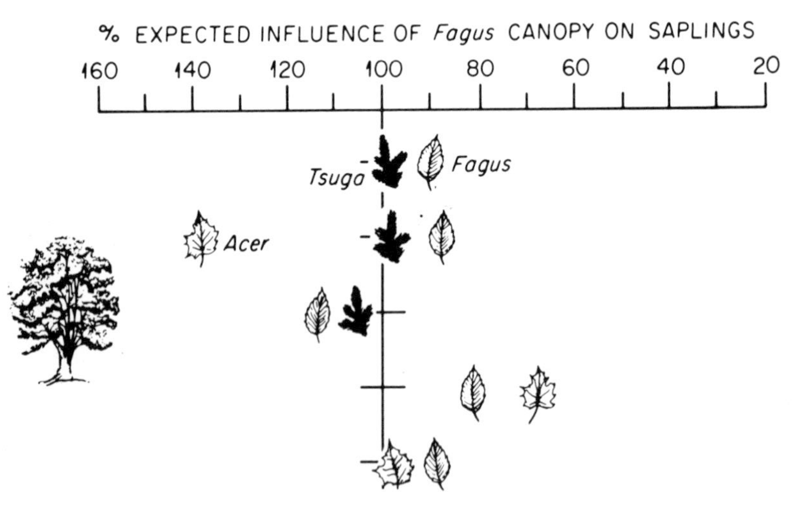

Figure 19.3. Distribution of species of saplings with respect to *Fagus* canopy influence. Leaf symbols represent sapling species. Each row is a sample. Horizontal axis is *Fagus* canopy influence as percentage of its expected value and is independently scaled for each sample. Each leaf is located at the mean value for all saplings of that species in that sample.

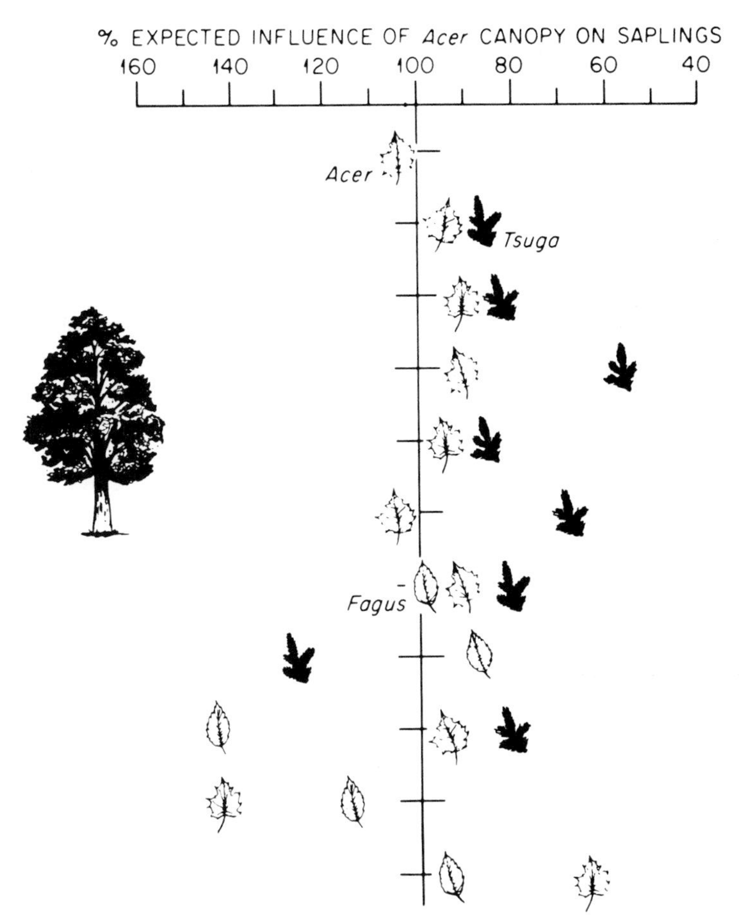

Figure 19.4. Distribution of species of saplings with respect to *Acer* canopy influence. Leaf symbols represent sapling species. Each row is a sample. Horizontal axis is *Acer* canopy influence as percentage of its expected value and is independently scaled for each sample. Each leaf is located at the mean value for all saplings of that species in that sample.

saplings are at greater maple canopy influence. Hemlock saplings appear to show some avoidance of maple canopy; in seven of eight cases they are found at lower maple canopy influence than any other sapling species.

These results support Poulson's contentions that maple canopy has stronger influence on sapling distribution than does beech canopy and that beech in the understory shows greater shade tolerance (especially notable beneath the dense canopy of hemlock). They also suggest that, when present, hemlock in the canopy exerts an even greater organizing force on the understory than do beech or maple. This has been suggested by work done on the herbaceous strata of forests in the Great Smoky

Mountains (Hicks 1978, 1980). The apparent inhibitory effect of maple canopy on hemlock reproduction may be an allelopathic interaction; Tubbs (1973, 1976) has found root exudates of maple seedlings to suppress growth in several species of conifers, though he did not test its effect on hemlock. Maple seedlings are extremely abundant beneath maple canopy in the stands studied.

These data can also be analyzed by techniques similar to those developed in the beech-maple reciprocal replacement study. Composite indices can be made for all pairs of canopy species, and pairs of sapling species can be compared in their response to these indices. All possible such comparisons involving beech, maple, and hemlock were made, using rank sum tests to determine whether the hypothesis of similar distribution of the two sapling species with respect to composite canopy influence should be rejected. In Fig. 19.5 the three tree symbols represent the canopy influence of the indicated species. The three lines between each pair of canopy species represent the possible pairs of sapling species and their relative response to the joint influence of the two canopy species. If sapling responses differ, the names of sapling species are written nearest the canopy species that they relatively prefer.

The relationships shown in Fig. 19.5 tend to support and clarify tendencies seen in earlier results. Comparison 1 in the figure is the beech-maple reciprocal replacement case more generally shown. In fact, for all comparisons involving beech canopy and maple saplings (1, 3, 4, and 6), maple saplings are found at values of the composite index indicating relatively greater influence of beech canopy. In comparisons involving beech saplings and maple canopy (1, 2, 7, and 8), beech saplings are found at greater maple canopy influence than other sapling species, except in one case (7), where differences in sapling species response were not significant. In most comparisons involving both saplings and canopy of maple (1, 3, 7, and 9) or of beech (1, 2, 4, and 5), saplings of these two species are found at relatively lower values for conspecific canopy influence. The exceptions to this trend occur when the comparison involves hemlock saplings and canopy (5 and 9), in which cases both species, relative to hemlock saplings, prefer conspecific canopy to that of hemlock. Hemlock saplings, in fact, in all comparisons involving hemlock canopy (5, 6, 8, and 9), are found at relatively great hemlock canopy influence.

Yellow birch and basswood are not particularly shade tolerant (Fowells 1965) and so do not appear frequently as suppressed seedlings or saplings. They do appear, though, to be capable of maintaining significant representation in the canopy. Yellow birch has been frequently described as a gap-phase species, reproducing successfully only in canopy openings where light levels are great enough and competitors small enough that its rapid growth rate allows it to reach the canopy (Forcier 1975). In this study few birch saplings were found, and they tended to be found at low total canopy influence. Since suppressed saplings are generally fewer

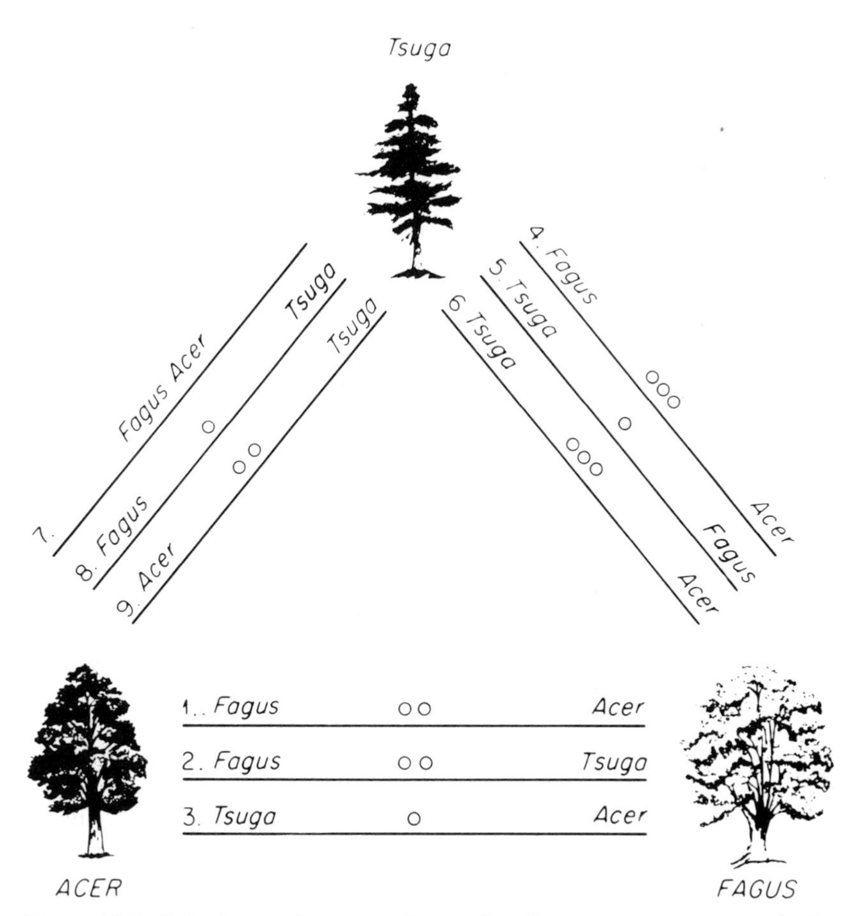

Figure 19.5. Pairwise species comparisons of sapling response to composite indices of canopy influence. The tree symbols represent canopy influence of indicated species. Each line represents a single pairwise comparison of sapling species with respect to the composite index of influence for the two canopy species connected by the line. All possible comparisons are represented. Where rank sum tests indicated significant differences in response of sapling species, sapling species names are shown nearest the canopy species that they, comparatively, favored. Each line represents several tests. Levels of significance at which the hypothesis of similar distribution of sapling species may be rejected are suggested by number of dots: one dot indicates rejection in at least three of four tests with $\alpha = 0.05$; two dots indicate rejection in all tests at $\alpha = 0.05$–0.005; three dots indicates rejection in all comparisons at $\alpha < 0.005$.

beneath hemlock canopy, it might be reasoned that hemlock gaps would relatively frequently be devoid of competitors for birch, and birch might more frequently succeed in hemlock gaps. This notion is supported by the fact that, while all other pairs of dominant species are negatively associated ($\alpha < 0.05$), hemlock and yellow birch are positively associated ($\alpha < 0.01$) in individual quadrats. Brown and Curtis (1952), Whittaker (1956), and Rogers (1978) also report that birch and hemlock are frequently found growing together. Runkle (1979) finds birch saplings to be more important in hemlock gaps than in gaps formed by any other species except itself and perhaps, tulip tree (*Liriodendron tulipifera*). Barden (1979) reports birch as the most frequent replacement of hemlock canopy trees (22 of 73 hemlock gaps observed) in the Great Smoky Mountains National Park. In his study beech is next in frequency as a replacer of hemlock (13 gaps).

Basswood, though very rarely found as seedlings or independent saplings in intact forests in this study, produces abundant and vigorous basal sprouts, which appear to succeed the parent stem in the canopy with high frequency (Fowells 1965). This tendency might be encouraged by subsidization of sprouts and the tendency of basswood trees to break off rather than uproot (Runkle 1979). Such literal self-replacement can be observed to have occurred for several generations of stems in many instances, frequently with an increase in number of stems. Individuals of basswoods, then, appear to have extremely long life expectancies, and they are capable of persisting on a single microsite for several generations of stems.

Through observation of distributional patterns of these five species in the understory, we might now suggest a complex of strategies and interactions that might lead to their coexistence in a stable community. Basswood shows a strong tendency toward self-replacement. The three shade-tolerant dominants have more probabilistic patterns of replacement, but in general hemlock may tend to replace itself, while maple and beech replace one another. Yellow birch is probably able to persist in the canopy through opportunistic use of occasional suitable gaps, which may, more frequently, be hemlock gaps. The patterns of replacement that our data suggest and their relative importances are illustrated in Fig. 19.6. These patterns suggest no strong directional trend, though relative importances might be expected to fluctuate somewhat over long periods without disturbance. The probabilities indicated here for various replacements might be altered somewhat by the mass action effect of more abundant reproduction by some species; a slight trend toward hemlock, suggested in Fig. 19.6, might be balanced by the greater abundance of maple and beech seedlings and saplings.

Other workers have studied patterns of canopy-understory interactions and replacement of canopy trees in various forest types. Runkle (1979) has, in addition to his previously mentioned work in Hueston Woods, studied old-growth stands of other forest types using Markov analyses of gap data to obtain transition probabilities and predictions of steady-state compositions. His findings support the hypothesis of reciprocal replacement of beech and maple in northern forests of low diversity, but in richer forests of the southern Appalachians most species tend to be self-

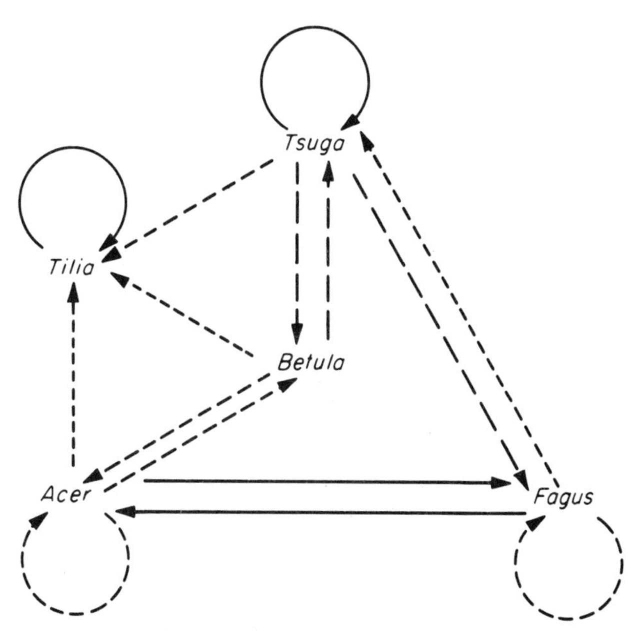

Figure 19.6. Diagram of patterns of canopy tree replacement as suggested by distributions of sapling species with respect to canopy species. The arrows are directed toward species being replaced. The solid lines indicate transitions that occur most frequently, dashed lines less common transitions, and dotted lines those that probably occur occasionally. Other possible transitions probably occur but more rarely.

replacing. Runkle also reports a slight tendency for large gaps to be colonized by species that, when dying, leave small gaps, and vice versa. This suggests a trend of large trees of shade-tolerant species being replaced by small gap-phase species, which are, in turn, replaced by shade-tolerant species. Predicted compositions at equilibrium are, in all cases, close to current compositions. Fox (1977) claimed reciprocal replacement phenomena occurred in several forest types including beech-maple (supported by Runkle's study and those reported here) and beech-hemlock (supported by neither Runkle's study nor ours). Forcier (1975) studied associations of various species as seedlings, saplings, and canopy trees in a late successional beech-birch-maple forest in New Hampshire. He found birch in the overstory to be associated with maple seedlings and saplings and maple overstory to be associated with beech seedlings and saplings. Both of these results are compatible with the results of studies described in this article. However, Forcier found beech canopy negatively associated with maple understory and most positively associated with beech root sprouts. Horn (1975b), in a Markov analysis of a successional woods in New Jersey, also suggests strong self-replacement of beech by means of root sprouts. Beech root sprouts are relatively more important in the northeastern part of the range of beech (Ward 1961), and it appears that this difference in reproductive strategy may be reflected in a difference in replacement patterns.

Summary and Conclusion

Many communities are prone to such frequent and severe disruption that a state recognizable as stability is rarely, if ever, reached (Connell and Orias 1964; Huston 1979). However, there exist continua of size and frequency of disturbance. In many forests and woodlands, disturbances are usually of such small size or low frequency that, on the scale of the single stand, these communities appear relatively stable.

In chronically disturbed systems the different adaptations of species to disturbance and succession are a major basis of species diversity (Loucks 1970). In more stable systems adaptations of species to different microsite conditions (including gap-phase effects) should be a major basis of species diversity. Effects of canopy species on reproductive microsites seem particularly important in maintaining both stability and diversity in stable forest communities, where this biologically induced differentiation might account for much of the diversity seen.

The driving factors of successional change, at least in late-secondary succession, have generally been thought to be rooted in modification of environment by existing organisms and in chance factors of seed source, disturbance history, substrate qualities, and so on (Drury and Nisbet 1973; Connell and Slatyer 1977). Work we have described shows that dynamics of species interaction and patterns of canopy tree replacement in forests we see as stable are controlled by essentially similar factors.

Though we have discussed some specific examples of biologically induced variety in microsites and different adaptations that could interact to produce dynamically stable communities, caution should be used in making generalizations about community dynamics and function. Indeed, in the relatively small range of forest types discussed in this article, patterns of canopy-understory interaction and of canopy replacement vary in kind among forest types and strategies vary among species. It is probably not justifiable to predict from our information that particular interactions such as reciprocal replacement will be important for species pairs other than beech and maple, or even that this pattern should be important throughout the joint range of these species. Among the several species we have studied, nearly as many distinct strategies appear to lead to persistence in the canopy of stable forests.

We would suggest, however, that the understanding of requirements and limitations of important species in early stages of their life cycles is generally necessary for adequate explanation and understanding of community dynamics. We have mentioned a number of mechanisms that could lead to coexistence of species in a stable community or in other ways influence community function and dynamics. None of these could have been suggested solely by observation of canopy patterns.

Chapter 20

Changes in Biomass and Production During Secondary Forest Succession

Robert K. Peet

Introduction

One of the most fundamental ways in which forests change during secondary succession is in energy relations (Odum 1969). However, a paucity of reliable data on forest development, coupled with frequent but uncritical speculation, has led to considerable confusion over long-term successional trends in forest biomass and production. The majority of forest biomass and production studies are of single forests and give little indication of the range of variation in a region or the factors responsible for that variation. A few studies have related biomass and production to environmental factors (e.g., Maruyama 1971; Waksman et al. 1975; Westman and Whittaker 1975; Whittaker et al. 1974; Whittaker and Niering 1975), but such studies remain the exception rather than the rule. While the forestry literature contains a wealth of information on merchantable yield as related to stand development and site conditions, few studies have examined changes in total stand biomass and production over time.

The slow rate of forest development is the primary reason for the paucity of studies of changes in biomass and production. Ideally long-term permanent sample plots should be studied, but the use of this approach depends on plot availability (e.g., Attiwell 1979; Cooper Chapter 21; Crow 1980). In areas where permanent plots are not available, investigators typically must construct artificial chronosequences from stands of differing developmental stages on similar sites (e.g., Long and Turner 1975; Ovington 1957; Switzer et al. 1966). While this approach is open to errors resulting from stochastic variation in initial conditions and stocking levels, it is often the only method available.

In this chapter I briefly review the literature, illustrate the most commonly encountered trends in forest development using studies I have conducted in the Rocky Mountains of Colorado and the piedmont of North Carolina, and attempt to show how several divergent patterns can be understood in terms of a single conceptual model based on tree population dynamics. In all cases I assume that forest development occurs in the absence of further major disturbance. Owing to limitations in available data, the discussion will be confined to above-ground biomass and production of trees.

Theories and Published Data

Biomass

By definition, biomass increases during forest succession from near zero to that found in mature forests. The details of this pattern of increase, while widely speculated upon, are not well established.

The simplest hypothesis is that biomass increases in a smooth, logistic fashion toward an upper limit fixed by site conditions. This is the pattern offered by Odum (1969) in his summary of successional trends, which was widely accepted at the time. Several studies present data that support this hypothesis. Forcella and Weaver (1977) reported asymptotic biomass accumulation for *Pinus albicaulis* forests of Montana, and MacLean and Wein (1976) presented evidence of similar forest development in Canadian *Pinus banksiana forests.* Holt and Woodwell (in Whittaker 1975b), working in the Brookhaven Forest, also found asymptotic forest regrowth.

Further evidence for an asymptotic or logistic biomass accretion model can be found for many even-aged stands. Both Switzer et al. (1966) and Kinerson et al. (1977) suggested that *Pinus taeda* stands add standing crop in a logistic manner, and Zavitkowski and Stevens (1972) provided similar evidence for *Alnus rubra* stands. However, such stands are mostly young and with maturation may break up, resulting in a drop in biomass.

A second hypothesis, well documented in the forestry literature (see Loucks 1970) and which incorporates stand breakup, is that biomass increases to a maximum in late successional stages, only to drop with development of the climax community. Bormann and Likens (1979a) proposed a "shifting mosaic" model, which accounts for this dip in biomass. In effect, their model is a stochastic version of the logistic yield model of Odum (1969) and others. Lacking a significant sample of old-growth forest, Bormann and Likens (1979a) employed JABOWA, the forest simulation model of Botkin et al. (1972a), to examine long-term changes in forest biomass. They, like Botkin and his co-workers, found biomass to peak after approximately 200 years and to drop to an asymptote by 350 years. The mechanism is intrinsic to the structure of the simulation model. JABOWA simulates growth in many 10 × 10 m plots which are then averaged. When forest growth begins all plots are synchronized with zero biomass. The plots remain nearly synchronized when peak biomass is reached after 200 years. Thereafter, plots start to experience the death of canopy trees in an irregular, stochastic manner, with the result being a dramatic drop in biomass for the affected plots. The drop in forest biomass results from the loss of synchrony of the individual plots; the eventual asymptote corresponds to an equilibrium in which plots are in all stages of recovery from disturbance. Invoking Watt's (1947) gap-phase model, Bormann and Likens suggested that the many small disturbances in the forest, when averaged, produce a pattern of first a peak and then a dip in biomass similar to that simulated using JABOWA.

A third hypothesis concerning biomass change during succession predicts an initial increase followed by a drop and then a subsequent recovery. Here, rather than a simple dip to an asymptotic climax level, biomass shows a series of damped oscillations around an equilibrium level. European foresters have long reported such damped oscillations. Ilvessalo's (1937) monograph on the dynamics of Finnish forests shows this result, as does Siren's (1957) somewhat more recent monograph. Plochmann (1956), working in the forests of Alberta, reported similar cycles but did not report damping.

A final hypothesis of forest biomass change during secondary succession is that after an initial peak, the quantity of living biomass that can be supported steadily decreases as available nutrients are tied up in dead organic matter (e.g., Bloomberg 1950, Strang 1973). This is primarily a phenomenon reported from boreal sites with organic soils.

As a means of partially unifying these disparate developmental patterns (Fig. 20.1), I propose a conceptual model closely related to the shifting mosaic model of Bormann and Likens. I suggest that the major patterns of biomass change can be understood in terms of time lags in regeneration and of synchrony of mortality. As with the shifting mosaic theory, an initial wave of regeneration dominates after disturbance. This is a period of increasing biomass (an aggrading system) during which tree growth and mortality are the dominant processes. The dense tree population is viewed as inhibiting regeneration through resource preemption (Connell and Slatyer 1977; Monsi and Oshima 1955). Failure of regeneration necessitates an eventual stagnation in biomass accumulation and eventually a drop. Regeneration starts only after natural thinning of the canopy has progressed to the point where resources released by mortality are no longer preempted by increased growth of existing trees (see Peet 1981; Peet and Christensen 1980a). Once the even-aged forest structure is lost, peak biomass cannot again be reached without outside disturbance. If for some reason regeneration had not been suppressed (no time lag) early in stand development, possibly because of low initial recruitment rates, the biomass drop might not have occurred.

The model I present differs from the present version of the shifting mosaic model in that this population-based model incorporates a greater range of observed patterns. The degree of time lag (see May 1976) introduced can modify the shape of the biomass accumulation curve from a simple logistic (no lag) through a simple peak (moderate lag) to damped or continued oscillations (long lag). Synchronous tree mortality would amplify the oscillations.

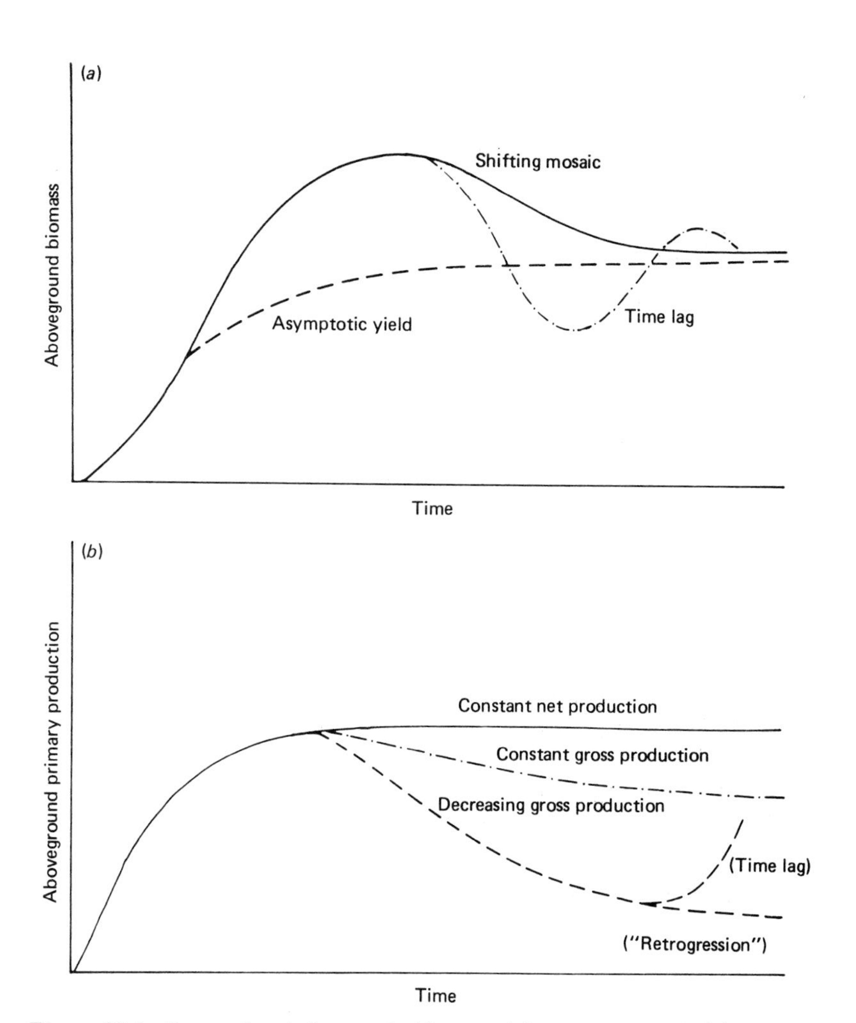

Figure 20.1. Successional changes in biomass (a) and production (b) as predicted by various models described in the text.

Production

During succession following forest removal, net primary production increases from zero to some positive value. If biomass lost through tree death and plant part shedding is subtracted, the net primary production should again approach zero during the climax or equilibrium stage (Kira and Shidei 1967; Odum 1969). This result is implicit in the definition of secondary succession. As death or shedding is not treated here as negative production, the net production at climax will be positive.

The classical interpretation of succession as formulated by Clements (1916) and others is that successive vegetation stages modify the environment, preparing the way for subsequent stages. This environmental amelioration should allow increased production. Certainly the increase of net production to an asymptote at climax is widely accepted for certain cases of classical xerarch succession. For secondary forest succession, however, only a few cases of asymptotic increase have been reported, and these only with limited data. Holt and Woodwell (in Whittaker 1975b) reported an uneven but asymptotic increase in net production for the Brookhaven Forest. Forcella and Weaver (1977) reported a slight but steady increase in net production of *Pinus albicaulis* forests over an age sequence from 50 to 650 years.

An alternative hypothesis is that a finite quantity of resources is available for plant growth each year, with the result that after an initial but brief recovery period, production is independent of stand age. This is, in effect, a generalization of the reciprocal yield law of Kira et al. (1953) to woody plants. Marks' (1974) work with *Prunus pensylvanica* is consistent with this hypothesis. He found net production in early successional *Prunus*-dominated communities to quickly reach the level of production of the mature forests of the region. In well-stocked stands the steady-state level was reached in as little as 4 or 5 years, with a somewhat longer recovery period needed where fewer seeds were available. Ovington (1957) reported near constant production for *Pinus sylvestris* plantations between 15 and 35 years old and for *Betula verrucosa* stands between 20 and 55 years of age (Ovington and Madgwick 1959). Similar results were reported by Van Cleve (1973) for *Alnus* stands in Alaska and by Meeuwig (1979) for piñion-juniper stands in Nevada.

The constant production concept applies more readily to gross production than to net production, at least if light or carbon is the limiting resource. It can be argued that as the quantity of living biomass in the forest increases, a steady decrease in net production should be expected because increasing amounts of respiration must be subtracted from constant gross production. Odum and Pinkerton (1955), Odum (1969), and Margalef (1963a) all theorized this shift in energy flow from production to maintenance with increasing system maturity.

A third hypothesis suggests that production drops more during the later portion of forest development than can be accounted for by increased respiration. Much evidence has been published supporting this view. In one of the earlier statements, Meyer (1937) suggested *Tsuga heterophylla* forests to peak in production after 30 years and to drop to an asymptote by 150 years. Rodin and Bazilevich (1967) cite several examples of major drops in forest production with age. They refer to forests of *Pinus* and of *Picea* on high quality sites in the subtaiga of European Soviet Union, both species reaching peak production between 40 and 50 years, only to drop to levels of around 25% of their maxima by 110 years. *Tilia* forests near Moscow and *Quercus* forests near Voronezh change in essentially the same way. Attiwill (1979) provides another example, with the basal area increment of *Eucalyptus obliqua* forests in Australia peaking around 40 years and dropping to near 25% of maximum by 65 years.

The drop in production during the latter phases of succession is variously attributed to developing forest structure, changing species composition, and changing nutrient availability. With respect to species composition, Horn (1971, 1974) has argued persuasively that the geometry of early successional trees is intrinsically more efficient for maximizing forest photosynthesis than the geometry of later successional species. Although the generality of Horn's theory has not been tested, it is consistent with production results from the southeastern United States, where early successional stands of species such as *Pinus taeda* or *Liriodendron tulipifera* are often 50% more productive than stands of old-growth *Quercus* or *Carya* forest on similar sites. Possible exceptions to Horn's model include tropical groups such as *Trema* and *Cecropia.* As Lieth (1974) has observed, the question of whether successional communities are more productive than climax ones is still unanswered for most successional series in the world.

Species-independent aspects of stand development can also be used to explain a late succession drop in production. The dynamics of most even-aged, postdisturbance forests are dominated by a thinning process; as dominant individuals in the initially dense population of trees grow, many smaller trees die. Over the short term, stand development can be viewed simply as decreasing tree density and increasing size. The relationship between stand density and productivity has long been debated. Moller (1947, 1954) and Baskerville (1965) are among those who suggest production to be independent of density; only at extreme densities should production fall. In developing his theory of density independence, Moller observed the amount of foliage per unit area to be constant. Further, he suggested that only in very high density stands would the quantity of nonphotosynthetic respiring area be sufficient to cause respiration to limit production. In contrast, Assman (1970) and others suggest that there is a narrow optimal density range over which production is maximal. If we accept Assman's view, the necessary consequence is a peak in stand production at the point when optimal density is reached during the thinning process.

An additional aspect of even-aged stand development that could limit production is tree senescence. As the original trees age, they lose their capacity for high levels of production, thus lowering stand production until such time as they are replaced by young, vigorous trees (Hellmers 1964; Kira and Shidei 1967).

The potential role of nutrient availability should also be mentioned. Vitousek and Reiners (1975, Gorham et al. 1979) have described how net ecosystem production (net primary production less mortality) should be a major determinant of nutrient losses from forest ecosystems. Nutrient losses can be hypothesized to have a negative feedback on primary production. Early in succession production should be relatively high due to a large pool of available nutrients. After this pool has been exhausted, production should be limited by nutrient inputs. When biomass (both living and dead) has reached a stable level and nutrients are not being sequestered in new biomass, production should be less limited by nutrients. Changes in available nutrient pools thus have the potential to partially counteract the tendency for a high peak in mid-stand development. More information is needed before the importance of nutrient limitation as a negative feedback can be adequately evaluated.

Observed changes in forest production during succession (Fig. 20.1) can partially be explained in terms of the time lag model applied earlier to changes in biomass. After disturbance, production should increase steadily to a plateau dictated by site conditions. The rate at which this plateau is approached will depend on the initial rate of tree establishment. Stands initially densely stocked will plateau in only a few years, whereas on extreme sites with low establishment rates, production will only slowly increase to an asymptote. On favorable sites, after the initial plateau, net production should drop at least gradually as nonphotosynthetic biomass builds up. The rate of drop will depend on the importance of density and tree senescence. During this period of thinning, regeneration will be minimal. When thinning reaches the point where existing trees can no longer successfully preempt resources released by tree mortality, regeneration will resume. If the loss of canopy trees were to occur over a relatively short period, the replacement of old, senescent trees by young, vigorous saplings could result in a marked increase in production and a subsequent slow drop. In contrast, if mortality were gradual, the second increase might never occur.

Case Studies

Two case studies drawn from my own work help illustrate the range of possible successional patterns in biomass and production, and the dependence of these patterns on underlying population processes. The first, based on work on the North Carolina piedmont, has the stronger data base in that permanent sample plots were used. The second, based on work in the Colorado Front Range, shows a greater diversity of patterns owing to a greater range of environmental conditions.

North Carolina Piedmont Forests

The original vegetation of the North Carolina piedmont was oak-hickory forest (Braun 1950; Peet and Christensen 1980b). Far from being homogeneous, the forests of this area take a diversity of forms, reflecting local topography and subtle edaphic variation (Braun 1950; Peet and Christensen 1980b). Since European settlement the piedmont has been intensively exploited for agriculture and for forest products with the result that the land now appears as a fine textured mosaic, the pieces representing various land-use histories. As described elsewhere in this volume (Christensen and Peet Chapter 15), this mosaic provides an ideal natural laboratory for studying secondary succession.

Together with O. P. Council, I have been examining changes in biomass and production during succession in the forests of Durham and Orange Counties, North Carolina. Data from permanent plots established in both the Duke Forest (Duke University) and the Hill Experimental Forest (North Carolina State University) have been available to us. Trees in these plots were numbered individually and measured for diameter at breast height (dbh) and usually total tree height (H) at roughly 5 year intervals. These data allowed us to estimate both biomass and net production at intervals over almost 50 years of plot history. Both mixed-aged hardwood forests and even-aged, old-field pine forests were studied.

To estimate above-ground tree biomass and production, we used the dimension analysis methodology developed by Whittaker and Woodwell (1968), and Whittaker and Marks (1975). Parabolic volume [PV $1/8\pi(\mathrm{dbh})^2 H$] was used as a biomass estimator, and estimated volume increment (EVI, change in PV per unit time) as an estimator for wood production, leaf and twig production being estimated by PV. Details of methodology are available in Peet and Council (1981). Use of these estimators rather than the more frequently used simple diameter at breast height allows partial compensation for changes in tree geometry with stand development.

Hardwood Forests

All piedmont forests have been disturbed by modern civilization in some major way. The hardwood permanent plots studied represent areas that appear to never have been ploughed and to have a species composition similar to what might have been found in the original forests. Doubtless, grazing and selective cutting were common before 1930, when the lands were acquired. Several forest types were included in the study, ranging from mesic alluvial forests to well-drained uplands to infertile uplands characterized by shrink-swell (montmorillonitic) clays. The range of production values observed was equivalent to the range of published production values for eastern North American deciduous forests (Peet and Council 1981). The shrink-swell clay site had production values as low as 450 g m^{-2} yr^{-1} (above-ground dry weight), less than the previous low values reported by Monk et al. (1970) for a mixed *Quercus-Carya* forest on the Georgia piedmont, or Whittaker's (1966) *Quercus rubra-Q. alba* type in the Smoky Mountains. A young alluvial forest was the most productive stand, which, with production averaging 1850 g m^{-2} yr^{-1} and with an initial production of 2100, was near the high of 2400 reported by Whittaker (1966) for a mesic *Liriodendron* forest in the Smoky Mountains.

Despite almost 50 years of records and a diversity of sites, within site production (Fig. 20.2) remained essentially constant for each site while biomass climbed at almost constant, site-specific rates. As these stands are undergoing thinning with little associated recruitment, the results support the view of Möller (1947), Baskerville (1965), and others that production is largely independent of density.

Rates of biomass accumulation (Fig. 20.3) similarly do not suggest a trend during the study period. In no case do the results suggest that

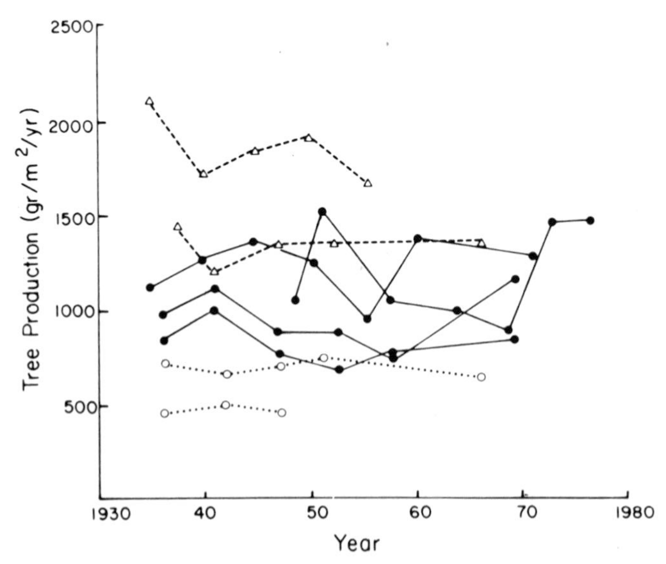

Figure 20.2. Above-ground tree production as a function of time for eight mixed-age hardwood permanent sample plots. Alluvial plots are indicated by open triangles, well-drained upland plots by solid circles, and shrink-swell clay dominated upland sites by open circles.

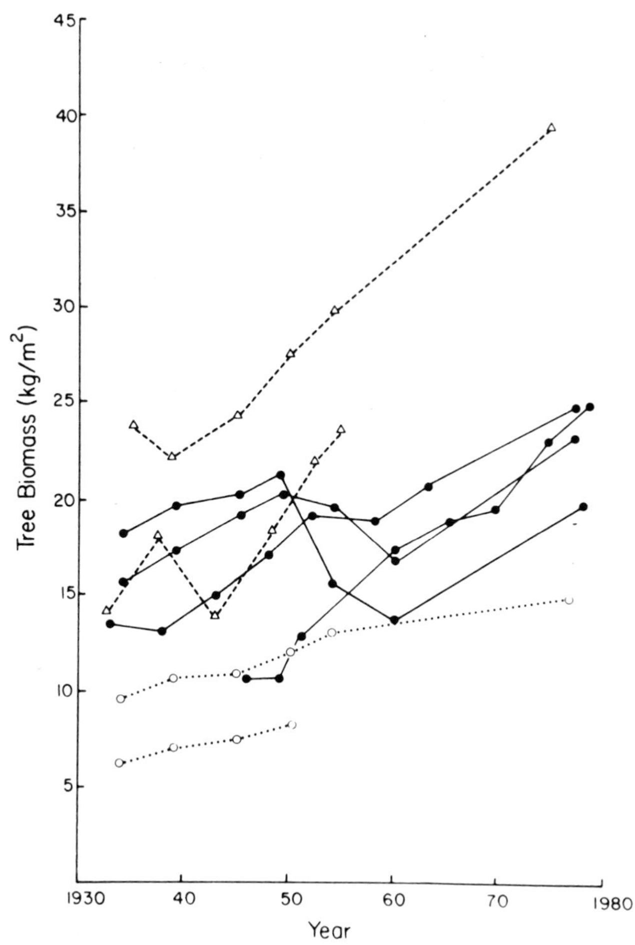

Figure 20.3. Above-ground tree biomass as a function of time for eight mixed-age hardwood permanent sample plots. Alluvial plots are indicated by open triangles, well-drained upland plots by solid circles, and shrink-swell clay dominated upland sites by open circles.

biomass in approaching an asymptote. Two of the well-drained upland stands dropped in biomass following a hurricane in 1954. In addition one very productive alluvial stand dropped in biomass during an early sampling interval as a result of mortality associated with extreme overcrowding. Otherwise, within each plot, biomass appears to have increased at a constant rate.

Whittaker and Likens (1973) observed that mature forests usually have biomass (metric tons/ha) in a range defined as (0.625 times production (g m^{-2} yr^{-1}) − 250) ± 125. With the exception of one plot on shrink-swell soils, all plots started the sampling period well below this range of biomass, and by the end of the sampling period an additional three plots had barely managed to reach the low bound of this range. If we accept Whittaker and Likens' generalization, the stands sampled appear sufficiently immature to account for failure to approach a biomass asymptote.

Pine Forests

Pine (*Pinus taeda*, *P. echinata*) often invades abandoned agricultural fields or other deserted lands of the piedmont. The resulting pine forests have long been recognized by foresters as highly productive. I consider here primarily a density series with plots having initial densities (8-year-old trees) from 25 trees per 0.1 acre (617/ha) to 1173 per 0.1 acre (28,985/ha).

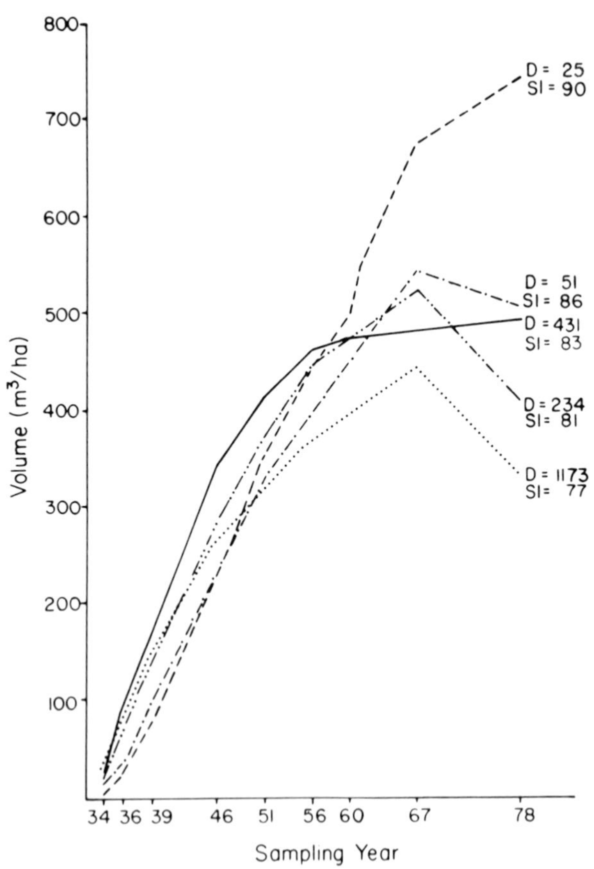

Figure 20.4. Volumes (PV) of old-field *Pinus taeda* stands of varying initial density as a function of age. Initially stands were ordered by density but after 44 years stands were ordered by site index. SI indicates site index (in feet). D indicates initial tree density (0.1 acre).

The denser plots (>6200/ha) had high production during the first interval, mostly in excess of 3000 g m^{-2} yr^{-1}, or three times the production of hardwood stands on similar sites. On an extremely favorable site elsewhere in the Duke Forest, Kinerson et al. (1977) found *Pinus taeda* production in excess of 4000 g m^{-2} yr^{-1}. By the time the trees in the present study were 25 years old, production had stabilized on all plots near 1700 g m^{-2} yr^{-1}, still about 1.7 times the production of hardwoods on similar sites. Consistent with Horn's (1974) theory, this evergreen, early successional invader species, *Pinus taeda*, greatly exceeds the production of potential climax species.

The 45-year period over which the pine density series was observed was one of tree growth and mortality. Very few new trees became established, and density decreased in a negative exponential manner, the rate being dependent on initial density (see Fig. 1 of Peet and Christensen 1980a). Despite this steady loss of individuals, total volume increment (EVI) remained nearly constant after age 25, suggesting production rapidly became independent (within extremes) of both stand age and density (Fig. 2 of Peet and Christensen 1980a). The result is consistent with the proposed generalization of the reciprocal yield relation of Kira et al. (1953) to woody plants.

The change in volume (PV) of *Pinus taeda* with time is shown in Fig. 20.4. Initially, when the trees were small, the density of stocking determined the volume present. With increasing plot age this relationship broke down, the new rank order of biomass having a one-to-one correspondence with site index (as predicted by depth of the A horizon and plasticity of the B horizon). However, because initial density was also correlated with site index, it is difficult to tell if the final order was primarily determined by site factors or by initial density.

Colorado Front Range Forests

Forests dominate the vegetation of the Colorado Front Range between 1600 and 3500 m elevation. Together with the numerous available topographic and moisture conditions, this elevational range assures a diversity of forest types. In addition, the forest vegetation of the Front Range reflects disturbance history. Fire, insect outbreaks, and wind have together produced a landscape mosaic made of patches differing in age and history. Forests near steady state are rare; the great majority of stands are in some stage of recovery (Peet 1978, 1981).

Although permanent sample plot data have not been available to me, the large number of sites undergoing natural recovery made it possible to deduce postdisturbance recovery patterns. Typical developmental sequences were constructed using temporary plots (0.1 ha) after comparison of roughly 20 plots for each major site type (Table 20.1). Note that in

Table 20.1. Developmental sequences of middle- and high-elevation forests in Rocky Mountain National Park, Colorado.

Stand	PV[a]	EVI[b]	EVI/PV	BA[c]	Density	Age[e]	Stand age
Pinus contorta forests at middle elevations (ca. 2900 m)							
A1	105.4	484.9	4.60	24.5	6154	36.7	45
A2	343.1	356.8	1.04	52.6	5783	94.0	110
A3	546.8	288.1	0.53	59.2	1347	203.2	260
A4	344.0	264.0	0.77	36.3	721	165.9	260
A5	272.4	218.8	0.80	30.9	1891	141.6	260
Picea, Abies forests near tree line (ca. 3400 m)							
B1	40.8	144.7	3.55	7.3	913	50.6	60
B2	100.6	194.0	1.93	19.0	2650	90.0	100
B3	256.6	302.2	1.18	34.9	1870	120.0	200
B4	415.0	323.8	0.79	55.3	2725	253.4	510

[a]PV, parabolic volume in m^3/ha, calculated as $1/8\pi\,(dbh)^2H$.
[b]EVI, estimated volume increment in $cm^3\ m^{-2}\ yr^{-1}$, calculated by subtracting sequential values of PV.
[c]BA, basal area in m^2/ha.
[d]Density, number of stems $\geqslant$2.5 cm dbh per ha.
[e]Age, mean age of all stems $\geqslant$2.5 cm dbh.

the case of the *Pinus contorta* forests, the last three stands, while at different developmental stages, are all the same chronological age. Differing rates of development can result from a number of factors, including initial density, disturbance history, minor site differences, and chance. Details of the ordering procedure are more fully explained in Peet (1980). For each site in the developmental sequences, parabolic volume and estimated volume increment were determined for each tree. Because reliable dimension analysis regression equations using PV and EVI as independent variables were not available, PV and EVI were summed for each plot and used to represent trends in biomass and production, respectively. The results closely approximate and vary linearly with those that would be obtained if dimension analysis equations were applied.

Consider first well-drained, middle-elevation sites dominated by *Pinus contorta* (Table 20.1). (*Pinus contorta* forests were selected because of their abundance, but similar results could be obtained in middle-elevation stands of *Picea*, *Abies*, or *Pseudotsuga*.) After a fire or other major disturbance, *Pinus* usually reestablishes quickly so that after perhaps 20 years a dense stand of trees is present. As with *Pinus taeda* in North Carolina, following canopy closure, competition from existing trees precludes further tree establishment; subsequent stand development is dominated by growth and mortality. The stand sequence shown in the first half of Table 20.1 illustrates the results. Volume and basal area both increase to maxima, which are followed by pronounced drops resulting largely from natural mortality in the absence of regeneration. Volume increment reaches an early maximum and then drops steadily. While part of this drop may be due to increased maintenance costs associated with high biomass, the drop continues unabated after biomass also starts to drop. After a sufficient decline in biomass and production in the original generation of trees, regeneration will again be possible. The synchrony and rate with which competition is released throughout the stand will determine whether biomass will simply drop to an asymptote as in the shifting-mosaic model of Bormann and Likens, or will undergo a series of damped oscillations which can last for several generations of trees. The more synchronous the fall in biomass, and the stronger the initial inhibition of regeneration, the stronger will be the subsequent oscillations.

A strikingly different type of forest development is found at highest elevations and on extremely dry sites (Table 20.1). On these sites, usually dominated by *Picea* and *Abies*, or *Pinus flexilis*, the rigorous environment greatly inhibits initial tree regeneration after disturbance. Some trees do become established, but often an excess of several hundred years is required before the area once again resembles a closed forest. Most regeneration is in the shelter of established trees. The critical result for our purposes is that tree basal area, production, and biomass all steadily increase to an asymptote over a several hundred year period.

Conclusions

The examples of secondary succession in the forests of the North Carolina piedmont and the Colorado Front Range illustrate the close interrelationship between tree population dynamics and changes in biomass and production. The period of initial establishment and canopy closure defines the recovery period for production. The subsequent buildup of biomass and the associated inhibition of regeneration by established trees introduces a time lag into the system. Although initially tree growth compensates for mortality, this process cannot continue indefinitely. Biomass and production will drop and regeneration will resume, with new trees filling gaps in the canopy.

The degree of time lag or length of interruption in regeneration will have a major impact of stand biomass and production. Where the time lag is minor, a single peak in biomass with a subsequent drop to an asymptote

is typical. Where lack of a time lag is coupled with slow initial regeneration, a simple asymptotic increase in biomass and production can be found. However, where the time lag in regeneration is marked, strong but damping oscillations can be expected in biomass and production.

The proposed time lag model does not incorporate all of the important factors influencing biomass and production during secondary succession. For example, Horn's explanation of decreasing production during succession based on changes in the geometry of the dominant tree species is probably important, yet it is not incorporated. Nutrient limitations are also important in many systems. The possible value of the time lag model is to be sought not in its completeness, but in its ability to explain a number of otherwise divergent or unrelated observations in a single theory. It will be successful to the extent that it provides increased conceptual unity or suggests improvements for quantitative models of forest growth.

Acknowledgments

This work was supported by National Science Foundation grants DEB 77-08743 and DEB 78-04043, and a grant from the North Carolina Energy Institute. The collaboration of Norman L. Christensen and Orin Pete Council is gratefully acknowledged.

Chapter 21

Above-Ground Biomass Accumulation and Net Primary Production During the First 70 Years of Succession in *Populus grandidentata* Stands on Poor Sites in Northern Lower Michigan

Arthur W. Cooper

Introduction

In recent years there have been many studies of biomass and primary production in forest ecosystems, but because forest ecosystems are long-lived, and because there are relatively few data sets from permanent sample plots or artificial chronosequences documenting changes in individual forest stands, few estimates have been made of change in biomass and net primary production over long periods in the life of a forest (Peet Chapter 20).

An opportunity to study such changes and to test the various hypotheses concerning changes in biomass and net primary production during succession (Peet Chapter 20) is available at the University of Michigan Biological Station (UMBS) in west central Cheboygan County, Michigan. Here bigtooth aspen (*Populus grandidentata*) forests now occur on soils previously occupied by pine, and ecological studies have been under way since 1910. Considerable data and several permanent sample plots established by F. C. Gates exist documenting changes in the biota of the region as it recovered from extensive logging and fire in the early 1900s.

During the summers of 1967–1969 and 1971 I carried out a program of research involving (1) dimension analysis of the canopy species of the bigtooth aspen ecosystem; (2) re-analysis of Gates' permanent and experimental burn plots; and (3) analysis of other sample plots in mature aspen stands. These data, together with remeasurement of the plots in the summers of 1973 and 1979, were combined to yield estimates of changes in biomass and net primary production over the first 70 years of succession in bigtooth aspen stands. The studies were confined to poor sites on deep, sandy soils, and the generalizations derived are valid only for those sites and soils. Furthermore, these stands were subjected to repeated, human-induced, intense fires and logging, which may have altered the soil organic matter and chemical content. Thus, regeneration may not be typical of that which followed presettlement fires. In spite of this possibility the stands represent unique opportunities for study of changes during one successional sequence.

Origin of the Aspen Forest

Sandy soils in northern lower Michigan are now occupied by pure stands of bigtooth aspen, but were originally covered with vast stands of large hemlock (*Tsuga canadensis*) and white and red pine (*Pinus strobus* and *P. resinosa*) (Gates 1930; Kilburn 1960a,b). Bigtooth aspen apparently was scattered throughout these forests, presumably as single clones in openings originating from disturbance. In the vicinity of the University of Michigan Biological Station pine and hemlock forests were cut between 1870 and 1880 with virtually all stands destroyed by 1890. Although hemlock was extirpated from the sandy uplands by logging and subsequent collecting of bark for tannin, enough scattered red and white pines remained to serve as a seed source for reestablishment of the species.

In the years immediately following the first logging, wildfires repeatedly swept across the landscape, preventing recovery of the forest, so that by the early 1900s most of the sandy soils supported only scrubby stands of bracken fern (*Pteridium aquilinum*), blueberry (*Vaccinium* spp.), and young bigtooth aspen root sprouts. By about 1920, substantial areas of sandy soils in western Cheboygan and adjacent eastern Emmet Counties had come under the management of UMBS and were thus protected from fire. Subsequently, forests dominated by bigtooth aspen developed over much of the area. Now these stands are dominated by bigtooth aspen with abundant white and red pine reproduction.

The late F. C. Gates recognized that these stands offered a unique opportunity and studied succession on them from 1930 to 1954. In cooperation with foresters at UMBS, he established a series of permanent sample plots and experimental burns designed to provide, through periodic remeasurement, a record of successional changes occurring on each major soil type in the region.

Other studies suggest that growth, community structure, and community composition differ dramatically as a response to differences in the properties of the sandy soils supporting bigtooth aspen (Benninghoff and Cramer 1963; Graham et al. 1963; Koerper and Richardson 1980). On good upland sites, with sandy loam soils, a clay layer near the surface, or the water table within 5 ft of the surface, bigtooth aspen basal area at maturity may exceed 28 m^2/ha and height at 30 years may range from 18 to

24 m (Graham et al. 1963). On poor aspen sites, with deep sandy soils, basal areas from 14 to 18 m^2/ha or less and heights of 9–15 m at 30 years are the rule (Graham et al. 1963). On good aspen sites, most tree reproduction and the herbaceous species are those of the northern hardwood forest (Benninghoff and Cramer 1963). On poor aspen sites, red oak (*Quercus rubra*) and red maple (*Acer rubrum*) occur with aspen in the canopy, and white and red pine reproduction is common. The herb layer is dominated by bracken, blueberry, and bush honeysuckle (*Diervilla lonicera*), and mosses are abundant.

Koerper and Richardson (1980) studied biomass and net annual primary production in mature (52–60 years) bigtooth aspen on good, intermediate, and poor sites at UMBS. The total above-ground biomasses of aspen were 171,565, 128,765, and 38,530 kg/ha at the good, intermediate, and poor sites, respectively. Corresponding above-ground net annual production values were 11,038, 7259, and 2925 kg/ha. The percentages of biomass components, except for leaves, generally were similar among sites. Poor site production per unit leaf weight was one-third that of the good site. Koerper and Richardson do not report biomass or net production data for other tree and herbaceous species associated with the aspen or changes in these properties over time.

All reproduction of bigtooth aspen is vegetative, by means of root and collar sprouts (Zahner and Crawford 1965). As a result, extensive clones of genetically identical individuals may arise. Clones often differ markedly in phenotypic features and growth rates. Thus, differences among sites due to soil conditions may be masked or accentuated by the differing growth rates of the clones occurring on them (Zahner and Crawford 1965). Individual clones also respond differently to silvicultural treatment (Garrett and Zahner 1964).

Site Description and Stand Sampling

Description of Study Areas

All areas involved in this study (Fig. 21.1) were located near UMBS on poor sites with very sandy soils. Studies were carried out on previously established permanent sample plots in aspen stands of varying age and in sample areas in the woods immediately surrounding these plots, or in similar stands of bigtooth aspen on other poor sites south and west of UMBS (Table 21.1).

The majority of the work was done near the center of Section 32, T37N, R3W, Cheboygan County. This area is a nearly level, morainic ridge covered with unstratified outwash. It has an elevation of 232–238 m mean sea level. Woodlands in this area are typically dominated by densely

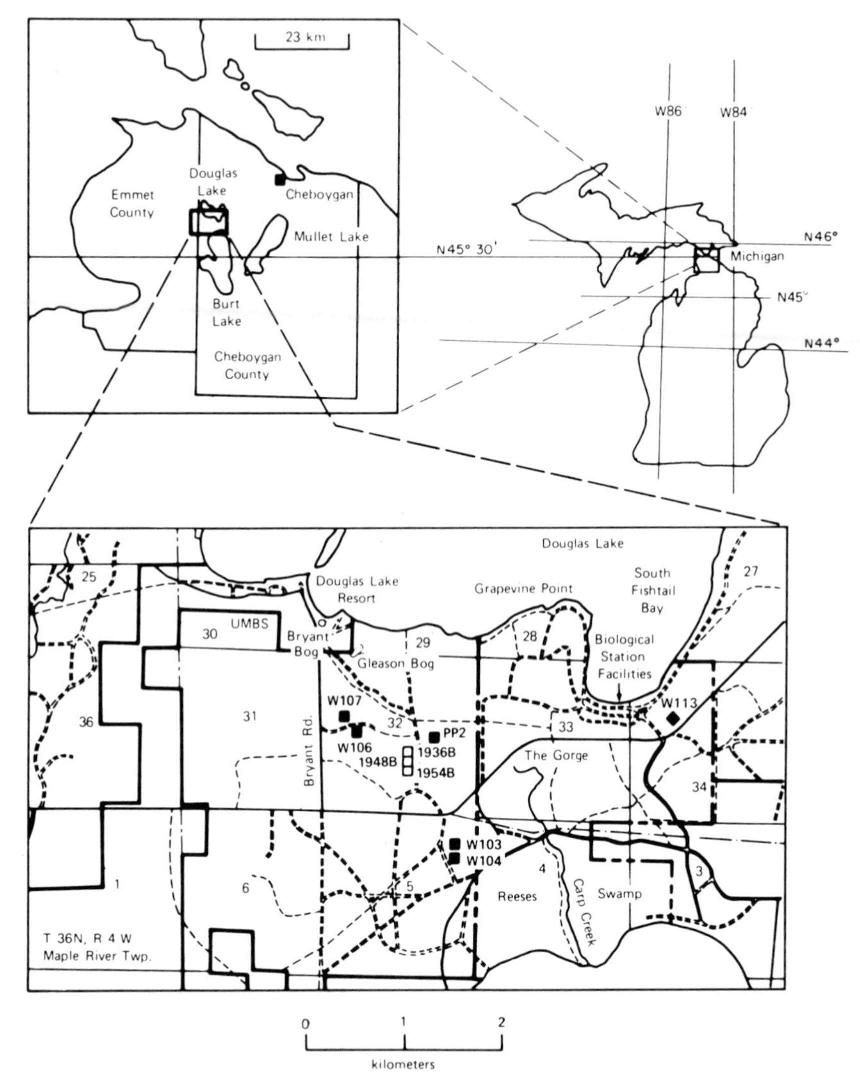

Figure 21.1. Location of study sites PP2, 36B, 48B, 54B, 103, 104, 106, 107, and 113 at UMBS. See text for description of specific sites. Plot sizes not to scale.

stocked, mature, bigtooth aspen forest with trees ranging from 10 to 18 cm dbh. Red oak and red maple are important subdominants and there is abundant reproduction of white and red pine. The herb layer is dominated by dense bracken averaging less than 0.7 m in height. Increment cores taken from dominant aspens in 1978 indicate an age of 66–67 years, suggesting that the last fire occurred in 1911. Because Kilburn (1960a) cited 1911 and 1923 as the years of the most recent fires at UMBS, 1911 was chosen as the date for the last fire and as the date of regeneration of all mature aspen stands studied in Section 32.

Several sites near the center of Section 32 were chosen for intensive study. The first, Permanent Plot 2 (PP2), is one of seven permanent 20.12 × 20.12-m study plots laid out in 1938 by F. C. Gates and W. H. Ramsdell (Table 21.1).

A second set of samples (Table 21.1) was obtained from three contiguous areas, each of about 1 1/2 ha, located just south of the center of Section 32, where Gates cleared and burned aspen forest in three different years. These areas, cleared and burned in 1936 (36B), 1948 (48B), and 1954 (54B), respectively, were designed to maintain various stages in aspen succession. All trees and brush were felled and piled so as to provide as complete a burn as possible. Photographs of the areas show that clearing and burning were very thorough and that all standing trees were cut. Although an effort was made to make these burns in relatively homogeneous aspen forest, the stands are not now homogeneous. The area of the 1948 burn seems to have several dense clones and thus has a somewhat higher density and basal area than other stands of similar age. Portions of the 1936 and 1954 burns are open with a low density of aspen, which could be due either to low density in the original forest or to hot spots created by burning brush that reduced the number of sprouts from the surviving aspen root systems. In addition, there is evidence in a portion of the 1936 burn of disturbance resulting from experimental work on aspen growth.

Data were collected from two 20.12 × 20.12-m plots established in each of these burns in 1957 (Farmer 1958), and from one 10 × 20-m plot laid out adjacent to each of the two larger plots in each of the three burns.

Several other sample areas in mature bigtooth aspen forest were located elsewhere on UMBS property (Table 21.1). Plots identical in size to those in PP2 and the burns were established. One area (113) originated in 1908. This area has a greater proportion of pine in the canopy and understory.

The pattern of change during succession was developed by treating these permanent sample plots as a series of samples over time, covering stands from ages 14 to 71. The method thus combines the permanent plot and artificial chronosequence methods (Peet Chapter 20).

Soils

Soils of all study areas are medium sands that have developed from glacial outwash sands (Spurr and Zumberge 1956). All are weakly to moderately developed spodosols. Occasional gravel pockets occur throughout, drainage is free to excessive, and there is no evidence of the bands of finer-textured material that are associated with intermediate and good aspen sites (Koerper and Richardson 1980). The soils of all sites are considered to be typical Rubicon sands, except for Plots 103 and 104, which lie close to a boundary between Rubicon and Graycalm sand (Mark Roberts, personal communication).

Table 21.1. Description and location of study sites, sample plots, and sources of data.

Sample area	Location and date of origin	Years sampled	Ages of samples	Plot sizes	Data gathered and years recorded
1954 Burn (54B)	Center, Sec. 32, T37, R3W. Last burned 1954.	1968 1971 1979[a]	14 17 25	2 plots, 20.12 x 20.12 m; 2 plots 10 x 20 m;	Tree diameter (1968, 71, 79); tree density (1968, 71, 79); herb clip samples (1968, 69); litter (1968, 69); litter fall (1969)
1948 Burn (48B)	Center, Sec. 32, T37N, R3W. Last burned 1948.	1968 1971 1979[a]	20 23 31	2 plots, 20.12 x 20.12 m; 2 plots 10 x 20 m	Tree diameter (1968, 71, 79); tree density (1968, 71, 79); herb clip samples (1968, 69); litter (1968, 69); litter fall (1969)
1936 Burn (36B)	Center, Sec. 32, T37N, R3W. Last burned 1936.	1968 1971 1979[a]	32 35 43	2 plots, 20.12 x 20.12 m; 2 plots 10 x 20 m	Tree diameter (1968, 71, 79); tree density (1968, 71, 79); herb clip samples (1968, 69); litter (1968, 69); litter fall (1969)

Permanent Plot 2 (PP2)	Center, Sec. 32, T37N, R3W. Last burned 1911.	1938[b] 1945[b] 1951[b] 1955[b] 1968 1971 1973[c] 1979[a]	27 34 40 44 57 60 62 68	1 plot, 20.12 x 20.12 m	Tree diameter (1938, 45, 51, 55, 68, 71, 73, 79); tree density (1938, 45, 51, 55, 68, 71, 73, 79); herb clip samples (1968, 69); litter (1968, 69); litter fall (1969)
103, 104	SE 1/4 of NE 1/4, Sec. 5, T36N, R3W. Last burned 1911.	1969 1973 1979[a]	58 62 68	1 plot, 20.12 x 20.12 in each area	Tree diameter (1969, 73, 79); tree density (1969, 73, 79); herb clip samples (1969)
106, 107	Western center, Sec. 32, T37N, R3W. Last burned 1911.	1969 1973 1979[a]	58 62 68	1 plot, 20.12 x 20.12 in each area	Tree diameter (1969, 73, 79); tree density (1969, 73, 79); herb clip samples (1969)
113[d]	SW 1/4 of NW 1/4, Sec. 34, T37N, R3W. Last burned 1908.	1969 1973 1979[a]	61 65 71	1 plot, 20.12 x 20.12	Tree diameter (1969, 73, 79); tree density (1969, 73, 79); herb clip samples (1969)

[a]Data from Mark Roberts, School of Forestry and Environmental Science, Duke University, Durham, North Carolina.
[b]Data in UMBS files.
[c]Data from C. J. Richardson, School of Forestry and Environmental Science, Duke University, Durham, North Carolina.
[d]The site of this plot was F. C. Gates "Set I," and was used from 1929 to 54 as a site for his class studies.

A typical Rubicon profile has a thin, black A1 horizon overlying a light brownish gray, strongly acid, albic A2 horizon. The B2 (spodic) horizon is typically encountered at about 15 cm; it is dark brown, weakly to moderately cemented, and averages about 30 cm in depth. The B3 horizon is yellowish brown in color, very friable, and usually somewhat deeper (up to 50 cm thick) than the B2 horizon. The light yellowish brown sandy parent material is usually encountered within 90 cm of the surface.

Tree Growth Data

Measurements included the diameter of each tree over 1.3 cm dbh (1.4 m) and a count of the number of tree seedlings (less than 0.3 m) and saplings (0.3–1.4 m). From these data density and basal area were calculated.

Biomass and Net Primary Production of Trees and Tree Reproduction

Above-ground biomass and net primary production were estimated by dimensional analysis (Newbould 1967; Whittaker and Woodwell 1968). Sampling of bigtooth aspen was done in July and early August of 1967, 1968, and 1969. All other species were sampled in July and early August of 1968. A total of 131 individuals of nine canopy species were felled and analyzed (Table 21.2). All trees were harvested from the area immediately surrounding PP2 and the experimental burns, except for a portion of the bigtooth aspen trees, which came from areas of similar forest on poor sites in the immediate vicinity of UMBS.

Individuals of bigtooth and trembling aspen (*Populus tremuloides*), juneberry (*Amelanchier* spp.), and beech (*Fagus grandifolia*) were felled and height and diameter at breast height measured. The entire tree was then separated into leaves and twigs, fruit (when present), live branches (wood plus bark), dead branches, and trunk wood plus bark. Each component was weighed and a subsample collected for determination of fresh-to-dry weight conversion factors. These samples were dried in a forced-draft oven to 80°C until change in weight had ceased. The leaf-twig component was separated into leaves and twigs by dividing subsamples, determining the average percentage of dry weight of each, and then multiplying the pooled weight by these percentages.

For red and white pine, red maple, red oak, and white birch (*Betula papyrifera*), the trunk was divided into 1-m sections after felling. The diameter of each branch was measured (just beyond basal swell), and one branch was randomly chosen from each section for detailed analysis. On each branch sampled, the fresh weight of leaves and twigs, dead branches, and live branches (wood plus bark) was determined and subsamples taken for determination of fresh-to-dry weight conversion factors. The fresh weight of the entire trunk was determined, its diameter at breast height and height measured, and samples for drying gathered. The samples were dried in the same manner as were those of the other species. In the case of the two pines, the needles were separated into current needles plus twigs and old needles (those on branches not produced during the current year).

No large white or red pines were included in the harvest sample. Rather than extend the equations derived in my study beyond the upper-diameter limits of trees measured in the plots, other data were sought. In the case of white pine, the raw data (Sollins and Anderson 1971) for 20 trees sampled at Coweeta, North Carolina, by Swank and Schreuder (1974) were combined with the six small trees collected at Douglas Lake to produce one set of equations. As Swank and Schreuder provided no data for dead branches, the Douglas Lake data had to be extended to larger trees. For red pine, six trees from Douglas Lake were pooled with similar data for ten trees from Minnesota provided by D. H. Alban to develop a single set of equations.

Component weights of seedlings, saplings, and of *Salix bebbiana* were determined by harvesting a representative sample of individuals and determining the mean weight of each component. Leaf-twig relationships were assumed to be similar to mature individuals.

Herbaceous Layer and Litter

Randomly placed square meter clip samples of the herbaceous layer were obtained in mid-August. In both 1968 and 1969, five plots were collected in each of the two sample areas in the 1936, 1948, and 1954 burns, and ten plots were collected from PP2. Plots collected in 1968 were eliminated from the 1969 sample. In addition, ten samples were taken in Plots 103, 104, 106, 107, and 113 in 1969. In each plot all herbaceous plants and shrubby species, such as *Diervilla* and *Vaccinium* spp., were clipped. Bracken was separated from all other plants and two categories, bracken and "other," recorded. The number of bracken fronds per square meter was also tallied. All materials were dried at 80°C to constant weight and results expressed in oven dry weight.

In the 1936, 1948, and 1954 burns and in PP2, litter was removed from the square meter from which the herbaceous materials were collected in 1968 and 1969. These samples were also dried to constant oven dry weight at 80°C. Finally, in 1969 each plot collected was staked, and, in mid-November, the freshly fallen litter was collected. These samples were used to estimate annual litter fall.

Table 21.2. Summary of dimension analysis data and regression coefficients for nine canopy species (no corrections made for bias[a]).

Species and component[b]	n	Diameter range (cm)	a[f]	b[f]	r^2	Standard error
Acer rubrum						
Leaves and twigs	14	0.8-15.2	4.7760	1.5636	0.98	0.24
Live branches	14	0.8-15.2	5.6797	1.9284	0.86	0.79
Dead branches	14	0.8-15.2	2.3009	2.9042	0.77	1.60
Bole	14	0.8-15.2	6.9399	2.2068	0.99	0.10
Amelanchier sp.						
Leaves and twigs	14	0.5-6.4	4.2232	1.1633	0.76	0.53
Live branches	14	0.5-6.4	5.6033	1.8879	0.94	0.40
Dead branches[c]	14	0.5-6.4	2.9488	1.9596	0.79	0.82
Fruit[c]	14	0.5-6.4	1.9129	1.4269	0.65	0.85
Bole	14	0.5-6.4	6.9787	1.9116	0.98	0.20
Betula papyrifera						
Leaves and twigs	12	1.5-14.0	4.7542	1.6388	0.89	0.41
Live branches	12	1.5-14.0	5.2630	2.1305	0.91	0.49
Dead branches	12	1.5-14.0	-0.4625	4.9073	0.83	1.61
Bole	12	1.5-14.0	6.6157	2.4266	0.99	0.14
Fagus grandifolia						
Leaves and twigs	10	0.3-3.1	4.5335	1.0288	0.76	0.49
Live branches	10	0.3-3.1	5.6670	1.6118	0.75	0.80
Dead branches	10	0.3-3.1	4.6884	2.6548	0.83	1.02
Bole	10	0.3-3.1	6.9187	1.5543	0.91	0.40
Pinus resinosa[d]						
Leaves	16	1.5-27.2	5.9654	1.4386	0.96	0.29
Live branches and twigs	16	1.5-27.2	5.5139	1.8017	0.95	0.38
Dead branches	16	1.5-27.2	0.8210	3.5025	0.92	0.96
Bole	16	1.5-27.2	6.5255	2.3529	0.98	0.31
Pinus strobus[e]						
Leaves	26	0.3-18.3	5.8343	1.5431	0.91	0.46
Live branches and twigs	26	0.3-18.3	6.1725	2.0218	0.95	0.45
Dead branches	6	0.3-5.1	1.0644	0.8608	0.78	0.53
Bole	26	0.3-18.3	6.6402	1.7038	0.95	0.38
Populus grandidentata						
Leaves and twigs	52	0.5-20.3	4.9714	1.3664	0.88	0.44
Live branches	52	0.5-20.3	5.1144	1.8981	0.93	0.47
Dead branches	51	0.5-20.3	4.2539	1.4087	0.77	0.68
Bole	42	2.5-20.3	6.5580	2.4283	0.99	0.14
Bole	10	0.2-2.5	6.4612	1.3543	0.93	0.22

Table 21.2. (*continued*)

Species and component[b]	n	Diameter range (cm)	a[f]	b[f]	r^2	Standard error
Populus tremuloides						
Leaves and twigs	5	0.5-8.1	4.9767	1.0935	0.95	0.31
Live branches	5	0.5-8.1	5.4544	1.5277	0.92	0.56
Dead branches	5	0.5-8.1	4.1325	1.9229	0.99	0.27
Bole	5	0.5-8.1	6.9579	1.8158	0.97	0.37
Quercus rubra						
Leaves and twigs	12	3.1-22.9	4.7204	1.7937	0.96	0.24
Live branches	12	3.1-22.9	4.6079	2.6567	0.91	0.54
Dead branches	12	3.1-22.9	2.4475	2.8384	0.90	0.64
Fruit[c]	12	3.1-22.9	-1.2086	3.0434	0.66	1.45
Bole	12	3.1-22.9	6.8995	2.3021	0.99	0.13

[a]Madgwick (1976).

[b]For bigtooth and trembling aspen, juneberry, and beech, regressions of the form $\log_e Y = \log_e a + B \log_e X$, where Y = dry weight of a component in grams and X = dbh in inches, were calculated using total weights of components determined directly from field analyses. For red and white pine, red oak, red maple, and white birch, regressions were first developed involving dry weight of branch components on branch diameter. These were then used to estimate leaf and branch component weights for each sample tree. The weights of each component were then summed for all branches of each sample tree and these summed weights, together with trunk weights, were regressed on sample tree diameters.

[c]Correction of +1 added to eliminate zero values.

[d]Includes 6 trees from Douglas Lake and 10 from Minnesota (see text).

[e]Includes 6 trees from Douglas Lake and 20 trees from Coweeta, North Carolina (see text).

[f]Regression coefficients for dbh in inches and weight in grams.

Biomass Accumulation and Net Production

The nine tree species can be placed into one of three groups: aspen, oak-maple, and pine. The aspens (bigtooth and trembling) reproduce entirely by root and collar sprouts, flushing immediately following fire with no subsequent reproduction by seed. Occasionally, sprouts arise from stems that die during succession; these are never numerous. The oak-maple group (red oak, red maple, juneberry, beech, white birch) also produces a flush of sprouts, chiefly of collar origin, immediately following fire. However, the seedlings of each of these species may also appear throughout the succession, a few of which may survive and reach reproductive maturity. Red maple and juneberry seedlings are the most abundant, followed by red oak, beech, and white birch. Red and white pines, on the other hand, reproduce only by seed and, except for an occasional wolf tree that survived the early fires, enter the succession during the first 15 years. Their number depends upon the proximity and age of potential seed trees and increases rapidly later in the successional sequence. Thus, a mature aspen forest contains aspens only of sprout origin, oaks, maples, and associated species of both sprout and seedling origin, and pines of seedling origin only.

Dimension Analysis

The results of dimension analyses (Table 21.2) are consistent with the results of similar analyses of other species in other areas and of bigtooth aspen in northern Michigan (Koerper and Richardson 1980) and adjacent areas (Zavitkovski 1971). In virtually all cases, the diameters of trees encountered in sample plots fell within the range of diameters of trees sampled in the dimension analysis work. Six bigtooth aspens exceeded 20 cm dbh in the 1968–1979 samples of PP2, and several more occurred in the other plots in the 55–71 year age group. The equations of Table 21.2 were extended to cover these individuals.

Biomass Accumulation

Trees

In order to estimate tree biomass in each sample plot, the weights derived from the dimension equations for trees equivalent to the diameters of those present in the plot were summed for each component of each species. Regressions were then applied to the data to describe the average change in biomass of a component during the time covered by the study (Fig. 21.2).

There is great variation in biomass of most of the components from plots of similar age (Fig. 21.2). However, in all cases the equation describing a given component had an F value significant at the 1% level or better. The variability in biomass data results primarily from differences in initial density of sprouts from the aspens and other deciduous species, secondarily from differences in rate and density of establishment of pine seedlings, and from whatever unknown differences there may have been in the treatment and climatic histories of the plots.

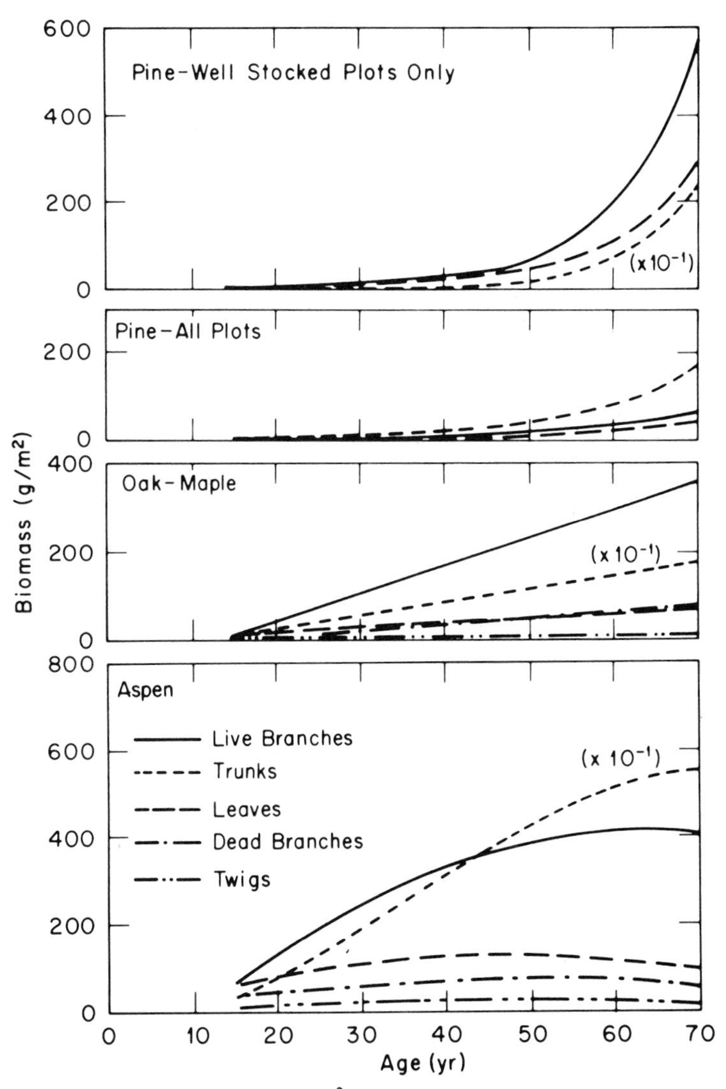

Figure 21.2. Estimated biomass (g/m^2) of aspen, oak-maple, and pine. See text for description of equations. Equations and raw data available from author.

Two equations ($Y = a + bX - cX^2$ and $Y = a + bX^2 - cX^3$) were fitted to the aspen data. Each gave a good empirical fit to the observed changes in biomass over time. Although it might be assumed that the equation with the cubic term, which produces a brief period of early slow growth and then a rapid rise to a maximum, would fit all of the aspen biomass data best, this did not prove to be the case. Actually, only the trunk data were best described by this equation, and the parabolic form fitted the remaining data best. The explanation for this fact appears to lie in the very rapid early growth of leaves and branches on aspen suckers and the slower growth of the stem of the sucker. Also, any early slow growth of leaves and branches that might have taken place occurred before the age of the youngest plots sampled in my study.

The biomass of aspen leaves, twigs, and dead branches increased rapidly to maximam of about 125, 12.8, and 73 g/m^2, respectively, by about the 50th year and declined slowly thereafter. Aspen live branches increased throughout the first 65 years, reached their peak standing crop of 417 g/m^2 in the 65th year, and declined thereafter. Although aspen trunk biomass continued to increase throughout the 70-year sequence and reached a maximum of about 5500 g/m^2 in the 70th year, growth slowed by the 70th year. These changes in aspen biomass reflect the rapid growth, natural thinning, and decline in rate of growth of the flush of aspen sprouts that arises after fire. Given the absence of aspen seedling reproduction, biomass can be expected to continue a slow decline to some small stable level similar to that assumed to be present in the original pine forests of the region (Kilburn 1960a,b).

The biomass for the oak-maple group (Fig. 21.2) increased linearly throughout the 70 years, but was quite variable, reflecting the variability in stocking of oak, maple, juneberry, beech, and white birch, particularly in the oldest sample plots. In the oak-maple group, by the 70th year leaf biomass was 70 g/m^2; twig biomass 4 g/m^2, live branch weight 365 g/m^2, trunk 1739 g/m^2; and fruit biomass 1.95 g/m^2.

The biomass of pine tree components increased rapidly from one sampling interval to the next, except for Plot PP2, where the increase in two intervals was small or slightly negative. The overall trend in PP2, however, was a rapid increase over the 40 years of sample data. Therefore, an exponential model was used to describe changes in pine biomass. When data from all plots were included together, the r^2 values were low (0.28–0.34), because of the three old sample plots in which only pine seedlings and saplings were present. However, when these plots were eliminated and the regressions recalculated, the r^2 values increased substantially (to 0.65–0.86), as did the predicted biomass values, of course. Thus, in describing pine biomass change, two patterns, one for pine in the aspen forest generally (all plots) and the other for pine in well-stocked portions of the forest (well-stocked plots only), were used. When all plots are considered, the maximum biomasses achieved in the 70th year were: leaves

39; twigs 3; live branches 60; dead branches 8; and trunks 164 g/m^2, respectively. For the well-stocked plots only, comparable values were: leaves 282, twigs 8, live branches 582, dead branches 35, and trunks 2443 g/m^3, respectively.

Herbaceous Layer

The standing crop of bracken declined with age of stand (Fig. 21.3), most rapidly during the first 20–25 years, in part a response to decrease in mean frond weight associated with change from sun to shade frond morphology.

The standing crop of other herbaceous species declined during the first 35 years from about 22 g/m^2 to less than 4 and then returned to about 12 by the 70th year. This pattern is a function of the decline of the species associated with burns and open areas (*Poa compressa*, *Solidago hispida*, *Vaccinium* spp.), which are virtually absent from the older sample plots, and eventual increase of the forest herbs such as *Gaultheria procumbens*, *Melampyrum lineare*, and *Oryzopsis asperifolia*.

Litter

Litter fall (Fig. 21.3) closely parallels development of the canopy. During the first 25 years, annual litter fall increased to about 200 g/m^2, and at age 65–70 is slightly in excess of 300 $g\ m^{-2}\ yr^{-1}$. Litter accumulation, therefore, took place most rapidly during the first 25 years of succession. The litter weight of a mature aspen forest, about 800 g/m^2, averages about 50% greater than that of young stands 20–25 years old, which have 400–500 g/m^2. Litter fall varied greatly. For example, one plot in the 1948 burn had a greater litter fall than any plot in PP2, whereas one plot each in the 1948 and 1936 burns had litter accumulated almost to the depth of that in PP2. The variability in litter fall in the young plots is undoubtedly associated with the variable density of stems and weight of foliage of aspen clones in those plots, whereas the variation in accumulation in older plots is probably due to long-term differences associated with the microsites that serve as foci for retention of fallen litter.

Production

Biomass data from the sample plots were used to estimate net primary production for aspen, oak-maple, pine, bracken, and other herbaceous components of the aspen forest. Values were estimated as follows (s = biomass in grams; t = time at beginning of interval; t + 1 = time at end of interval):

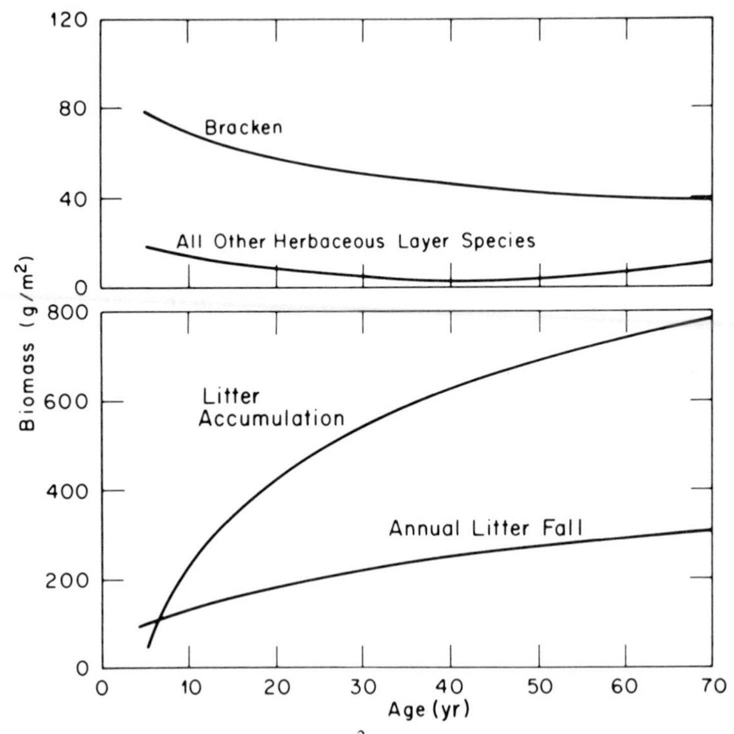

Figure 21.3. Estimated biomass (g/m^2) of bracken and other herbaceous layer species, and estimated weight (g/m^2) of litter accumulation and annual litter fall. Equations and raw data are available from the author.

1. The leaf and twig production for aspen and oak-maple and fruit production for oak-maple were calculated as the average standing crop in the plot for the interval $[(s_t + s_{t+1})/2]$.

2. The leaf production for pines was calculated in the same way except that the average standing crop at each sample interval was multiplied by 0.3, the approximate percentage of the total weight of needles that was produced anew each year $[(s_t + s_{t+1})/2(0.3)]$.

3. The production of live and dead branches together and trunks (wood and bark) for all groups was calculated as the total increase in weight over a sampling period divided by the number of years in the interval $[(s_{t+1} - s_t)/n]$. Negative values were treated as zero.

4. The standing crop of bracken and other herbaceous species (Fig. 21.3) was assumed to represent annual production.

Regressions of productivity values on time, employing the same models used to describe standing crop, were made for three components: leaves and twigs, live and dead branches, and trunks (Fig. 21.4). The total net annual productivity values were calculated as the sum of the estimated

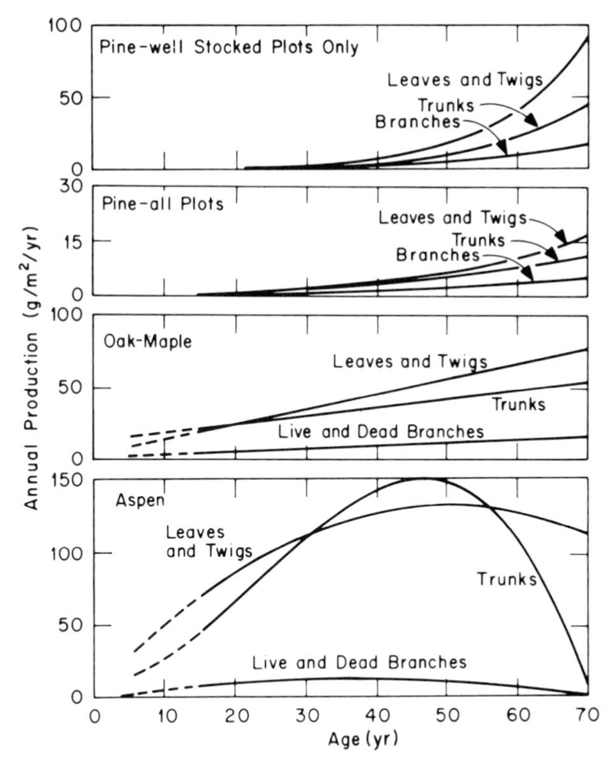

Figure 21.4. Estimated net annual production (g m^{-2} yr^{-1}) of aspen, oak-maple, and pine. Equations and raw data are available from the author.

values for individual components (in the case of species groups) or as the sum of the species groups for total community values. Because these values are not corrected for trees that died during sample intervals (mortality), they are therefore low estimates of production.

The net production of aspen leaves and twigs (Fig. 21.4) increased gradually to about 120 g m^{-2} yr^{-1}, by the 35th year, and did not decline below that level until after the 65th year. The maximum leaf and twig production was 132 g m^{-2} yr^{-1} in the 50th year. Trunk production increased rapidly during the first 30 years to about 110 g m^{-2} yr^{-1}, more slowly thereafter to a maximum of 151 in years 40–45, and declined to 11 by the 70th year. Branch production was low, never exceeding 12 g m^{-2} yr^{-1}, and peaked earlier (years 30–40) than production of leaves and trunks.

Net primary production in the oak-maple group reflected the linear models used to describe it (Fig. 21.4). Leaf and twig production rose to 76 g/m^2 by year 70. Branch, trunk, and fruit production reached 15, 52, and 2 g m^{-2} yr^{-1}, respectively, over the same interval.

The net primary production of all pine components (Fig. 21.4) rose rapidly after years 35–40. In the well-stocked plots, leaf and twig production rose from 11 g/m^2 in year 45 to 91 in the 70th year. Comparable values for branches and trunks are 4 and 16 and 8 and 45, respectively. The values for all plots were much lower, with leaf and twig, branch, and trunk production reaching only 17, 5, and 11 $g\ m^{-2}\ yr^{-1}$, respectively, in the 70th year.

The standing crop and production data reported here compare favorably with those of Koerper and Richardson (1980), who reported a total above-ground standing crop for a 60-year-old bigtooth aspen stand on a poor site at UMBS of 3853 g/m^2. Although their distribution of biomass is similar to that I found in 18–23-cm trees, the absolute values for trunks (5000–5500 g/m^2) and live branches (400–420g/m^2) found for trees of the same age in my study (Fig. 21.3) are higher. This difference can be explained by the fact that the basal area of aspen on Koerper and Richardson's poor site was substantially lower (11.4 m^2/ha) than the 15.7-m^2/ha average basal area of aspen in my 19 samples of mature (50+ years) aspen forest. Aspen also made up a smaller percentage of stand basal area in Koerper and Richardson's stand than in mine.

The pattern of above-ground net primary production for bigtooth aspen forest predicted for the first 70 years of development by the models used in this study (Fig. 21.5) shows that tree leaf and twig production rose gradually to about 200 $g\ m^{-2}\ yr^{-1}$ at about years 50–55. Thereafter, in the plots well stocked with pine, it continued to increase, reaching 280 by the 70th year. However, when all the pine plots are considered, leaf production remained constant at about 200 through the 70th year. Branch production reached 25 $g\ m^{-2}\ yr^{-1}$ in year 45, thereafter rising to 31 in the dense pine plots and declining to an average of 20 in all pine plots. Trunk production reached a maximum of 203 $g\ m^{-2}\ yr^{-1}$ in the 50th year and declined thereafter. In the plots well stocked with pine it declined to 108, whereas the average for all the plots declined to 74 $g\ m^{-2}\ yr^{-1}$ by the 70th year. The total net primary production for all vascular plants for the entire sequence, including bracken and other herbaceous species, increased very rapidly for the first 5–10 years to over 200 $g\ m^{-2}\ yr^{-1}$ and then less rapidly to about 450 by the 40th year. The rapid rise in production of the first 5 years is a function of the rapid growth of bracken and weedy species and of aspen suckers in response to full sunlight and whatever nutrients are made available as a result of fire. Because no data were available for the first 10 years of aspen growth, there was no evidence of the rapid increase in growth during the first 3–5 years, which, in turn, is usually followed by a decline in growth due to competition and natural thinning (Graham et al. 1963). The data available do, however, portray the rapid growth of the stand during the years 10–20 reported by Graham et al. (1963). After the 40th year, the total net production estimate is influenced by the stocking of pine. In areas where there are few pines, production declines slowly from a peak of 465 $g\ m^{-2}\ yr^{-1}$ in the year 40–50 to 354 by year 70 (although a correction for stem mortality might alter this

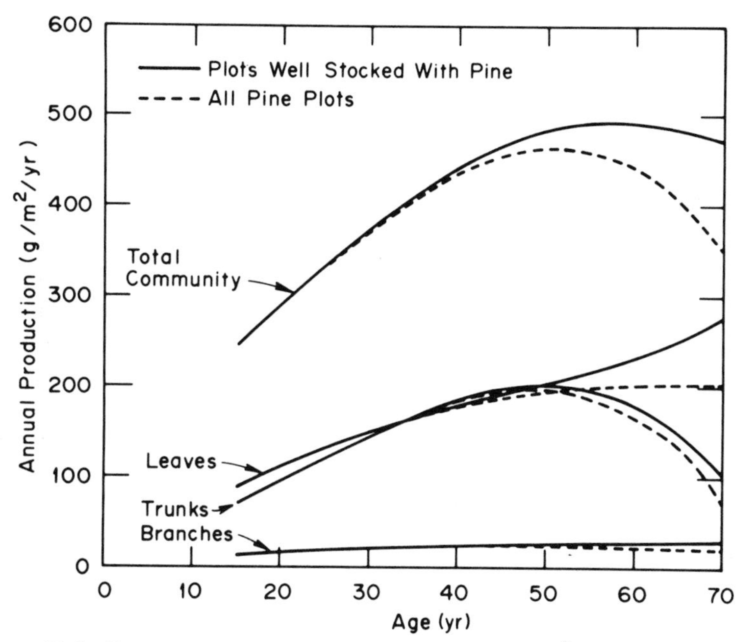

Figure 21.5. Total net annual aboveground biomass (g/m^2) and net annual primary production (g m^{-2} yr^{-1}) for leaf and twig, branch, stem, and herbaceous layer components. The values are calculated at 5-year intervals from data in Fig. 21.4.

conclusion). In areas where stocking of pines is higher, the total net primary production is about 125 g m^{-2} yr^{-1} greater, and it remains approximately constant at nearly 500 for the last 20 years of the study period.

Discussion

This study represents one of the longest unbroken records of change in biomass and net community production available from stands of forest vegetation in North America. It offers an opportunity to examine ecological theory dealing with succession against documented events.

The aspen forests studied at UMBS, although appearing homogeneous, are in reality a mosaic of stands of differing composition and varying growth rates. The even-aged, old-growth bigtooth aspen population is the "matrix" within which three major sorts of variation occur. First, the aspens themselves, because of variations in morphology and growth of individual clones, differ in density, form, and growth rate. Second, red oak and red maple, because of their origin as stump sprouts, occur in clumps and patches. Third, pine stocking varies greatly both in density and species composition. In areas where small pines become established early in the succession, where good seed sources were available, or where mature pines were established before the fires ceased, pines are now well stocked and are growing rapidly. In other areas lacking an initial seed source, pines are just now becoming established. Except in areas where red pine seed sources are present, white pine is the more common pine, despite the fact that red pine appeared to be more abundant than white pine in the original forest on sandy soils (Kittredge and Chittenden 1929; Kilburn 1960b).

The forests range from open stands of aspen of poor form and slow growth mixed with a few oaks and maples and having no large pines to closed stands of aspen of good form and better growth with abundant oak, maple, and some canopy-sized pines. Stands exist showing all degrees of intermediacy between these extremes.

A mature bigtooth aspen forest on sandy soil at UMBS has an aspen above-ground standing crop ranging from 5.5 to 6.5 kg/m^2 and an annual production range for aspen of 150–300 g m^{-2} yr^{-1}; a pine standing crop ranging from 0.4 to 1.4 kg/m^2 and production of 30–110 g m^{-2} yr^{-1}; and an oak-maple biomass of 1.5–2.7 kg/m^2 and production of 90–180 g m^{-2} yr^{-1}. The total aboveground biomass ranges from slightly less than 7 to nearly 10 kg/m^2; annual tree productivity ranges from 300 to 500 g m^{-2} yr^{-1}.

These values will hold only until mortality begins to have a significant impact on the mature aspen trees; then major changes will take place. The data clearly indicate that the 70-year-old bigtooth aspen stands at UMBS are in this period of transition and probably have been so for at least 10 years (Fig. 21.6). The net annual production of aspen is now declining rapidly, and aspen standing crop has reached its maximum level. Increased mortality of mature aspen will further decrease production and will cause the standing crop to begin to decline. On the other hand, the standing crop and net annual production of pine and oak-maple are increasing. Whereas net annual production is now about 140 g m^{-2} yr^{-1} each for aspen, pine, and oak-maple, within 10–20 years the production of the forest will be almost entirely confined to pine and oak-maple. This transition will be the most evident event of the next two decades.

The differing productivities of aspen, pine, and oak-maple combine to produce what is a temporary "steady state" (Fig. 21.6). Where pines are well stocked, total above-ground community productivity (including herbaceous species), although now declining slightly, has essentially stabilized at about 475–500 g m^{-1}, the apparent maximum for mature bigtooth aspen forest on poor soils at UMBS. On the other hand, in areas where few pines are present, total community production is apparently declining, rather than remaining stable (Fig. 21.5).

These trends will continue until one or more new species become dominant. Where pines are well stocked, they will clearly be the next dominant. In areas where they are not, it is likely that there will be a

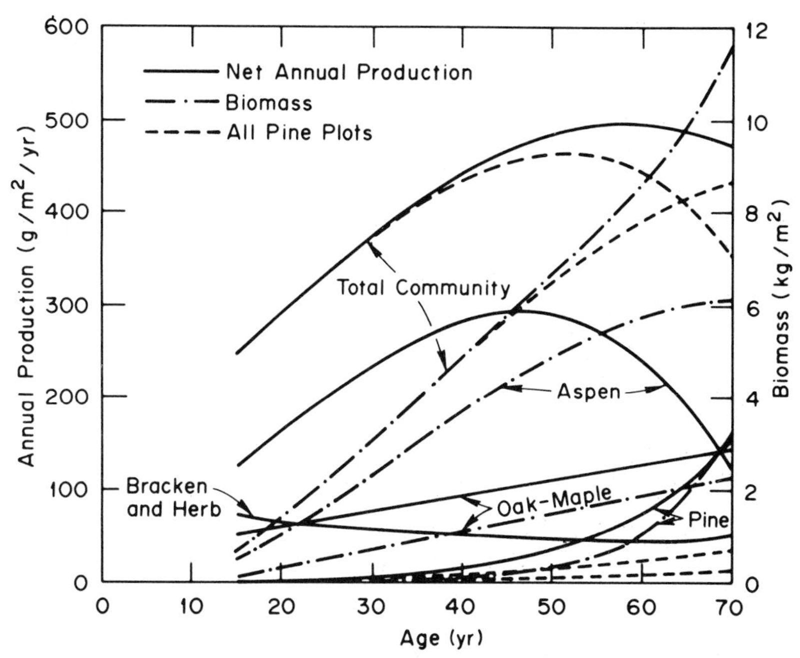

Figure 21.6. Total annual aboveground biomass (g/m^2) and net annual primary production (g m^{-2} yr^{-1}) for species groups. The values are calculated at 5-year intervals from data in Fig. 21.4.

period of unpredictable duration when the oaks and maples are dominant before pine finally assumes dominance. In those few areas where pines are poorly stocked and where there are few oaks and maples, the future is unclear. In such areas, the canopy will probably become much more open than it is now, and it is entirely possible that aspen suckers arising from the roots of recently dead canopy trees may be able to survive and mature under such conditions. In this way, small patches of aspen dominance may persist when the remainder of the forest has been converted to pine or mixed pine and oak.

Without information on tree mortality, it is difficult to determine which of the productivity patterns outlined by Peet (Chapter 20) are displayed by this sere. Time lags in pine reproduction and establishment are clearly present, but the extent to which pine growth is suppressed by aspens is unknown. A peak in productivity followed by a decline is apparent in those sites where pine establishment is slow, however.

Acknowledgments

I wish to acknowledge the following assistance: R. Zahner for information on aspen silvics and growth and for data from UMBS files; R. Sabo and M. F. Rebar for help with all phases of the field work, particularly the dimension analysis felling program; G. L. Wheeler for programming the analyses of tree dimensions; B. L. Ward for laboratory analyses and programming; M. F. Cutting for leaf area determinations; C. J. Richardson and M. Roberts for advice and information relating to their on-going studies of aspen forest at UMBS; D. H. Alban for data from ten red pine trees harvested at Pike Bay Experimental Forest, Minnesota; W. L. Hafley for assistance with statistical analyses; and D. M. Gates and the late F. K. Sparrow, directors of UMBS during the period the work was done, for assistance and use of station facilities. This research was partially supported by a grant from the Department of Energy to David M. Gates, Contract No. DE-AC02-EV10091.

Chapter 22

Below-Ground Processes in Forest Succession

Kermit Cromack, Jr.

Introduction

In many forests, most energy utilization and nitrogen transformations of carbon compounds take place below ground (Harris et al. 1980; Fogel 1980). This is true throughout secondary succession. The above-ground patterns of changes in species composition and vegetative structure with forest succession are accompanied by and indicative of major qualitative and quantitative changes below ground. One stage in succession may be affected by prior stages through processes that occur below ground, but the extent of this influence is little known and is of considerable current research interest.

In this chapter, important below-ground ecosystem components and processes will be reviewed, and then, through selected examples, the importance of specific components and processes in different stages of forest succession will be discussed. Some inferences also will be made concerning effects of particular below-ground components or processes upon later stages of succession. Pacific Northwest forests will be emphasized as the primary source of illustrative cases.

Below-Ground Ecosystem Components and Processes

The below-ground portion of forest ecosystems includes the following general components: (1) plant roots and associated microflora which function in the uptake of nutrients and water; (2) heterotrophic microorganisms and soil animals active in decomposition and nutrient cycling processes; (3) bacteria specialized in transformations of certain elements such as N, S, Fe, and Mn; (4) organisms pathogenic or parasitic on other biota; (5) organic matter in all stages of decay; and (6) the mineral soil.

Early discussions of ecological succession recognized the importance of below-ground components in ecosystem organic matter turnover and nutrient cycling (Darwin 1881; Russell 1921; Waksman 1932), but only in recent decades have adequate methods been developed to measure in the field root, microbial, and soil animal biomass and processes related to their activities (Parkinson et al. 1971; Phillipson 1971; Postgate 1971; Russell 1973; Dickinson and Pugh 1974; Lohm and Persson 1977; Marshall 1978; McNabb and Geist 1979; Soderstrom 1979; Swift et al. 1979; Batistic et al. 1980; Clareholm and Rosswall 1980; Powell et al. 1980). Concurrently, new methods have evolved for fractionating soil organic matter, minerals, and organomineral complexes that are less destructive chemically than classical robust chemical extraction of soil humic fractions (Greenland and Ford 1964; Spycher and Young 1977; Turchenek and Oades 1979). Both stable and radioactive isotopes also have facilitated the interpretation of key below-ground processes related to physical-chemical-biological below-ground components (Jenkinson 1971, 1977; Clark 1977; Paul and van Veen 1978; Ladd and Amato 1980). The work of Jenny (1941, 1961, 1980) and Jenny et al. (1968) has provided the framework for a broad synthesis of above- and below-ground ecosystem components through analysis of state factor equations for soil and ecosystem development.

The root systems of most higher plant families are colonized by fungal symbionts known as mycorrhizal fungi, which effectively replace root hairs in uptake of nutrients and water. In exchange, these fungi utilize plant photosynthate as their major energy source (Harley 1972, 1975; Bowen 1973). Due to their greater surface area per unit of biomass, mycorrhizae permit a more energy-efficient exploitation of the soil pore space and soil particle surface contact exchange than do root hairs (Richards 1974; Harley 1975; Nye and Tinker 1977; Ruehle and Marx 1979).

Two major functional types of mycorrhizal associations have evolved, the ectomycorrhizae and the endomycorrhizae. Ectomycorrhizae are characterized by a dense surface mantle of hyphae sheathing the root surface. Endomycorrhizae lack a characteristic fungal mantle, and they penetrate root cells with morphologically distinct structures (Harley 1972). Ectomycorrhizae occur mainly in plant families such as the Pinaceae, Betulaceae, and Fagaceae in the northern hemisphere, in the tropical Caesalpinoideae and Dipterocarpaceae, and in tree genera such as *Eucalyptus* in Australia.

Endomycorrhizae colonize roots of most gymnosperms other than the Pinaceae, most herbaceous angiosperms, including grasses, and many shrub and tree species of angiosperms such as Ericaceae, Leguminosae, and the Rosaceae. Depending upon the relative dominance of particular plant species during different stages of succession, it is possible to shift from dominance by endomycorrhizal plants to dominance by ectomycorrhizal species.

The decomposition of organic residues and recycling of nutrients are carried out by heterotrophic bacteria, fungi, and soil animals. Eukaryotic microflora such as fungi are most important in the decay of complex structural carbon substrates. Fungi can transport substances internally through hyphal networks stretching over considerable distances and connecting chemically diverse microsites (Alexander 1961; Hudson 1968; Swift et al. 1979). Certain fungal hyphae are specialized for the penetration of structural organic surfaces over long time periods, whereas unicellular prokaryotes such as bacteria form colonies by binary fission of individual cells on surfaces. Fungi thus exploit pore space more effectively, while bacteria colonize solid surfaces, including dead fungal hyphae (Parr et al. 1967; Swift et al. 1979).

Prokaryotes such as actinomycetes and other bacteria have considerable enzymatic diversity that enhances their ability to break down complex carbon substrates. Though bacterial biomass may be somewhat less than that of fungi in forest soils, they can contribute nearly as much to total soil carbon mineralization as can fungi (Clareholm and Rosswall 1980). Prokaryotes are important, for example, in the conversion of nitrogenous organic compounds to ammonia. Prokaryotes also are essential in nitrogen fixation both as free-living species and as symbionts of higher plants (Walker 1975). Chemically autotrophic prokaryotes oxidize inorganic compounds of N, S, Fe, and Mn; their activities, coupled with those of organisms that reduce compounds of these elements, can significantly influence relative availability of these elements in soil solution, their gain or loss by leaching from the rooting zone, and their volatilization from the soil surface. It is important to point out, as will be discussed later, that significant changes in activity by these prokaryotes are associated with major forest disturbances and with particular below-ground processes occurring in given stages of primary and secondary succession.

Soil animals, both microscopic and macroscopic, interact synergistically with microbes in decomposition processes. The animals not only alter litter chemically but also physically, converting it to smaller particles that have a greater surface-to-volume ratio. These animals derive a substantial portion of their nutrition from microbes embedded in litter and from internal microbial flora (Swift et al. 1979). Soil animals also mix soil substrates, transport microbial spores, and enhance microbial growth by their "grazing" effects (Van der Drift and Witkamp 1959; Parkinson et al. 1979; Swift et al. 1979). Many invertebrates excrete readily available forms of nitrogen such as ammonia as waste products. Protozoa may cause the direct release of mineralized phosphorus from bacteria (Cole et al. 1978). Until recently, vertebrates other than burrowing ones were not thought to have much influence on below-ground processes. Now it is realized that many small mammals derive substantial food resources from the fungal sporocarps of mycorrhizae, as well as from the sporocarps of saprophytic fungi. This discovery has led to important new hypotheses concerning mammalian effects on succession processes (Maser et al. 1978a,b).

Soil pathogens and parasites influence directly or indirectly a number of below-ground processes. Viruses, for example, infect both prokaryotic and eukaryotic organisms. Fungal diseases of plant roots can alter both species composition and relative density of tree populations (McCauley and Cook 1980). Soil animals such as nematodes injure roots by grazing on feeder roots. It has been hypothesized that they may help inoculate roots with mycorrhizae, in some cases as a beneficial compensation for their root feeding. Fungal parasites may invade senescent roots and leaves and function as initial decomposers (Waid 1957; Swift et al. 1979).

Vegetative Succession in Pacific Northwest Forests

Descriptions of vegetative changes during secondary succession in Pacific Northwest communities dominated by Douglas-fir (*Pseudotsuga menziesii*) have been provided by Zavitkovski and Newton (1968a), Dyrness (1973), Franklin and Dyrness (1973), and by Franklin and Hemstrom (Chapter 14). In these forests, substantial accumulations of coarse woody

Table 22.1. Annual organic matter and N inputs by above- and below-ground litter components in a young *P. menziesii* stand in western Oregon (Fogel and Hunt 1979, Fogel 1980).

Component	kg ha^{-1} yr^{-1} Organic matter	N
Foliage	2410	14.
Branches	270	0.4
Boles	137	0.1
Total above ground:	2817	14.5
Mycorrhizal root (host tissue)	9157	55[a]
Mycorrhizae (fungal sheath)	6104	36[a]
Total below ground:	15261	91

[a]N concentrations assumed to be the same in both components for budget given. Fungal sheath would probably have a higher N concentration if separate data were taken (Fogel 1980, pers. comm.).

debris persist in stands of various ages and successional stages (Franklin and Waring 1980; Franklin and Hemstrom Chapter 14). The theoretical basis for dominance by evergreen conifers during the later stages of succession in the Pacific Northwest has been presented by Waring and Franklin (1979). Succession following major fires will often result in the presence of early seral species such as red alder (*Alnus rubra*) and varnishleaf ceanothus (*Ceanothus velutinus*), which are capable of fixing nitrogen (Zavitkovski and Newton 1968a,b). Plants such as alder can function in both a primary and secondary successional role in areas such as landslides, snow avalanche tracts, and glacial moraines (Lawrence 1951; Crocker and Major 1955; Ugolini 1968). Early successional herbaceous species such as fireweed (*Epilobium angustifolium*) also come in very densely (Dyrness 1973); such species have been hypothesized to be important in the nutrient conservation of disturbed sites (Marks and Bormann 1972; Vitousek and Reiners 1975).

Root and Mycorrhizal Dynamics in Forest Succession

In several forest ecosystems root death has been recognized as contributing a major input to soil organic matter (SOM), generally exceeding above-ground litter fall (Harris et al. 1980). Associated processes such as turnover of mycorrhizal fungi are also major contributors to below-ground SOM dynamices (Fogel and Hunt, 1979). Though root and mycorrhizal turnover have not been investigated through entire sequences of secondary succession, direct evidence of substantial changes in SOM and nitrogen caused by early successional nitrogen-fixing species (Tarrant and Miller 1963; Youngberg and Wollum 1976) strongly implies significant below-ground SOM inputs in early as well as later successional stages.

The below-ground turnover of SOM and nitrogen by ectomycorrhizae in a young-growth Douglas-fir ecosystem has been found to be substantially greater than corresponding above-ground components (Fogel and Hunt 1979; Fogel 1980, pers. comm.). Organic matter inputs directly attributable to Douglas-fir were 5.4 times greater for the below-ground than the above-ground components. Mycorrhizae (fungal sheath) contributed 40% of the total below-ground organic matter input (Table 22.1). For nitrogen, the below-ground components contributed 86% of the total tree system nitrogen input to the soil (Table 22.1).

In an old-growth Douglas-fir ecosystem Sollins et al. (1980) estimated that nearly 71% of organic matter inputs to the soil were attributable to fine roots, but mycorrhizal root tissue and fungal sheaths were not included (Table 22.2). Dead boles made up 45% of total organic matter inputs in the old-growth ecosystem in contrast to less than 1% of the inputs in the young Douglas-fir stand reported by Fogel and Hunt (1979).

Nitrogen inputs from fine roots in the old-growth Douglas-fir ecosystem contributed 45% of total nitrogen inputs to the soil; if bole litter inputs are omitted, fine root nitrogen input would have contributed 50% of the total. In these comparisons, I have omitted precipitation nitrogen input which in our area is about 2 kg ha^{-1} yr^{-1} (Sollins et al. 1980). For the old growth I included an estimate of 2.1 kg ha^{-1} yr^{-1} of nitrogen from lichen litter fall (Sollins et al. 1980).

Endomycorrhizae or ectomycorrhizae dominate different stages of succession. In the Pacific Northwest, for example, endomycorrhizae dominate early successional brush communities where ceanothus dominates above ground (Rose, 1980). Later successional stages are dominated by Douglas-fir and other conifer species which are ectomycorrhizal, although some residual understory species remain endomycorrhizal. Even where dominance of functional types of mycorrhizae may not change appreciably, as in coniferous forests with climax species not in the Pinaceae, pioneer species such as red alder can have mycorrhizae that are host specific (Molina 1979; Trappe 1979). Alders, and others in the family Betulaceae, are predominantly ectomycorrhizal (Harley 1972).

Some plant species colonizing disturbed sites have less dependence on mycorrhizae than do mature forest species, at least in lowland tropical forests (Janos 1980). In the Pacific Northwest, a recent experimental study of the mycorrhizal pioneer species, ceanothus, showed significantly greater growth and higher rates of nitrogen fixation by vesicular-arbuscular (VA) endomycorrhizal plants than non-mycorrhizal controls (Rose and Youngberg 1981).

Table 22.2. Annual organic matter and N inputs by above- and below-ground litter components in an old-growth *P. menziesii* stand in western Oregon.[a]

Component	kg ha^{-1} yr^{-1} Organic matter	N
Foliage	2500	14.1
Lichens	121	2.1
Branches	1100	3.8
Boles	7000	3.9
Total above-ground:	10721	23.9
Fine roots ($\leqslant$5 mm diam)	3400	21.0
Coarse roots (>5 mm diam)	1400	1.3
Total below-ground:	4800	22.3

[a]Sollins et al. 1980.

Rooting patterns can change substantially in Pacific Northwest forest stands of different ages, even in those dominated by the same species. For example, a 23-year-old Pacific silver fir (*Abies amabilis*) stand had a greater percentage of mycorrhizal and fibrous textured roots in the A horizon than did a 180-year-old stand (Vogt et al. 1980). This pattern reflects significant changes in root resource allocation between the mineral soil, traditionally considered the true below-ground realm, and the forest floor. It also cautions against estimating ecosystem organic matter nutrient budgets for forest floor solely from above-ground litter inputs.

Dynamics of Below-Ground Nutrients and Soil Organic Matter

Nitrogen Fixation

Early successional species of plants that fix nitrogen symbiotically in the Pacific Northwest include both legumes and nonlegumes. Nonlegumes generally contribute greater amounts of fixed nitrogen (Table 22.3), but some perennial legumes can continue to fix nitrogen for more than 15 years. (McNabb et al. 1976) and may be important in long-term nitrogen accretion.

Long-term field studies have shown that nitrogen fixation associated with red alder and ceanothus can increase total soil nitrogen by 33% to 39%, respectively (Tarrant and Miller 1963; Youngberg and Wollum 1976). Data from experimentally established plantations following a wildfire in old-growth Douglas-fir indicate that the presence of a nitrogen fixing species in secondary succession can influence rate of stand development, including emergence of dominants. The total stand wood volume for a mixed Douglas-fir and red alder stand was nearly twice the volume of an adjacent pure stand of Douglas-fir at age 48 years (Miller and Murray 1978). There were approximately 66 uniformly distributed "superdominants" per hectare in the 38–46-cm diameter size class in the mixed stand and none in the pure Douglas-fir stand.

Nitrogen fixation in leaf litter substrates by free-living bacteria can be influenced by the stage of succession. Nitrogen fixation does not occur in leaf litter from red alder, since its litter is already rich in that element. Needle litter from Douglas-fir, a later successional species, serves as a substrate for nitrogen fixation, but the process appears to be absent from needle litter of western hemlock (*Tsuga heterophylla*), a "climax" species (Silvester 1980, pers. comm.). Litter layer nitrogen fixation is found in New Zealand forests also, indicating independent evolution of similar fixation processes (Silvester 1978).

Table 22.3. Representative data for symbiotic nitrogen fixation in the Pacific Northwest.

Species	Location	Annual nitrogen fixation (kg ha^{-1} yr^{-1})	Reference
Alnus rubra	Coast Range, Oregon	320	Newton et al. (1968)
	Western Cascades, Washington	40	Tarrant and Miller (1963)
	Cedar River, Washington	85	Cole et al. (1978)
Ceanothus velutinus	Western Cascades, Oregon	108	Youngberg and Wollum (1976)
	Eastern Oregon	70	Youngberg and Wollum (1976)
Lupinus sericeus	Eastern Oregon	7.2	McNabb et al. (1976)
Lupinus leucophyllus	Eastern Oregon	10.5	McNabb et al. (1976)

Successional Changes in Litter Substrate Quality

Litter substrate quality as measured by carbon compounds and the C:N ratio can influence litter decomposition rates (Jensen 1929; Waksman 1932; Alexander 1961; Cromack and Monk 1975; Fogel and Cromack 1977; Meentmeyer 1978; Sharpe et al. 1980; Aber and Melillo 1980). Successional species in the Pacific Northwest differ in carbon and nitrogen litter substrate quality (Table 22.3) and litter of early successional species generally decays more rapidly.

Preliminary evidence suggests that below-ground root litter differs in litter substrate quality for several of the species listed in Table 22.4 (Cromack et al. 1979a). Chemical composition of live fine roots suggests that early successional species such as red alder and ceanothus should have faster decomposition rates of root litter (Cromack et al. 1979a).

Successional Changes in Soil Organic Matter and Nitrogen Capital

Changes in soil nitrogen have direct effects upon soil inorganic nitrogen availability (Keeney 1980). Studies suggest a strong correlation between available inorganic nitrogen and total soil nitrogen (Heilman 1974). Disturbances to the vegetation and soil influence relative magnitude of nitrogen losses from leaching (Vitousek et al. 1979) or gaseous products of denitrification and/or partial NH_4^+ oxidation (Bremner and Blackmer 1979).

A number of soil properties can be altered by the addition of nitrogen from biological fixation and result in long-term effects on the below-ground ecosystem. These include: (1) soil bulk density; (2) soil pore size and pore distribution; (3) cation exchange capacity (CEC); (4) soil aggregate size and aggregate strength. Addition of SOM to forest soils from nitrogen-fixing species can affect soil structure (Tarrant and Miller 1963). Such structural changes influence water-holding capacity. The addition of fixed nitrogen would lower the C:N ratio and thus increase the availability of inorganic nitrogen in soil solution and on cation exchange sites. Addition of SOM would also increase the CEC of the soil. Since changes in SOM have effects on soil physical properties, new methods of studying soil structure by means of image analysis, for example, should facilitate work on effects of plant succession on soil physical properties (Bullock and Thomasson 1979).

Changes in soil aggregate structure are important in influencing root growth and relative access to available nutrients by roots (Greenland 1979, Hewitt and Dexter 1979). Young and Spycher (1979) hypothesized that soil aggregate structure is an important factor regulating turnover of carbon and nitrogen in soil as does the chemical nature of SOM. Soil mi-

Table 22.4. Nutrient composition and C substrate quality of foliage and needle litter of both early and late successional species in the H. J. Andrews Forest (additional data available from author).

Species	N (%)	C (%)	C:N ratio	Lignin (%)[a]	Cellulose (%)[a]	ADF (%)[a]	Annual weight loss (%)
Epilobium angustifolium[b]	0.65	46	71	10.1	14.9	74.6	52
Alnus rubra[b]	2.1	15	21	9.4	9.9	80.6	51
Ceanothus velutinus[b]	0.81	45	55	10.0	11.0	77.3	48
Acer circinatum[b,c]	0.70	44	63	8.5	14.7	73.9	59
Rhododendron macrophyllum[c]	0.20	51	255	10.3	16.0	73.7	25
Castanopsis chrysophylla[c]	0.32	50	156	22.8	21.1	55.9	37
Pseudotsuga menziesii							
Green needles[d]	0.80	50	63	21.8	13.2	64.2	25
Senescent needles[d]	0.50	50	100	24.1	14.6	59.5	20
Seed cones[d]	0.24	49	204	44.2	34.1	21.2	8
Twigs[d]	0.15	49	327	43.4	30.0	26.7	9
Bark[d]	0.19	56	294	58.6	12.6	23.0	3

[a] After Van Soest (1963). ADF is the acid detergent fraction.
[b] Generally occupy sites as early successional species; *C. velutinus* and *A. rubra* fix N.
[c] Present as both early and late successional species.
[d] Predominant litter components of old-growth *Pseudotsuga* communities.

croflora such as fungi can cause soil aggregation to occur (Clough and Sutton 1978). Hence, availability of nutrients to plants may alter with changes in aggregate structure on both a short-term and a long-term basis during succession.

Considerable interest exists in understanding the specific mechanism of nitrogen accumulation in the soil from decomposition of litter components of nitrogen-fixing species. Models of SOM humification and soil organomineral complexes have postulated mechanisms of amino acid incorporation into the various SOM fractions (Anderson 1977, 1979; Spycher and Young 1979; Young and Spycher 1979; Ladd and Amato 1980). The residence time of both the "light" density SOM fraction as well as the "heavy" organomineral fractions may be controlled by changes in environmental factors and in rates of inputs to the various SOM fractions from the different species present in succession. Future studies of interactions between SOM, nitrogen, inorganic soil mineral particles and soil aggregates will be necessary to understand the changes in SOM and nitrogen in different stages of succession.

Coarse Woody Debris as a Special Organic Matter Fraction

Great accumulations of coarse woody debris in Douglas-fir ecosystems persist through several stages of succession due to the large size of the trees and their slow decomposition rate (Franklin and Waring 1980). Amounts of coarse woody debris range from 70 to 580 metric tons/ha in old-growth forests in the region (Grier and Logan 1977; Grier 1978; Franklin and Waring 1980). Fires can create more coarse woody debris than they destroy (Franklin and Hemstrom Chapter 14).

Coarse woody debris is part of the above-ground litter fraction, but its large accumulation and long residence time (Maser et al. 1979) lead to considerable interaction with the below-ground fractions. Coarse woody debris serves as an organic seedbed for forest tree species (Jones 1945; Franklin and Hemstrom Chapter 14) and as a major habitat for invertebrates and vertebrates (Elton 1966; Maser et al. 1979; Triska and Cromack, 1980) which feed on microbial and animal tissues within or near decaying logs. Logs provide physical cover as well.

Many mycorrhizal fungi produce spores that are easily spread by wind and water; well-known sporocarps such as mushrooms and puffballs are adapted for this purpose. Other mycorrhizae produce either single spores or enclosed spore-containing structures so that dispersal by a wide variety of both invertebrates and vertebrates is necessary for the proper reinoculation of sites (McIlveen and Cole 1976; Maser et al. 1978a). Thus, microbial and animal cocolonization of sites may be essential to the presence or absence of certain species or assemblages of species in different successional stages.

Small mammals are important vectors for dissemination of spores from mycorrhizal fungi with hypogeous sporocarps (truffles) (Maser et al. 1978a,b). Following the logging and burning of clear cuts, changes in small mammal populations can occur due to changes in availability of hypogeous fungi associated with the previous stand (Maser et al. 1978b). Since logs provide protective cover as well as other habitat functions, residual coarse woody debris can supply a physical link between different stages of forest succession by serving as a dispersal pathway for small mammals carrying undigested mycorrhizal fungal spores from adjacent undisturbed areas (Maser et al. 1979).

Furthermore, coarse woody debris functions as an important SOM component for colonization by mycorrhizal fungi (Harvey et al. 1976, 1978, 1979). Species such as Douglas-fir and western hemlock have mycorrhizal roots which are capable of colonizing wood in an advanced decay stage (Harvey et al. 1976). The community of small mammals and decomposer invertebrates inhabiting logs and snags is important in processing of mycorrhizal fungal spores, some of which will inoculate roots colonizing decaying wood. The management of coarse woody debris is therefore an important consideration in intensive forest management for fiber and fuel (Maser et al. 1978b, 1979). Complete removal of wood debris might alter forest species composition, or at least the relative abundance of certain species.

Interactions of Below-Ground Ecosystem Components: An Integrated Example

The substantial turnover of SOM and nitrogen by mycorrhizae in a young Douglas-fir stand (Fogel and Hunt 1979; Fogel 1980) is one indication of the magnitude of below-ground dynamics in Pacific Northwest forests. The intense colonization of soil by mycorrhizae is also important in processes such as soil weathering and release of nutrients (Graustein et al. 1977; Cromack et al. 1979b). The hypogeous basidiomycete mycorrhiza *Hysterangium crassum* spores found in rodent digestive tracts (Maser et al. 1978a) suggest that the mammals are a principal vector of the fungus. Thus, some Douglas-fir mycorrhizal species are likely spread by small mammals, which may have used residual wood as a dispersal mechanism into young stands. The young Douglas-fir stand discussed above (Table 22.1) had several species of hypogeous mycorrhizae (Fogel 1976). One species, *H. crassum*, occupied 9.6% of the A soil horizon to a depth of 10 cm. The degree of soil weathering and intensity of colonization was similar to that in fungal soil mats observed in coniferous stands in both Finland and North America (Hintikka and Naykki 1967; Fisher 1972).

Acknowledgments

Support for the writing of this paper came from NSF Ecosystems grant DEB-77-06075. Publication No. 1531 from the Forest Research Laboratory, Oregon State University, Corvallis, Oregon.

Chapter 23

Fire and Succession in the Conifer Forests of Northern North America

Miron L. Heinselman

Introduction

Ecosystem theories must encompass the conclusions now emerging from studies of fire and vegetation in fire-dependent northern conifer forests. Such forests comprise more than half the present forest area of North America, including most of the forests that have never been altered through logging or land clearing. Furthermore, vast areas are still influenced by natural lightning-fire regimes, and it is possible to study directly the role of fire in controlling vegetation mosaics and ecosystem function.

In this chapter the relation of fire to vegetation change in conifer dominated forests is reviewed for three regions: the boreal forest, the Great Lakes-St.Lawrence-Acadian forests, and the subalpine forests of the Rocky Mountains (see Rowe 1972 and Küchler 1966 for the regions and vegetation types included).

Fire Regimes and Fire Cycles in Northern Conifer Forests

The kind of fire history that characterizes an ecosystem is called its *fire regime*. Its elements are (1) *fire type* and *intensity* (e.g., crown fires, severe surface fires, and light surface fires); (2) *size* of typical significant fires; and (3) *frequency* or *return intervals* for specific land units.

The average time required for a fire regime to burn over an area equivalent to the total area of an ecosystem is its *fire rotation* (Heinselman 1973) or *fire cycle* (Van Wagner 1978). Fire is really not that orderly, however, because each physiographic site tends to have its own return interval, and fire is a semirandom process. Some areas are skipped for long periods, while others may burn two or more times during a rotation.

In the main boreal forest the dominant regime is high-intensity crown fires or severe surface fires of very large size—often more than 10,000 ha, and sometimes more than 400,000 ha. Such fires kill entire stands and, hence, result in total stand replacement. Fire rotations are shorter in the drier regions of western Canada and Alaska, where they average 50–100 years, than in the high precipitation areas of eastern Canada, where they may average 200 years or more. Fire cycles are also longer near the tree line in the subarctic spruce-lichen woodlands than in the closed boreal forests. Cycles of 100–150 years typify the lichen woodlands of the western boreal, while in the east cycles may average 150–300 years. In some jack pine (*Pinus banksiana*) and lodgepole pine (*Pinus contorta*) forests in western Canada surface fires of low intensity occur at intervals of 25 years or so, without killing whole stands. The longest cycles in the western boreal probably are in floodplain white spruce (*Picea glauca*), where they may be 300 years long. Documentation of regimes can be found in Lutz (1956); Barney (1971); Rowe and Scotter (1973); Viereck (1973, 1981); Rowe et al. (1974, 1975); Simard (1975); Donnelly and Harrington (1978); Heinselman (1978, 1980); Johnson (1979); Rowe (1979); and Stocks and Hartley (1979).

In the Great Lakes-St.Lawrence-Acadian forests there were several distinct regimes. In the north, enclaves of boreal forest had stand-killing crown fire regimes similar to the full boreal. Red and white pine (*Pinus resinosa, P. strobus*) forests on xeric landforms in the west had combinations of moderate-intensity surface fires at intervals of 20–40 years, with more intense fires killing much of the stand at intervals of 150–200 years. Farther east, or on mesic sites, white pine forests had regimes of crown fires or intense surface fires that brought in new age classes in longer cycles—perhaps 200–300 years. "Jack pine barrens" on sand plains had regimes of light surface fires, with cycles as short as 15–30 years. Large peatlands in Minnesota supporting black spruce (*Picea mariana*) had crown fire regimes at cycles of 150–200 years. Documentation of regimes can be found in Marschner (1930); Spurr (1954); Frissell (1973); Heinselman (1973, 1974, 1978, 1980); Swain (1973); Cwynar (1977); and Wein and Moore (1977).

The subalpine forests of the Rocky Mountains formerly had regimes that varied with elevation, physiography, and vegetation. In lodgepole pine forests near the lower ecotone with Douglas fir (*Pseudotsuga menziesii*) savanna, a common regime was one of moderate intensity surface fires of medium size (40–400 ha), with fire cycles averaging about 25 years. Such fires killed portions of stands—producing multiaged structures. In the middle subalpine, mixed forests of lodgepole pine, Engelmann spruce (*Picea engelmannii*), and subalpine fir (*Abies lasiocarpa*) generally experienced high-intensity crown fires or severe surface fires, in cycles of 50–200 years. Most such fires failed to reach the tree line, but some burned up into tundra. Documentation of regimes can be found in Clements (1910); Habeck (1970); Wellner (1970); Day (1972); Habeck and Mutch (1973); Houston (1973); Loope and Gruell (1973); Brown (1975); Arno (1976); Gabriel (1976); and Tande (1979).

Current anthropogenic changes in fire regimes are easily summarized: (1) in the unexploited boreal forests of northern Canada and Alaska natural lightning fire regimes still prevail, either because fire control has not been attempted, or it has had little effect, and human-caused ignitions make only small contributions to the total area burned; and (2) in the Rockies, the exploited boreal forests, and the Great Lakes-St.Lawrence forests, fire control, logging, land clearing, roads, and urban areas have so greatly modified natural fire regimes that they are relevant chiefly in the management of nature reserves and in the light they shed on original ecosystem function.

Organic Matter Accumulation, Permafrost, and Paludification

In the north, organic matter decomposition on the forest floor may not keep pace with the accumulation of moss and humus, particularly on cold or wet sites. This potential for net accumulation of biomass has long been recognized in northern peatlands (Gorham 1957; Sjörs 1959; Heinselman 1963, 1970, 1975), but its significance on mineral soils and its relation to fire are just now being studied. Unfavorable conditions for bacterial and fungal decay and incomplete nutrient cycling are the causes. Throughout the north, forest fires were a major factor in periodically reducing organic layers. The details of nutrient cycle-fire interactions are still being worked out, but the outlines seem clear (Heilman 1966; Ohmann and Grigal 1979; Viereck 1981; Van Cleve and Viereck, Chapter 13). The feathermosses (*Pleurozium schreberi*, *Hylocomium splendens*, *Hypnum crista-castrensis*) and *Sphagnum* are the major humus builders, and they form continuous carpets in many conifer stands 40 or 50 years after fire. Such a tieup of nutrients in moss and organic layers runs counter to expected trends in temperate forest successions. Maximum nutrient cycling may come midway in postfire vegetation sequences (Chapin and Van Cleve 1981), and the accumulation of biomass may continue indefinitely in the absence of fire (Viereck 1981). This is one of the factors that may cause species diversity to peak early in postfire vegetation sequences. In late stages there is degradation of site quality and depauperization of the community.

Permanently frozen ground (*permafrost*) is discontinuous but widespread in the boreal forests of northwestern Canada and interior Alaska (Brown 1970). The depth of summer thaw (the *active layer*) varies from 30 to 200 cm or more. Fires that kill whole stands cause increased surface soil temperatures and thawing because of overstory removal, changes in albedo, and removal of the insulating organic layer (Viereck and Dyrness 1979; Viereck 1981, in press). Increased soil temperatures combined with nutrient release cause increases in productivity. Higher temperatures and deeper thawing peak out only 10–15 years into the post-fire period, however, after which increasing tree cover and moss layers cause a gradual reversion to preburn conditions. Prolonged fire-free intervals cause a rise in permafrost tables, impeded soil drainage, and site degradation beyond that attributable to stand maturity. Instead of succession leading toward stable, diverse, and productive ecosystems, the trend is toward depauperate black spruce muskeg (Drury 1956; Viereck and Dyrness 1979; Viereck 1981; Van Cleve and Viereck Chapter 13).

On gently sloping mesic to wet terrain, long fire-free periods may allow so much organic matter to accumulate that actual peat is formed, and mineral soil sites become permanent peatland. This process is called *paludification* (Heinselman 1975). Vast regions have already been paludified in the last 5000 years, particularly the Hudson Bay Lowlands (Sjörs 1959, 1961), the glacial Lake Agassiz plain in Minnesota, Ontario and Manitoba (Heinselman 1963, 1970; Zoltai 1971), the Ontario Clay Belt, the lowlands of major northern rivers such as the Kuskokwim and Tanana (Drury 1956), and several areas in Newfoundland. Periodic fires have apparently burned away increments of peat from time to time, and, through their stimulation of nutrient cycling, have retarded peat accumulation. The trend of vegetation change with organic matter accumulation is toward depauperate black spruce raised bogs, or string bogs and ribbed fens, depending on drainage patterns, access to mineral-bearing waters, and other factors. Excluding fire accelerates paludification and vegetation changes that can only be seen as "retrogession" in the context of normal successions (Heinselman 1973, 1975).

Fire Characteristics and Ecological Effects

Fires vary in many ways that influence vegetation sequences. Three basic fire types occur: (1) crown fires, (2) surface fires, and (3) ground fires (fires in organic layers or in peat). Each type may vary in intensity (rate of burn and energy released) and area burned. Fire characteristics are relevant to revegetation because: (1) fire type and intensity control the kill of trees, shrubs, herbs, mosses, and lichens; (2) they control the opening of serotinous (closed) cones and seed dispersal in jack pine and lodgepole pine; (3) they control survival of stored seeds in the cones of jack pine, lodgepole pine, and black spruce, and of viable seeds of other conifers; (4) they control the kill of stored seeds and vegetative propagules in the organic soil and in the mineral soil; (5) they control the release of the carbon and nutrients in the vegetation and organic layers.

The state of the vegetation affects the fire, particularly by influencing the amount and kind of fuel. The species, age, and density of trees, shrubs, herbs, mosses, and lichens are important variables that influence fuel characteristics. For example, the Sphagnaceae are almost always moist and rarely burn, but the feathermosses dry out and are readily burnable. These mosses are the major contributors to the organic layer beneath spruce, fir, and some pine stands. Such layers may vary from a few centimeters to more than 1 m in thickness. The long-crowned conifers such as balsam fir (*Abies balsamea*), subalpine fir, northern white cedar (*Thuja occidentalis*), and the spruces often have flammable branches near the ground that serve as *fire ladders* to move flames into the canopy. Such trees often "torch off" even when they are scattered in stands that lack these characteristics e.g., red pine or aspen (*Populus*). Arboreal lichens (*Usnea* spp., *Alectoria* spp.) also facilitate crowning. Stand age is important because old stands contain dead timber and may have been subject to insect, disease, and parasite attack, or windfall. In eastern North America the spruce budworm (*Choristoneura fumiferana*) causes heavy mortality in balsam and white spruce—particularly in old stands with a high fir component. Mistletoes (*Arceuthobium*) are important parasites on lodgepole pine and black spruce. They kill trees, and their "witches brooms" are significant fuel. Barkbeetles (mostly *Dendroctonus*) cause mortality in old lodgepole pine and Engelmann spruce. All of these factors alter the abundance and arrangement of fuels.

The climate for a year or more preceding fires influences ecological effects because large, high-intensity burns often occur during extended droughts (Haines and Sando 1969). Fire history records indicate that most of the area burned in a region is generated by a few large fires, which occur at 10–20 year intervals (Heinselman 1973; Tande 1979). Extended drought also sets the stage for fires that can consume most of the organic layer.

Season of burn is important. It determines growth stage at the time of burning, important in the abilities of the plant to survive or to reproduce if killed; it determines the season in which initial reseeding will occur, and the timing of vegetative regeneration; it influences the extent of drying of organic layers. The winter-dry, summer-wet climatic patterns of northern North America influence seasonal effects. The ground is snow covered from November through April, or even from mid-October through May, and fires are only possible when the ground is bare. Snow melt recharges surface soil moisture reserves in April or May, but spring fires still occur because the cured leaves, needles, and grasses from the previous summer dry out quickly during even short spring dry periods. Spring fires are common in the southern boreal and Great Lakes region, but such fires seldom burn much of the organic layer because it is still cold and wet. Crown fires are possible at this time. Summer fires come at a time when soil moisture reserves may be much lower, and organic layers may become almost completely dried after significant droughts. Therefore, summer fires are much more likely to consume these layers. In normal years rains occur frequently through the summer, so it is chiefly in years of unusual drought that major summer fires occur.

Lightning ignitions also vary seasonally. In general, lightning peaks in midsummer, and lightning fires are much less common in early spring or in fall. The annual influx of thunderstorms into the boreal forests of western Canada occurs as a seasonal wave, entering from the south with the invasion of moisture-laden pacific air into the cold, dry arctic air. This invasion just reaches the subarctic spruce-lichen woodlands in midsummer, and then quickly recedes southward (Johnson and Rowe 1975; Rowe 1979). Most thunderstorms are accompanied by enough rain to extinguish lightning fires, but it is the strikes that come during "dry" storms in a drought that ignite major fires (Schroeder and Buck 1970).

Weather at the time of a fire is crucial to its effects. The key variables are humidity, wind speed, wind direction with respect to fuels and land forms, temperature, and sky conditions. Once a fire is under way, rate of spread, fire-induced winds, whirls, convection columns, and "spotting" can influence the intensity and fire size.

Fire barriers such as lakes, streams, barren rock, wetlands, snowfields and the timberlines of mountains can repeatedly influence fire paths. Seasonally nonflammable vegetation such as aspen stands may also check spread. However, long-distance "spotting" due to flying brands can overcome wide barriers when winds are high and fire convection columns strong.

The interfire interval is a regulator of fuel and fuels characteristics. Short periods obviously allow little time for new fuels to grow. However, if stand kill was good but fuel consumption poor in the last fire, there may be important carryover fuels. Trees killed by fire are seldom consumed (commonly only the needles, twigs and loose bark burn, even in crown fires), and so snag stands can furnish fuel for later fires. In the north snags last 20–50 years. Young, dense conifer stands will support crown fires under proper drying, humidity, and wind, and so young stands are not immune to fire either (jack pine, lodgepole pine, black spruce, and red pine burn well at only 15–20 years). Thus, "readiness to burn" is not a simple function of time since the last fire, but varies widely with stand conditions, weather, and fire behavior.

The total area of a burn influences revegetation from seed sources outside the burn. This effect is modified by the shape and alignment of the burn with respect to land forms, and by unburned inclusions. These factors can be important, but on many burns the regeneration comes largely from propagules originating within the burn.

The degree to which organic layers are consumed is important in determining revegetation patterns for the following reasons:

1. Exposure of mineral soil seedbeds favors the establishment of conifers (LeBarron 1939, 1944; Place 1955; Horton 1956; Ahlgren and Ahlgren 1960; Horton and Bedell 1960; Fowells 1965; Rowe 1981; Viereck 1981).
2. Burning recycles nutrients locked up in the organic layer. These layers may contain the largest pool of unavailable nutrients in the system (Ohmann and Grigal 1979; Zasada et al. 1979; Viereck 1981; Van Cleve and Viereck Chapter 13).
3. Burning of the organic layer increases soil temperatures—an important factor in nutrient mobilization.
4. Burning of organic layers causes thawing of permafrost through removal of insulating moss and exposure of darkened soils to solar radiation.
5. Soil surface temperatures are lower in mineral soils than in blackened organic soils. Black organic soils may reach temperatures lethal to seedlings or become so dry that seedlings die (LeBarron 1944; Place 1955; Zasada 1971).
6. Complete burning of the organic layer kills most of the seeds of plants that rely on seedbanks in the organic layer. Partial burning may enhance or not injure the germination of such plants (Cushwa et al. 1968; Moore and Wein 1977).
7. Depth of burning influences vegetative regeneration from plant structures in the organic layer (Ahlgren and Ahlgren 1960; Buckman 1964; Ohmann and Grigal 1979; Rowe 1981; Viereck 1981).

Plant Strategies for Coping with Fire

Noble and Slatyer (1977) proposed a set of species' "vital attributes" associated with persistence following fire, another set covering strategies for community-imposed stresses ("tolerance"), and a third set covering life histories. Tables 23.1 and 23.2 present classifications of northern plants according to the Noble and Slatyer attributes, based on Rowe (1981), the literature, and my field experience.

The canopy storage of seeds in closed or semiclosed cones is one of the best-known fire adaptations. The cones of most jack pine and many lodgepole populations are of the true serotinous (fully closed) type—the cone scales being sealed with resins that melt at the temperatures attained in crown fires or intense surface fires (Clements 1910; Horton 1956; Beaufait 1960). Some populations of both species are open coned, however, and these populations may be associated with regions having a history of light surface fires (Lotan 1976; Schoenike 1976). Black spruce has persistent, but only semiclosed cones, and seed dispersal occurs gradually over several years (LeBarron 1939; Wilton 1963; Zasada et al 1979; Viereck 1981).

Many northern plants have *highly dispersed propagules*—mostly wind-transported seeds or spores (Table 23.1). Moss spores are airborne, and these plants are the principal ground cover in many northern forests. *Ceratodon purpurea*, *Funaria hygrometrica*, and *Pholia nutans* invade immediately after fire, especially on moist sites where the organic layer is consumed. *Polytrichum* spp. and the liverwort *Marchantia polymorpha* are also abundant on such sites. *Pleurozium schreberi*, *Hylocomium splendens*, *Hypnum crista-castrensis*, and some species of *Dicranum* do not appear until after a tree canopy is established.

Seeds stored in the humus are important in the prompt revegetation of many burns. Several early postfire species germinate and flower profusely in the first 1–3 years after fire, and then bank their seeds in the organic layer and do not reappear until the next fire. The clearest examples are *Corydalis sempervirens*, *Gernanium bicknellii*, *Aralia hispida*, and *Polygonum cilinode* (Ahlgren 1960; Ohmann and Grigal 1979; Viereck and Foote 1979; Rowe 1981; Viereck 1981). Deep-burning ground fires theoretically might kill all of the seeds of such species, but even on the most severely burned sites some usually appear. There may be a northward decrease in the abundance and viability of seeds in organic layers (Cushwa et al. 1968; Johnson 1975; Moore and Wein 1977; Rowe 1981; Viereck 1981).

Vegetative reproduction may occur from root suckers, root crown and basal stem sprouts, rhizomes, tubers, bulbs, and in lichens from fragments. Suckers are important in the postfire reproduction of quaking aspen (*Populus tremuloides*), bigtooth aspen (*P. grandidentata*) and balsam poplar (*P. balsamifera*). Deep burning of the organic layer may kill much of the *Populus* root network, but even on severely burned sites some suckering is usual. Root crown or stem sprouts are important in the regeneration of *Betula*, *Acer*, *Fraxinus*, *Salix*, *Alnus*, *Corylus*, *Vaccinium*, *Ledum*, *Arctostaphylos*, and other woody genera. Deep burning of the organic layer can nearly eliminate the vegetative reproduction of *Alnus*, *Corylus*, *Salix*, and the Ericaceae (Buckman 1964; Rowe 1981). The regeneration of lichens is not well understood, but vegetative reproduction from fragments occurs. Caribou, moose, and other mammals may be important in dispersing ground lichens (Hale 1961; Viereck 1981).

Most northern trees are not fire resistant. All of the spruces, firs, cedars, hemlocks (*Tsuga*), aspens, birches (*Betula*), maples (*Acer*), and ash (*Fraxinus*) are very sensitive, even as large trees. Even moderate surface fires kill most individuals. The only significant exceptions are red and white pine, and to a lesser degree jack and lodgepole pine. Red and white

Table 23.1. Mode of disseminule-based reproduction of some northern forest species, classified according to Noble and Slatyer's (1977) descriptors.

Mode of reproduction	Communal relationship: I = Intolerant	T = Tolerant	R = Tolerant with Requirement
D = highly dispersed propagules	*Pinus resinosa*	*Abies balsamea*	*Circaea alpina*
	Pinus banksiana (open-cone ecotypes)	*Abies lasiocarpa*	*Corallorhiza* spp.
	Pinus contorta (open-cone ecotypes)	*Picea engelmannii*	*Goodyera repens*
	Populus tremuloides v	*Picea glauca*	*Monotropa uniflora*
	Populus grandidentata v	*Picea rubens*	*Dicranum rugosum*
	Populus balsamifera v	*Pinus strobus*	*Dicranum scoparium*
	Larix laricina	*Thuja occidentalis* v	*Dicranum* spp.
	Betula papyrifera v	*Tsuga canadensis*	*Hylocomium splendens*
	Fraxinus nigra v	*Aster macrophyllus*	*Hypnum crista-castrensis*
	Salix spp. v	*Linnaea borealis*	*Pleurozium schreberi*
	Epilobium angustifolium v	*Mitella nuda*	*Sphagnum* spp.
	Agrostis scabra	*Vaccinium vitis-idaea*	
	Calamagrostis canadensis v	*Lycopodium obscurum* v	
	Carex spp.	*Lycopodium annotinum*	
	Ceratodon purpureus	*Lycopodium clavatum*	
	Polytrichum juniperinum		
	Polytrichum piliferum		
	Pohlia nutans		
	Funaria hygrometrica		
	Marchantia polymorpha		

S = storing propagules in soil	*Prunus pennsylvanica*	*Cornus stolonifera*
	Shepherdia canadensis	*Gaultheria hispidula*
	Symphoricarpos	*Gaultheria procumbens*
	Rubus strigosus	*Ribes triste*
	Aralia hispida	*Ribes glandulosum*
	Corydalis sempervirens	*Rubus chamaemorus*
	Geranium bicknellii	*Rubus pubescens*
	Geocaulon lividum	*Viburnum edule*
	Polygonum cilinode	*Amelanchier* spp.
C = storing propagules in canopy	*Pinus banksiana*	*Picea mariana* v
	Pinus contorta	

[a]v = also capable of vegetative reproduction. Note: *Picea mariana* cones are persistent, but semiclosed. Seed dispersal is slow, but continuous. *Picea mariana* and *Thuja occidentalis* can reproduce vegetatively by layering, but not if burned.

Table 23.2. Mode of vegetative regeneration or survival of some northern forest species, classified according to Noble and Slayter's (1977) descriptors.

Vegetative attribute	Communal relationships		
related to persistence	I = Intolerant	T = Tolerant	R = Tolerant with requirement
V = Able to resprout even if burned in juvenile stage, but above ground stems killed in adult or juvenile stage	*Populus tremuloides*	*Acer rubrum*	*Usnea* spp.
	Populus grandidentata	*Acer spicatum*	*Alectoria* spp.
	Populus balsamifera	*Alnus crispa*	*Cladonia (Cladina) alpestris*[b]
	Betula papyrifera	*Alnus rugosa*	*Cladonia (Cladina) mitis*[b]
	Betula pumila	*Corylus cornuta*	*Cladonia (Cladina) rangiferina*[b]
	Betula glandulosa	*Cornus canadensis*	*Cladonia (Cladina) uncialis*[b]
	Fraxinus nigra	*Rubus chamaemorus*	*Peltigera apthosa*[b]
	Rosa acicularis	*Aralia nudicaulis*	*Stereocaulon paschale*[b]
	Salix glauca	*Clintonia borealis*	
	Salix planifolia	*Maianthenum canadense*	
	Salix bebbiana	*Pyrola secunda*	
	Arctostaphylos uva-ursi[a]	*Pyrola virens*	
	Ledum groenlandicum[a]	*Epilobium angustifolium*	
	Vaccinium angustifolium[a]	*Pteridium aquilinum*	
	Vaccinium myrtilloides[a]	*Lycopodium obscurum*	
	Vaccinium vitis-idaea[a]	*Equisetum sylvaticum*	
	Vaccinium uliginosum[a]		
	Apocynum androsaemifolium		
	Calamogrostis canadensis		
	Polytrichum commune		
	Polytrichum juniperinum		
	Polytrichum piliferum		

W = Adults virtually unaffected, but juveniles killed and cannot resprout	*Pinus resinosa* *Pinus strobus*[a]

[a]Tolerance intermediate: they thrive in open, but can persist under moderate canopy.
[b]These lichens reproduce vegetatively, but do not invade fresh burns. They are intolerant of canopy shade. Reproduction is from fragments.

pine can withstand moderate surface fires because of their thick bark and clear trunks once they attain the small tree stage, but seedlings and saplings are vulnerable.

Growth rates and age of reproductive maturity set the interfire intervals that still permit reproduction from seed after stand-killing fires. Longevity is important for the opposite reason—it sets a maximum interfire interval for reproduction of fire-requiring species. Interestingly, this interval is *much longer* than the probable natural fire cycle over almost the entire range of each species (Table 23.3). Theoretically, the shade-tolerant, K-selected specialists ("climax" species) should have greater longevities, slower growth rates, and later reproductive maturities. There is a trend in this direction, but balsam fir—the principal tolerant conifer east of Alberta—is notoriously short-lived, and invests heavily in early reproduction. However, tolerance is an advantage in local environments where fire cycles are long, and that is where tolerant species usually occur.

Many northern trees have broad niches. For example, the natural geographical ranges of quaking aspen, paper birch (*Betula papyrifera*), jack pine, black spruce, and white spruce are extremely large, they occupy a wide range of soils and land forms; they mingle freely with many different associates in various regional and local contexts. All but white spruce also occur in areas with long fire cycles, short cycles, and in areas with predominantly crown fires as well as some with surface fires. Such species may be less vulnerable because of their broad niches. Aspen and birch can reproduce both vegetatively and from seed after fire. Black spruce reproduces well from seed after fire, but if fire fails to return for long periods, it can reproduce vegetatively by layering. Such multiple strategies allow survival under many circumstances. Many northern shrubs, herbs, mosses, and lichens have circumpolar ranges and exhibit multiple strategies. Conversely, a few boreal species seem to have "small" niches and are relegated to habitats where generalists are less successful. For example, northern white cedar is very shade tolerant and long-lived (Table 23.3), but is easily eliminated by fire and thus occupies only sites where interfire intervals are long. If fire (and deer) were eliminated, it might increase (Grigal and Ohmann 1975).

Vegetation Change Following Fire

In the north most natural forests are either maturing following the last fire or being instantly recycled by the next. Primary successions occur locally on the floodplains of rivers (Van Cleve and Viereck Chapter 13), but the vegetation changes following fire are the major ones that we must understand to interpret northern landscapes. This brief regional review of selected postfire sequences indicates the range of vegetation changes present.

Closed Boreal Forests

Postfire Establishment Period

Stand-killing crown fires or severe surface fires are the most common fire types in boreal forests. The postfire establishment period lasts no more than 10 years on most sites. There are two patterns, depending on depth of burn of the organic layer.

Where the organic layer is not deeply burned, tree seedling establishment is often poor. Adequate rainfall in the germination period and the first few summers is critical. White and black spruce and jack and lodgepole pine are especially vulnerable on dry, blackened organic soil. Suckers of aspen and sprouts of paper birch are prolific in the first year wherever fire-killed parent trees exist. Black spruce may filter in for 5 years, but most pines are established in the first 2 years. Shrubs, herbs, and grasses adapted for vegetative reproduction resprout in the first postfire season. The "fire herbs" originating from seed banks (*Corydalis*, *Geranium*, *Aralia*, *Polygonum*) proliferate rapidly and then disappear after 4–8 years. The fire liverworts and mosses appear the first year and disappear by the fifth or sixth, but are not as abundant as on sites where the organic layer is consumed. The feathermosses, *Dicranum*, and *Sphagnum* are eliminated wherever the moss layer has burned. The principal ground lichens are also eliminated except in unburned open areas. In summary, the following picture emerges. Most of the community is reestablished quickly, with the exception of the feathermosses, *Sphagnum* and the lichens. The fidelity of the tree stratum to the burned stand depends on seedbed conditions and rainfall for pine and black spruce. Aspen and birch are likely to increase at their expense. Decreases or near elimination of white spruce, balsam, subalpine fir, and northern white cedar are frequent. Unless the fire occurred in the fall of a seed year, no seed will be available except from unburned patches or the edge of the burn (Dix and Swan 1971; Viereck and Foote 1979).

Where the organic layer is largely consumed by fire, black spruce, jack pine, and lodgepole pine usually maintain or increase their numbers relative to the preburn stand. Dense monotypes often result. Aspen from sucker origins may be much reduced, but aspen and birch seedlings may be abundant if rainfall in the first season is heavy. White spruce, balsam, or cedar may also seed in. The vegetative reproduction of several herbs and of such shrubs as *Alnus*, the *Vacciniums*, and *Ledum* is greatly reduced. Seedbank species are much reduced, although some reproduction usually occurs. The invaders, *Marchantia*, *Funaria*, *Ceratodon*, and *Epilobium* are abundant on sites with suitable moisture regimes. The feather-

Table 23.3. Life history data for principal northern conifer forest trees.[a]

Species	Juvenile growth rate	Adult growth rate[b]	Age effective seed production starts (yr)	Ordinary life span without fire (yr)[c]	Maximum longevity (yr)[d]	Propagules lost without fire (yr)[e]	Shade tolerance	Vegetative propagation after fire
Pinus banksiana	Fast	Medium	15-20	150-200	250	270	Low	No
Pinus contorta	Fast	Medium	15-20	250-350	500	520	Low	No
Pinus resinosa	Medium	Fast	30-60	300	400	400	Low	No
Pinus strobus	Medium	Fast	30-50	300-350	500	500	Medium	No
Picea mariana	Medium	Slow	15-30	180-250	300	na	Medium	No
Picea glauca	Slow	Fast	30-50	200-250	300	na	High	No
Picea rubens	Slow	Medium	30-50	300	400	na	High	No
Picea engelmannii	Slow	Medium	30-50	400	500+	na	High	No
Abies balsamea	Slow	Fast	25-35	150	200	na	High	No
Abies lasiocarpa	Slow	Medium	30-50	230	300?	na	High	No
Tsuga canadensis	Slow	Medium	30-50	400-500	500+	na	High	No
Thuja occidentalis	Slow	Slow	30-40	300	500+	na	High	No
Larix laricina	Fast	Fast	30-40	150-200	250-330	300	Low	No
Populus tremuloides	Fast	Fast	25	120-160	200	na	Low	Yes
Populus grandidentata	Fast	Fast	25	100-140	150?	na	Low	Yes
Populus balsamifera	Fast	Fast	25	130-150	200	na	Low	Yes
Betula papyrifera	Fast	Fast	25	130-200	230	na	Low	Yes
Fraxinus nigra	Medium	Medium	30-40	150	250?	na	Low	Yes
Acer rubrum	Medium	Slow	30-40	100-150	180?	na	Medium	Yes

[a]Data from Fowells (1965), Viereck and Little (1972), and present author's files.

[b]All species grow slowly in the northern fringe of the boreal forest or near timberline if present in such environments. Other age-related attributes may also change with latitude and elevation.

[c]Life spans and maximum longevities given are for the principal range of the species in the boreal, northern Great Lakes-St.Lawrence and subalpine forest regions. Life spans are variable, and much shorter for some species nearer the limits of their ranges.

[d]Age at which all individuals originating after a given fire would be dead.

[e]na, not applicable. These species are either tolerant enough to reproduce from seed without fire, capable of vegetative reproduction by layering, or have such widely dispersed seeds that they could r ecolonize areas temporarily lost.

mosses and lichens do not reappear in the first decade. If only white spruce was present, and seed trees are eliminated over large areas, an initial aspen establishment of seedlings may occur, and the reestablishment of white spruce will be slow (Van Cleve and Viereck Chapter 13). The fidelity of stand composition to that of the burned stand will be high if the canopy-storage pines or black spruce were prominent, but may be low if white spruce, balsam, cedar, or aspen were dominant (Dix and Swan 1971).

Canopy Development Period

The canopy development period lasts from about year 11 to year 50. Where stocking of spruce, fir, jack pine, or lodgepole pine is heavy, crown closure and increasing shade permit expansion of the feathermosses and *Dicranum* between years 20 and 40. This often marks the principal change in composition that might be seen as a successional stage. Not all stands develop a feathermoss carpet, but most stands of black spruce do, as do most well-stocked jack pine-black spruce or lodgepole pine-black spruce mixtures—especially on cool, moist sites. So also do many white spruce forests, and many balsam-spruce mixtures (Damman 1964; Dix and Swan 1971; Viereck 1981). These moss layers consist primarily of *Pleurozium schreberi*, *Hylocomium splendens*, *Hypnum crista-castrensis*, and two or three *Dicranums*. Species coverages vary with moisture regimes, nutrient status, overstory density, and region. In general, *Hylocomium* and *Hypnum* are more abundant on moister, more fertile sites, and *Pleurozium* is more abundant on drier, less fertile sites. Nevertheless, they are not very exacting in site requirements, except for the necessity of a coniferous overstory. These mosses are the great humus builders of the north, covering hundreds of millions of hectares as the principal ground layer of closed coniferous stands from Newfoundland to Alaska, and southward into northern Minnesota, Michigan, and Maine. They first become noticeable about 10 or 15 years after fire, and high coverages often develop within 50 years. The ground lichens, *Cladonia*, *Stereocaulon*, and *Peltigera* may appear in open pine stands and in open areas within spruce forests during this period.

Where the initial stand of trees includes more than one species, sorting into dominance classes occurs. If black spruce is mixed with jack pine, lodgepole pine, aspen, or birch, most spruce are soon overtopped. Spruce persists, however, and many stands develop an overstory-understory structure suggesting succession. This situation may also develop with white spruce, balsam, or northern white cedar if these species seed in with the pines or aspen and birch. They are less well equipped to reseed after fire, however, and frequently do arrive later. Herbivores also influence stand composition. Moose selectively eliminate aspen, birch, and fir, snowshoe hares may thin out almost any of the common trees, and beaver frequently eliminate all aspen and birch within 100–200 m of waterways (Bendell 1974; Kelsall et al. 1977).

Mature Stand Period

Fire usually terminates stand life during the mature stand period, if not earlier. I set this period at 51–120 years, but maturity varies with species and locale. As the stand matures, moss carpets thicken and spread and arboreal lichens develop on conifer branches. Toward the end of the period small stand openings narrow, and ground lichens are replaced by feathermosses. Where monotypes of jack pine, lodgepole pine, black spruce, white spruce, or aspen exist, they continue into late maturity, but where balsam is available, seedlings and saplings may begin to appear in these monotypes. Spruce budworm attacks often develop in extensive stands with a high initial balsam component, and heavy mortality usually results. White spruce trees are injured, but many survive. Black spruce is unaffected (Blais 1954, 1958, 1965; Batzer 1969).

Senescence Period

A final senescence period, marked by death of the first-generation postfire trees may begin between years 120 and 200 in those rare situations where stands are not burned earlier. In the east where fire cycles are longer, more stands occur in the senescence stage. In the west they occur chiefly on floodplains, in barren rocky areas, along lakeshores, on islands, and in other habitats that offer protection from fire. Some old stands also escape as long, narrow "stringers" that are missed through the vagaries of fire (Rowe and Scotter 1973; Viereck 1973; 1981; Rowe et al. 1975). If balsam or white spruce are available as replacement species, they may enter the main canopy during this stage. In the absence of fire, the demise of old stands may come quickly in a windstorm or budworm outbreak, or it may occur gradually through attrition by mistletoes, insects, and fungi. Most trees have a potential longevity that greatly exceeds the natural fire cycle, however, and at least a few individuals usually persist until the next fire. The increase of arboreal lichens, thickening moss carpets, ladder fuels, old loose-barked birch trees, and down timber make such stands very susceptible to fire. Where stand breakdown occurs without fire, aspen suckers, birch sprouts, and balsam or spruce seedlings restock the area.

In much of the northwestern boreal forest, white or black spruce are the only trees on many sites, and they often occur in pure stands through site adaptations or the vagaries of past fires. Lutz (1956) thought that white spruce was the climax in interior Alaska, and that it could replace black spruce on sites that escaped fire for long periods. However, Viereck (1981) has seen no evidence of such replacement. Thus, in vast regions of the northern boreal forest there simply are no replacement species.

Subarctic Spruce-Lichen Woodlands

This vegetation consists of stunted spruce interspersed with treeless areas covered by lichens, dwarf shrubs, and herbs. Black spruce, white spruce, and tamarack (*Larix laricina*), and rarely paper birch, jack pine, or balsam, are the only trees. Variants occur from Labrador to Alaska. The role of fire has been much studied because of controversy over the effects of fire on the winter range of the barren ground caribou and over new pipelines and other developments (Ritchie 1962; Scotter 1964; Rowe 1970, 1979; Rowe and Scotter 1973; Bergerud 1974; Rowe et al. 1974; Johnson and Rowe 1975, 1977; Kelsall et al. 1977).

A postfire sequence in black spruce in the Mackenzie Valley near Inuvik, North West Territories is described by Black and Bliss (1978). They found no replacement of shrubs and herbs because most plants simply resprout and soon regain prefire prominence. Stage 1, lasting from 1 to 20 years, is dominated by the early mosses and liverworts, and by herb-shrub-grass cover of *Epilobium*, *Salix*, *Betula glandulosa*, *Arctogrostis*, and *Calamagrostis canadensis*. Black spruce seedlings are obscured by the vegetation. Stage 2 lasts from 20 to 120 years. It is dominated by early *Cladonia* and *Peltigera* lichens, with increasing moss cover of *Aulocomnium* and *Hylocomium* with age. *Vaccinium uliginosum*, *Petasites frigidus*, and *Empetrum nigrum* increase later. Black spruce becomes a short tree by 120 years. Stage 3 is characteristic of rare forest stringers 120–200 years from fire occasionally found along lakes and rivers. Black spruce may reach heights of 5–7 m, and as a canopy develops, the *Cladonia* lichens and *Vaccinium vitis-idaea* increase. A very rare fourth stage, 200–300 years from fire, consists of old, dying black spruce undergoing self-replacement through layering, with dominance of the lichens *Cladonia alpestris*, *C. mitis*, and *C. rangiferia*, and a low cover of *Cetraria*.

Maikawa and Kershaw (1976) described another sequence on drumlins southeast of Great Slave Lake. Here the woodland is dominated by spruce and *Stereocaulon paschale*. Phase 2, the *Cladonia* phase, from years 20 to 60, sees a *Polytrichum* mat invaded by *Cladonia alpestris*, *C. unciallis*, and other lichens. Phase 3, the spruce-*Stereocaulon* phase, from years 60 to 130, is the mature lichen woodland. Fire normally recycles the system in this phase, or earlier according to Johnson and Rowe (1975), who studied the same region—the wintering ground of the Beverley caribou herd. Some areas, however, are skipped by fire long enough to enter Phase 4, the spruce-moss woodland, in which canopy closure allows the feathermosses to dominate. Thus, maintenance of lichen woodlands is dependent on periodic large fires at a frequency of about 100 years, and the caribou range is not threatened by fire at its natural cycle. More than 99% of the annual area burned in this region is due to lightning ignitions.

Great Lakes-St.Lawrence-Acadian Conifer Forests

I shall discuss just two kinds of forests in this complex region: (1) "near-boreal" conifer forests, and (2) red pine-white pine forests.

"Near-Boreal" Conifer Forests

From Minnesota and Ontario east to New Brunswick there are large areas of jack pine, black spruce, fir-spruce, and aspen-birch-fir forests not unlike those of the eastern boreal forests. A 3- or 4-year postfire establishment period sets the stage for most subsequent vegetation changes. Most fires are crown fires or high-intensity surface fires that kill the aboveground vegetation.

Postfire Establishment Period. Two recent fires in the Boundary Waters Canoe Area Wilderness of Minnesota (BWCAW) demonstrate the importance of initial floristics, season of burn, and organic matter consumption in determining stand composition. The Little Sioux Fire of May, 1971, was a human-caused spring fire after a short drought. The Roy Lake Fire of August, 1976, was a lightning-caused summer fire after prolonged drought. Both study areas were on ridges with thin, bouldery soils over granitic bedrock, both had organic layers varying in thickness from 3 to 20 cm, and both supported stands dominated by jack pine and black spruce, with some aspen and paper birch, and variable understories of balsam. I shall compare two jack pine stands from the Little Sioux with three conifer stands from Roy Lake (Table 23.4; Ohmann and Grigal 1979).

Both Little Sioux stands occurred in an area previously burned in 1864, supporting jack pine communities 107 years old. Both burned in an intense crown fire, but because the organic layer was still wet below the surface, several centimeters of unburned organic matter remained (Ohmann and Grigal 1979). On the Roy Lake Fire, three burned stands, each with different prefire composition and stand age, were sampled in July, 1979 (Table 23.4). Stand 1, on Sea Gull Lake, was a 175-year-old mixed conifer stand dating from an 1801 burn; Stand 2, on Grandpa Lake, was a 101-year-old upland black spruce area dating from an 1875 fire; and Stand 3, on Saganaga Lake, was a 73-year-old jack pine stand dating from a 1903 burn. All burned in high-intensity crown fires. The organic layer was almost entirely consumed.

Comparison of prefire data with third-year regeneration data (Table 23.4; Ohmann and Grigal 1979) leads to the following conclusions: Jack pine and black spruce reproduced abundantly on mineral soil on all sites at Roy Lake. Jack pine also reproduced well on the Little Sioux, although not as abundantly as at Roy Lake, but black spruce failed almost com-

Table 23.4. Comparison of prefire composition of three conifer stands burned in crown fires, August, 1976, with regeneration three seasons later, Roy Lake Fire, Boundary Waters Canoe Area Wilderness, Minnesota.[a,b]

Species	Stand 1: 175-year-old mixed conifers			Stand 2: 101-year-old black spruce			Stand 3: 73-year-old jack pine		
	Prefire stand		Regeneration 1979:	Prefire stand		Regeneration 1979:	Prefire stand		Regeneration 1979:
	Stocking (trees/ha)	Basal area (m^2/ha)	Stocking (trees/ha)	Stocking (trees/ha)	Basal area (m^2/ha)	Stocking (trees/ha)	Stocking (trees/ha)	Basal area (m^2/ha)	Stocking (trees/ha)
Pinus banksiana	99	5.2	27,227	13	0.5	4010	3100	32.5	241,250
Picea mariana	812	8.1	3636	1388	21.0	45,208	516	3.2	6650
Abies balsamea	58	1.5	—	402	0.5	—	86	0.3	—
Pinus resinosa	—	—	—	13	4.6	—	—	—	—
Populus tremuloides	49	0.6	79,136	—	—	26,510	—	—	71,600
Populus grandidentata	—	—	90	—	—	—	—	—	—
Betula papyrifera	141	1.2	—	52	0.6	208	—	—	—
Total: All Species	1159	16.6	110,089	1868	27.2	75,936	3702	36.0	319,500

[a]This late summer fire burned-off all of the organic layer.
[b]See text of this chapter for descriptions of fire and burned sites. Data taken by graduate class under supervision of M. L. Heinselman, July, 1979. J. C. Almendinger compiled these data.

pletely. Close inspection shows that most jack pine seedlings became established locally where the organic layer was thin or absent. Some of the failure of black spruce may have been due to seed kill by the intense crown fire, but similar intensities at Roy Lake resulted in heavy spruce reproduction. There was major seeding in of aspen on mineral soils at Roy Lake, although some of these high aspen counts are due to clumps of suckers near fire-killed trees. No comparable seeding of aspen occurred in the Little Sioux jack pine stands. The future of aspen seedlings at Roy Lake is unclear. Many may eventually be eliminated by beaver, hares, and moose, or suppressed by competing jack pine on the drier sites. Moose increased rapidly on the Little Sioux burn, and are now heavily browsing aspen and shrubs (Peek 1974). Balsam, the major potential replacement species on both fires, was eliminated in all stands. However, field observations show that seed trees exist nearby in unburned enclaves, along lakeshores, streams, bog margins, and at the perimeter of the burn. Given a century for reseeding, balsam too will probably regain its former status. Thus, fire will probably have changed the tree stratum little in the end, except for decreases of black spruce on the Little Sioux.

The shrub and herb strata in the Little Sioux stands virtually recovered to prefire conditions within 5 years (Ohmann and Grigal 1979). Vegetative reproduction of many shrubs and herbs and the germination of seeds stored in the organic layer occurred because the organic layer was so lightly burned. The fire ephemerals, *Corydalis*, *Geranium bicknellii*, *Polygonum cilinode*, and *Aralia hispida*, flowered profusely, but by 1975 were much reduced. *Prunus pennsylvanica* became a major tall shrub. Even such plants as *Clintonia borealis* and *Lycopodium obscurum*, not usually considered "pioneers," reproduced vegetatively in abundance. Only the moss and lichen ground layers were temporarily eliminated. As Ohmann and Grigal (1979) summarized it, "The major change in stand structure was lowering of the tree canopy from a former height of 10 to 20 m to 0.1 to 0.2 m after the first growing season following the fire."

At Roy Lake, the shrub, herb, and ground layer revegetation was substantially different. The much greater exposure of mineral soil led to abundance of the fire invaders, *Marchantia*, *Funaria*, *Polytrichum*, *Polygonum*, and *Epilobium*, but the fire ephemerals that depend on stored seeds were much less common. Most of the herbs and shrubs that reproduce vegetatively were also much reduced. Thus, by removing the organic layer at Roy Lake, fire greatly increased the stocking of black spruce and jack pine, allowed more seeding of aspen, and greatly reduced the competition from the shrubs and herbs that reproduce vegetatively or from seed stored in the organic layer. Ahlgren (1959, 1960) reported similar early revegetation and tree seedling establishment.

Canopy Development Period. The canopy development period lasts from about year 11 to year 40. By the 40th year crown closure has occurred in all but the most open stands. If present, jack pine, aspen, and birch usually assume dominance, and black spruce, balsam fir, white spruce, and northern white cedar fall into intermediate or subcanopy positions. Monotypes are common, especially of jack pine, aspen, or black spruce. If a feathermoss groundlayer is to appear, it usually is well developed in 40–50 years. The *Cladonia* lichens appear in rocky openings about 30 years after fire, especially in spruce or pine stands. Shrubs such as *Alnus*, *Corylus*, *Salix*, and *Prunus pennsylvanica* decrease in vigor, and some disappear.

Mature Stand Period. The mature stand period lasts from year 40 to year 100 or even 150, depending on species and sites. In presettlement times fire usually recycled the sequence in this period, if not before. Aspen and jack pine often persist as intact mature stands for 90–130 years in the BWCAW, but if fire fails to return, the senescence of dominant individuals begins on a significant scale at about 100 years. Balsam fir and occasionally cedar seedlings and saplings may become abundant in mature stands of other species. When the early postfire stand contained a high percentage of fir, it reaches maturity in 60–80 years, and by 100 years a spruce budworm outbreak has often killed most of the fir. In black spruce and some jack pine and fir stands, feathermoss carpets build thick mats. As stand openings close, lichens are gradually replaced by mosses and herbs.

Senescence Period. The senescence period begins at 100–150 years and lasts until fire returns. In the BWCAW, a forest dating from a 1727 fire still contains living jack pines—now more than 250 years old. I have been unable to find any stands lacking at least a few remnants of a postfire overstory. Little by little, however, the original overstory dies, and after 200 years only a few old jack pine, aspen, or spruce tower above the balsam saplings, shrubs, and jumbled fallen logs that characterize such old stands. Before fire exclusion was attempted, such stands undoubtedly were rare. Now several are present in the BWCAW on 1727, 1755, 1759, and 1801 burns. Succession to balsam might be "forced" by fire exclusion, but a spruce budworm outbreak has already killed the balsam in most such areas.

Red and White Pine Forests

The red and white pine forests of the Lake States and eastward were clearly of fire origin (Maissurow 1935, 1941; Spurr 1954; Horton and Bedell 1960; Frissell 1973; Heinselman 1973; Swain 1973, 1978; Cwynar 1977, 1978). At least from Michigan westward, pine stands on fresh to dry sites were subject to repeated light-to-moderate surface fires, probably at intervals of 20–40 years. Such fires consumed the litter and killed back shrubs and invading tolerant conifers, but only scarred or left uninjured pines more than 30 or 40 years old. Perhaps once every 100–200 years a higher-intensity fire killed out patches of mature pines. The resulting openings then were recolonized by a new age class. This hypothetical sequence has been verified by age and fire scar analyses in virgin pine in Itasca Park, Minnesota (Spurr 1954; Frissell 1973), the BWCAW (Heinselman 1973), and Algonquin Park, Ontario (Cwynar 1977). Stands with such fire regimes exhibited little vegetation change. Both white and red pine were involved, although this kind of history was more prevalent in red pine. Since neither of these pines has persistent or closed cones, and only occasional good seed years, a regime that does not eliminate all seed trees is essential. Succession was impossible because tolerant invaders were periodically elminiated.

On mesic sites and eastward, white pine had a different history. Here, higher-intensity fires occurred, but at very long intervals—perhaps 200–300 years or more. The physiograhic situation where such stands occurred evidently favored incomplete stand kills, and thus seed sources remained. Seeding in may have required 10–30 years, but such stands were essentially even aged. During the long interfire intervals significant invasion by balsam, white spruce, northern white cedar, and in the east also hemlock (*Tsuga canadensis*) often occurred. White pine is so long lived, however, that these tolerant species rarely replaced it before fire returned (Heinselman 1973). Where fire delayed too long, such stands may have been destroyed by blowdowns that initiated a whole new stand, perhaps of different composition (Henry and Swan 1974).

Rocky Mountain Subalpine Forests

The lodgepole pine, aspen, and Engelmann spruce-subalpine fir forests of the northern Rockies have histories similar to those of the boreal forests. Stand-killing crown fires or high-intensity surface fires at cycles of 50–200 years were the most common regimes at the higher elevations (Clements 1910; Wellner 1970; Day 1972; Gabriel 1976; Habeck 1976; Tande 1979). In a study of three large, intense burns in Idaho and Montana, Lyon and Stickney (1976) found that most shrubs and herbs reproduced from on-site sources, including vegetative reproduction from rhizomes, root crowns, and underground stems, and germination from seeds or fruits stored on site. Their studies indicate that, "Practically all plants that survived the fire reestablished within the first year. Data also suggest that virtually all species that contributed significantly to early vegetal cover were established in the first year." Lyon and Stickney (1976) conclude: "Forest succession in the Northern Rocky Mountains is not an autogenic process in which initial seral plants modify the site to their own exclusion and permit the establishment of interseral and eventually climax species."

Vast areas in the Rockies still support nearly pure stands of lodgepole pine. Careful stand origin and fire scar studies, such as those of Tande (1979), demonstrate that most of these stands at the higher elevations are nearly even aged and can be dated to specific fire years. Fire simply recycles the whole community. On mesic sites and at upper elevations, Engelmann spruce and subalpine fir are often available as invaders (Habeck and Mutch 1973). However, the work of Day (1972) and Tande (1979) indicates that even most spruce-fir stands are of first generation, postfire origin. Day notes that many such stands still contain 220–240-year-old lodgepole pine that date the last fire. Tande found similar evidence. Apparently Engelmann spruce often seeds into burns along with lodgepole pine on suitable sites, but the slower-growing spruce is overtopped by lodgepole for many years.

Day proposed the following hypothetical four-phase history for mixed stands. In Phase 1 (55 years after fire), pine is still dense and even aged, but beginning to decline, and spruce exists as an understory about one-quarter the height of the lodgepole pine. Subalpine fir is sparse, but seeding in slowly. In Phase 2 (155 years after fire), spruce is beginning to dominate the declining pine canopy, and fir is developing an aggressive all-aged understory. In Phase 3 (255 years after fire), pine is dying, and spruce dominates. By now fir is about four times as numerous as spruce, but subordinate. As the pines die, spruce fills the gaps. Phase 4 (355 years after fire) is conjectural because until fire control began, fire almost always recycled the system before this stage. In this phase, however, virtually all pines would be eliminated, and the stand would gradually develop into a ragged, all-aged spruce-fir stand. The work of Gabriel (1976), Arno (1976), and Tande (1979) confirms that fire was so prevalent that few stands ever persisted long enough to eliminate all first generation pine and/or spruce.

At lower elevations many lodgepole stands are multiaged and parklike. In Jasper, Tande (1979) found that such stands had been subject to regimes of light surface fires at intervals of 10–40 years. Houston (1973) found similar evidence in Yellowstone, as did Arno (1976) and Gabriel (1976) in Montana. Some of these pine stands have open cones (Lotan 1976). Succession could not proceed far in stands subject to such short-interval fires.

Is Lightning-Caused Fire Random or Time Dependent?

It is not clear how the probability of a significant lightning-caused fire increases with time since fire for a given area. The questions are: (1) Is fire probability closely linked to fuel accumulation, vegetation changes since fire, and increasing tree mortality with stand age, or (2) Is fire largely a random process, dependent on random lightning ignition patterns, and the vagaries of winds and weather during the burn? Two related questions are involved: (1) Does the probability of a successful ignition increase with time? (2) How do probabilities vary with time for fire intensity and size? Nobody doubts that there is at least a short period after a major fire when the probability of a reburn is reduced. But is that period only a few years, or is there a continuing increase in probability for ignition, intensity, and size?

The arguments for rising probability are:

1. Total fuel increases with vegetation regrowth after fire, and biomass does not peak for at least 100 years.
2. The change from bare mineral soil or charred organic matter on the forest floor to continuous carpets of flammable feathermosses or lichens is time dependent.
3. The fuel content of the organic layer increases for at least 50–100 years, and on some sites indefinitely.
4. Arboreal lichens are time-related fuels.
5. "Ladder fuels" increase with time.
6. In aspen-birch forests flammability increases with stand age where there are parallel increases in the conifer component.
7. Mortality increases with stand age due to insects, diseases, and wind. Dead timber may be more prone to ignition.
8. Changes in fuels chemistry over time may increase flammability (Mutch 1970).

The arguments for randomness are:

1. Lightning strike patterns may be random, or at least unrelated to stand age.
2. Fire paths and spread rates may be largely controlled by wind direction and speed, and by changing temperature and humidity. In major droughts most vegetation burns.
3. "Spotting" due to flying brands makes it possible for fires to leap barriers and nonflammable vegetation.
4. Fires create fuels (by killing timber), and such fuels peak early in stand life.
5. Fire cycle models that assume randomness fit actual stand origin data fairly well (Van Wagner 1978). Johnson (1979), however, has developed a similar model that includes terrain and vegetation factors.
6. Even a cursory examination of aerial photographs or actual stands in the boreal forest shows that many stands are reburned at short intervals.
7. Fuel studies indicate that many dense young conifer stands burn well in crown fires—sometimes better than old stands.

8. The early reproductive maturity of many northern trees suggests environments with either very short fire cycles or much randomness.

The best estimate at this time is that the probability of ignition is higher in old stands, and that "blowups" into major fires are more likely in old stands. However, once a major fire is under way it clearly may burn many age classes. The facts are not all in, but answers are needed to understand vegetation changes.

Fire Patterning of Vegetation Mosaics

The vegetation of northern regions is a patchwork of forest-age classes and plant communities resulting from a long history of fires of various sizes and shapes. These patches are often visible from aircraft or on aerial photographs. What one sees are the outlines of new burns, and of large areas of trees of specific age classes, each area dating from a different fire. It is a record of the cycling of ecosystem units by fire over the centuries.

Maps of such age-class mosaics have been made in two large forest reserves: the BWCAW (Heinselman 1973), and Jasper Park (Tande 1979). The map units are *stand origin fire years*, i.e., patches on the landscape containing forest stands dating from specific fire years. Fire years are determined either from historical records or by tree ring dating of fire scars and dominant trees. Stand origin maps help one understand factors that influence fire movements and patch sizes and shapes. The outlines of large former fires become evident, as do the skipped areas and short-cycle reburnings. Such maps are invaluable in studies of natural fire rotations and the factors that control time-related vegetation changes. For example, topography and natural fire barriers are often seen to influence fire frequency. Certain areas are swept repeatedly by large fires, while others with better protection are more likely to support old stands and species not well adapted to short cycles.

Knowledge of the scale of the vegetation units generated by fire is fundamental to resolving our views of ecosystem dynamics. If the patches in a mosaic are large, then studies designed to look at system processes must encompass enough patches to sort out the time-related factors that explain differences between patches. This is even important in evaluating the nutrient dynamics of watersheds (Vitousek and Reiners 1975). Patch sizes do vary in major ways, depending on topography, regional climates, lightning fire frequencies, and other factors. Single fire patches may cover as much as 500,000 ha in western Canada. There are also many far smaller patches, however. Small patches are created by two factors—skipped areas and small fires. Rowe (1979) studied the relation of fire size to total area burned in the Mackenzie Valley. He found that for any one year, only 1% of the fires exceeded 40,470 ha, but those few fires burned

50% of the total area. Five percent of the fires exceeded 4047 ha and burned 95% of the total. Maximum patch sizes in the BWCAW are smaller, but some exceed 60,000 ha. Patch size distributions by fire years over large regions tell us much about the periodicity of major droughts. Distributions can be studied by random point sampling, using helicopters for ground truth checks, instead of by stand-origin mapping if the interest is in determining the spatial distribution of year classes, and in fire cycles (Rowe et al. 1974, 1975; Johnson and Rowe 1975, 1977; Van Wagner 1978; Johnson 1979). The problem of randomness as opposed to system-regulated fire occurrence is related to factors that control maximum patch size, variability in sizes, and geographic distribution over time. We need to understand the pattern diversity generated by natural fires on a continental basis.

The vegetation over large regions may be nearly in dynamic balance, even though on a smaller scale the individual patches are all undergoing change. As White (1979, p. 283) puts it,

> There may be a characteristic and relatively stable distribution of land into various disturbance, regrowth and old age classes. On this larger scale the landscape may be in a steady-state, as long as (the) disturbance regime remains constant.

Loucks (1970) and Wright (1974) have made similar arguments. The ongoing paludification of certain northern landscapes raises questions about long-term stability, however.

Our view of diversity in northern ecosystems also depends on scale and pattern renewal effects. Many northern forests are monotypes in the tree stratum. In many communities the ground layer at given phases of the postfire vegetation sequence may also be low in diversity. Thus, within-site diversity is low. Also, because of the often large size of fires, the patches showing low diversity may be large. However, if we stand back and look at vast regions, a different picture emerges. As Rowe (1979) puts it, "Fire continually creates between-site diversity, renewing the spatial mosaic of community types and age classes." It is this between-site diversity that accommodates most of the diversity that exists in northern ecosystems.

Implications for Succession Theory

The traditional meaning of succession is a consistent, unidirectional sequence of changes over time in the species composition of a community, accompanied by parallel structural and functional changes (Drury and Nisbet 1973a). McIntosh (1980) makes the additional point that such a sequence must be demonstrated on the same site. Many writers also link succession to progression toward steady states (climax), and to community-driven changes.

Most of the vegetation changes in northern fire-dependent forests do not fit these concepts. Most species, and even most individuals become established in the first few years after fire. Apparently many individual are not eliminated from the site. Their above-ground parts are killed, but they reappear at once through vegetative reproduction. Many arise from seed stored in the canopy or in the organic layer. Many stands regenerate to a composition almost identical to that of the burned stand, but sometimes there are changes because propagules are eliminated by fire, or through an influx of new propagules. Egler (1954) drew attention to the importance of initial floristics in old-field successions, and the situation after fire is similar. Furthermore, many of the visually impressive changes in stands as they mature simply reflect different growth rates of species. As Viereck (1981) notes,

> The use of the term succession is questioned because following fire in the black spruce type most of the species that were on the site reinvade directly and there is no replacement of large numbers of species. Differential growth rates only make it appear that there are several different stages before a mature stand is reached.

Or as Methven et al. (1975) put it,

> Fires in the boreal forest always result in the reestablishment of forest . . . . Seeding-in is completed quickly, and all individual young trees capable of taking part in stand development are present from the start. There is no succession in the normal sense of the term, only cycling of the forest by fire.

In the north, white spruce and balsam have long been considered the key climax species, an idea traceable to Cooper's (1913) classic studies on Isle Royale. However, in many forests balsam does not occur, and in northwestern Canada and Alaska, where white and black spruce are the principal conifers, white spruce does not replace black spruce. Thus in many far northern forests there are no replacement species. In eastern Canada and the Great Lakes region, balsam is short-lived and subject to spruce budworm outbreaks that decimate its populations before it can replace the intolerant postfire dominants. Also, with normal fire cycles balsam is periodically reduced before replacement of pine, aspen, or birch can occur. Furthermore, in much of the north many forests regenerate as virtual monotypes, with no possibility of species replacement before fire occurs.

There is, however, a clear temporal sequence in the moss and lichen ground layers in many northern forests. The feathermosses, *Dicranum*, *Sphagnum*, and the lichens are temporarily eliminated by fire. *Marchantia* and several ephemeral mosses reinvade quickly, but the feathermosses, *Dicranum*, and *Sphagnum* appear when a canopy is reestablished. The feathermosses have a definite shade requirement (Tamm 1964), and their expansion clearly requires a conifer overstory. Mosses are significant because they accumulate thick layers of raw humus, and in some forests much of the carbon fixation occurs in the moss stratum. Many lichens may have requirements that must be met before they reinvade new burns, but they also give way to the feathermosses when a tree canopy is established. Viereck (1981) agrees that, "The cryptogams follow a more typical succession pattern than do the vascular plants . . . . They seem to have an orderly sequence of species invading and being eliminated over time following fire."

McIntosh (1980b) points out that a mosaic of successional habitats generated by random or periodic disturbance should not be confused with a sequence on a single site. The problem, though, has always been to separate spatial heterogeneity from real changes on a specific site, because investigators cannot follow single sites long enough to answer these questions. In northern forests there are two different kinds of changes under way: (1) vegetation changes on the same site over time—these are the changes within patches that occur after initial establishment following fire; (2) different outcomes during the establishment period, on different sites. It is in the first category that we can see any succession. The sequences in moss and lichen layers are the most noteworthy. Often there is no succession in the tree stratum, either because the first generation trees start out simultaneously, there are no replacement species, or fire returns too soon.

The second category is the source of confusion. Site variations that selectively favor different species and variations in fire factors cause different species mixes on different land forms. The variability of the fire process also causes different outcomes on similar sites. When we look at a large region and see the mix of age classes and species generated by site effects, fire history, and actual time-related factors, it is hard to sort out any real on-site successions. Checking stand ages and fire scar dates would answer many questions, but most earlier studies did not use these techniques.

Recent northern studies have mostly concentrated on fire regimes, fire history, species regeneration patterns, and fire effects on ecosystem processes. Most workers have been unable to identify steady-state communities. As Dix and Swan (1971) noted,

> The nature of the fire-free forest is a matter of conjecture about which no evidence is available . . . . Succession does not seem to be important. It seems more important that the area has probably undergone an infinite number of fire disturbances through time . . . . Any attempt to fit the vegetation into the mold of a climax concept would be unreal . . . and unjustified.

Is the explanation for vegetation changes then to be found just by adding up the autecologies of species? The situation does not seem that simple. There are changes following fire that are tied to species interactions, and to changes in the vegetation and system function over time. Many of these changes, however, do not fit the concept of succession. It may be better to simply study them for what they are, and not try to fit them into an ill-suited theoretical framework.

For example, the phenomenon of moss growth, organic matter accumulation, and nutrient tieup during long interfire periods needs to be understood. Viereck (1981) notes that,

> Forest floor biomass increases greatly once the tree canopy is established and the feathermosses are abundant . . . . On most of the cooler, more moist sites the forest floor continues to thicken and accumulate material. The criteria for a climax ecosystem that the net increment of biomass should equal zero is not attained.

In the far north, permafrost encroachment, site degradation, and depauperization of vegetation often result. On some land forms, even without permafrost, bog formation occurs—hardly "succession." Fire exclusion could remove what balance exists, causing increased biomass accumulations and more paludification.

Any general model of vegetation change must accommodate the system interactions and patch dynamics of fire-dependent northern forests. We need models that will simulate the vegetation, fuel, hydrologic, edaphic, nutrient cycle, and biotic changes that occur under alternative fire regimes and cycles. We also need models to handle the spatial questions of patch size variation, reburn patterns, and other area-dependent variables that should be looked at with random and time-dependent ignition and intensity assumptions. The collaboration of fire ecologists, fuels and fire behavior specialists, and model builders is needed. A start has been made by Horn (1976), Van Wagner (1978), Cattelino et al. (1979), and Johnson (1979), but comprehensive theories that link vegetation changes with fire regimes and ecosystem processes are still needed. We tend to look at large forest fires as "natural disasters." However, large fires may play a major role in maintaining the diversity and productivity of northern ecosystems. We need to understand the consequences of modifying fire regimes or of fire exclusion for the large ecosystems of the north.

The studies reviewed in this chapter show the following:

1. Most significant fires are large-scale crown fires or high-intensity surface fires, causing total stand replacement at cycles of 50–200 years.
2. Fire characteristics and drought strongly influence ecological effects.
3. Many postfire vegetation changes can be accounted for by prompt species establishment from canopy-stored seeds, organic layer seedbanks, vegetative regeneration, or wind-blown seeds from nearby, and by different rates of growth and senescence of the initial plants.

4. Establishment of feathermoss carpets depends on the development of a conifer canopy.

5. Moss cover contributes to organic layer accumulations, incomplete nutrient cycling, permafrost encroachment, depauperization of communities, and bog formation.

6. Fires consume organic layers and recycle nutrients, retarding such trends.

7. There is no unidirectional development of vegetation, and there are no steady-state communities.

8. The vegetation is a mosaic of patches of even-aged forest, each patch dating from the last fire but varying in species composition with site factors, propagule sources, and fire variables.

9. It is not clear to what extent patch sizes, turnover rates and locations are random as opposed to time dependent.

10. Models are needed to simulate patch dynamics and system interactions under varying fire regimes.

Chapter 24

Vegetation Change in Chaparral and Desert Communities in San Diego County, California

Paul H. Zedler

Introduction

Arid region vegetation types exhibit patterns of change that do not always conform to existing theories of succession. This is certainly true for the chaparral and desert shrub communities of the southwestern United States in part because of the important role of extreme environmental events and natural disturbances. In the chaparral the problem stems from the overriding importance of fire in determining readily observable vegetation change. The tendency has been either to view fire as preventing chaparral from reaching true climax condition (Horton 1950, cited from Hanes 1971) or to hold that the vegetation is climax because fire is a part of the prevailing climate (Clements 1928). The latter view has been given formal status by Hanes (1971), who proposed the term "autosuccession" to describe the special case of succession in the chaparral in which fire neither causes extinction nor initiates invasion, but rather reestablishes species that are already present.

Like the chaparral, the desert has been said to absorb disturbance without compositional change. Muller (1940) suggested deserts were a kind of "superclimax" because of this extreme resilience. Shreve (1951) did not believe that succession was an important process in the Sonoran desert. He felt that species sorted out by climatic zone, substrate, and slope position and that successional stages were not identifiable. It seems that later investigators have given tacit agreement to this view, since the literature of secondary succession in the desert is scanty (Richardson 1977). More recent treatments of vegetation change in the desert have often indicated problems in applying successional concepts to deserts (Noy-Meir 1973) or have ignored succession entirely.

Many studies have, of course, contributed to our knowledge of vegetation change in the desert and chaparral, but the ideas of succession as applied in less arid regions have not provided a useful framework for organizing what has been learned. This may be because vegetation change in arid regions is either less important or qualitatively different compared to change in humid regions. It may also be that successional theory is inadequate, a suggestion finding increasing favor (Niering and Goodwin 1974; White 1979), and which I shall attempt to further in this chapter. It is time to include vegetation change in arid regions within successional theory, instead of focusing attention on those special cases (mostly from humid, forested, regions) that are the most pleasing basis for theoretical structures (McIntosh 1980b).

Examples of short-term vegetation change from the chaparral and desert of San Diego County, California, will first be discussed without particular reference to successional concepts, and some generalizations about vegetation change in these shrub communities will then be suggested. I shall then discuss the cause of the difficulties in applying successional concepts in arid vegetation. Finally, an amended view of succession will be proposed.

The Regional Setting

The vegetation and physical environment of San Diego County and Southern California have been described in a number of recent publications (e.g., Thrower and Bradbury 1977; Miller in press) and only a brief summary will be given here. For present purposes it is sufficient to divide the vegetation into three zones: (1) coastal zone dominated by varying mixtures of summer-deciduous (i.e., "coastal sage scrub" types) and evergreen sclerophyllous shrubs (i.e., "chaparral" types); (2) montane zone dominated by mixed oak-conifer forest but with a significant admixture of chaparral; and (3) desert zone in the rain shadow of the mountains dominated by open Sonoran desert scrub. The studies reported on here are being carried out in the chaparral and in the desert.

Precipitation gradients in this region are relatively steep. At the coast, annual rainfall is about 25 cm. This rises to a maximum of about 100 cm at the higher elevations of the mountains, and declines to about 8 cm in the desert just beyond the eastern boundary of San Diego County about 100 km from the ocean. Rainfall in all zones is highly variable, but with the usual tendency for greater relative variation with increasing aridity. At San Diego, a 129-year weather record has a mean fall to spring precipitation of 24.9 cm and a coefficient of variation of 41%. At El Centro, about 50 km east of the San Diego County boundary, the mean annual precipitation based on a 44-year record is 6.0 cm with a coefficient of variation of

64%. At San Diego the maximum recorded rainfall season is 66 cm (7.5 times greater than the minimum), while at El Centro the maximum annual rainfall is 17.3 cm (27 times greater than the minimum).

The coastal strip has a Mediterranean climate with mild temperatures and a prolonged summer drought. The desert zone has less equitable temperatures, with severe drought possible at any season, but typically most severe in May through July. Unlike the chaparral, the easternmost portions of the desert in San Diego County receive a significant proportion of their total rainfall in the late summer and autumn.

Fire is an important factor in the chaparral, as it probably has been ever since the climate has had a prolonged dry period. Under certain conditions fires can burn hundreds of square kilometers, and there is no significant area in San Diego County that has remained unburned for much more than a century. Humans have certainly altered both the spatial and temporal pattern of fire, but it is a matter of debate whether the overall effect has been to increase or decrease the average length of time between fires. Fires occur occasionally in the desert, but it is not an important factor except at the higher elevations where plant biomass is sufficient to provide cover dense enough to carry a fire. Fire was probably even less important in the desert vegetation in presettlement times.

Vegetation Change in *Cupressus forbesii* Dominated Chaparral in San Diego County

Cupressus forbesii, Tecate cypress, is a small tree that occurs in scattered populations near the coast over a 450 km range from just below Los Angeles, California, south into Baja California, Mexico. Three discrete populations occur in San Diego County, all in association with chaparral between elevations of 275–1200 m. Like other members of its genus, Tecate cypress shows a marked tendency to occur on unusual substrates, and in San Diego County it is found on soils derived from ultramafic gabbroic rocks and Jurrasic metavolcanics. Where cypress occurs it is locally dominant, forming dense stands that are or were originally coextensive with large expanses of chaparral. Although cypress is often classified as a small tree, ecologically cypress woodlands are a phase of chaparral vegetation.

Chaparral Shrub Life Histories

The shrub species found with Tecate cypress are mostly evergreen sclerophyllous, though some elements of the summer-deciduous sage scrub vegetation are also present and locally dominant in the near vicinity of cypress stands. Although many articles (e.g., Vogl 1968; Hanes 1971; Biswell 1974; Keeley 1981) have been written about the response of chaparral species to fire, our quantitative knowledge is still limited, and even the qualitative understanding deficient in many areas. This is understandable since the species diversity of the chaparral is relatively high, and many species, like Tecate cypress, have very local distributions. Geographic variation within species can also be large and ecologically significant. It is important to note that the information to be presented in this article is based entirely on observations within San Diego County.

Vegetation change depends on the life history traits of the component species, and a qualitative description of the important traits (Table 24.1) will help to illustrate the following points with respect to community response to fire:

1. All of the species recover rapidly from fire.
2. There is wide variation among the species with regard to the relative dependence on the survival of vegetative parts and the germination of stored seeds.
3. Some species, including important chaparral dominants, do not always exploit the postfire conditions to expand their populations, and appear to require other or additional conditions for population growth.

The treatment of life history traits (Table 24.1) is modified from Zedler (1977), and may be compared to a more general one in Keeley (1981). Species are characterized according to five traits that relate to the response of the species to fire and to conditions between fires. The traits were determined by field observations in San Diego County, mostly in or near stands of Tecate cypress over the last 8 years. Considerable quantitative data support some of the classifications, while others are based on less substantial evidence. On the whole, the table should be considered tentative, and will certainly be subject to revision and refinement as data and experience accumulate.

The traits considered in the table are:

1. Fire related mortality, which is the proportion of individuals (disregarding seed) killed by wildfires. A species in which every individual present before the fire resprouts would therefore be considered to have no fire-related mortality.
2. Germination of stored seeds after the fire. This is the average abundance of seedlings produced from seeds present on the site before the fire, relative to the prefire abundance of individuals of the species.
3. Germination in the fire free interval is the observed tendency of a species to germinate seeds in older closed stands or unburned and undisturbed areas, that is, the ability to establish new individuals without fire.
4. Longevity is the estimated maximum life span of dominant, well-placed individuals when they are unburned. Species that can live beyond 100 years are considered long-lived.

Table 24.1. Life history traits of shrub species found associated with or near stands of *Cupressus forbesii* in southern San Diego County, California.[a,b]

Species	Mortality from fire	Germination after fire	Germination in fire-free interval	Longevity	Colonization disturbed area
Group 1. Obligate seeders					
Ceanothus greggii	VH	H	N	M-H	N-VL
Ceanothus tomentosus	VH	VH	N	M	N-VL
Cupressus forbesii	VH	VH	N-VL	H	VL
Group 2. Sprouters-seeders					
Adenostoma fasciculatum	L-M	H	N-(M)	M-H	L-M
Arctostaphylos glandulosa	N-L	L-H	N?	H	N-L?
Pickeringia montana	L-M	L		M?	L-M?
Rhus laurina	N-L	M	N?	M-H	M?
Rhus ovata	N-L	M	L?	H	L?
Group 3. Sage-scrub species					
Artemisia californica	(M)-H	H	M-H	L-M	M-H
Eriogonum fasciculatum	H	H	M-H	L-M	H
Lotus scoparius	H	H	M?	L	H
Salvia apiana	L-H	H	M?	L-M	M
Salvia mellifera	H	H	M	L-M	M?
Group 4. Sprouters, non-seeders					
Cercocarpus minutiflorus	L-M	N	N-M	H?	L?
Cneoridium dumosum	L	N		H?	N?
Heteromeles arbutifolia	N-VL	N	N-M	H	N?
Quercus dumosa	VL	N	L?	H	N?
Xylococcus bicolor	L	VL	VL?	H	N

[a]Modified from Zedler (1977).
[b]Symbols indicate degree of expression of the particular trait on an increasing scale from N (for nil or none), to L (low or few), M (intermediate), to H (high or many); V stands for very.

5. Ability to colonize disturbed areas (other than burns) is based on the frequency of occurrence of a species on road cuts, old fields, fuel breaks, etc. It is an estimate of the invasive potential of a species.

In the table, these traits are scored by a four-point scale: N for none or negligible, L for few or low, M for intermediate, and H for many or high. A V for very indicates extremes.

For ease of discussion, the species may be placed into four groups on the basis of their similarity in qualitative traits, though the groups are certainly not internally homogeneous. Detailed quantitative data would probably show that each species is unique, and reveal nearly continuous variation in all of the traits listed. Idealized population dynamics for the four groups are shown in Fig. 24.1. Each group is discussed in turn below.

The first group consists of species that are the most influenced by fire, and corresponds to the "obligate seeding" life history, as described by other authors (Hanes 1971). These species are completely killed by fire and reestablish in the first year or two from seed. They rarely establish seedlings except after fire because only after fire do environmental conditions stimulate a large proportion of the dormant seed to germinate. In *Cupressus forbesii* this is because the seeds are retained in the cones until fire causes nearly simultaneous seed release. Between fires release is primarily from damaged or old disintegrating cones.

Most obligate seeding species, like the other two listed here, have long-lived seeds that accumulate in the soil. Although it has been said that fire is necessary to stimulate the germination of some species, seedlings of *Ceanothus greggii* and other obligate seedling species are often common in bulldozed but unburned chaparral. However, in stands that develop without significant human intervention, seedlings of these species are rare except after fire, and the species populations are locally even aged and regionally a mosaic of different-aged populations dating to past fires. Obligate seeding species may thus be said to be fire dependent. Although this life history probably evolved in response to fire (Wells 1969), it is also very sensitive to variations in fire pattern. The expected pattern of population change would be one of sharp increases after fire and an intervening period during which there is mortality but little establishment (Fig. 24.1a).

Species in the second group (sprouters-seeders) both resprout after fire and typically establish large numbers of seedlings after fire. They suffer varying degrees of mortality from fire (Plumb 1961, Keeley and Zedler 1978), but a large proportion of the population will survive even the most intense fire by sprouting either from burls, the base of the stem, or the roots. Seedling establishment after fire is also usually high. Like the obligate seeders, most species in this group rarely establish seedlings in the interval between fires. *Adenostoma fasciculatum* is able to do so, but survival of the seedlings beyond a single year seems to occur only in areas recovering from human disturbance or severe overgrazing by domestic livestock.

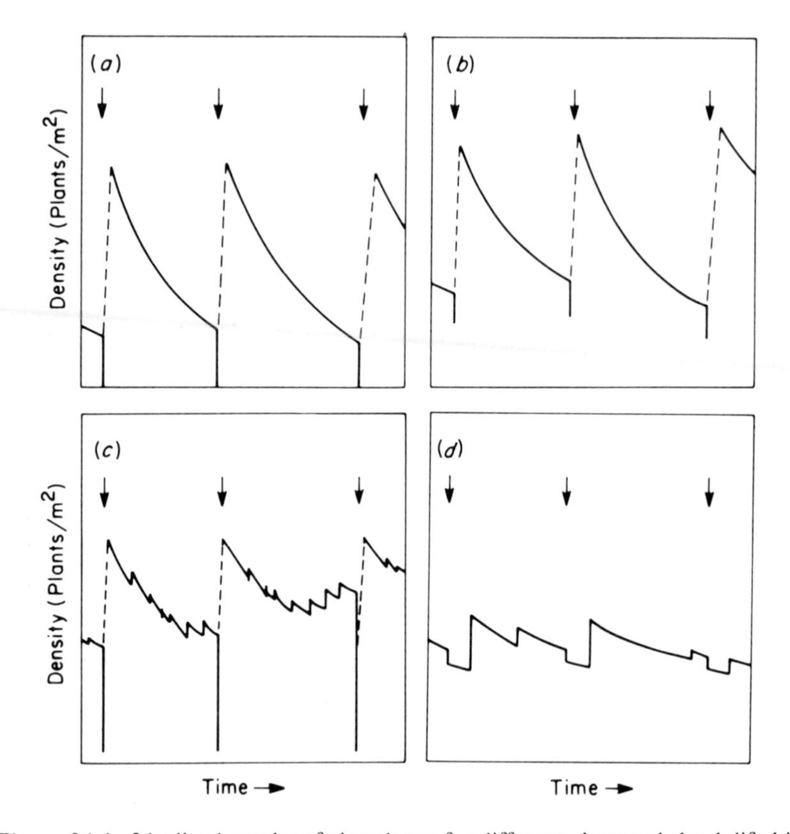

Figure 24.1. Idealized graphs of abundance for different chaparral shrub life history types during a period encompassing three fires: (a) obligate seeders; (b) sprouters-seeders; (c) sage scrub type; (d) sprouters, nonseeders. The vertical arrows indicate the occurrence of fires.

The longevity of all of these species is relatively great and the pattern of population change (Fig. 24.1b) is assumed to be one of small to moderate declines in density resulting from fire, followed by a sharp pulse of establishment after the fire. As in the obligate seeding species, there is a decline in numbers between fires as mortality occurs, but because seedling densities are not usually as high, there is less thinning as individual size increases. The expected age structure of genets in areas undisturbed except for fire would consist of discrete age classes tracing back to past fires.

The first two life history types are those usually emphasized in discussion of chaparral succession, and because of the capacity of species in these groups to regenerate after fire, they conform to the fire-induced autosuccession view. Not all species in the chaparral and coastal scrub are so clearly fire dependent, however. Species in the remaining groups would be less directly affected if fire were excluded from the chaparral.

The third group (sage-scrub species) includes species that reach their greatest abundance in the coastal sage scrub vegetation at lower elevations or on sites with high insolation. All except *Lotus scoparius* are of minimal importance within areas dominated by *Cupressus forbesii*, though they commonly occur in the vicinity of cypress stands at lower elevations, and are among the dominant species in the drier portions of the region. They resemble the obligate seeding species because all establish seedings after fire and generally do not resprout. They are relatively short-lived, and often appear structurally decrepit after 30 years. Unlike the obligate seeders, they also commonly establish seedlings in the intervals between fires. This combination of traits gives species in this group a higher potential for population expansion in open but fire free conditions, and they often invade ruderal areas such as roadsides and abandoned fields.

The expected pattern of population change in relation to fire depends on the degree to which the species resprouts and the length of time between fires. For a species with some vegetative survival a pattern like that shown in Fig. 24.1c would occur if the intervals between fire are relatively long, allowing some establishment of new seedlings in the fire-free interval. The age structure of genets in a population could vary from even-aged to nearly all aged, depending on fire history.

The fourth group (nonseeding sprouters) consists of species that resprout after fire but usually do not establish seedlings in proportion to their prefire abundance. Compensating for this surprising inability to exploit what seem to be ideal conditions for seedling establishment is a very low rate of mortality from fire (Keeley and Zedler 1978). The individuals are not immortal, however, and the question arises as to how the populations are maintained.

The lack of seedlings of *Quercus dumosa* and *Cercocarpus minutiflorus* in burns evidently results from the fact that these species have fruits that are not well adapted to long-term survival in the soil or to resisting fire. Nor is the timing of seed production optimal with respect to the occurrence of wildfires. Both species typically disperse their fruits in the autumn, and, in fire years, there is a high probability that wildlife would consume nearly all the fruit of the current season before it was dispersed.

It is not obvious why the other species in the fourth group do not normally establish seedlings in burns. They possess fruits that are similar to those of species that establish abundant seedlings, and at least *Xylococcus bicolor* does establish a few seedlings. Since our observations cover a limited time, the possibility that low densities of seedlings in burns is unrepresentative of the normal behavior of species in this group cannot be ruled out. Unusually high rates of seed predation either before or after dispersal was a possible explanation. Nevertheless, the failure of seedling establishment in these species has been noted over a wide area and in several burns, and it is unlikely that this is a rare form of population behavior.

Granting then that there are species that do not exploit postfire conditions to establish seedlings, two extreme views of the population regulation in this group seem possible. First, the populations may be varying about a steady-state condition in which the relatively low mortality rates are balanced by a low but consistent rate of seedling establishment. Because very few seedlings would have to survive to replace the annual mortality, it would be extremely difficult to measure the addition of new individuals to the population without extensive sampling. At the other extreme, the populations of these species could be replenished by occasional flushes of establishment with intervening periods in which establishment is nil. If this model applies, then one would expect to observe occasional incidents of very high seedling establishment, probably triggered by some factor other than fire.

Observations so far suggest that the second model is more likely to hold. In the wet winters of 1977–1978 and 1978–1979 seedlings of three of the five species (*Q. dumosa*, *C. minutiflorus*, *H. arbutifolia*) were observed in unusually great abundance. In fact, no seedlings of the last two species had been observed in the previous five years. The most dramatic response was by *H. arbutifolia*, which is a common but not dominant species in the Southern California chaparral. In 1977–1978, the third year of recovery after fire on Tecate Mountain, abundant seedlings were found in unburned chaparral and a large number in the adjacent burned area. The survival rate of these seedlings was low, but the observations suggest that *H. arbutifolia* populations are renewed during occasional periods of heavier than average rainfall. Seedlings of *Cercocarpus minutiflorus*, previously absent, were noted on Tecate Mountain in the same year.

Seedlings of *Q. dumosa* can occasionally be found in older stands, but usually at very low densities. Until 1977–1978 most of the seedlings found appeared to be several years old. In the 1977–1978 growing season, however, clumps of seedlings evidently arising from small mammal caches were abundant at some localities on Otay Mountain.

Abundant germination stimulated by high rainfall can also occur in burns in special circumstances. A controlled burn was conducted in December of 1977 at a site a few kilometers from the Guatay Mountain cypress site. The burn was conducted in an area of heavy oak cover, after the dispersal of the current year's acorn crop. In the spring a very dense population of seedlings was present, probably because the acorns, presumably buried by small mammals, were well protected from the heat of the fire. It is not clear how representative this situation is of natural fire conditions, but it shows the potential for the interaction of fire and climate to produce flushes of germination. The evidence, however, favors the idea that the establishment of new individuals is probably not dependent on fire, and is certainly not continuous. The ages of genetic individuals in the population would not be uniformly distributed.

The predicted long term behavior of species in this group is shown in Fig. 24.1d. A fire results in a small amount of mortality and a large reduction in biomass. The individuals lost in fires and from other causes between fires are replaced in the fire-free interval by discontinuous periods of establishment. Depending on the species these periods may be very few or more common, but it is very likely that seasons of higher than normal rainfall are important for all of them, and that establishment occurs primarily between fires rather than after fires.

Consideration of the species in more detail and the listing of additional species would reveal still more variety in population behavior. Most notable of the special cases is *Yucca whipplei*. Much of the above-ground biomass of this monocarpic species survives fire, but flowering and seedling establishment seem not to be dependent on fire, and populations are probably all aged (Keeley 1981). A number of subshrubs (e.g., *Haplopappus squarrosus*) share with the species of the third group an ability to spread between fires (Zedler 1977).

The survey of life history traits makes it clear that fire could substantially affect the abundance of any single species, and the relative balance among species. There is a considerable range, however, in sensitivity to fire, that is, maximum change in abundance expected from any one fire. The obligate seeding species, for example, are highly vulnerable both to fire occurring before the onset of reproduction or so late that there has been a depletion of seed reserves. The fourth group is on the other end of the scale of sensitivity, since neither mortality nor seedling establishment appears to be much affected by fire. Fire effects on the fourth group are largely indirect; they are of the result of effects on populations of competitors, predators, and parasites, and changes in soil properties and stand microclimate. Hence, it is possible to overemphasize the controlling importance of fire in the chaparral.

The species of the chaparral show a range of life history types from long-lived and slow to expand in open conditions to short-lived and capable of invading. Despite this, a consideration of the seminatural condition in which only fire, minimal grazing, and climatic variation play a role does not suggest any clear-cut successional relations among species. It is obviously wrong, however, to say that there is no population change.

Population Changes in *Cupressus forbesii*

Because all of the shrub species of the chaparral are adapted to recover from fire, it is expected that a single fire will have a low probability of causing the local extinction of species. Because of the significantly different life histories of the shrubs, however, changes in relative numbers and biomass are inevitable. For the system to approach equilibrium, a minimal requirement would seem to be a uniform length of time between fires. When fire frequency is variable or shifts to a new average value, the result could be large changes in the vegetation. In extreme cases, the pattern of fires could lead to species extinction. Recent changes in populations of *Cupressus forbesii* (Tecate cypress) show that this kind of change does occur.

Tecate cypress possesses serotinal cones that are opened by fire. This trait allows the determination of seed availability before fire, and estimated seed rain after fire. Since at least the base of even small individuals remains after fire, the density of trees present before fire can also be measured. It is therefore possible to evaluate initial establishment in burned stands by comparing seedling densities to seed input and prefire stand densities.

Recent fires in extreme southern San Diego County on Tecate and Otay Mountains have given an opportunity to observe stand recovery. In 1975 a fire burned nearly all of the area mapped in Fig. 24.2. At the time of the fire there was a complex mosaic of past fires, which could be partly mapped by the use of fire history records, ring counts of cypress and other woody species, and aerial photographs. The 1975 fire burned stands that had burned previously in about 1880, about 1923, 1944, and 1965. In 1976 a fire burned an area at the base of Otay Mountain about 15 km west of the area in Fig. 24.3. This location had previously burned in 1943.

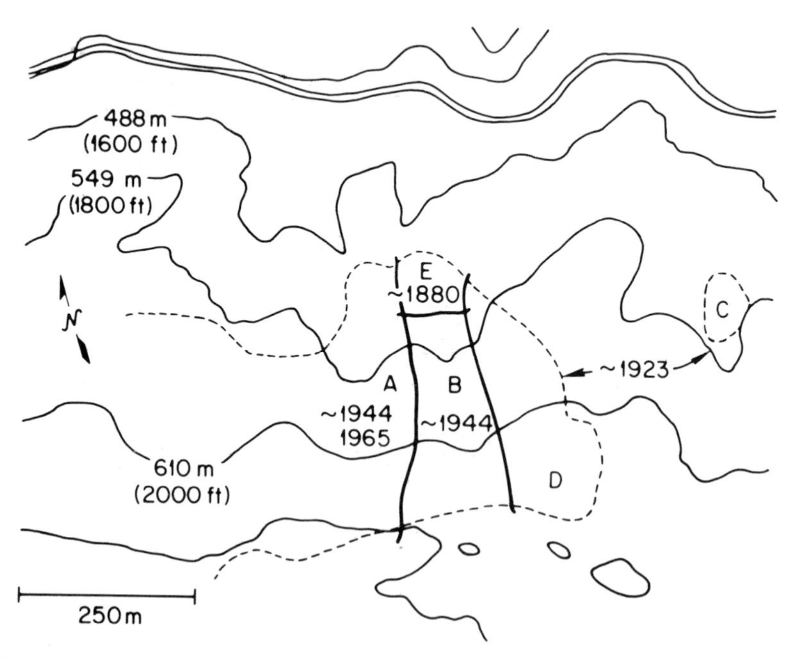

Figure 24.2. Fire history of a *Cupressus forbesii* stand on the north face of Tecate Mountain, San Diego County, California. The area depicted lies about 2 km north of the United States-Mexico border and 6 km northwest of the town of Tecate. Essentially all of the area pictured was burned in 1975. The times of earlier fires are indicated on the map. Dating of burned remains also indicates that all five stands were burned in about 1880.

Thus, these two wildfires, both of human origin, burned stands of five different ages. Data for stands A and B on Tecate Mountain were collected in the unburned stands in 1972. Sampling to assess recovery from fire began on Tecate Mountain in the winter of 1976 and on the Otay Mountain Mocogo site in the winter of 1977.

The cone counts in the stands (Table 24.2) showed that the average number of cones per tree increases with stand age, as would be expected, since this is primarily a function of tree size. Assuming equal probability of seedling establishment, individuals in older stands would seem to have a higher probability of establishing offspring in the event of a fire.

This is borne out by the data (Table 24.3) on stand recovery for the Tecate Mountain stands burned in 1975, and the Mocogo stand burned in 1976. The number of seedlings per tree is higher in the stands that were at least 52 years old at the time of the fire. These stands have reestablished at a density several times higher than the prefire density. The stands that were 33 years and younger at the time of the fire have established fewer

Table 24.2. Age of *Cupressus forbesii* stands at the time of the last fire, and number of cones per tree for each stand.

	Age of time of last fire (yr)	Cones/tree
Tecate A	10	0.8
Tecate B	32	3.7
Otay Mocogo	33	9.9
Tecate C	52	21.7
Tecate D	52	29.9
Tecate E	~95	33.5

Table 24.3. Various measures of stand recovery for burned *C. forbesii* stands.[a]

	Age at time of fire	Preburn trees/m^2	Seedings/cone	Seedlings/ preburn tree	Seedlings/ m^2
Tecate A	10	0.05	0.33	0.28	0.02
Tecate B	32	8.90[b]	0.25	0.07	0.60
Otay Mocogo	33	1.70	0.03	0.33	0.57
Tecate C	52	1.21[b]	0.22	4.25	5.15
Tecate D	52	0.36	0.51	15.30	5.51
Tecate E	95	0.44	0.13	4.25[c]	1.87[c]

[a]Seedling data are for May-June 1976 at the close of the first growing season, except for Tecate E where the figures are based on data from May 1978 and corrected for mortality to approximate 1976 data.

[b]Based on sample data from 1972; other values based on burned remains.

[c]Estimates of 1976 values calculated from 1978 sample data; 17% mortality is assumed.

seedlings than there were mature trees before the fire. The average tree in these stands failed to replace itself, and there has been a significant drop in density.

Absolute densities show that the decline is also detectable in the number of trees per unit area. If stands of cypress were to be burned every 33 years, these data suggest that near extinction would result after three or four fires.

The prediction of extinction by fire is given empirical support by further data from Tecate Stand B taken from Zedler (1977), and independently reported on by Armstrong (in Vogl et al. 1977). This stand burned in 1944, as did Stand A, but was burned again in 1965 (Fig. 105). Because of the 1972 sample data, it was possible to reconstruct the history of this stand (Table 24.4; Zedler 1977). The 1972 sample gave seedling counts as of that date as well as counts of dead trees remaining after the 1965 fire, and the 1976 data gave seedling density after the 1975 fire. The density at other times was estimated by assuming mortality rates comparable to those seen in stands sampled after the 1975 fire. In 1944 just before the fire, the density was probably at least 1 tree/m^2. Reestablishment after the 1944 fire was relatively strong, since the minimal estimate of the 1965 density based on the burned remains was greater than one tree/m^2. The observed density of seedlings established after the 1965 fire and surviving to 1972 was only 0.05/m^2, an order of magnitude drop in abundance. Surprisingly, there was no significant decline in abundance in the next fire, but this may be due to input of seed from surrounding areas. These data suggest that cypress populations fluctuate in response to varying intervals between fire, and support the idea that a run of fires separated by short periods could lead to extinction of the species.

Cone and seed production depend on factors other than age alone, however. Tecate cypress responds dramatically to site factors, and there is large variation in average tree size and hence cone production within stands. Predicated rates of decline based on stand averages will mask the spatial differences. The effects of site factors is explored below.

Because of the long summer drought, soil depth and slope position are factors of importance in cypress growth. A common pattern is for trees at the bottom of slopes and along drainages to be significantly larger than the trees upslope. Transects taken across a topographic gradient in the Mocogo stand on Otay Mountain illustrate this trend. The average diameter and average density of the prefire stand, averaged by 5-m segments, vary with relative elevation (Fig.24.3). The average diameter of trees declines upslope, then inreases at the top of the stand, where the cypress gives way to other species of shrubs. Average density is low at the bottom, rises to a peak in the middle, and drops again at the top of the cypress stand. The pattern is one of decreasing average tree size slightly complicated by interaction between number and size. The largest diameters are at the bottom of the slope, where conditions are most favorable,

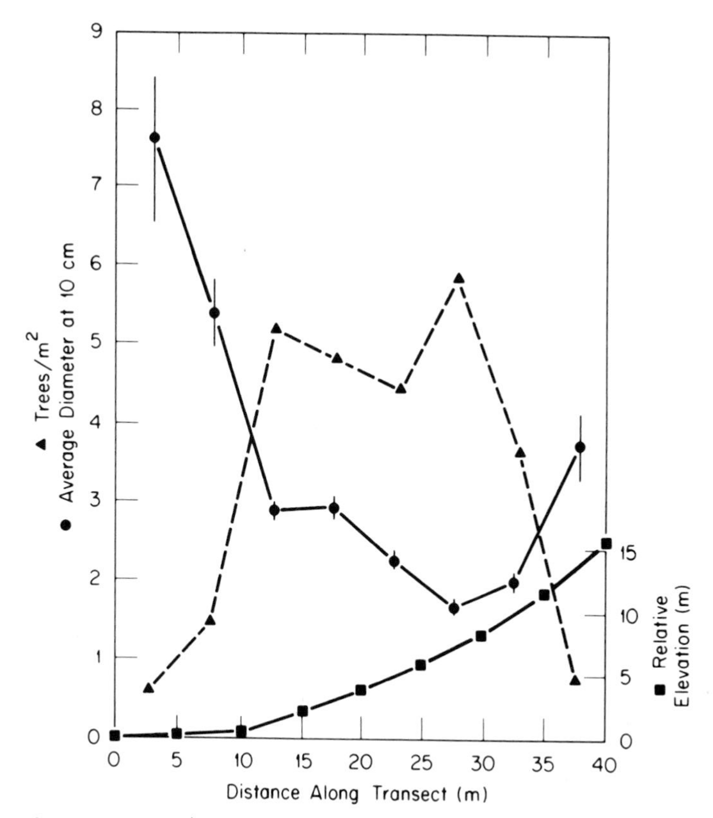

Figure 24.3. Density (triangles) and average diameter (solid circles) of dead Tecate cypress trees along the topographic gradient in the Otay Mocogo stand. Data are averages for 5-m segments. The relative elevation is measured from the zero point of the transect, which is at the edge of a dry wash that contains no perennial vegetation.

and at the top of the stand, where physical conditions are less favorable, but where intraspecific competition is reduced.

The number of cones (Fig. 24.4) declines upslope as a result of reduced tree size and lowered density. Paralleling this trend is a decreasing density of seedlings established in 1977 after the 1976 fire. The ratio of seedlings to preburn trees (Fig. 24.4b) puts these data in the same terms as those of Table 23.1. Recall that overall this stand showed a decline in density. It is apparent (Fig. 24.4b) that the decline is most severe upslope and that the lower slope positions have recovery rates comparable to those of older stands. The break-even point is approximately where the terrace next to the stream ends. These data show that capacity for regeneration is site dependent and that spatial factors must be considered in predicting local extinction. In the example, the population with lowest slope position could persist 33 years between fires, assuming that these

Table 24.4. Density of *Cupressus forbesii* in Tecate A.[a]

Year	Time since last fire (yr)	Density of C. forbesii individuals (number/m^2)
(1944)	~64	(~1.0)
– Fire 1944 –		
(1945)	0.5	(>1.5)
1965	21	>1.04
– Fire 1965 –		
1966	0.5	(~0.07)
1972	7	0.05
– Fire 1975 –		
1976	0.5	0.02

[a]Figures in parentheses are estimates, other data by direct measurement.

data are representative, while the upslope portions of the stand could be maintained only by dispersal from the lower slopes. Since periods between fires undoubtedly varied in the past, the historical pattern may have been one of small-scale extinction and reinvasion from refuges where age at first reproduction is lower and seed production per tree higher.

Vegetation Change in *Larrea-Ambrosia* Desert

The desert vegetation of eastern San Diego County differs floristically and physiognomically from the chaparral. Rainfall is lower and the vegetation much more open, with the result that fire is rare. However, like the chaparral, shrub populations observed at any one time are likely to lack seedlings or young individuals, and consist almost entirely of larger, older plants. It is obvious that as in the chaparral, species do not establish seedlings every year, and in most years the populations of most species are stable or declining. The restocking of populations takes place only infrequently, and requires higher than average soil moisture levels.

An unusually heavy storm caused by a tropical hurricane moved through Southern California in September of 1976. Rainfall was heavy in the desert and catastrophic in the higher mountains. Severe flooding was caused in the larger washes, and smaller drainages were substantially altered. As expected, the rain was followed by the most abundant seedlings establishment of shrubs seen in many years.

In a study conducted jointly with Thomas A. Ebert, a study site was established about 10 km west of Ocotillo, California to follow the fate of shrub seedlings. The site occupies a nearly level area in a broad alluvial valley, and is dissected by an anastomosing network of small drainages.

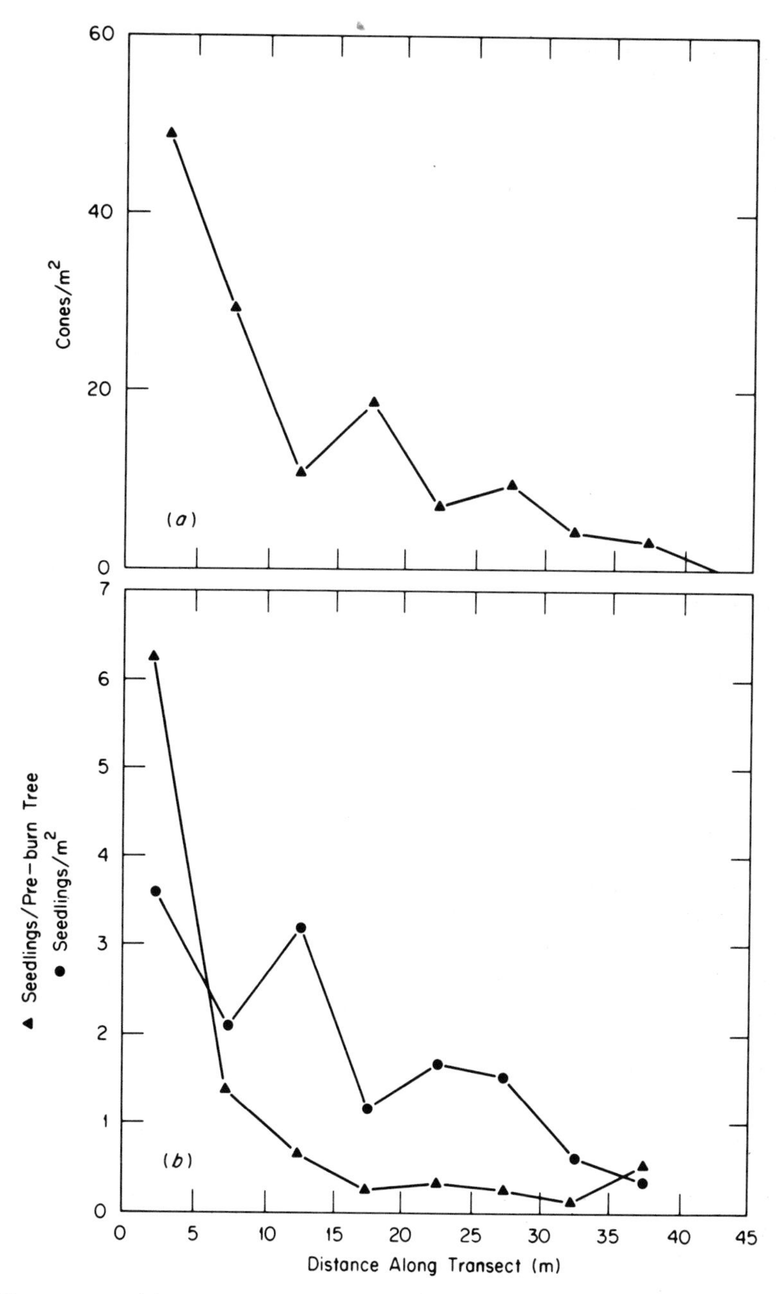

Figure 24.4. (a) Cones remaining on dead *Cupressus forbesii* trees expressed on an area basis plotted along the topographic gradient in the Otay Mocogo stand. (b) Density of cypress seedlings (solid circles) and the ratio of seedling density to preburn tree density (solid triangles) along the topographic gradient. Ratios greater than one indicate that the density of seedlings is greater than the density of preburn trees. In both graphs distance along the transect increases upslope (see Fig.24.3). Data are averages for 5-m meter segments.

The mean annual rainfall in the area is about 10 cm, while the September storm produced 7.6 cm of rain.

Permanent plots were established within an area of approximately 5 ha. Seedlings were counted and measured in a total of 850 m^2 of permanent plots, and the density of established individuals estimated in belt transects that had a total area of 6240 m^2. Estimating the density of individuals was a problem in some species, but reasonable estimates could be obtained for all but *Agave deserti.*

A portion of the resulting data (Zedler and Ebert, unpublished data) are given in Table 24.5, where the densities of seedlings and mature plants are compared. The relative establishment rates of the shrub species present varied greatly, and fell into three groups. In the first group, the establishment of seedlings was high, and proportional to the abundance of mature plants. A double log plot of seedling density against adult density

Table 24.5. Seedling and mature plant densities recorded in fall 1976 at a site near Ocotillo, California.

Species	Established plants (number/ha)	Seedlings (number/ha)
Group 1. Present as adults, common as seedlings		
Ambrosia dumosa	1151	34900
Encelia farinosa	462	2449
Hymenoclea salsola	83	2168
Larrea divaricata	122	765
Opuntia (echinocarpa?)	135	625
Dalea schottii	19	472
Simmondsia chinensis	12	179
Bebbia juncea	10	No data, but present[a]
Group 2. Present as adults, no seedlings recorded		
Agave deserti	Abundant[b]	0
Fouquieria splendens	6	0
Krameria parviflora	26	0
Opuntia bigelovii	77	0
Asclepias subulata	2	0
Group 3. Rare or absent as adults, abundant as seedlings		
Dalea spinosa	<1	714
Acacia greggii	0	612
Hyptis emoryi	<1	357
Beleperone californica	<1	153

[a]Seedlings of Bebbia were initially thought to be herbs and not recorded.

[b]Because of clone formation, no satisfactory estimate of individuals could be made of Agave. Its relative cover is about 10%.

revals a significant linear relationship. Within this group of species no major readjustments in relative abundance are to be expected as a result of this storm.

The five species that failed to establish any seedlings in the study plots (Group 2 of Table 24.5) differ considerably in life history traits. *Opuntia bigelovii* (jumping cholla) spreads vegetatively by the rooting of disarticulated joints, and is thought to set limited numbers of viable seeds. Its failure to establish seedlings is expected. *Agave deserti*, another succulent, does produce viable seed and also reproduces vegetatively. Individuals appear to have a very long average life span, but the rate of vegetative spread is slow. *Agave deserti* seeds are evidently not well adapted to survival in the soil, and reproduction seems likely to depend upon the coincidence of high seed production followed in a short time by a period of high soil moisture. Since the heavy rain reported in September 1976 followed a series of rather dry years, the seedbank of *A. deserti* was presumably very low. Jordan and Nobel (1979) estimate that *A. deserti* has established seedlings only once in the last 17 years in their study area in Riverside County, California.

Fouquieria splendens (ocotillo) produces at least some seeds in most years, but the seeds lack the necessary physical properties to enable them to survive for long in the soil. Like *A. deserti*, *F. splendens* probably requires a sequence of conditions to stimulate germination, most probably heavy winter rain for seed production, followed by heavy summer rain. We cannot account for the failure of germination in the other two species, *Krameria parviflora* and *Asclepias subulata* (Table 24.5).

Fouquieria splendens and *Agave deserti* seem to be similar to the sprouting but nonseeding species in the chaparral (Group 4), in that they are long-lived, stable elements of the vegetation that have specialized requirements for germination that limit their capacity to exploit opportunities for population expansion.

The third group of species (Table 24.5), found in reasonable abundance as seedlings in the vegetation but absent as adults, are all most common in wash areas subject to periodic flooding. Since the species all occur upstream from the study area, it is presumed that seeds were carried in with flood waters. As a result, small populations of wash species have become established within the study area. Despite very high seedling mortality, some individuals persist, and small groups of well-established *Dalea spinosa* and *Acacia greggii* seem likely to produce stands of mature individuals where none existed before.

Even this limited study reaffirms the importance of physical factors in the desert. It seems unjustified to conclude, however, as some have, that competition is of less importance than in other systems. During the most severe droughts, it may be true that survival of many individuals is independent of the surrounding plants, and it may also be true that establishment and growth in the initial phases of a favorable period also have little to do with competitive interaction. Nonetheless, as the vegetation declines from peak periods of growth, individuals in close proximity must suffer increased mortality and reduced growth (Noy-Meir 1973). In this process, the larger, well-established individuals have an overwhelming advantage, and so it will seem that competition has done little to alter the composition of the perennial vegetation. As with chaparral fire, the effect of competition will be difficult to measure in the short run, but immensely important in the long run.

Discussion

Comparison of Vegetation Change in the Desert and Chaparral

There are underlying similarities in vegetation change in the chaparral and desert. In both systems most species of woody plants establish new individuals only in some years and usually in response to extreme events; fire in the chaparral, and heavy rain in both. In the two systems the response of different species to the same extreme event is varied, but in neither community is local extinction or invasion likely to result from a single natural event except on a very small scale (1–100 m^2). In both systems, excluding agricultural clearing, massive earth movement, and urbanization, sequences of extreme events are generally more significant than individual events. The timing and occurrence of fires influence the increase and decrease of chaparral shrub populations. Some shrub species, especially in the desert, will not establish seedlings unless seed production and seedling establishment are made possible by the proper timing of rains. After site and average climatic factors have been accounted for, the species composition and stand structure at a particular place will be dependent on the past pattern of extreme events. Southern California is not unique in this, the significance of the pattern of extreme events in structuring arid and semiarid ecosystems has been noted for other areas (Hall et al. 1964; Jackson 1968; Noy-Meir 1973; Bykov 1974).

In both the chaparral and desert shrub communities, the range of shrub life history types includes species that are long-lived, stable elements (e.g., *Larrea divaricata*, *Quercus dumosa*), and shorter-lived, more volatile ones (e.g., *Salvia mellifera*, *Ambrosia dumosa*). The long-lived species have a good prospect of short-term survival once they are well established, almost regardless of the pattern of environmental variation, and form a semipermanent matrix around which less stable species vary. Displacement of these long-lived, stable species must be a slow process unless there is a drastic change in the prevailing climate, or an outbreak of a pathogen or predator. If conditions were to alter slightly and affect the relative ability of species to establish new individuals, it would be difficult in a single human life span to observe that directional change was underway.

Logical deduction from the above points suggests that local composition in chaparral and desert communities is continually, but not necessarily conspicuously, shifting in response to changes in climatic conditions and varying patterns of extreme events. This is supported by Hastings and Turner's (1965) study of long-term vegetation change in Arizona. Whether or not the small-scale changes noted in the decline of cypress in the chaparral and the local invasion of *Dalea spinosa* and *Acacia greggii* in the desert are indicative of future trends is unknown. They are, however, examples of the way in which change is brought about, not by massive extinction but by gradual shifts in population boundaries. The expansion and contraction about refuges, hypothesized for the cypress, may also be a common pattern. It is certain that no detailed understanding of vegetation change in arid regions is possible without considering spatial factors. The evidence argues for combining studies of pattern with population studies (Greig-Smith 1979).

The examples presented do not exclude the possibility that there are significant gap-phase cyclical patterns in the communities, and at least one has been suggested for the Sonoran desert (Yeaton 1978). The existence of such patterns would not invalidate any of the points made, but it would add additional complexity to the process of change, and reinforce the need for careful attention to spatial aspects.

The importance of fire in the chaparral, and its insignificance over most of the desert, is the most obvious distinction between the systems. Chaparral species have features that presumably could only evolve in an environment with frequent fire, but striking as these adaptations are, from a population point of view the difference between species that experience fire and those that do not is not really very great. The behavior of a desert species that must wait until the next period of prolonged surface soil moisture to establish new individuals is much like that of a chaparral species that faces a small probability of establishment between fires. In both systems there are long periods during which populations are stable or declining, interspersed with brief periods of very heavy seedling establishment.

Is Vegetation Change in the Chaparral and Desert To Be Considered Succession?

Succession is usually defined as a predictable change in species composition over time (Drury and Nisbet 1973). The model case is that of secondary succession in the deciduous forest, in which there is a "disturbance" followed by a series of predictable changes that are initiated by the invasion of "pioneer" species, and sustained by subsequent invasions, ending in an equilibrium climax state. It is said that one can identify the stages in the progression toward the equilibrium state.

The chaparral and desert deviate from the ideal case in two fundamentally important ways:

1. Single extreme events that would qualify as "disturbance" in the deciduous forest produce little, or no, compositional change. Arid systems are, to use Hanes' (1971) term, autosuccessional.
2. When there is compositional change, it does not seem part of a predictable undirectional successional sequence.

Desert and chaparral communities are visualized as changing randomly in response to buffeting from external factors with no apparent direction. It is not possible to use the species composition of the perennial woody vegetation in either ecosystem as a measure of the time since the last fire, flood, or drought, although relative abundance is strongly influenced by these events, as described above. The strongest patterns in terms of composition are those of substrate, aspect, and slope position.

The awkward discrepancies might be explained in two ways:

1. Vegetational change in arid regions is fundamentally different from that in humid forests, and it is therefore necessary either to accept that succession does not occur or to formulate a separate set of concepts.
2. Successional concepts and terminology may be inadequate.

I believe that the second explanation is the correct one. Arid ecosystems, at the level of generalization appropriate to a theory of vegetation change, are not qualitatively different, but exhibit in a stronger and less avoidable form features that are characteristic of all ecosystems.

We may consider the first of the two problems raised above, the "autosuccessional" nature of arid systems. In "well-behaved" ecosystems, extinction is followed by a series of invasions. Since the sequence of invaders is said to be predictable, it follows that they are drawn from a pool of species that is stable over long time intervals. Obviously, if one of the species that is expected to participate in the succession goes extinct in the region and is unavailable for invasion, the predictability is lost. Successional "predictions" are based on assumptions about availability of plant propagules, and therefore about the spatial distribution of species outside the area to which the predictions apply. In reality, these predictions apply only to a large-scale system with consistent composition. At this large scale, single extreme events do not change the species composition (Zedler and Goff 1973).

Clearly the "well-behaved" case and the chaparral and desert differ only in scale, not in process. In the chaparral, fire causes the death (local extinction) of some plants that may be replaced with local invasion. But at a scale of 1–10 ha, however, no compositional change is expected. In the deciduous forest, a fire may cause local extinction of a species at a scale of 1–10 ha or more. On a quantitative basis the scale is significantly different, and important natural history features that determine the means of species reestablishment will also be different; conceptually, however, the process is identical, since both systems are self-reproducing.

It is also questionable if arid systems are more "unpredictable" than "well-behaved" ones. In discussing this point, it is convenient to refer to a diagram (Fig. 24.5). Regardless of the complexity of natural ecosystems, the understanding of ecosystem change reduces to being able to predict or explain changes in the properties of biological entities. Because of the importance of spatial effects, the system within which the properties will be measured must be well defined. For an acceptably general theory, the entities chosen will be species or other collections of organisms of evolutionary significance. Which properties of the biological entities are measured is of crucial importance, and will vary with the purpose of a particular study, but it is certain that any comprehensible and workable approach will deal with a small number of the virtually infinite number of properties that could be considered. A minimal set of properties and entities might be the total number and average size or weight of

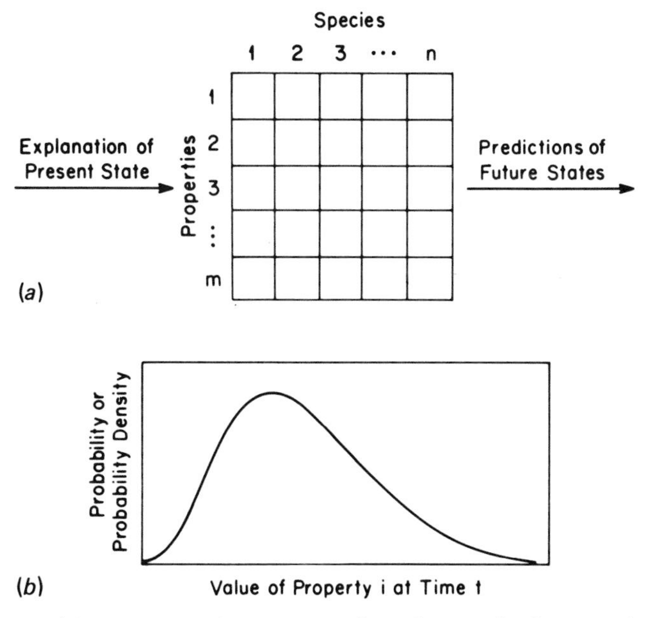

Figure 24.5. (a) Diagrammatic representation of a matrix that contains values for m properties for each of n species. The matrix is a description of the ecosystem at a particular time. (b) A hypothetical probability distribution of predicted states for a particular ecosystem property at a future time, illustrating the fact that realistic prediction for natural systems is necessarily stochastic.

all taxonomic species. The present state of the biota can be represented by one such matrix with values for each property of interest. It is clear that multidimensional matrices will be needed for most realistic applications, but this does not alter the logic.

A fully devloped theory of vegetation change should be able to work from the present state of the ecosystem as defined in a matrix of properties to predictions of future states and explanations of the present and past states, to make use of the distinction supported by Mayr (1961). Such a mature theory will specify which knowledge outside of the information in the matrix itself is necessary to be able to make predictions and indicate the procedures for deriving the future states and for assigning error bounds to the predictions.

To make predictions, it is obviously necessary to make assumptions about features of the environment, and in particular, it is necessary to base predictions on assumptions of future environmental variation, including extreme events like fire, flood, and drought. It is possible to assume that a particular pattern will hold and to make predictions that are deterministic with respect to those conditions, or to assign probabilities to different patterns and obtain in that way a probability distribution of future states (Fig. 24.5b). The latter approach is necessary for realistic prediction.

With this model in mind, it is apparent that there is little difference in principle in predicting change in arid ecosystems and in well-behaved ecosystems like the deciduous forest, for which successional theory is supposed to provide predictability. In both cases it is necessary to consider environmental variation, including extreme events, over the period of prediction. In the more realistic stochastic case, predictions in both instances will take the form of probability distributions for the particular property of interest (Fig. 24.5). One measure of predictability would be the variance of these probability distributions, and it is not obvious that these variances will be less for systems that are conventionally thought of as having successional "predictability."

When it is said that successional theory is predictive, what is actually meant is that if environmental variation is maintained within certain bounds, and if certain extreme events are assumed not to occur within the period of prediction, the composition change is expected to follow patterns that have been recorded in the past. If the same restrictions are applied in arid regions, no doubt an equal or better degree of predictability could be claimed; but deterministic projections of this kind that exclude so-called disturbance factors are so patently ridiculous in arid regions that few have chosen to make predictions on this basis.

Much of the difficulty in fitting vegetation change in arid regions in with successional theory arises because of the fixation on climax. In terms of the model given above, climax is a subset (assumed to be small) of the total possible set of possible outcomes when vegetation undergoes change in an environment free of events and patterns of variation considered to be disturbances long enough to achieve stability of species composition. Which subset will be designated as conforming to climax depends on which sorts of environmental patterns and events a particular investigator feels are "abnormal." In the real world a certain area after any interval of time, no matter how long, may end with the vegetation within the climax subset or in some other totally different state. The climax condition may not be the most probable state of the vegetation, and in some cases it may be purely hypothetical. Speculating about what vegetation would be like if it were to develop under specified conditions is the basis of scientific land management, and therefore obviously worthwhile, but there seems to be no reason justifiable from a theoretical point of view for the inordinate concern for "disturbance" free vegetation development. There are strong arguments for discarding the climax concept (Harper 1977a, White 1979).

It is necessary to reformulate concepts of vegetational change in a framework that relegates climax to its proper place, and recognizes the importance of environmental variation, and especially extreme events, in shaping ecosystems. The simplest remedy is to make succession synonymous with change in species composition. The basis for measuring successional change is then the special case of the matrix in Fig. 24.5 in which the biological units are species and the only property is presence or absence. Succession can be said to have taken place in an interval if there has been change in the presence-absence vector. The degree of change in this vector will depend on the size and geographic location of the area being considered, and this will require explicit introduction of spatial considerations into successional theory. We may then proceed with studies of vegetation change and seek to understand its causes.

Taking presence-absence as the basis of succession leaves a great deal of vegetation change outside the definition, but this seems desirable. If one wishes to call changes in relative abundance or biomass succession, then it will be necessary either to call practically all change succession, or to adopt a more detailed and by no means clear-cut definition. For example, one could define succession as changes in the rank order of dominance, and then go on to define how dominance is to be measured. It seems preferable to keep succession simple, and resign all other change below presence-absence to the category of vegetation development.

In conclusion, it is not necessary to redefine succession in order to explain vegetation change in chaparral and desert, but it is necessary to view the ecosystem as the product of a long history of catastrophes great and small, and to consider the importance of spatial scale (Whittaker and Levin 1977). As shown here, this is essential in the arid shrub lands of southern California, and others have argued the need for this view in other systems (e.g., Raup 1957; cited in White 1979; Gilbert 1959; Loucks 1970; Heinselman Chapter 23) and very generally (e.g., Wright 1974; Vitousek and Reiners 1975; Grubb 1977). Because of the difficulty which ecology has in escaping from the "tyranny of the particular," we should be sceptical of generalization about ecosystem changes that are made without reference to the properties of the component species or the features of the environment in which they are found.

Acknowledgements

Special thanks are due Thomas A. Ebert who shared equally in the collection of the desert data and their interpretation. F. Glenn Goff stimulated my interest in succession and many of his thoughts are incorporated. The paper was written while in residence at the School of Plant Biology, University College of North Wales, and benefitted from discussions with staff and visitors. The work was supported by National Science Foundation grants DEB 76-19742 and DEB 79-13424.

Chapter 25

Is Succession a Viable Concept in African Savanna Ecosystems?

B. H. Walker

Introduction

The savannas of Southern and East Africa have five major characteristics that together distinguish them from other ecosystems:

1. Both the herbaceous and woody layers contribute significantly to primary production, and they generally occur in an irregular mosaic.
2. The vegetation is spatially very heterogeneous.
3. They support, in their natural state, a high biomass of large ungulates.
4. Rainfall is very variable and, as a consequence, so is primary production.
5. Fires occur at irregular intervals, the frequency decreasing with aridity, from virtually annual in some of the wetter savannas (>750 mm rainfall per annum) to nil at the driest extremes. In the moister savannas, the vegetation would be closed woodland in the absence of fire.

Change is also a characteristic feature of the savannas, and the question that concerns us here is how much of the change can be regarded as succession. This chapter is in two parts; first, a review of the evidence of change in savannas and the factors that determine it, and second, in the light of this information, a discussion on how best to interpret it.

Determinants of Savannas

Water

The overriding factor giving rise to the structure of the vegetation in African savanna regions is the soil water balance (Walter 1971; Tinley 1977; Walker and Noy-Meir 1982). In general terms this is a function of the

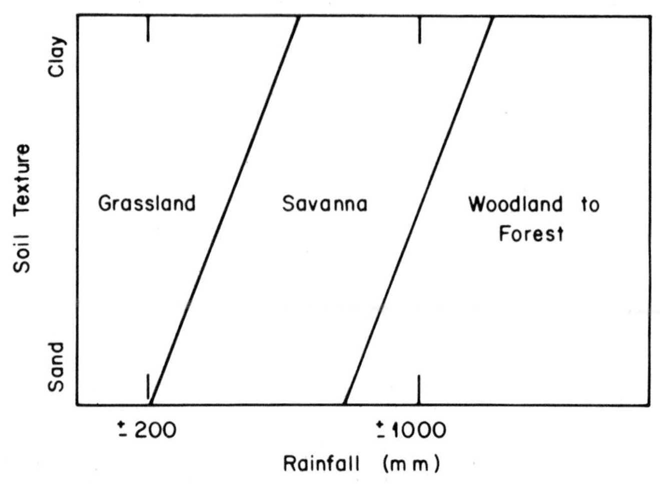

Figure 25.1. The distribution of savanna vegetation in relation to grassland and woodland, according to rainfall and soil type.

rainfall and soil type, and gives rise to the situation depicted in Fig. 25.1. Within this general pattern, however, major differences in vegetation result from differences in the soil water balance, which are due to different patterns of drainage and to the presence or absence of a pan horizon, and the depth of this horizon below the soil surface (Tinley 1977). At shallow depths such an horizon leads to a perched water table, with hydromorphic soils, in which woody vegetation does not grow. Changes in the soil water balance, due to erosion or changes in base levels of the water table, lead to marked changes in the vegetation of a region, as discussed more fully later on.

Walter (1971) put forward the hypothesis that, in semiarid savannas with free drainage, the existence of a more or less stable mixture of grass and woody vegetation is due to a two-layer rooting zone. Grasses use water from only the upper (topsoil) layer, in which they are superior competitors for water, and the woody vegetation makes use of both topsoil and deeper subsoil water. Owing to the pattern of rainfall some water always gets through to the lower layer. Walker and Noy-Meir (1982) have analyzed the dynamics of the two vegetation components, based on this hypothesis, and the results of their model support the existence of a stable mixture of grass and trees, provided that (1) there is sufficient rainfall for some to always penetrate through to the subsoil, and (2) there is insufficient rainfall for the maintenance of a woodland.

The failure of woody vegetation in savannas to develop into closed woodland is therefore a result of either too much or too little water. Woody plants constantly attempt to invade the grassy areas and the conditions under which they succeed are discussed later.

Nutrients

Soil nutrient differences (themselves influenced by moisture status) strongly influence the species composition of both woody and herbaceous vegetation, their productivity, and vertical structure. Bell (1982) presents evidence to show that high-nutrient status soils give rise to vegetation of high food quality, which in turn leads to a high biomass of smaller herbivores, which, through their grazing and browsing pressure, keep the vegetation in a low-biomass, shortened form. In areas of low nutrient soils (mainly southern Africa) the biomass of plants is high, limited by available resources (water and nutrients) rather than by herbivores. Bell argues that this gives rise to a "community succession" from high plant biomass used by low animal biomass to low plant biomass used by high animal biomass. Use of the term "community succession," though, is confusing in this case, and is probably better replaced by "a community continuum." The change occurs spatially, but the gradation does not occur as a dynamic sequence on one site.

Further evidence for the effect of nutrients on the dynamics of savannas comes from Sinclair's (1975) work in the Serengeti, in which he showed that nitrogen (as protein) levels in the grass, during a short, critical dry season period, were a major limitation to herbivore populations. Botkin et al. (1981) have reviewed the evidence for regulation of large mammalian herbivore populations in the savannas, and conclude that the population dynamics of these herbivores are linked to ecosystem dynamics through nutrient cycling. The birth and death rates are affected by these mineral cycles, while in turn the storage and flux of nutrients is influenced by the living vegetation and animals.

Within the herbaceous layer nutrients directly affect species composition. In undisturbed grassland the amount of available soil nitrogen is generally low (c.f. Bate and Gunton 1982). Several workers (e.g., Mills 1966, Roux 1969) have recorded that applications of high levels of N (especially as N-NO_3) lead to a reversal in succession, until the climax species have all been replaced by a few pioneer grasses and forbs.

The bulk of the available nutrients exist in the top 10–25 cm of soil, and there is a steep decline with depth. This implies that the vegetation (particularly the grass) is rapidly taking up released nutrients, and preventing them from being leached further down. A hypothesis that I am currently testing is that should the grass layer be severely reduced, there will be a downward movement of nutrients into the lower root zone, where the woody vegetation has exclusive access to them. This will further enhance the effects of reduced competition for water, leading to an increase in woody plants and making a return to a high grass-to-woody plant ratio a difficult and lengthly process.

Nutrient availability is further complicated by the activity of the vegetation. There is some evidence (Munro 1966; Meikeljohn 1968) that grasses, in particular the late successional species, secrete root toxins that inhibit nitrifying bacteria. This would accord with the fact that the "climax" grasses are replaced by early successional species when NO_3 is applied. The evidence is not, however, unequivocal. Purchase (1974a,b) claims that this type of inhibition does not exist and that the reduced amounts of NO_3 are a result of inhibition of nitrification due to low PO_4 levels.

Herbivores

Perhaps the most distinguishing feature of African savannas is the high diversity and biomass of their large-mammal fauna. It is an old fauna and has long influenced the plants with which it evolved. Cumming (1982) has reviewed the evidence for the influence of herbivores on the vegetation, geomorphology, and soil nutrient status of savannas and concludes that, at a local level, the dynamics of savannas are closely tied in to herbivore activities.

The most striking, and probably best-known effects are those induced by elephants. The radical changes that they have brought about in the woody vegetation-to-grass ratio have been recorded from many areas (e.g., Buechner and Dawkins 1969; Laws 1970; Wing and Buss 1970; Anderson and Walker 1974; Thompson 1975). Anderson and Walker (1974) have described how the species composition of the woody vegetation has been changed by them. Interpretation of current changes due to elephants are difficult, as they are influenced by the recent massive increase in human occupation, and the compression effect that this has had on elephant densities. These changes represent, today, a trend to open, grassy vegetation when elephant numbers have declined and fire been prevented, though this has been followed in some areas (for example, parts of the Chizarira plateau in Zimbabwe) by the development of thick scrub. Caughley (1976) has postulated a ± 200 year stable limit cycle of elephants and woodlands. This plant-based evidence for peaks of elephant activity is open to other interpretations, though, since the area has experienced a long human occupation. It is, nevertheless, an interesting hypothesis, and suggests that in these particular areas the vegetation is in a continuous state of flux, and is alternately undergoing succession and disturbance.

Large grazing herbivores are markedly selective in their feeding habits, and this leads to a number of effects. Area-selective grazing leads to enhancement of the differences between favored and unfavored communities or sites. The less an area is used, the more moribund it becomes, and the less attractive it is to animals. This same effect occurs at a species level within any particular community. Such species-selective grazing can lead to two different results, depending on the stocking rate.

At high stocking rates, favored species are overgrazed and may disappear from the community. At low stocking rates, favored species receive optimum treatment, and plants of the unfavored species become moribund, and the species decreases in abundance.

Light defoliation of a grass tuft, especially of the older leaves, has a stimulating effect on productivity. Kelly and Walker (1976) have observed this for the sward as a whole, and McNaughton (1979a,b) has recorded it for individual plants. The increase in production of the sward at low stocking rates is due both to this stimulation of the grass plant, and to the increased rate of nutrient cycling that results from the physical effects of animals in converting standing dead material to litter in contact with the soil surface. The interaction between herbivores and nutrients has already been discussed, and the community continuum from high plant biomass-low animal biomass to low plant biomass-high animal biomass (Bell 1982) finds its counterpart, at least in East Africa, in the sequence of changes that are induced by the grazing behavior of a seasonal succession of herbivores (Vesey Fitzgerald 1960; Bell 1970) on a particular site. This annual cycle of events does not constitute any long-term change which could be regarded as succession. According to Botkin et al. (1981), however, the net effect of the large herbivores is to act as a nutrient retention mechanism, and without their presence there would be a marked shift in the nutrient status of the soils and vegetation, and therefore a change in the vegetation itself.

The effects of browsing herbivores on woody plants are seldom, if ever, as marked as the effects of grazers on grass. This is largely because only a relatively small proportion of the total biomass of most browse species is available as food. Nevertheless, as demonstrated by Taylor and Walker (1978), the density of small woody plants can be significantly reduced by browsers and this effect can carry through to a reduced density of all age classes.

Fire

The role of fire in the dynamics of savanna vegetation has been discussed and disputed for many years. Avoiding for the moment its overall role, we can make the following observations on its particular effects.

Natural fires are a common phenomenon, at least in the savannas of southern Africa. West (1971) quotes eight recorded observations of lightning-induced fires, and unconfirmed verbal reports from various sources over many years substantiate this claim. It is, however, important to note that a number of West's recorded fires were put out by rain associated with the lightning. This suggests that the majority of successful, naturally induced fires would have occurred as a result of the dry thunderstorms that are characteristic of the hot, October period just prior to the onset of the main rainy season. Both West (1971) and Phillips (1974) estimated that fires set by humans began at least around 55,000–50,000 years ago, and more recent unpublished data extend it at least as far back as around 100,000 years ago.

Owing to their long evolution with fire, most savanna plants are adapted to it, and the proportion of species that are easily killed is very low. Such species nevertheless exist, and there is a continuum from those that are highly susceptible to those that are extremely resistant. The relative abundances of the two extremes on any site is a fair reflection of the fire history of that site. Most species are easily killed as seedlings, and their resistance increases with age, up to the point where they become senescent, and again susceptible.

Among the woody plants, some species are adapted to withstand fire by such characteristics as a thick, corky bark (e.g., *Uapaca* spp., *Cussonia* spp.), and others die back to ground level, but regrow vigorously by coppicing. The former occur on all types of soil, but the latter are more characteristic of deep, sandy soils, where the vegetation has a high root-to-shoot biomass. Rushworth (1975), for example, has described the rapid redevelopment of the woody vegetation on Kalahari sand deposits, following fire. If the fires are sufficiently intense and frequent, the end result is that all the trees and shrubs will have been eliminated, and only "underground trees" survive. These are species that grow mostly on sandy soils, and have large, extensive underground woody systems, and produce only leaves above ground (e.g., *Parinari capensis*, *Dichapetalum cymosum*, *Lannea edulis*, *Combretum platypetalum* var. *oatesii*, *Pygmaeothamnus zeyheri*).

The general effect of regular or frequent fires on the maintenance of savannas depends on the average rainfall (and therefore production). In the drier savannas it is harmful, in that it destroys litter. This leads to a marked reduction in water infiltration (Kelly and Walker 1976) and increased evaporation. In the wetter savannas, dead plant material (particularly grass) accumulates as top hamper, slowing down nutrient cycling and shading new growth. If fire is very infrequent under such conditions, it has a disastrous effect when it does occur.

There is ample evidence to show that fire interrupts the development of vegetation toward the climax (West 1971), and as Phillips (1930) has pointed out, vast areas of Africa remain in a nonclimax state as a result of fire. The degree to which fire influences vegetation structure and composition depends on the fuel load, and on the type and time of the fire. For example, Trapnell (1959) showed that, in the miombo vegetation around Nadola in Zambia, the dominant tree species *Brachystegia spiciformis* and *Julbernardia globiflora* were killed by frequent burning. Yet in the burning trials at Grasslands research station in Zimbabwe, with a lower rainfall, these same species, though reduced in height, were not killed (Strang

1974). West (1965) claims that this is due to the more intense fires that occurred at Ndola as a result of the greater fuel load.

In both these trials, early burning (i.e., burning as soon after the end of the rains as possible) had little effect on the trees, and in fact it can lead to an increase in woody vegetation. Burning at the end of the dry season destroys the woodland. The reasons for this are twofold. Early burning has little effect on dormant trees, but causes the grasses to break dormancy, and a flush of growth occurs, which depletes their root reserves. The amount of growth that takes place, however, is insufficient for replenishment of the reserves, and the new leaves die before the next rainy season begins. The woody vegetation breaks dormancy from August onwards, some two to three months before the rains begin. Grasses only begin growth with the onset of the rains around November. A fire in October therefore occurs at the time of maximum susceptibility for the trees (when they are undergoing rapid upward translocation of reserves and the phloem cells are most active) but when the grasses are still dormant. The resulting flush of grasses then persists until the rains come, and growth can continue.

Burning during the mid dry season leads to a profuse flowering of many spring forbs and grasses (West 1971), some of which flower only after fire. Frequent fires in the wetter savannas lead to a simplification of the grass sward, and the development of what is generally regarded as a fire subclimax grassland (with some trees). These grasslands are typically dominated by tall species of *Hyparrhenia* and by *Hyperthelia dissoluta* (commonly known as "thatch" grass). In the semiarid savannas frequent fires reduce grass vigor and lower the grass cover, to the point where a fire will no longer take hold. This often coincides with the stage at which the perennial grass species are replaced by annuals.

As a final observation, the effects of fire interact strongly with the effects of herbivores, and much of the information that has emerged from the many burning trials in Africa is severely limited in its usefulness as a result. Both large herbivores and insects are strongly attracted to the new flush of growth that follows a fire, and their effects must be included with those of the fire. As an example, Trollope (1974) has shown that fire and herbivores (goats) together can lead to the elimination of shrubs, where either one alone could not.

Dynamics of Savannas

The structure of any particular area of savanna results from the interaction of all four of the major determinants: water, nutrients, herbivores, and fire. The changes that they bring about occur concurrently at different time scales, and in order to get some perspective of savanna dynamics, it is useful to consider the changes at each scale.

Long-Term Changes

According to Cole (1982) most of the savannas of southern Africa are associated with old planation surfaces and represent a legacy of the vegetation that flourished during the Tertiary Period when, under hot, wet conditions, laterization was active. Under subsequent drier conditions, many species were lost, plant cover deteriorated, and the top A horizon of soil was removed and deposited elsewhere to form extensive sand regions, such as the Kalahari sand deposits.

Geomorphological evolution of the landscape gives rise to different combinations of relief and drainage and of soil depth, and, since the last glacial period (which was a pluvial period in Africa), a shortening of the rainy season and drainage through erosion channels has led to the development of increasingly xerophytic savanna communities. There is constant evidence of active drainage (often enhanced by overgrazing and burning), though shorter-term climatic fluctuations tend to hide this long-term trend.

A long-term cycle of soil development and erosion occurs over much of Africa. Tinley (1977) describes such a typical sequence of events, beginning with sandy, hydromorphic soils, which were laid down in a littoral, freshwater fan environment. They are strongly leached, leading to alluviation and deposition of clay to form a hardpan layer. This layer results in poor drainage and therefore (owing to a seasonally perched water-table) grassland. The water does not drain laterally, owing to the existence of primary base levels, and changes in these lead to incision of the fan, increased aeration, invasion by woody vegetation, and the development of ferricrete from the iron-rich clay pan. Incision and headward extension of nick points then leads first to exposure of the ferricrete (which is cracked and jointed), and finally the subpan clays are exposed, and invaded by a completely different vegetation community, the sands having been redeposited elsewhere.

Medium-Term Changes

Medium-term changes occur again largely as a result of modifications to the soil water balance. The same basic process of incision and development of drainage channels can occur over a shorter time scale, and lead to invasion by woody vegetation. A further type of change is due to the effect of woody vegetation itself modifying the site. Trees dry out the soil, and once they have managed to invade seasonally wet soils (for example, during dry periods or by gradual extension from termite mounds), they perpetuate themselves (Tinley 1977). Once the trees are removed, the seasonally flooded soils prevent redevelopment of woodlands, suggesting that the successional development of the woodland in the humid savannas is very uneven, and advances erratically when climatic fluctuations give rise to a sufficiently long period of favorable soil conditions.

At the dry extreme of the semiarid savannas the reverse is true. Only when there is a sufficiently long period of wetter years, allowing the development of a root system that is extensive enough to carry the trees through a drier year, is there a successful establishment of woody vegetation.

Thus far, we have considered only the changes occurring in developed vegetation. Classical primary succession does, however, occur. It is most common, as a medium-term phenomenon, on the alluvial deposits of rivers in the low-altitude savannas. In these regions the river courses have very large catchments (having risen on the high plateaus of the mid-continent) and carry large silt loads.

The course of succession on these alluvia, though, bears little resemblance to commonly held ideas. The very first invader of newly exposed sand banks is frequently the tree *Acacia albida*, a large and long-lived species. The herbaceous vegetation at this stage varies according to the water regime, but plays little part in any biotic reaction. As the alluvium becomes older, other woody species manage to invade. A good example of such a succession is on the alluvium of the mid-Zambezi River, in the Mana Pools Game Reserve (Zimbabwe). The downstream section (which is the most recent) is dominated by a woodland of *A. albida* and *Trichelia emetica*, with an understory of tall, coarse grasses. The older, upstream section is much more diverse, with many other tree species (but still including *A. albida*), a developed shrub layer, and a much more diverse herbaceous component. The time span for the development of this sequence is difficult to judge, but is in the order of several hundreds of years.

Short-Term Changes

Three main types of changes occur over short time scales (from 5 to 100 years), which I have dubbed stochastic, successional, and (for want of a better word) interactive.

Stochastic

This class of changes includes all of those unpredictable, episodic events, which have a sudden, marked effect on the composition and structure of the savanna, and from which the savanna then recovers over a few to many years afterwards. Many of the apparent anomalies in savanna structure are undoubtedly due to such changes. Without a knowledge of these major past events, it is sometimes impossible to interpret the present (cf. also Heinselman Chapter 23; Zedler Chapter 24). Four examples of such disruptive events are rinderpest, major droughts, frost, and insect eruptions.

According to the time scales used in this chapter, rinderpest was, strictly speaking, a medium-term change, since the last outbreak occurred around the turn of the century. It resulted in much more than a decimation of at least all the large ungulates, and the current status and trend in, for example, the wildebeest population on the Serengeti plains can be attributed to this event (Sinclair and Norton-Griffiths 1979).

Major droughts occur at least a few times each century, and their effects persist for many years. During the droughts, hydromorphic grasslands are invaded by woody vegetation, which may or may not persist. On shallow lithosols many of the more susceptible tree species (e.g., *Commiphora* spp.) die. There is a general reduction in perennial grasses, and a concommitent increase in annual grasses when the drought is over.

Frost is patchily common, though generally light, over much of southern Africa. In 1968 unusually severe frosts occurred, including a sudden, two-day period in November, well after tree growth had commenced. Many thousands of large animals and countless birds perished, and the woody vegetation over vast areas was severely reduced in general height and in density. The effects in some areas are still discernible today.

Insect eruptions of many kinds are common throughout the savannas. They vary in magnitude from the massive locust outbreaks (which are gradually being eliminated) to more frequent and less extensive eruptions of more or less host-specific larvae of moth species such as *Gonimbrasia epimethea*, which attacks the tree *Colophospermum mopane*, and *Cirina fonda*, which attacks the tree *Burkea africana*. In some years, over areas of thousands of square kilometers, these tree species (and others like them with their own particular larvae) are completely defoliated. They may reflush and be defoliated again twice or even three times in one season.

Stochastic events of the kind outlined above give rise to effects that are not easily interpreted when encountered at some future time. They may also, however, cause changes that initiate succession, and in this way are related to the second type of short-term changes.

Successional

A typical feature of the semiarid savannas is invasion of grassland or open savanna by species of *Acacia* (e.g., *A. karroo*), usually on well-drained, shallow, medium-textured soils. They are vigorous when young, but are short-lived (about 50 years or even less). If protected from fire, these stands of trees gradually give way to longer-lived, broad-leaved species, many of them with fleshy fruits, which were brought in by birds. Once in this state, the composition of the vegetation remains fairly constant, unless it is burned. Fire reduces the density of trees and eliminates fire-sensitive species and, if sufficiently frequent and intense, eventually eliminates all the broad-leaved species. The stage is then set for the next invasion by *Acacia*.

Secondary succession after clear felling and cultivation is extremely widespread in the wetter savannas and woodlands, which comprise the main agricultural region of the subcontinent. A typical example is given by Strang (1974), for the miombo region in Zimbabwe. In many respects, the pattern of events is very similar to those for the northeastern forests of the United States, as described by Bormann and Likens (1979a,b). The main features are as follows. There is a rapid increase in basal area of woody vegetation up to about 50 years, whereafter it is slow. The density of trees increases up to 20 years, and then decreases. The herbaceous layer decreases up to 20 years and then remains constant. The height of trees initially increases rapidly up to 25 years, and then slowly increases until only very few trees above 13 m are encountered. With regard to species changes the most important feature is that two tree species (*Brachystegia spiciformis* and *Julbernardia globiflora*) remain characteristic dominants throughout the period. The total number of species increases (due to release from fire) for about 20 years. In the herbaceous layer the grass species change from typical early successional to the taller, "climax" grasses and then back again to the early species as the trees develop.

Succession after fire is influenced by the fact that only exceptionally are trees killed by fire. If the above-ground parts are destroyed, or severely reduced, root suckering leads to rapid redevelopment. Many studies (cf. the earlier section on fire) have demonstrated that woody species density remains much the same in regularly burned as opposed to unburned sites, and it is only the basal area and height that changes. Few data are available on the effects of fire on nutrient dynamics, but what evidence there is suggests that exchangeable bases in the soil are affected very little, but perhaps decrease slightly (Strang 1974). Depending on fire frequency, nitrogen in the soil remains much the same or also decreases, together with soil organic matter (Brookman-Amissah et al. 1980).

In general, therefore, it appears that fire may change the rate of any medium- or long-term successional trend, but not the direction. Its main effects are, on a regional basis, to maintain a higher grass-to-woody vegetation ratio than would otherwise exist, and to shift the ratio of fire-resistant to susceptible species in favor of the former.

Interactive

I come now to those short-term changes that do not fit easily into any existing theories of succession based on the amelioration or removal of some overriding factor. They comprise many of the changes occurring on normal well-drained sites over the extensive semiarid regions of Africa, and are a result of the interactions of variation in rainfall, herbivore pressure, nutrients and fire. It is more useful to consider the changes of this type in terms of a general theory of community dynamics, rather than to attempt their description in the framework of a successional trend.

For any successional explanation to be viable it is necessary that all of the vegetation stages claimed to be part of the succession can occur on any one site. For example, Boughey's (1963) hypothesis of the changes in the various "secondary communities" in the Wankie National Park, while plausible in many respects, includes communities that, from some of my own unpublished work, occur on inherently different sites, and could not be transformed one into the other.

The following account of savanna dynamics is based on analytical models, using insufficient field data, and is therefore admittedly still speculative. It seems, nevertheless, to be a plausible explanation of observed changes.

Extending the earlier hypothesis of semiarid savanna on normal well-drained soils as a stable joint equilibrium of grass (G) and woody vegetation (W), maintained by a split resource (soil water), Walker and Noy-Meir (in press) analyzed the dynamics of G and W based on existing knowledge of the factors influencing each. The results indicated that in the absence of herbivores, and provided that the state of the vegetation did not directly influence water input to the soil, such a single equilibrium was in fact the only outcome. Noy-Meir (in press) then included the effects of herbivores and showed that these can lead to multiple equilibria, and Walker et al. (1981) have developed a quantitative model which indicates that semiarid savannas on soils prone to surface capping are likely to have two stable equilibria (the joint G-W combination, and W alone) separated by an unstable G-W equilibrium.

The reasons for this are as follows. Owing to its exclusive access to sub-soil water the W zero-isocline in G-W space appears as in Fig. 25.2. No matter how much grass there is, some water always gets through to the subsoil and W cannot be eliminated by high G. At high W however, G is eliminated. On medium to heavy texture soils a surface cap develops when the surface is bare, and this results in up to a tenfold reduction in rainfall infiltration (Kelly and Walker 1976). For various reasons input to the subsoil is affected less, and the net effect is that at low levels of grass cover there is an inflection in the G zero isocline, resulting in the upper stable G-W equilibrium and the lower stable W equilibrium. The position of the separatrix, which passes through the unstable equilibrium and divides the G-W space into two domains of attraction, depends on the relative speeds at which G and W change.

The really interesting part of these dynamics which provide a somewhat novel view of savanna succession is demonstrated in Fig. 25.3. The zero isocline for G moves toward the origin as grazing pressure increases. Whether or not it can be reduced to below the W isocline depends on the

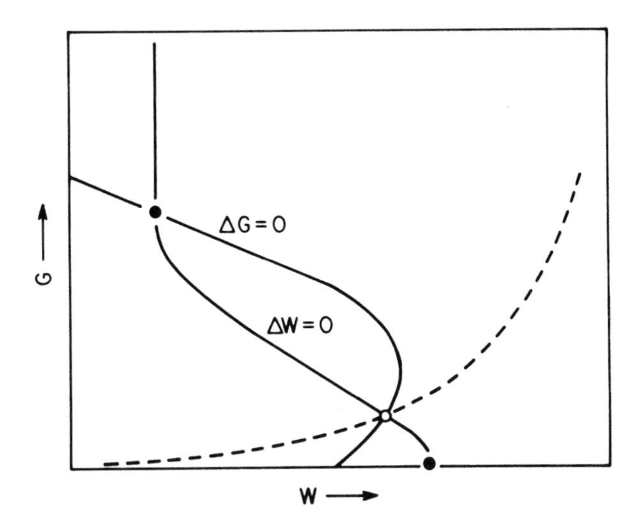

Figure 25.2. The zero isoclines for grass (G) and woody vegetation (W) in G-W space, giving rise to the two stable equilibria (●) and the unstable equilibrium (○). The dashed line indicates the position of the separatrix. See text for explanation.

minimum amount of G at which herbivore intake is unrestricted, and the grazing cutoff level (G_0), below which no grazing is possible. Walker et al. (1981) have discussed this in terms of changes in the resilience of the system (*sensu* Holling 1973). A high level of G_0 can only develop under heavy grazing pressures. It is achieved by species with large underground reserves, rhizomatous and stoloniferous growth, unpalatable and very coarse green parts, etc. Such plants are poor competitors with erect, leafy bunch grasses which channel most of their net growth into further photosynthetic tissue, and they therefore require at least intermittent periods of heavy grazing in order to persist (e.g., McNaughton 1979b). In the absence of herbivore pressure, the vegetation changes to a highly productive state but with much reduced resilience. Under heavy or "over"grazing the resulting changes, regarded everywhere as retrogressive succession or degradation, lead to a state with lower production by less "desirable" species, but one that is highly resilient to further grazing pressure.

The many examples of badly eroded savanna that have become scrub thicket, or where even the W has gone owing to severe erosion, indicate situations where the herbivores were able to graze the G down into the lower domain. Once in the lower, W-equilibrium state, it is remarkedly difficult to return the system to its original savanna condition. Grass is, of course, never entirely eliminated, owing to spatial and temporal heterogeneity in the amount of W, which provides small pockets where G

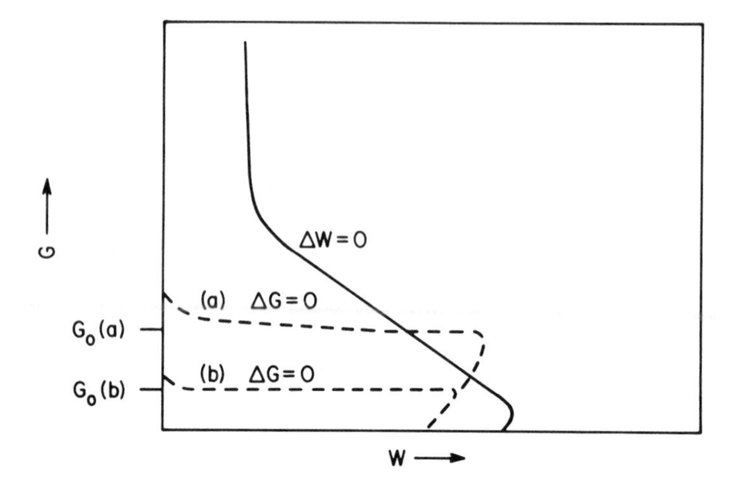

Figure 25.3. Differences in the position of the grass zero isocline under the same very high stocking rate: (a) with a high level of G below which no grazing can occur (G_0); and (b) with a lower level of G_0. In (a) both stable equilibria are retained. In (b) the only possible outcome is woodland or thicket.

may survive or invade. But it is insignificant and insufficient to carry a fire. Over a fairly long period of time (as yet unknown) intra-W competition will reduce the density of W, and leaf drop and accumulation of dead twigs, etc. will lead to the redevelopment of a topsoil with improved infiltration properties. Provided that grazing herbivores are excluded, the system may then redevelop, since the inflection in the G isocline will be much weakened. A strong browsing pressure will enhance the return, but over most of the savanna ranching areas browsers have been eliminated or reduced to insignificant levels.

A further mechanism whereby herbivores and fire influence savanna dynamics, and lead to different communities, can develop as a result of spatial heterogeneity. A possible example, which is currently being investigated, comes from the South African Savanna Ecosystem Project at Nylsvley in the northern Transvaal. This savanna occurs on infertile sands with a rainfall of about 600 mm/yr. The tree layer is dominated by *Burkea africana*, and the herbaceous layer by a coarse, unpalatable grass, *Eragrostis pallens.* Within this general vegetation are small patches (2–10 ha) dominated exclusively by *Acacia* spp. and the palatable grass *Cenchrus ciliaris.* All of these patches contain evidence of past human occupation, but they have been abandoned now for at least 50 years. Some have soil inherently different from the Burkea savanna, but others are very similar and apparently could have been *Burkea* in the past. Aerial photographs show that they have not changed at all over at least the last 20 years, and the following hypothesis has been put forward (Walker and Noy-Meir 1982) as an explanation for this remarkably stable situation. It is based on differences in nutrient import, and nutrient cycling and retention, induced by herbivores.

Burkea and *E. pallens* are well adapted to infertile sands, and under such conditions outcompete *Acacia* and *Cenchrus*. Under high soil nutrient levels *Acacia* and *Cenchrus* are dominant. With the establishment of a village, the existing *Burkea* vegetation was removed and cattle were concentrated on the area, being penned at night and herded close by during the day. The resulting concentration of nutrients, imported via dung and urine, led to the establishment of *Acacia* and *Cenchrus* when the village was abandoned. Wild and domestic herbivores show around a fivefold preference for these sites as opposed to the *Burkea* savanna (Zimmerman 1978), and though they move through the latter, eating off what is palatable, they graze the *Cenchrus* far more and concentrate nutrients in the *Acacia* areas. The soil effect on nutrient dynamics is depicted in Fig. 25.4. In the *Burkea* savanna there is a low stable equilibrium concentration of nutrients, where annual input (NI) is balanced by net annual uptake plus losses (NL). Any increase in input is leached away, and any reduction in soil nutrient level is gradually restored through precipitation input or weathering. The high nutrient levels in the *Acacia* patches are maintained by the concentration effect of the herbivores, which sustains a high input. There is also less leaching, plant growth (and therefore nutrient uptake) is much increased, and the vegetation has a higher nutrient content (which increases its palatability). The result is a high equilibrium concentration of nutrients. If nutrients are reduced for some reason, the herbivore concentration effect will raise them back to the level above, where further inputs are leached out.

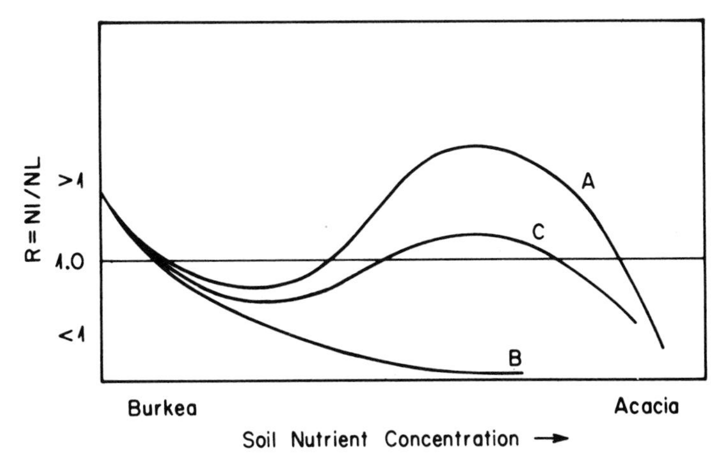

Figure 25.4. Soil nutrient recruitment curves for the suggested relationship between the *Burkea* and *Acacia* vegetation at Nylsvley in the northern Transvaal, South Africa: NI, nutrient inputs; NL, nutrient losses (uptake + leaching); A, high herbivore biomass; B, no herbivores; C, intermediate herbivore biomass.

Should the herbivores be reduced, nutrient import will decline, palatability will decrease, and the nutrient recruitment curve will move from A toward C, and finally (in the absence of herbivores) to B, as in Fig. 25.4. The middle, unstable equilibrium where the curve crosses the $R = 1$ line is a threshold point, and once the nutrient level drops below this point, it will continue to decline, as the vegetation changes toward the *Burkea* nutrient equilibrium. Even if the herbivore levels are reestablshed once the nutrient level is below the threshold, the attraction of the vegetation is so reduced that nutrient inputs will remain low and nutrient leaching will exceed inputs until the lower equilibrium is attained. A single, sudden increase in nutrients in *Burkea* savanna to above the threshold is unlikely to move the system to the upper equilibrium, as the *Burkea* vegetation is unlikely to be sufficiently palatable and attractive to animals (though this has not been tested). The high levels need to be maintained long enough for the vegetation to change to *Acacia.*

This hypothesis is admittedly speculative, but it agrees with a number of facts and offers an explanation for the existence of only the pure *Burkea* and pure *Acacia*, with no mixtures. It is also of interest that the presence of *Cenchrus ciliaris* in small patches, in many other parts of southern Africa, has been associated with archaeological sites that iron-age inhabitants abandoned several hundred years ago, and is now used to locate such sites (T. Huffman 1980, pers. comm.). If the hypothesis is true, the *Acacia* areas are maintained by a concentration effect, which means that there is some maximum proportion of the area that can be kept in this form.

A very similar, but less permanent, phenomenon occurs as a result of fire. Where large areas of savanna are lightly used by herbivores for several years, the less favored parts accumulate moribund growth, and become less and less used. The fuel load and state of the fuel (old, dry material) increase the probability of fire, and eventually it burns. The new flush of growth is highly palatable and induces a concentration of animals, which keeps the grass short, and maintains nutrient cycling. The effect can persist for some years until a fire elsewhere attracts the animals away.

The first two examples of interactive changes (the grass-woody and *Burkea-Acacia* changes) both represent ecosystems with multiple equilibria. It is neither useful nor even possible to attempt to represent these changes in the framework of succession. The systems are better understood if one sees them as having two domains of attraction around two stable equilibria. The equilibrium points will move around within the domains of attraction in response to climatic variation, fluctuations in animal numbers, etc., and the boundary between them will likewise shift. The position of the system itself will also be moving within one of the domains, and at any one time will be attracted toward the current position of the associated equilibrium. Given some fairly major disturbance (such as one of the stochastic events considered earlier), the relative positions of the system and the boundary may alter such that the system is now within the other domain of attraction, and the system will then move toward the second stable equilibrium.

In the case of the grass-woody balance, once the system has changed and is approaching its new physiognomic equilibrium, further changes may occur in species composition, and the latter changes may well be successional. In the *Burkea-Acacia* example (if, indeed, it is shown to be true), the concept of secondary succession cannot be invoked at any stage.

The third example, of accumulating moribund growth leading to fire, induced palatability, and maintenance of palatable vegetation by grazing more closely represents a cyclic phenomenon. Again it cannot be interpreted within the framework of succession. However, if the burn is severe, and the interval between fires is sufficiently long, then succession (as described previously) may occur.

Conclusions

Within the savannas of Southern and East Africa various different kinds of changes are occurring concurrently at different time scales. Some are directional, some cyclic and others are unpredictable. Seldom is any one type so predominant as to give rise to a single, clear picture of change over a large region. The situation more closely resembles an apparently confusing mixture of communities of various sizes, arranged in different patterns at different scales, and each of which may be changing in both an orderly and disorderly fashion, at different rates. This same general conclusion has been reached by others, and Tinley (1977), for example, recognized three kinds of concurrent multidirectional succession with feedbacks between them, namely substrate (either geomorphic surface replacement of edaphic changes *in situ*), biotic (either spatial, or *in situ* within a community), and evolutionary including long-term climatic and diastrophic (sic) changes.

In answer, therefore, to the question posed in the title, yes, succession is a viable concept in savannas, but it is not clear-cut and obvious, of the classical kind. It is multidirectional, occurring over different time scales, and is confounded by equally important nondirectional changes of other kinds. Within established natural savanna (i.e., excluding the secondary succession after cultivation), the most frequently occurring and easily discernible successional changes are short-term changes due to fire, followed by those due to changes in the soil water balance. However, when the other types of interactive changes are more fully understood, their significance, relative to the better-known successional changes, may well increase.

References

Abbott, H. G., and T. F. Quink. 1970. Ecology of ecosystem white pine seed caches made by small forest mammals. Ecology 51:271–278.

Aber, J. S., D. B. Botkin, and J. M. Melillo. 1978. Predicting the effects of different harvesting regimes on forest floor dynamics in northern hardwoods. Can. J. For. Res. 8:305–315.

Aber, J. S., D. B. Botkin, and J. M. Melillo. 1979. Predicting the effects of different harvesting regimes on productivity and yield in northern hardwoods. Can. J. For. Res. 9:10–14.

Aber, J. S., and J. M. Melillo. 1980. Litter decomposition: Measuring relative contributions of organic matter and N to forest soils. Can. J. Bot. 58:416–421.

Abraham, B. J. 1980. Spatial and temporal changes in spider communities. Ph.D. Thesis. Utah State University, Logan.

Ahlgren, C. E. 1959. Some effects of fire on forest reproduction in northeastern Minnesota. J. For. 57:194–200.

Ahlgren, C. E. 1960. Some effects of fire on reproduction and growth of vegetation in northeastern Minnesota. Ecology 41:431–445.

Ahlgren, I. F., and C. E. Ahlgren. 1960. Ecological effects of forest fires. Bot. Rev. 26:483–533.

Ahti, T. 1959. Studies on the caribou lichen stands of Newfoundland. Ann. Bot. Soc. Zool. 'Vanamo' Fenn. 30 N:4, 44 pp.

Alexander, M. 1961. Introduction to Soil Microbiology. Wiley, New York. 472 pp.

Allen, D. L. 1952. Wildlife Management. Boy Scouts of America, North Brunswick, New Jersey. 94 pp.

Amman, G. 1977. The role of the mountain pine beetle in lodgepole pine ecosystems: Impact on succession. pp. 3–18. In W. J. Mattson (ed.), The Role of Anthropods in Forest Ecosystems. Springer-Verlag, New York.

Amundson, D. A., and H. E. Wright. 1979. Forest changes in Minnesota at the end of the Pleistocene. Ecol. Monogr. 49:1–16.

Anaya-Lang, A. L. 1976. Considerations sobre el potencial alelopatico de la vegetacion secundria. pp. 428–556. In A. Gomez-Pompa, C. Vazquez-Yanes, S. del Amo, and A. Butanda (eds.), Regeracion de Selvas. ÇECSA, Mexico, D.F.

Anaya-Lang, A. L., and M. Ravalo. 1976. Alelopatia en plantas superiores. pp. 388–427. In A. Gomez-Pompa, C. Vazquez-Yanes, S. del Amo, and A. Butanda (eds.), Regeracion de Selvas. CECSA, Mexico, D.F.

Andersen, D. C., J. A. MacMahon, and M. L. Wolfe. 1980. Herbivorous mammals along a montane sere: Community structure and energetics. J. Mammal. 61:500–519.

Andersen, S. T. 1970. The relative pollen productivity and pollen representation of north European trees, and correction factors for tree pollen spectra. Dan. Geol. Unders. Ser. 11, 96, 99 pp.

Anderson, D. W. 1977. Early stages of soil formation on glacial till mine spoils in a semi-arid climate. Geoderma 19:11–19.

Anderson, D. W. 1979. Process of humus formation and transformation in soils of the Canadian Great Plains. J. Soil Sci. 30:77–84.

Anderson, G. D., and B. H. Walker. 1974. Vegetation composition and elephant damage in the Sengwa wildlife research area, Rhodesia. J. S. Afr. Wildlife Management Assoc. 4:1–14.

Anderson, J. M. 1978. The organization of soil animal communities. pp. 15–23. In V. Lohm and T. Persson (eds.), Soil Organisms as Components of Ecosystems. Ecol. Bull. 25. Swedish Natural Science Research Council, Stockholm.

Anderson, S. T. 1967. Tree pollen rain in a mixed deciduous forest in South Jutland (Denmark). Rev. Palaeobotan. Palynol. 3:267–275.

Anon. 1968. Review of Australia's water resources—monthly rainfall and evaporation. Bur. Meteorol., Melbourne.

Anon. 1973. A resource and management survey of the Cotter River catchment. Report prepared for the Forest Branch, Department of Capital Territory by the Consultancy Group, Department of Forestry, Australian National University, Canberra.

Anon. 1975. Climatic averages, Australia. Bur. Meteorol., Canberra.

Applequist, M. B. 1941. Stand composition of upland hardwood forests as related to soil type in the Duke Forest. M.S. Thesis. Duke University, Durham, North Carolina.

Archibold, O. W. 1979. Buried viable propagules as a factor in post-fire regeneration in northern Saskatchewan. Can. J. Bot. 57:54–58.

Arno, S. F. 1976. The historical role of fire on the Bitterroot National Forest. USDA For. Serv. Res. Note INT-187. Intermountain Forest and Range Experiment Station. Ogden, Utah. 29 pp.

Ashe, W. W. 1897. Forests of North Carolina. N.C. Geol. Survey Bull. 6:139–224.

Ashe, W. W. 1911. Chestnut in Tennessee. Tennessee Geol. Survey Ser. 10B. 35 pp.

Ashton, D. H. 1979. Fire in tall open forests (wet schlerophyll forests). In A. M. Gill, R. H. Groves, and I. R. Noble (eds.), Fire in the Australian Biota. Proc. MAB Conf. Fire in Australian Biota, Canberra, ACT. (in press).

Ashton, P. S. 1978. Crown characteristics of tropical trees. pp. 591–615. In P. B. Tomlinson and M. H. Zimmermann (eds.), Tropical Trees as Living Systems. Cambridge University Press, Cambridge, England.

Assman, E. 1970. The principles of forest yield study: Studies in the organic production, structure increment, and yield of forest stands. Pergamon Press, Oxford. 506 pp.

Atkin, D. 1980. The age structures and origins of trees in a New Jersey floodplain forest. Ph.D. Thesis. Princeton University, Princeton, New Jersey.

Attiwill, P. M. 1979. Nutrient cycling in a *Eucalyptus obliqua* (L'Herit.) Forest. III. Growth, biomass, and net primary production. Aust. J. Bot. 27:439–458.

Aubreville, A. M. A. 1938. La foret coloniale: Les forets de l'Afrique occidentale francaise. Ann. Acad. Sci. Colon., Paris 9:1–245.

Aubreville, A. M. A. 1947. Les brousses secondaires eu Afrique equatoriale, Côte d'Ivoire, Cameroun. A.E.F. Bois For. Trop. 2:24–35.

Auchmoody, L. R. 1979. Nitrogen fertilization stimulates germination of dormant pin cherry seed. Can. J. For. Res. 9:514–516.

Ayala, F. J., and T. Dobzhansky. 1974. Studies in the Philosophy of Biology: Reduction and Related Problems. University of California Press, Berkeley. 390 pp.

Bach, C. E. 1980. Effects of plant diversity and time of colonization on an herbivore-plant interaction. Oecologia 44:319–326.

Bailey, R. E. 1972. Vegetation history of northwest Indiana. Ph.D. Thesis. Indiana University, Bloomington, Indiana.

Bailey, R. E. 1977. Pollen stratigraphy, Wintergreen Lake. In Handbook for Paleoecology Field Trip in Central Lower Michigan. Ecological Society of America, AIBS, East Lansing, Michigan.

Bailey, R. L., and T. R. Dell. 1973. Quantifying diameter distributions with the Weibull function. For. Sci. 19:97–104.

Ball, L. C. 1924. Report on oil prospecting near Tewantin. Queensland Gov. Mining J. 25:354–62.

Barden, L. S. 1979. Tree replacement in small canopy gaps of a *Tsuga canadensis* forest in the Southern Appalachians, Tennessee. Oecologia 44:141–142.

Barden, L. S. 1980. Tree replacement in a small canopy gaps of a cove hardwood forest in the southern Appalachians. Oikos 35:16–19.

Barney, R. J. 1971. Wildfires in Alaska—Some historical and projected effects and aspects. pp. 51–59. In Fire in the Northern Environment. Symposium Proc. (Fairbanks, Alaska, 1971).

Barney, R. J., and K. Van Cleve. 1973. Black spruce fuel weight and biomass in two interior Alaska stands. Can. J. For. Res. 3:304–311.

Barney, R. J., K. Van Cleve, and R. Schlentner. 1978. Biomass distribution and crown characteristics in two Alaskan *Picea mariana* ecosystems. Can. J. For. Res. 8:36–41.

Barrett, L. E., and A. A. Downs. 1943. Hardwood invasion of pine forests of the Piedmont plateau. J. Agr. Res. 67:111–127.

Bartholomew, W. V., J. Meyer, and H. Laudelout. 1953. Mineral nutrient inmobilization under forest and grass fallow in the Yangambi (Belgian Congo) region. Publ. Inst. Etude Agron. Congo Belgue Ser. Sci. No. 57. 27 pp.

Baskerville, G. L. 1965. Dry-matter production in immature balsam fir stands. For. Sci. Monogr. 9:42 pp.

Bate, G. C., and C. Gunton. 1982. Nitrogen in the Burkea savanna. In B. J. Huntley and B. H. Walker (eds.), Ecology of Tropical Savannas. Springer-Verlag, Heidelburg. (in press).

Bates, M. 1960. Ecology and evolution. pp. 547–568. In S. Tax (ed.), Darwin and

after Darwin, Vol. I. University of Chicago Press, Chicago.
Batistic, L., J. M. Sarker, and J. Mayaudon. 1980. Extraction, purification and properties of soil hydrolases. Soil. Biol. Biochem. 12:59–63.
Batzer, H. O. 1969. Forest character and vulnerability of balsam fir to spruce budworm in Minnesota. For. Sci. 6:194–199.
Batzli, G. O. 1975. The role of small mammals in arctic ecosystems. pp. 243–268. In F.B. Golley, K. Petrusewicz, and L. Ryszkowski (eds.), Small Mammals: Their Productivity and Population Dynamics. International Biological Programme, Vol. 5. Cambridge University Press, New York.
Bazzaz, F. A. 1979. The physiological ecology of plant succession. Ann. Rev. Ecol. Syst. 10:351–371.
Bazzaz, F. A., and S. T. A. Pickett. 1980. The physiological ecology of tropical succession: A comparative review. Ann. Rev. Ecol. Syst. 11:287–310.
Beard, J. S. 1944. Climax vegetation in tropical America. Ecology 25:127–158.
Beard, J. S. 1945. The progress of plant succession of the Soufriere of St. Vincent. J. Ecol. 33:1–9.
Beard, J. S. 1955. The classification of tropical American vegetation types. Ecology 36:89–100.
Beard, J. S. 1976. The progress of plant succession of the Soufriere of St. Vincent: Observations in 1972. Vegetatio 31:69–77.
Beattie, A. J., D. C. Culver, and R. J. Pudlo. 1979. Interactions between ants and the diaspores of some common spring flowering herbs in West Virginia. Castanea 44:177–186.
Beattie, A. J., and N. Lyons. 1975. Seed dispersal in *Viola* (Violaceae): Adaptations and strategies. Am. J. Bot. 62:714–722.
Beatty, S. Personal communication. Department of Ecology and Systematics, Cornell University, Ithaca, New York. April 25, 1980.
Beaufait, W. R. 1960. Some effects of high temperatures on the cones and seeds of jack pine. For. Sci. 6:194–199.
Bedard, J., M. Crete, and E. Audy. 1978. Short-term influence of moose upon woody plants of an early seral wintering site in Gaspe Peninsula, Quebec. Can. J. For. Res. 8:407–415.
Bell, R. H. V. 1970. The use of the herb layer by grazing ungulates in the Serengeti. In A. Watson (ed.), Animal Populations in Relation to Their Food Resources. Brit. Ecol. Soc. Symp. No. 10:111–123. Blackwells, Oxford.
Bell, R. H. V. 1982. The effect of soil nutrient availability on community structure in African ecosystems. In B. J. Huntley and B. H. Walker (eds.), Ecology of Tropical Savannas Springer-Verlag, Heidelburg. (in press).
Bendell, J. F. 1974. Effects of fire on birds and mammals. pp. 73–138. In T. T. Kozlowski and C.E.Ahlgren (eds.), Fire and Ecosystems. Academic Press, New York.
Benninghoff, W. S., and K. J. Cramer. 1963. Phytosociological analysis of aspen communities on three site classes for *Populus grandidentata* in western Cheboygan County, Michigan. Vegetatio 11:253–264.
Bergerud, A. T. 1974. Decline of caribou in North America following settlement. J. Wildl. Manage. 38:757–770.
Bergthorsson, P. 1962. Preliminary notes on past climate of Iceland. (Data presented to NCAR-USAF). Conference on the Climate of the 11th and 12th Centuries. Aspen, Colorado.

Bernabo, J. C., and T. Webb, III. 1977. Changing patterns in the Holocene pollen record of northeastern North America: A mapped summary. Quat. Res. 8:69–96.

Bevege, D. I. 1978. Biomass and nutrient distribution in indigenous forest ecosystems. Dept. For. Queensland Tech. Pap. No. 6.

Billings, W. D. 1938. The structure and development of old field shortleaf pine stands and certain associated properties of the soil. Ecol. Monogr. 8:437–499.

Birks, H. H. 1973. Modern macrofossil assemblages in lake sediments in Minnesota. pp. 9–15. In H. J. B.Birks and R. C. West (eds.), Quaternary Plant Ecology. Blackwell, Oxford.

Birks, H. J. B. 1976. Late-Wisconsonian vegetational history at Wolf Creek, central Minnesota. Ecol. Monogr. 46:395–429.

Birks, H. J. B., and S. Pegler. 1980. Identification of *Picea* pollen of late Quaternary age in eastern North America: A numerical approach. Can. J. Bot. 58:2043–2058.

Biswell, H. H. 1974. Effects of fire on the chaparral. pp. 321–364. In T. T. Kozlowski and C. E. Ahlgren (eds.), Fire and Ecosystems. Academic Press, New York.

Black, R. A., and L. C. Bliss. 1978. Recovery sequence of *Picea mariana-Vaccinium uliginosum* forests after burning near Inuvik, Northwest Territories, Canada. Can. J. Bot. 56:2020–2030.

Blais, J. R. 1954. The recurrence of spruce budworm infestations in the past century in the Lac Seul area of northwestern Ontario. Ecology 35:62–71.

Blais, J. R. 1958. The vulnerability of balsam fir to spruce budworm attack in northern Ontario, with special reference to the physiological age of the tree. For. Chronol. 34:405–422.

Blais, J. R. 1965. Spruce budworm outbreaks in the past three centuries in the Laurentide Park, Quebec. For. Sci. 11:130–138.

Bledsoe, L. J., and G. M. Van Dyne. 1971. A compartment model simulation of secondary succession. pp. 479–511. In B. C. Patten (ed.), Systems Analysis and Simulation in Ecology, Vol. 1. Academic Press, New York.

Bliss, L. C. 1971. Arctic and alpine plant life cycles. Ann. Rev. Ecol. Syst. 2:405–438.

Bliss, L. C. 1975. Tundra grasslands, herblands, and shrublands and the role of herbivores. Geosci. Man 10:51–79.

Bliss, L. C., G. M. Courtin, D. L. Pattie, R. R. Riewe, D. W. A. Whitfield, and P. Widden. 1973. Arctic tundra ecosystems. Ann. Rev. Ecol. Syst. 4:359–399.

Bloomberg, W. G. 1950. Fire and spruce. For. Chronol. 26:157–161.

Blum, K. E. 1968. Contributions toward understanding of vegetational development in the pacific lowlands of Panama. Ph.D. Thesis, Florida State University, Tallahassee, Florida.

Bocock, K. L. 1963a. The digestion and assimilation of food by *Glomeris.* pp. 137–148. In J. Doeksen and J. van der Drift (eds.), Soil Organisms. North-Holland Publ., Amsterdam.

Bocock, K. L. 1963b. Changes in the amount of nitrogen in decomposing leaf litter of sessile oak. J. Ecol. 51:555–566.

Bond, W., M. Ferguson, and G. Forsyth. 1980. Small mammals and habitat structure along altitudinal gradients in the southern Cape Mountains. S. Afr. J. Zool. 15:34–43.

Bormann, F. H. 1953. Factors determining the role of loblolly pine and sweetgum in early old field succession in the Piedmont of North Carolina. Ecol. Monogr. 23:339–358.

Bormann, F. H., and G. E. Likens. 1979a. Pattern and Process in a Forested Ecosystem. Springer-Verlag, New York. 253 pp.

Bormann, F. H., and G. E. Likens. 1979b. Catastrophic disturbance and the steady state in northern hardwood forests. Am. Sci. 67:660–669.

Botkin, D. B. 1976. The role of species interactions in the response of a forest ecosystem to environmental perturbation. pp.147–171. In B. C. Patten (ed.), Systems Analysis and Simulation in Ecology, Vol. IV. Academic Press, New York.

Botkin, D. B., S. Golubek, B. Maguire, B. Moore, III, H. J. Morowitz, and L. B. Slobodkin. 1979. Closed regenerative life support systems for space travel: Their development poses fundamental questions for ecological science. pp. 3–12. In R. Holmquist (ed.), Life Sciences and Space Research. Vol. XVII. Pergamon Press, Oxford.

Botkin, D. B., J. F. Janak, and J. R. Wallis. 1970. Rationale, limitations and assumptions of a northeast forest growth simulator. IBM J. Res. Develop. 16:101–116.

Botkin, D. B., J. F. Janak, and J. R. Wallis. 1972. Some ecological consequences of a computer model of forest growth. J. Ecol. 60:849–872.

Botkin, D. B., J. M. Melillo, and L. S. Wu. 1981. How ecosystem processes are linked to large mammal population dynamics. In C. Fowler (ed.), Population Dynamics of Large Mammals. Wiley, New York. (in press).

Boughey, A. S. 1963. Interactions between animals, vegetation and fire in Southern Rhodesia. Ohio J. Sci. 63:193–209.

Bourdeau, P. 1954. Oak seedling ecology determining segregation of species in Piedmont oak-hickory forests. Ecol. Monogr. 24:297–320.

Bourliere, F., and M. Hadley. 1970. The ecology of tropical savannas. Ann. Rev. Ecol. Syst. 1:125–152.

Bowen, G. D. 1973. Mineral nutrition of ectomycorrhizae. pp. 151–205. In G. C. Marks and T. T. Kozlowski (eds.), Ectomycorrhizae. Academic Press, New York.

Bowen, G. D. 1980. Coping with low nutrients. In J. S. Pate and A. J. McComb (eds.), The Biology of Australian Native Plants. University Western Australia Press, Perth. (in press).

Bowler, J. M., G. S. Hope, J. N. Jennings, G. Singh, and D. Walker. 1976. Late Quaternary climates of Australia and New Guinea. Quat. Res. 6:359–394.

Brahmachary, R. L. 1980. On the germination of seeds in the dung balls of the African elephant in the Virunga National Park. Rev. Ecol. (Terre Vie) 34:139–142.

Bratton, S. P. 1976a. The response of understory herbs to soil depth gradients in high and low diversity communities. Bull. Torrey Bot. Club 103:165–172.

Bratton, S. P. 1976b. Resource division in an understory herb community: Responses to temporal and microtopographic gradients. Am. Nat. 110:679–693.

Braun, E. L. 1950. Deciduous forests of eastern North America. Hafner Press, New York. (Reprinted 1974) 596 pp.

Bremner, J. M., and A. M. Blackmer. 1979. Effects of acetylene and soil water content on emission of nitrous oxide from soils. Nature 280:380–381.

Brewer, R., and P. G. Merritt. 1978. Windthrow and tree replacement in a climax beech-maple forest. Oikos 30:149–152.
Brewer, R., and C. H. Thompson. 1979. Morphology of two sub-tropical podzols formed on a siliceous dune in coastal Queensland. pp. 179–180. Proc., Committee IV and V, Int. Soil Sci. Soc. CLAMATROPS Conf. (August 1977), Kuala Lumpur, Malaysia.
Brian, M. V. 1978. Production Ecology of Ants and Termites. International Biological Programme 13. Cambridge University Press, Cambridge, England.
Bridge, B. J. (personal communication). Division of soils, CSIRO, Brisbane, Queensland, Australia. April 25, 1980.
Brinkman, W. A. R. 1976. Surface temperature trend for the northern hemisphere—Updated. Quat. Res. 6:355–358.
Broecker, W. S., and J. Van Donk. 1970. Insulation changes, ice volumes and the ^{18}O record in deep sea cores. Rev. Geophys. Space Phys. 8:169–198.
Brookman-Amissah, J., J. B. Hall, M. D. Swaine, and J. T. Attakorah. 1980. A reassessment of a fire protection experiment in northeastern Ghana savanna. J. Appl. Ecol. 17:85–100.
Brown, J. H., D. W. Davidson, and O. J. Reichman. 1979. An experimental study of competition between seed-eating desert rodents and ants. Am. Zool. 19:1129–1143.
Brown, J. K. 1975. Fire cycles and community dynamics in lodgepole pine forests. pp. 428–456. In B. M. Baumgartner (ed.), Management of Lodgepole Pine Ecosystems. Washington State University Extension Service, Pullman, Washington.
Brown, R. J. E. 1970. Permafrost in Canada—Its Influence on Northern Development. University of Toronto Press, Toronto. 234 pp.
Brown, R. T., and J. T. Curtis. 1952. The upland conifer-hardwood forests of northern Wisconsin. Ecol. Monogr. 22:217–234.
Brubaker, L. B. 1975. Post-glacial forest patterns associated with till and outwash in Northcentral Upper Michigan. Quat. Res. 5:499–527.
Bryson, R. A., and F. K. Hare. 1974. Climates of North America. Elsevier, Amsterdam. 420 pp.
Buckman, R. E. 1964. Effects of prescribed burning on hazel in Minnesota. Ecology 45:626–629.
Budowski, G. 1959. The ecological status of fire in tropical American lowlands. Bol. Mus. Cienc. Nat. pp. 113–127.
Budowski, G. 1961. Studies on forest succession in Costa Rica and Panama. Ph.D. Thesis, Yale University, New Haven, Connecticut.
Budowski, G. 1965. Distribution of tropical American rain forest species in the light of successional process. Turrialba 15:40–42.
Budowski, G. 1970. The distinction between old secondary and climax species in tropical Central American lowland forests. Trop. Ecol. 11:44–48.
Buechner, H. K., and H. C. Dawkins. 1961. Vegetation change induced by elephants and fire in Murchison Falls National Park, Uganda. Ecology 42:752–766.
Bullock, P., and A. J. Thomasson. 1979. Rothamsted Studies of Soil Structure. II. Measurement and characterization of macroporosity by image analysis and comparison with data from water retention measurements. J. Soil. Sci. 30:391–413.
Burgess, I. P., A. Floyd, J. Kikkawa, and V. Pattemore. 1975. Recent develop-

ments in the silviculture and management of sub-tropical rainforest in New South Wales. Proc. Ecol. Soc. Aust. 9:74–84.

Burgess, P. F. 1970. An approach towards a silvicultural system for the hill forests of the Malay peninsula. Malay. For. 33:126–134.

Bykov, B. A. 1974. Fluctuations in the semi-desert and desert vegetation of the Turanian plain. pp. 245–251. In R. Knapp (ed.), Vegetation Dynamics. W. Junk, The Hague.

Cain, S. A. 1935. Studies on virgin hardwood forest: III. Warren's Woods, a beech-maple climax forest in Berrien County, Michigan. Ecology 16:500–513.

Cain, S. A. 1939. The climax and its complexities. Am. Midland Nat. 21:146–181.

Campbell, W. A. 1938. Preliminary report on decay in sprout northern hardwoods in relation to timber stand improvement. USDA For. Serv. Occas. Pap. 7. Northeastern Forest Experiment Station, Upper Darby, Pennsylvania. 8 pp.

Canham, C. D. 1978. Catastrophic windthrow in the hemlock-hardwood forest of Wisconsin. M.S. Thesis, University of Wisconsin, Madison, Wisconsin.

Castro, A. R., and S. Guevara S. 1976. Viabilidad de semillas en muestrans de suelo almacenado de "Los Tuxtlas," Veracruz. pp.233–249. In A. Gomez-Pompa, C. Vazquez-Yanes, S. del Amo and A. Butanda (eds.), Regeneracion de Selvas. CECSA, Mexico, D.F.

Cates, R. G., and G. H. Orians. 1975. Successional status and the palatability of plants to generalized herbivores. Ecology 56:410–418.

Cattelino, P. U., I. R. Noble, R. O. Slatyer, and S. R. Kessell. 1979. Predicting the multiple pathways of plant succession. Environ. Management 3:41–50.

Caughley, G. 1976. The elephant problem—An alternative hypothesis. E. Afr. Wildlife J. 14:265–283.

Chapin, F. S., III. 1980. The mineral nutrition of wild plants. Ann. Rev. Ecol. Syst. 11:233–260.

Chapin, F. S., III., and K. Van Cleve. 1981. Plant nutrient absorption and retention under differing fire regimes. pp. 303–324. In H. A. Mooney N. L. Christensen, J. E. Lotan, and W. A. Reiners (eds.), Fire Regimes and Ecosystem Properties. USDA Gen. Tech. Rep. WO-26, Washington, D.C.

Chettleburgh, M. R. 1952. Observations on the collection and burial of acorns by jays in Hainault Forest. Brit. Birds 45:359–364.

Chew, R. M. 1974. Consumers as regulators of ecosystems: An alternative to energetics. Ohio J. Sci. 74:359–370.

Chew, R. M. 1978. The impact of small mammals on ecosystem structure and function. pp. 167–191. In D. P. Snyder (ed.). Small Mammal Populations. Special Publ. No. 5. Pymatuning Laboratory of Ecology. University of Pittsburgh, Pittsburgh, Pennsylvania.

Churchill, E. D., and H. C. Hanson. 1958. The concept of climax in arctic and alpine vegetation. Bot. Rev. 24:127–191.

Clareholm, M., and T. Rosswall. 1980. Biomass and turnover of bacteria in a forest soil and a peat. Soil Biol. Biochem. 12:49–57.

Clark, F. E. 1977. Internal cycling of ^{15}N in shortgrass prairie. Ecology 58:1322–1333.

Clayton, H. H. 1944. World weather records. Smithsonian Misc. Collect. 79:1–1199.

Clayton, H. H., and F. L. Clayton. 1947. World weather records 1931–1940.

Smithsonian Misc. Collect. 105:1–646.
Clements, F. E. 1910. The life history of lodgepole burn forests. USDA For. Sci. Bull. 79. 56 pp.
Clements, F. E. 1916. Plant succession: An analysis of the development of vegetation. Carnegie Inst. Pub. 242. Washington, D.C. 512 pp.
Clements, F. E. 1928. Plant succession and indicators. Wilson, New York. 453 pp.
Clements, F. E. 1936. Nature and structure of the climax. J. Ecol. 24:252–284.
Clough, K. S., and J. C. Sutton. 1978. Direct observation of fungal aggregates in sand dune soil. Can. J. Microbiol. 4:333–335.
Coaldrake, J. E. 1962. The coastal sand dunes of southern Queensland. Proc. R. Soc. Queensland 72:101–116.
Coaldrake, J. E. 1968. Quaternary climates in north east Australia and some of their effects. Proc. R. Soc. Queensland 79:5–18.
Coile, T. S. 1940. Soil changes associated with loblolly pine succession on abandoned land of the Piedmont plateau. Duke Univ. School For. Bull. 5. 85 pp.
Coile, T. S. 1949. Effect of soil on the development of hardwood understories in pine stands of the Piedmont plateau. Soil Sci. Soc. Am. Proc. 1949:350–352.
Cole, D. W. 1977. Ecosystem dynamics in the coniferous forest of the Willamette Valley, Oregon, U.S.A. J. Biogeogr. 4:181–192.
Cole, D. W., S. P. Gessel, and J. Turner. 1978. Comparative mineral cycling in red alder and Douglas-fir. pp. 327–336. In D. G. Briggs, D. S. DeBell, and W. A. Atkinson (eds.), Utilization and Management of Alder. USDA For. Serv. Gen. Tech. Rep. PNW-70. Pacific Northwest Forest and Range Experiment Station, Portland, Oregon.
Cole, D. W., and M. Rapp. 1980. Elemental cycling in forested ecosystems. pp. 341–409. In D. E. Reichle (ed.), Dynamic Properties of Forest Ecosystems. Cambridge University Press, Cambridge, England.
Cole, M. M. 1982. The influence of soils, geomorphology and geology on the distribution of plant communities in savanna ecosystems. In B. J. Huntley and B. H. Walker (eds.), Ecology of Tropical Savannas. Springer-Verlag, Heidelburg. (in press).
Colinvaux, P. A. 1973. Introduction to ecology. John Wiley and Sons, New York. 621 pp.
Connell, J. H. 1978. Diversity in tropical rain forests and coral reefs. Science 199:1302–1309
Connell, J. H. 1979. Tropical rain forests and coral reefs as open non-equilibrium systems. pp. 141–163. In R. M. Anderson, B. O. Turner, and L. R. Taylor (eds.), Population Dynamics. Blackwell, Oxford.
Connell, J. H., and E. Orias. 1964. The ecological regulation of species diversity. Am. Nat. 98:399–441.
Connell, J. H., and R. O. Slatyer. 1977. Mechanisms of succession in natural communities and their role in community stability and organization. Am. Nat. 111:1119–1144.
Cook, R. F. 1977. Raymond Lindeman and the trophic dynamic concept in ecology. Science 193:22–26.
Coombe, D. E. 1960. An analysis of the growth of *Trema guineensis.* J. Ecol. 48:219–231.
Coombe, D. E., and W. Hadfield. 1962. An analysis of the growth of *Musanga cecropioides.* J. Ecol. 50:221–234.

Cooper, W. S. 1913. The climax forest of Isle Royale, Lake Superior, and its development. Bot. Gaz. 15:1–44, 115–140, 189–235.
Cooper, W. S. 1926. The fundamentals of vegetational change. Ecology 7:391–413.
Corner, E. J. H. 1932. Wayside trees of Malaya. Government Printing Office, Singapore.
Costin, A. B. 1954. A study of the ecosystems of the Monaro Region of New South Wales. Government Printer, Sydney. 657 pp.
Covington, W. W., and J. D. Aber. 1980. Leaf production during secondary succession in northern hardwoods. Ecology 61:200–204.
Cowles, H. C. 1899. The ecological relations of the vegetation on the sand dunes of Lake Michigan. Bot. Gaz. 27:95–117, 167–202, 281–308, 361–369.
Crafton, W. M., and B. W. Wells. 1934. The old field prisere: An ecological study. J. Elisha Mitchell Sci. Soc. 49:225–246.
Craig, A. J. 1970. Vegetational history of the Shenandoah Valley, Virginia. Geol. Soc., Spec. Pap. 123:283–297.
Craig, A. J. 1972. Pollen influx to laminated sediments: A pollen diagram from northeastern Minnesota. Ecology 53:46–57.
Crocker, R. L., and J. Major. 1955. Soil development in relation to vegetation and surface age at Glacier Bay, Alaska. J. Ecol. 43:427–448.
Cromack, K., Jr., C. C. Delwiche, and D. H. McNabb. 1979a. Prospects and problems of nitrogen management using symbiotic nitrogen fixers. pp. 210–223. In J. C. Gordon, C. T. Wheeler, and D. A. Perry (eds.), Symbiotic Nitrogen Fixation in the Management of Temperate Forests. Oregon State University Press, Corvallis, Oregon.
Cromack, K., Jr., P. Sollins, W. C. Graustein, K. Speidel, A. W. Todd, G. Spycher, C. Y. Li, and R. L. Todd. 1979b. Calcium oxalate accumulation and soil weathering in mats of the hypogeous fungus *Hysterangium crassum.* Soil Biol. Biochem. 11:463–468.
Cromack, K., Jr., and C. D. Monk. 1975. Litter production, decomposition and nutrient cycling in a mixed hardwood watershed and a white pine watershed. pp. 609–624. In F. G. Howell, J. B. Gentry, and M. H. Smith (eds.), Mineral Cycling in Southeastern Ecosystems. CONF-740513. Technical Information Center, Oak Ridge, Tennessee.
Cromartie, W. J., Jr. 1975. The effect of stand size and vegetational background on the colonization of cruciferous plants by herbivorous insects. J. App. Ecol. 12:517–534.
Crow, T. R. 1980. A rainforest chronicle: A 30-year record of change in structure and composition at El Verde, Puerto Rico. Biotropica 12:42–55.
Crow, T. R., and D. F. Grigal. 1980. A numerical analysis of arborescent communities in the rain forest of the Luquillo Mountains, Puerto Rico. Vegetatio 40:135–146.
Crow, T. R., and P. L. Weaver. 1977. Tree growth in a moist tropical forest of Puerto Rico. USDA For. Serv. Res. Pap. ITF-22. Institute of Tropical Forestry, Rio Piedras, Puerto Rico. 18 pp.
Cumming, D. M. 1982. The influence of large herbivores on savanna structure in Africa. In B. J. Huntley and B. H. Walker (eds.), Ecology of Tropical Savannas. Springer-Verlag, Heidelburg. (in press).
Curtin, M. E. 1978. Tree succession redux. B.A. Thesis. Princeton University,

Princeton, New Jersey.
Cushing, E. J. 1967. Late Wisconsin pollen statigraphy and the glacial sequence in Minnesota. pp. 59–88. In E. J. Cushing and H. E. Wright, Jr. (eds.), Quaternary Paleoecology. Yale University Press, New Haven, Connecticut.
Cushwa, C. T., R. E. Martin, and R. L. Miller. 1968. The effects of fire on seed germination. J. Range Management 21:250–254.
Cwynar, L. C. 1977. The recent fire history of Barron Township, Algonquin Park. Can. J. Bot. 55:1524–1538.
Cwynar, L. C. 1978. Recent history of fire and vegetation from laminated sediment of Greenleaf Lake, Algonquin Park, Ontario. Can. J. Bot. 56:10–21.
Dahl, E. 1975. Stability of tundra ecosystems in Fennoscandia. pp. 231–236. In F. E. Wielgolaski (ed.), Fennoscandian Tundra Ecosystems, Part 2, Animals and systems analysis. Ecological Studies, 17. Springer-Verlag, New York.
Damman, A. W. H. 1964. Some forest types of Central Newfoundland and their relation to environmental factors. For. Sci. Monogr. 8. 62pp.
Dansgaard, W., S. J. Johnsen, H. B. Clausen, and C. C. Langway, Jr. 1971. Climatic record revealed by the Camp Century ice core. pp.37–56. In K. K. Turekian (ed.), The Late Cenozoic Glacial Ages. Yale University Press, New Haven, Connecticut.
Darwin, C. 1881. The formation of vegetable mould through the action of worms with observations on their habits. John Murray, London. 326 pp.
Dash, M. C., and J. B. Cragg. 1972. Selection of microfungi by Enchytraeidae (Oligochaeta) and other members of the soil fauna. Pedobiologia 12:282–286.
Daubenmire, R. 1968. Plant Communities. Harper and Row, New York. 300 pp.
Davidson, D. W. 1977a. Species diversity and community organization in desert seed-eating ants. Ecology 58:711–724.
Davidson, D. W. 1977b. Foraging ecology and community organization in desert seed-eating ants. Ecology 58:725–737.
Davis, M. B. 1963. On the theory of pollen analysis. Am. J. Sci. 261:897–912.
Davis, M. B. 1967. Pollen deposition in lakes as measured by sediment traps. Geol. Soc. Am. Bull. 78:849–858.
Davis, M. B. 1968. Pollen grains in lake sediments: Redeposition caused by seasonal water circulation. Science 162:796–799.
Davis, M. B. 1969a. Palynology and environmental history during the Quaternary Period. Am. Sci. 57:317–332.
Davis, M. B. 1969b. Climatic changes in southern Connecticut recorded by pollen deposition at Rogers Lake. Ecology 50:409–421.
Davis, M. B. 1973. Redeposition of pollen grains in lake sediments. Limnol. Oceanogr. 18:44–52.
Davis, M. B. 1974. Holocene migrations of ecotones: Continuing northward migration of tree species throughout the Holocene caused changes in community composition and migrations of ecotones. pp.18–21. In Abstracts, Am. Quat. Assn. 3rd Biennial Meeting. Madison, Wisconsin.
Davis, M. B. 1976. Pleistocene biogeography of temperate deciduous forests. Geosci. Man 13:13–26.
Davis, M. B. 1978. Climatic interpretation of pollen in Quaternary sediments. pp. 35–51. In D. Walker and J. D. Guppy (eds.), Biology and Quaternary Environments. Aust. Acad. Sci. Canberra A.C.T.
Davis, M. B. 1981. Outbreaks of forest pathogens in Quaternary History. Proc.

IV Intern. Conf. Palynology, Lucknow (1976–1977) 3:216–227.
Davis, M. B., L. B. Brubaker, and T. Webb, III. 1973. Calibration of absolute pollen influx. pp. 9–27. In H. J. B. Birks and R.G.West (eds.), Quaternary Plant Ecology. Blackwell, Oxford.
Davis, M. B., and E. S. Deevey. 1964. Pollen accumulation rates: Estimates from late-glacial sediment of Rogers Lake. Science 145:1293–1295.
Davis, M. B., and J. C. Goodlett. 1960. Comparison of the present vegetation with pollen-spectra in surface samples from Browington Pond, Vermont. Ecology 41:346–357.
Davis, M. B., R. W. Spear, and L. C. K. Shane. 1980. Holocene climate of New England. Quat. Res. 14:240–250.
Davis, R. B., T. E. Bradstreet, R. Stuckenrath, and H. W. Borns, Jr. 1975. Vegetation and associated environments during the past 14,000 years near Moulton Pond, Maine. Quat. Res. 5:435–465.
Davis, R. B., and T. Webb, III. 1975. The contemporary distribution of pollen in eastern North America: A comparison with vegetation. Quat. Res. 5:395–434.
Davis, R. B. 1967. Pollen studies of near-surface sediments in Maine lakes. pp. 143–173. In E. J. Cushing and H. E. Wright, Jr. (eds.), Quaternary Paleoecology. Yale University Press, New Haven, Connecticut.
Davis, R. B. 1974. Stratigraphic effects of tubificids in profundal lake sediments. Limnol. Oceanogr. 19:466–488.
Day, R. J. 1972. Stand structure, succession, and use of southern Alberta's Rocky Mountain forest. Ecology 53:472–478.
Dayton, B. R. 1966. The relationship of vegetation to Iredell and other Piedmont soils in Granville County, North Carolina. J. Elisha Mitchell Sci. Soc. 82:108–117.
Deevey, E. S. 1949. Biogeography of the Pleistocene. Part I. Europe and North America. Geol. Soc. Am. Bull. 60:1315–1416.
del Amo R. S. 1978. Crecimiento y regeneracion de especies primarias de selva alta perennifolia. Tesis de Doctorado. Facultad de Ciencias, UNAM, Mexico. 259 pp.
Delcourt, H. R. 1978. Late Quaternary vegetation history of the eastern Highland Rim and adjacent Cumberland Plateau of Tennessee. Ph.D. Thesis. University of Minnesota, Minneapolis.
Delcourt, H. R. 1979. Late Quaternary vegetation history of the eastern Highland Rim and adjacent Cumberland Plateau of Tennessee. Ecol. Monogr. 49:255–280.
Delcourt, P. A. 1980. Goshen Springs: Late Quaternary vegetation record from southern Alabama. Ecology 61:371–386.
Delcourt, P. A., and H. R. Delcourt. 1977. The Tunica Hills, Louisiana-Mississippi late-glacial locality for spruce and deciduous forest species. Quat. Res. 7:218–237.
Delcourt, P. A., H. R. Delcourt, R. C. Brister and L. E. Lackey. 1980. Quarternary vegation history of the Mississippi Embayment. Quat. Res. 13:111–132.
DeLiocourt, F. 1898. De l'amenagement des Sapinieres. Bul. Soc. Forest Franch-Conte Belfort, Besancon. pp. 396–409.
Della-Bianca, L., and D. F. Olson, Jr. 1961. Soil-site studies in piedmont hardwood and pine hardwood upland forests. For. Sci. 7:320–329.
Dickinson, C. H., and G. J. F. Pugh. 1974. Biology of Plant Litter Decomposition (2 vol.). Academic Press, London. 775 pp.

Diftler, N. 1947. The invasion of hardwoods in pine stands as related to soil texture on the upland soil types of the lower Piedmont plateau region of Durham County and its adjacent area. Unpublished M.S. Thesis. Duke University, Durham, North Carolina.

Dix, R. L., and J. M. A. Swan. 1971. The roles of disturbance and succession in upland forest at Candle Lake, Saskatchewan. Can. J. Bot. 49:657–676.

Donnelly, R. E., and J. B. Harrington. 1978. Forest fire history maps of Ontario. Canada Dept. Environ., Can. Forestry Serv., Forest Fire Res. Institute, Aviation and Fire Management Centre, Sault St.Marie, Ontario (7 large maps, text).

Doyle, T. W. 1980. FORICO: Gap dynamics model of the lower montane rain forest in Puerto Rico. M.S. Thesis. University of Tennessee, Knoxville.

Drury, W. H. 1956. Bog flats and physiographic processes in the upper Kuskokwim River region, Alaska. Gray Herbarium, Harvard Univ. Pub. No. 178. 130 pp.

Drury, W. H., and I. C. T. Nisbet. 1971. Interrelations between developmental models in geomorphology, plant ecology, and animal ecology. Gen. Syst. 16:57–68.

Drury, W. H., and I. C. T. Nisbet. 1973. Succession. J. Arnold Arbor. 54:331–368.

Duffey, E. 1978. Ecological strategies in spiders including some characteristics of species in pioneer and mature habitats. Symp. Zool. Soc. Lond. 42:109–123.

Dunbar, M. J. 1960. The evolution of stability: Natural selection at the ecosystem level. pp. 99–109. In T. W. M. Cameron (ed.), Evolution, Its Science and Doctrine. University of Toronto Press, Toronto.

Duncan, W. H. 1941. The study of root development in three soil types in the Duke Forest. Ecol. Monogr. 11:141–164.

DuRietz, G. D. 1930. Classification and nomenclature of vegetation. Svensk. Bot. Tidskr. 24:489–503.

Dyrness, C. T. 1973. Early stages of plant succession following logging and burning in the western Cascades of Oregon. Ecology 54:57–69.

Egler, F. E. 1954. Vegetation science concepts. I. Initial floristic composition—a factor in old-field vegetation development. Vegetatio 4:412–417.

Egler, F. E. 1977. The nature of vegetation. Its management and mismanagement. Connecticut Conservation Association, Bridgewater, Connecticut. 527 pp.

Ellison, L. 1960. Influence of grazing on plant succession of grasslands. Bot. Rev. 26:1–78.

Elton, C. S. 1966. The Pattern of Animal Communities. Methuen, Co., Ltd., London. 432 pp.

Emanuel, W. R., H. H. Shugart, and D. C. West. 1978a. Spectral analysis and forest dynamics: Long-term effects of environmental perturbations. pp. 195–210. In H. H. Shugart (ed.), Time Series and Ecological Processes. Society of Industrial and Applied Mathematics, Philadelphia, Pennsylvania.

Emanuel, W. R., D. C. West, and H. H. Shugart, Jr. 1978b. Spectral analysis of forest model time series. Ecol. Modeling 4:313–326.

Emerson, A. E. 1960. The evolution of adaptation in population systems. pp. 307–348. In S. Tax (ed.), Darwin and After Darwin. Vol. I. University of Chicago Press, Chicago.

Emiliani, C. 1972. Quaternary hypsithermals. Quat. Res. 2:270–273.

Ettershank, G., N. Z. Elkins, P. F. Santos, W. G. Whitford, and E. F. Aldon. 1978. The use of termites and other soil fauna to develop soils on strip mine spoils. USDA For. Serv. Rocky Mt. For. Range Exp. Sta. Res. Note RM-361, Fort Collins.

Ewel, J. 1971. Experiments in arresting succession with cutting and herbicides in four tropical environments. Ph.D. Thesis. University of North Carolina. Chapel Hill, North Carolina.

Ewel, J. 1976. Litter fall decomposition in a tropical forest succession in eastern Guatemala. J. Ecol. 64:293–308.

Faegri, K., and J. Iverson. 1975. Textbook of Pollen Analysis. Manksgaard, Copenhagen.

Farmer, R. E., Jr. 1958. Some effects of prescribed burning following clearcutting in poor site aspen. M.S. Thesis. University of Michigan, Ann Arbor.

Farnworth, E. G., and F. B. Golley. 1974. Fragile Ecosystems. Evaluation of Research and Applications in the Neotropics (Rep. Inst. Ecol.). Springer-Verlag, New York.

Feibleman, J. K. 1954. Theory of integrative levels. Brit. J. Philos. Sci. 5:59–66.

Feinsinger, P. 1978. Ecological interactions between plants and hummingbirds in a successional tropical community. Ecol. Monogr. 48:269–287.

Ferrell, W. K. 1953. Effect of environmental conditions on survival and growth of forest tree seedlings under field conditions in the Piedmont region of North Carolina. Ecology 34:667–688.

Finley, R. B., Jr. 1969. Cone caches and middens of *Tamiasciurus* in the Rocky Mountains region. pp. 233–273. In J. K. Jones, Jr. (ed). Contributions in Mammalogy. University of Kansas, Lawrence.

Fisher, R. F. 1972. Spodosol development and nutrient distribution under Hydnaceae fungal mats. Soil Sci. Soc. Am. Proc. 36:492–495.

Fogel, R. 1976. Ecological studies of hypogeous fungi. II. Sporocarp phenology in a western Oregon Douglas-fir stand. Can. J. Bot. 54:1152–1162.

Fogel, R. 1980. Mycorrhizae and nutrient cycling in natural forest ecosystems. New Phytologist 86:199–212.

Fogel, R. 1980. Personal communication. Herbarium, University of Michigan. Ann Arbor, Michigan.

Fogel, R., and K. Cromack, Jr. 1977. Effect of habitat and substrate quality on Douglas-fir litter decomposition in western Oregon. Can. J. Bot. 55:1632–1640.

Fogel, R., and G. Hunt. 1979. Fungal and arboreal biomass in a western Oregon Douglas-fir ecosystem: Distribution patterns and turnover. Can. J. For. Res. 9:245–256.

Foote, M. J. 1976. Classification, description and dynamics of plant communities following fire in the taiga of interior Alaska. USDA For. Serv., Final Rep. (BLM Fire effects study). 211 pp.

Forbes, S. A. 1880. On some interactions of organisms. Bull. Ill. State Lab. Nat. Hist. 1:1–18.

Forcella, F., and T. Weaver. 1977. Biomass and productivity of the subalpine *Pinus albicaulis-Vaccinium scoparium* association in Montana. Vegetatio 35:95–105.

Forcier, L. K. 1975. Reproductive strategies and co-occurrence of climax tree species. Science 189:808–810.

Foster, H. D., and W. W. Ashe. 1908. Chestnut oak in the southern Appalachians. USDA For. Serv. Circ. 105. 27 pp.

Foster, R. B. 1980. Heterogeneity and disturbance in tropical vegetation. Chap. 5, pp. 75–92. In M. E. Soule and B. A. Wilcox (eds.). Conservation Biology. Sinauer, Sunderland, Massachusetts.

Fowells, H. A. 1965. Silvics of Forest Trees of the United States. USDA Handbook No. 271. U.S. Gov. Printing Office. Washington, D.C.

Fox, J. F. 1977. Alternation and coexistence of tree species. Am. Nat. 111:69–89.

Fox, J. F. 1978. Forest fires and the snowshoe hare-Canada lynx cycle. Oecologia 31:349–374.

Franklin, J. F., and C. T. Dyrness 1973. Natural vegetation of Oregon and Washington. USDA For. Serv. Tech. Rep. PNW-8. Pacific Northwest Forest and Range Experiment Station, Portland, Oregon. 417 pp.

Franklin, J. F., and R. H. Waring. 1980. Distinctive features of the northwestern coniferous forest. pp. 59–86. In R. H. Waring (ed.). Forests: Fresh Perspectives from Ecosystem Analysis. Proc. 40th Biol. Colloq. Oregon State University Press. Corvallis, Oregon.

Franklin, J. F., K. Cromack, Jr., W. Denison, A. McKee, C. Maser, J. Sedell, F. Swanson, and G. Juday 1981. Ecological characteristics of old-growth forest ecosystems in the Douglas-fir region. USDA For. Serv. Tech. Rep. PNW-118. Pacific Northwest Forest and Range Experiment Station, Portland, Oregon. 48 pp.

Frey, D. G. 1953. Regional aspects of the late-glacial and post-glacial pollen succession of the southeastern North Carolina. Ecol. Monogr. 23:289–313.

Fries, M. 1962. Pollen profiles of late pleistocene and recent sediments from Weber Lake, northeastern Minnesota. Ecology 43:295–307.

Frissell, S. S. 1973. The importance of fire as a natural ecological factor in Itasca State Park, Minnesota. Quat. Res. 3:397–407.

Fritts, H. C., G. R. Lofgren, and G. A. Gordon. 1979. Variations in climate since 1602 as reconstructed from tree rings. Quat. Res. 12:18–46.

Fujimori, T. 1972. Primary productivity of a young *Tsuga heterophylla* stand and some speculations about biomass of forest communities on the Oregon coast. USDA For. Serv. Res. Pap. PNW-123. Pacific Northwest Forest and Range Experiment Station, Portland, Oregon. 11 pp.

Futuyma, D. J., and F. Gould. 1979. Association of plants and insects in a deciduous forest. Ecol. Monogr. 49:33–50.

Gabriel, H. W. 1976. Wilderness ecology: The Danaher Creek drainage, Bob Marshall Wilderness, Montana. Ph.D. Thesis, University of Montana, Missoula, Montana.

Garcia-Gutierrez, A. 1976. Algunos aspectos del ciclo de vida de dos especies arboreas tropicales de diferentes estados sucesionales (*Bernoullia flammeay Ochroma lagopus*). pp. 594–640. In A.Gomez-Pompa, C. Vazquez-Yanes, S. del Amo and A. Butanda (eds.). Regeneracion de Selvas. CECSA, Mexico, D.F.

Garrett, P. W., and R. Zahner. 1964. Clonal variations in suckering of aspen obscurves effect of various clear-cutting treatments. J. For. 62:749–750.

Gates, F. C. 1930. Aspen association in northern lower Michigan. Bot. Gaz. 90:233–259.

Gauch, H. G., R. H. Whittaker, and T. R. Wentworth. 1977. A comparative study of reciprocal averaging and other ordinations. J. Ecol. 61:237–249.

Ghent, A. W., D. A. Fraser, and J. B. Thomas. 1957. Studies of regeneration of forest stands devastated by the spruce budworm. For. Sci. 3:184–208.

Ghiselin, M. T. 1974. The Economy of Nature. University of California Press, Berkeley. 346 pp.

Gholz, H. L. 1979. Limits on aboveground net primary production, leaf area, and biomass in vegetation zones of the Pacific Northwest. Ph.D. Thesis. Oregon State University, Corvallis, Oregon.

Gibbs, H. S. 1971. Nature of Paleosols in New Zealand and their classification. pp. 229–245. In D. H. Yaalon (ed.), Paleopedology. Israel University Press, Jerusalem.

Gilbert, J. M. 1959. Forest succession in the Florentine Valley. Proc. R. Soc. Tasmania 93:129–151.

Gleason, H. A. 1917. The structure and development of the plant association. Bull. Torrey Bot. Club 43:463–481.

Gleason, H. A. 1927. Further views on the succession concept. Ecology 8:299–326.

Gleason, H. A. 1939. The individualistic concept of the plant association. Am. Midland. Nat. 21:92–110.

Gleason, H. A., and M. I. Cook. 1927. Plant ecology of Porto Rico. Scientific Surv. of Porto Rico and the Virgin Islands. New York Academy of Science. Vol. 7. 96 pp.

Gliessman, S. R., and C. H. Muller. 1972. The phytotoxic potential of bracken, *Pteridium aquilinum* (L.). Kuhn. Madrono 21:299–304.

Goff, F. G., and D. C. West. 1974. Canopy-understory interaction effects on forest population structure. For. Sci. 21:98–108.

Golley, F. B. 1977. Ecological Succession. Dowden, Hutchinson and Ross, Stroudsburg, Pennsylvania. 373 pp.

Golley, F. B., J. T. McGinnis, R. G. Clements, G. L. Child, and M. J. Duever. 1975. Mineral Cycling in a Tropical Moist Forest Ecosystem. University of Georgia Press, Athens.

Gomez-Pompa, A. 1971. Possible papel de la vegetacion secundaria en la evolucion de la flora tropical. Biotropica 3:125–135.

Gomez-Pompa, A., and C. Vazquez Yanes. 1974. Studies on the secondary succession of tropical lowlands: The life cycle of secondary species. Proceedings of the First International Congress of Ecology. W. Junk, The Hague. pp. 336–342.

Gomez-Pompa, A., C. Vazquez Yanes, S. del Amo, and A. Butanda. 1976. Investigaciones sobre la regeneracion de selvas altas en Veracruz, Mexico. Compania Editorial Continental, S. A. Mexico. 640 pp.

Gomez-Pompa, A., C. Vazquez Yanes, and S. Guevara. 1972. The tropical rain forest: A nonrenewable resource. Science 177:762–765.

Goodman, R. B. 1930. Conditions essential to selective cutting in northern hardwoods and hemlock. J. For. 28:1070–1075.

Gorham, E. 1957. The development of peatlands. Quart. Rev. Biol. 32:145–166.

Gorham, E., P. M. Vitousek, and W. A. Reiners. 1979. The regulation of chemical budgets over the course of terrestrial ecosystem succession. Ann. Rev. Ecol. Syst. 10:53–84.

Gottsberger, G. 1978. Seed dispersal by fish in the inundated regions of Humaita, Amazonia. Biotropica 10:170–183.
Graham, S. A., R. P. Harrison, Jr., and C. E. Westell, Jr. 1963. Aspens: Phoenix Trees of the Great Lakes States. University of Michigan Press, Ann Arbor. 272 pp.
Grant, W. E., and N. R. French. 1980. Evaluation of the role of small mammals in grassland ecosystems: A modelling approach. Ecol. Modeling 8:15–37.
Graustein, W. C., K. Cromack, Jr., and P. Sollins. 1977. Calcium oxalate: Occurrence in soils and effect on nutrient and geochemical cycles. Science 198:1252–1254.
Graves, R. C. 1960. Ecological observations on the insects and other inhabitants of woody shelf fungi (Basidiomycetes: Polyporaceae) in the Chicago area. Ann. Entomol. Soc. Am. 53:61–78.
Greeley, W. B., and W. W. Ashe. 1907. White oak in the southern Appalachians. USDA For. Serv. Circ. 105. 27 pp.
Greenland, D. J. 1979. The physics and chemistry of the soil-root interface: Some comments. pp. 83–98. In J. L. Harley and R.S.Russell (eds.), The Soil-Root Interface. Academic Press, New York.
Greenland, D. J., and G. W. Ford. 1964. Separation of partially humified organic materials from soils by ultrasonic dispersion. Trans. 8th Int. Soil Sci. Congr., Bucharest 3:137–148.
Greenland, D. J., and J. M. L. Kowal. 1960. Nutrient content of the moist tropical forest of Ghana. Plant and Soil 12:154–174.
Greig-Smith, P. 1979. Patterns in vegetation. J. Ecol. 67:755–779.
Grene, M. 1967. Biology and the problem of levels of reality. New Scholasticism 41:427–449.
Grier, C. C. 1978. A *Tsuga heterophylla-Picea sitchensis* ecosystem of coastal Oregon: Decomposition and nutrient balances of fallen logs. Can. J. For. Res. 8:198–206.
Grier, C. C., and R. S. Logan. 1977. Old-growth *Pseudotsuga menziesii* communities of a western Oregon watershed: Biomass distribution and production budgets. Ecol. Monogr. 47:373–400.
Grigal, D. F., and L. F. Ohmann. 1975. Classification, description, and dynamics of upland plant communities within a Minnesota wilderness area. Ecol. Monogr. 45:389–407.
Griggs, R. F. 1946. The timberlines of northern America and their interpretation. Ecology 27:275–289.
Grime, J. P. 1979. Plant Strategies and Vegetation Processes. Wiley, Chichester, England. 222 pp.
Grubb, P. J. 1977. The maintenance of species richness in plant communities. The importance of the regeneration niche. Biol. Rev. 52:107–145.
Gruger, E. 1972. Pollen and seed studies of Wisconsonian vegetation in Illinois, U.S.A. Geol. Soc. Am. Bull. 83:2715–2734.
Guevara, S., and A. Gomez-Pompa. 1972. Seeds from surface soils in a tropical region of Veracruz, Mexico. J. Arnold Arbor. 53:312–335.
Gulmon, S. L., P. W. Rundel, J. R. Ehleringer, and H. A. Mooney. 1979. Spatial relationships and competition in a Chilean desert cactus. Oecologia 44:40–43.
Gutierrez, L. T., and W. R. Fey. 1975. Simulation of successional dynamics in ecosystems. pp. 73–82. In G. Innis (ed.), New Directions in the Analysis of

Ecosystems. Sim. Counc. Proc. Ser. Vol. 5, No. 1. Simulation Councils, La Jolla, California.

Gutierrez, L. T., and W. R. Fey. 1980. Ecosystem succession. A general hypothesis and a test model of a grassland. MIT Press, Cambridge, Massachusetts. 231 pp.

Habeck, J. R. 1970. Fire Ecology Investigations in Glacier National Park. Dept. Botany, Univ. Montana, Missoula, Montana. 80 pp.

Habeck, J. R. 1976. Forests, fuels and fire in the Selway-Bitterroot Wilderness, Idaho. Proc. Tall Timbers Fire Ecology Conf. 14:305–353.

Habeck, J. R., and R. W. Mutch. 1973. Fire-dependent forests in the Northern Rocky Mountains. Quat. Res. 3:408–424.

Haines, D. A., and R. W. Sando. 1969. Climatic conditions preceding historically great fires in the North Central Region. USDA For. Serv. Res. Pap. NC-34, North Central Forest Experiment Station, St. Paul, Minnesota. 19 pp.

Hale, M. E. 1961. Lichen Handbook. Smithsonian Institution, Washington, D.C. 178 pp.

Halevy, G. 1975. Effects of gazelles and seed beetles (Bruchidae) on germination and establishment of Acacia species. Israel J. Bot. 23:120–126.

Halfon, E. 1979. Theoretical Systems Ecology. Academic Press, New York. 516 pp.

Hall, E. A. A., R. L. Specht, and C. M. Eardley. 1964. Regeneration of the vegetation on Koonamore vegetation reserve. Aust. J. Bot. 12:205–264.

Hall, N., R. D. Johnston, and G. M. Chippendale. 1970. Forest trees of Australia. Australian Government Printing Service, Canberra. 334 pp.

Hall, W. L. 1907. The waning hardwood supply and the Appalachian forests. USDA For. Serv. Circ. 116. 16 pp.

Halle, F., R. A. A. Oldeman, and P. B. Tomlinson. 1978. Tropical trees and forests, an architectural analysis. Springer-Verlag. Berling.

Hammond, A. L. 1972. Ecosystem analysis: Biome approach to environmental science. Science 175:46–48.

Hanes, T. L. 1971. Succession after fire in the chaparral of Southern California. Ecol. Monogr. 41:27–52.

Harcombe, P. A. 1972. Plant succession and ecosystem recovery. Ph.D. Thesis, Rice University, Houston, Texas.

Hardeman, W. D. 1966. Geologic map of Tennessee, 1:250,000. Tenn. Dept. Conserv., Div. Geol., Nashville, Tennessee.

Harley, J. L. 1972. The Biology of Mycorrhizae. Leonard Hill, London. 334 pp.

Harley, J. L. 1975. Problems of Mycotrophy. pp. 1–24. In F.E.Sanders, B. Mosse, and P. B. Tinker (eds.), Endomycorrhizae. Academic Press, London.

Harper, J. L. 1977a. Population Biology of Plants. Academic Press, London. 892 pp.

Harper, J. L. 1977b. The contribution of terrestrial plant studies to the development of the theory of ecology. pp. 139–157. In C. E. Goulden (ed.), Changing scenes in the Life Sciences, 1776–1976. Special Publ. 12, Acad. Nat. Sci., Philadelphia.

Harris, W. F., D. Santantonio, and D. McGinty. 1980. Dynamic belowground ecosystems. pp. 119–129. In R. H. Waring (ed.), Forests: Fresh Perspectives from Ecological Analysis. Proc. 40th Biol. Colloq. Oregon State University Press, Corvallis, Oregon.

Harrold, A. G. Personal communication. Noosa Head, Queensland, Australia. February 14, 1980.
Hartshorn, G. S. 1978. Tree falls and tropical forest dynamics. pp. 617–638. In P. B. Tomlinson and M. H. Zimmermann (eds.), Tropical Trees as Living Systems. Proc. 4th Cabot Symp. Harvard Forest (1976) Cambridge University Press, Cambridge.
Harvey, A. E., M. F. Jurgensen, and M. J. Larsen. 1978. Seasonal distribution of ectomycorrhizae in a mature Douglas-fir/larch forest soil in western Montana. For. Sci. 24:203–208.
Harvey, A. E., M. J. Larsen, and M. F. Jurgensen. 1976. Distribution of ectomycorrhizae in a mature Douglas-fir/larch forest soil in western Montana. For. Sci. 22:393–398.
Harvey, A. E., M. J. Larsen, and M. F. Jurgensen. 1979. Comparative distribution of ectomycorrhizae in soils of three western Montana forest habitat types. For. Sci. 25:350–358.
Hastings, J. R., and R. M. Turner. 1965. The changing mile. University of Arizona Press, Tucson, Arizona. 317 pp.
Hatley, C. L., and J. A. MacMahon. 1980. Spider community organization: Seasonal variation and the role of vegetation architecture. Environ. Entomol. 9:632–639.
Haukioja, E. 1975. Importance criteria in birch defoliators. pp. 189–194. In F. E. Wielgolaski (ed.), Fennoscandian Tundra Ecosystems, Part 2, Animals and Systems Analyses. Ecological Studies, Vol. 17. Springer-Verlag, Berlin.
Hawk, G. M. 1979. Vegetation mapping and community description of a small western Cascade watershed. Northwest Sci. 53:200–212.
Hawkins, P. J. 1975. Forest management of Cooloola State Forest. Proc. Ecol. Soc. Aust. 9:328–333.
Hays, J. D., T. Saito, H. D. Opdyke and L. H. Burckle. 1969. Pliocene-Pleistocene sediments of the equatorial pacific: Their paleomagnetic, biostratigraphic and climatic record. Geol. Soc. Am. Bull. 80:1481–1514.
Heilman, P. E. 1966. Change in distribution and availability of nitrogen with forest succession on north slopes in interior Alaska. Ecology 47:825–831.
Heilman, P. E. 1974. Effect of urea on nitrification in forest soils in the Pacific Northwest. Soil Sci. Soc. Am. Proc. 38:664–667.
Heilman, P. E., H. W. Anderson, and D. M. Baumgartner 1979. Forest soils of the Douglas-Fir Region. Washington State University Cooperative Extension Service, Pullman, Washington. 298 pp.
Heinselman, M. L. 1963. Forest sites, bog processes, and peatland types in the Glacial Lake Agassiz Region, Minnesota. Ecol. Monogr. 33:327–374.
Heinselman, M. L. 1970. Landscape evolution, peatland types, and the environment in the Lake Agassiz Peatlands Natural Area, Minnesota. Ecol. Monogr. 40:235–261.
Heinselman, M. L. 1973. Fire in the virgin forests of the Boundary Waters Canoe Area, Minnesota. Quat. Res. 3:329–382.
Heinselman, M. L. 1974. Interpretation of Francis J. Marschner's map of the original vegetation of Minnesota. Text on reverse of The Original Vegetation of Minnesota—map by F. J. Marschner, 1930; (Pub. in color by USDA For. Serv., North Central Forest Expt. Sta., through U.S. Govt. Printing Office, Washington, D.C.).

Heinselman, M. L. 1975. Boreal peatlands in relation to environment. pp. 93–124. In A. D. Hasler (ed.), Coupling of Land and Watnsen, J. E. Lotan, and W. A. Reiners
(eds.), Fire Regimes and Ecosystem Properties. USDA Gen. Tech. Rep. WO-26, Washington, D.C.
Hellmers, H. 1964. An evaluation of the photosynthetic efficiency of forests. Quart. Rev. Biol. 39:249–257.
Hemstrom, M. A. 1979. A recent disturbance history of forest ecosystems at Mount Rainier National Park. Ph.D. Thesis. Oregon State University, Corvallis, Oregon.
Henry, J. D., and J. M. A. Swan. 1974. Reconsrnsen, J. E. Lotan, and W. A. Reiners (eds.), Fire Regimes and Ecosystem Properties. USDA Gen. Tech. Rep. WO-26, Washington, D.C.
Hellmers, H. 1964. An evaluation of the photosynthetic efficiency of forests. Quart. Rev. Biol. 39:249–257.
Hemstrom, M. A. 1979. A recent disturbance history of forest ecosystems at Mount Rainier National Park. Ph.D. Thesis. Oregon State University, Corvallis, Oregon.
Henry, J. D., and J. M. A. Swan. 1974. Reconstructing forest history from live and dead plant material—An approach to the study of forest succession in southwest New Hampshire. Ecology 55:772–783.
Hesselman, H. 1917. Studier over saltpeterbildningen i naturliga jordmaner och dess betydelse i vaxtekologist auseende. Medd. Statens. Skogsforkningsinst. 12:297.
Hewitt, J. S., and A. R. Dexter. 1979. An improved model of root growth in structured soil. Plant Soil 52:325–343.
Hicks, D. J. 1978. A niche analysis of estival herb communities of Cove Forest. M.S. Thesis. Cornell University, Ithaca, New York.
Hicks, D. J. 1980. Intrastand distribution patterns of southern Appalachian cove forest herbaceous species. Am. Midl. Nat. 104:209–223.
Hill, M. O. 1979. DECORANA. Cornell Ecology Program, Cornell University, Ithaca, N.Y. 51 pp.
Hinds, W. T., and G. M. Van Dyne. 1980. Abiotic subsystem. pp. 11–57. In A. I. Breymeyer and G. M. Van Dyne (eds.), Grasslands, Systems Analysis and Man, International Biological Programme, Vol. 19. Cambridge University Press, Cambridge, England.
Hintikka, V., and O. Naykki. 1967. Notes on the effects of the fungus *Hydnellum ferrugineum* (Fr.) Karst. on forest soil and vegetation. Commun. Inst. For. Fenniae 62:1–22.
Holdridge, L. R. 1967. Life Zone Ecology, rev. ed. Tropical Science Center, San Jose, Costa Rica. 206 pp.
Holling, C. S. 1973. Resilience and stability of ecological systems. Annu. Rev. Ecol. Syst. 4:1–23.
Holman, T. 1949. The invasion of hardwoods in pine stands as related to soil characteristics of the lower piedmont plateau of North Carolina. M.S. Thesis, Duke University, Durham, North Carolina.
Holmes, R. T., R. E. Bonney, Jr., and S. W. Pacala. 1979. Guild structure of the Hubbard Brook bird community: A multivariate approach. Ecology 60:512–520.

Hopkins, M. S., J. Kikkawa, A. W. Graham, J. G. Tracy, and L. J. Webb. 1977. An ecological basis for the management of rainforests. pp.57–66. In R. Monroe and N. C. Stevens (eds.), The Border Ranges: A Land-use Conflict in Regional Perspective. Royal Society of Queensland, St. Lucia, Queensland.
Horn, H. S. 1971. The adaptive geometry of trees. Monogr. Pop. Biol. 3. Princeton University Press, Princeton, New Jersey. 144 pp.
Horn, H. S. 1974. The ecology of secondary succession. Ann. Rev. Ecol. Syst. 5:25–37.
Horn, H. S. 1975a. Forest succession. Sci. Am. 232(5):90–98.
Horn, H. S. 1975b. Markovian properties of forest succession. pp.196–211. In M. L. Cody and J. M. Diamond (eds.), Ecology and Evolution of Communities. Harvard University Press, Cambridge, Massachusetts.
Horn, H. S. 1976. Succession. pp. 187–204. In R. M. May (ed.), Theoretical Ecology. Blackwell, Oxford.
Horn, H. S. 1979. Adaptation from the perspective of optimality. pp. 48–61. In O. T. Solbrig, S. Jain, G. B. Johnson and P.H.Raven, Topics in Plant Population Biology. Columbia University Press, New York.
Horton, J. S. 1950. Effect of weed competition upon survival of planted pine and chaparral seedlings. USDA For. Serv. Res. Note 72. Pacific Southwest Forest and Range Experiment Station, Berkeley, California. 6 pp.
Horton, K. W. 1956. The ecology of lodgepole pine in Alberta and its role in forest succession. Forestry Branch Tech. Note 45. Canadian Department of Northern Affairs and National Resources, Montreal. 29 pp.
Horton, K. W., and G. H. D. Bedell. 1960. White and red pine ecology, silviculture, and management. Forestry Branch Tech. Note 124. Canadian Department of Northern Affairs and National Resources, Montreal. 185 pp.
Hosner, J. F., and D. L. Graney. 1970. The relative growth of three forest tree species on soils associated with different successional stages in Virginia. Am. Midland Nat. 84:418–427.
Hough, A. F. 1932. Some diameter distributions in forest stands of northwestern Pennsylvania. Ecology 17:9–28.
Hough, A. F., and R. D. Forbes. 1943. The ecology and silvics of forests in the plateaus of Pennsylvania. Ecol. Monogr. 13:299–320.
Houston, D. B. 1973. Wildfires in northern Yellowstone National Park. Ecology 54:1111–1117.
Howard, J. A. 1970. Aerial Photo Ecology. Faber and Faber, London. 252 pp.
Howe, H. F., and G. A. Vande Kerckhove. 1979. Fecundity and seed dispersal of a tropical tree. Ecology 60:180–189.
Hudson, H. J. 1968. The ecology of fungi on plant remains above the soil. New Phytol. 67:837–874.
Huffman, T. Personal communication, Department of Archaeology, University of the Witwatersrand, Johannesburg, South Africa. March 3, 1980.
Hull, D. 1978. A matter of individuality. Phil. Sci. 45:335–360.
Huston, M. 1979. A general hypothesis of species diversity. Am. Nat. 113:81–101.
Hutchinson, G. E. 1942. Addendum to R. L. Lindeman. The Trophic-Dynamic Aspect of Ecology. Ecology 23:417–418.
Ilvessalo, Y. 1937. Pera-Pohjolan luonnon normaalien metsikoiden kasvu ja kehitys. Metsatieteellisen Tutkimuslaitoken Julkaisuja 24:1–168.

Innis, G. S. 1976. Reductionist vs. whole system approaches to ecosystem studies. pp. 31–36. In S. A. Levin (ed.), Ecological Theory and Ecosystem Models. Office of Ecosystem Studies, The Institute of Ecology, Indianapolis, Indiana.

Innis, G. S., and E. F. Cheslak. 1975. Systems ecology: An introductory course sequence. pp. 25–34. In G. S. Innis (ed.). New Directions in the Analysis of Ecological Systems. Sim. Counc. Proc. Ser. Vol. 5, No. 1. Simulation Councils, La Jolla, California.

Isaac, L. A. 1943. Reproductive habits of Douglas-fir. Charles Lathrop Pack Forestry Foundation, Washington, D.C. 107 pp.

Iversen, J. 1969. Retrogressive development of a forest ecosystem demonstrated by pollen diagrams from fossil mor. Oikos Suppl. 12:35–49.

Jackson, W. D. 1968. Fire, air, water and earth—an elemental ecology of Tasmania. Proc. Ecol. Soc. Aust. 3:9–16.

Jacobson, G. L. 1980. The use of small paired basins and the selection of sites for palynological studies, with special reference to an example from Minnesota. p. 186. In 5th Intern. Palynol. Conf. (Cambridge) Abstracts.

Janos, D. P. 1980. Vesicular-arbuscular mycorrhizae affect lowland tropical rain forest plant growth. Ecology 61:151–162.

Janssen, C. R. 1967. A comparison between the recent regional pollen rain and the sub-recent vegetation in four major vegetation types in Minnesota (USA). Rev. Palaeobot. Palynol. 2:331–342.

Janzen, D. H. 1971. Seed predation by animals. Ann. Rev. Ecol. Syst. 2:465–492.

Janzen, D. H. 1975. Ecology of Plants in the Tropics. The Institute of Biology, Studies in Biology No. 58. Edward Arnold, London. 66 pp.

Janzen, D. H. 1977. Why fruits rot, seeds mold, and meat spoils. Am. Nat. 111:691–713.

Janzen, D. H. 1978. Seeding patterns of tropical trees. pp. 83–128. In P. B. Tomlinson and M. H. Zimmerman (eds.), Tropical Trees as Living Systems. Proc. 4th Cabot Symp. Harvard Forest (1976), Cambridge University Press, Cambridge, England.

Jehne, W., and C. H. Thompson. 1981. Endomycorrhizae in plant colonization and phosphorus uptake on coastal sand dunes at Cooloola, Queensland. Aust. J. Ecol. (in press).

Jelgersma, S. 1962. A late-glacial pollen diagram from Madelia, south-central Minnesota. Am. J. Sci. 260:522–529.

Jenkinson, D. S. 1971. Studies on the decomposition of ^{14}C labelled organic matter in soil. Soil Sci. III:64–70.

Jenkinson, D. S. 1977. Studies on the decomposition of plant material. V. The effects of plant cover and soil type on the loss of carbon from ^{14}C labelled ryegrass decomposing under field conditions. J. Soil Sci. 28:209–213.

Jenny, H. 1941. Factors of Soil Formation. McGraw-Hill, New York. 281 pp.

Jenny, H. 1961. Derivation of state factor equations of soils and ecosystems. Soil Sci. Soc. Am. Proc. 25:385–388.

Jenny, H. 1980. Soil Genesis with Ecological Perspectives. Ecological Studies, Vol. 37. Springer-Verlag, New York. 560 pp.

Jenny, H., R. J. Arkley, and A. M. Schultz. 1969. The pigmy forest-podsol ecosystem and its dune associates of the Mendocino coast. Madrono 20:60–74.

Jenny, H., A. E. Salem, and J. R. Wallis. 1968. Interplay of soil organic matter and soil fertility with state factors and soil properties. pp. 5–37. In Study Week on Organic Matter and Soil Fertility. Pontificae. Acad. Sci. No. 32, Wiley, New York.

Jensen, H. L. 1929. On the influence of the carbon-nitrogen ratios of organic material on the mineralization of nitrogen. J. Agr. Sci. 19:71–82.

Joern, A. 1979. Feeding patterns in grasshoppers (Orthoptera: Acrididae): Factors influencing diet specialization. Oecologia 38:325–347.

Johnson, E. A. 1975. Buried seed populations in the subarctic forest east of Great Slave Lake, Northwest Territories. Can. J. Bot. 53:2933–2941.

Johnson, E. A. 1979. Fire recurrence in the subarctic and its implications for vegetation composition. Can. J. Bot. 57:1374–1379.

Johnson, E. A., and J. S. Rowe. 1975. Fire in the subarctic wintering ground of the Beverley Caribou herd. Am. Midland Nat. 94:1–14.

Johnson, E. A., and J. S. Rowe. 1977. Fire and vegetation change in the western subarctic. Canadian. Department of Indian Affairs and Northern Development ALUR 75-76-61. 58 pp.

Johnson, M. D. 1975. Seasonal and microseral variations in the insect populations on carrion. Am. Midland Nat. 93:79–90.

Johnston, D. W., and E. P. Odum. 1956. Breeding bird populations in relation to plant succession on the Piedmont of Georgia. Ecology 37:50–62.

Jones, E. W. 1945. The structure and reproduction of the virgin forest of the North Temperate zone. New Phytol. 44:130–148.

Jordan, C. F. 1969. Derivation of leaf area index from quality of light on forest floor. Ecology 50:663–666.

Jordan, C. F., J. R. Kline, and D. S. Sasscer. 1972. Relative stability of mineral cycles in forest ecosystems. Am. Nat. 106:237–253.

Jordan, P. W., and P. S. Nobel. 1979. Infrequent establishment of seedlings of *Agave deserti* (Agavaceae) in the northwestern Sonoran desert. Am. J. Bot. 66:1079–1084.

Juliano, J. P. 1940. Viability of some philippine weed seeds. Philippine Agr. 29:312–326.

Kapp, R. O. 1977. Late Pleistocene and post-glacial plant communities of the Great Lakes Region. pp. 1–27. In R. C. Romans (ed.), Geobotany. Plenum, New York.

Kartawinata, K. 1977. Biological changes after logging in lowland dipterocarp forest. Bull. Kebun Raya 3:29–34.

Kasanaga, H., and M. Monsi. 1954. On the light transmission of leaves and its meaning for the production of matter in plant communities. Jpn. J. Bot. 14:304–312.

Keay, R. W. J. 1960. Seeds in forest soil. Nigerian Forest Infor. Bull. 4:1–4.

Keeley, J. E. 1981. Reproductive cycles and fire regimes. pp. 231–278. In H. A. Mooney, N. L. Christensen, J. E. Lotan, and W. E. Reiners (eds.), Fire Regimes and Ecosystem Properties. USDA For. Serv. Gen. Tech. Rep. WO-26, Washington, D.C.

Keeley, J. E., and P. H. Zedler. 1978. Reproduction of chaparral shrubs after fire: A comparison of the sprouting and seeding strategies. Am. Midland Nat. 99:142–161.

Keeney, D. R. 1980. Prediction of soil nitrogen availability in forest ecosystems:

A literature review. For. Sci. 26:159–171.
Keever, C. 1950. Causes of succession an old fields of the Piedmont, North Carolina. Ecol. Monogr. 20:231–250.
Kellman, M. C. 1969. Some environmental components of shifting cultivation in upland Mindanao. J. Trop. Geogr. 28:40–56.
Kellman, M. C. 1970. Secondary plant succession in tropical montane Mindanao. Research School of Pacific Studies Department of Biogeography. Geomorphology Publication. BG/2. Australia.
Kelly, R. D., and B. H. Walker. 1976. The effects of different forms of land use on the ecology of a semi-arid region in south-eastern Rhodesia. J. Ecol. 64:553–576.
Kelsall, J. P., E. S. Telfer, and T. D. Wright. 1977. The effects of fire on the ecology of the boreal forest with particular reference to the Canadian north: A review and selected bibliography. Occasional Paper 32. Canadian Wildlife Service, Montreal. 58 pp.
Ker, J. W., and J. H. G. Smith. 1955. Advantages of the parabolic expression of height-diameter relationships. For. Chron. 31:235–246.
Kerfoot, W. C. 1974. Net accumulation rates and the history of cladoceron communities. Ecology 55:51–61.
Kilburn, P. D. 1960a. Effect of settlement on the vegetation of the University of Michigan Biological Station. Pap. Mich. Acad. Sci., Arts Let. 45:77–81.
Kilburn, P. D. 1960b. Effects of logging and fire on xerophytic forests in northern Michigan. Bull. Torrey Bot. Club 87:402–405.
Kinerson, R. S., C. W. Ralston, and C. G. Wells. 1977. Carbon cycling in a loblolly pine plantation. Oecologia 29:1–10.
King, J. E., and E. H. Lindsay. 1976. Late Quaternary biotic records from spring deposits in western Missouri. pp. 63–78. In W. R. Wood and R. B. McMillan (eds), Prehistoric Man and his Environments. A Case Study in the Ozark Highland. Academic Press, New York.
Kira, T., H. Ogawa, and K. Shinozaki. 1953. Intraspecific competition among higher plants. 1. Competition-density-yield interrelationships in regularly dispersed populations. J. Inst. Polytech., Osaka City Univ. D4:1–16.
Kira, T., and T. Shidei. 1967. Primary production and turnover of organic matter in different forest ecosystems of the western pacific. Jpn. J. Ecol. 17:70–87.
Kitchell, J. F., R. V. O'Neill, D. Webb, G. W. Gallepp, S. M. Bartell, J. F. Koonce, and B. S. Ausmus. 1979. Consumer regulation of nutrient cycling. BioScience 29:28–34.
Kittredge, J., and A. K. Chittenden. 1929. Oak forests of northern Michigan. Mich. Agr. Exp. Sta. Spec. Bull. 190. East Lansing, Michigan. 47 pp.
Klopsch, M. Personal communication, Forest Science Department, Oregon State University, Corvallis, Oregon. May 25, 1980.
Knight, D. H. 1975. A phytosociological analysis of species-rich tropical forest on Barro Colorado Island, Panama. Ecol. Monogr. 45:259–284.
Knuchel, H. 1953. Planning and control in the managed forest. Oliver & Boyd, Edinburgh. 360 pp.
Kochummen, K. M. 1966. Natural plant succession after farming in Sg. Kroh. Malay. For. 29:170–181.
Koerper, G. J., and C. J. Richardson. 1980. Biomass and net annual primary production regressions for *Populus grandidentata* on three sites in northern lower

Michigan. Can. J. For. Res. 10:92–101.

Korstian, C. F., and T. S. Coile. 1938. Plant competition in forest stands. Duke Univ. School For. Bull. 3. 125 pp.

Kozlowski, T. T. 1949. Light and water in relation to growth and competition of Piedmont forest tree species. Ecol. Monogr. 19:207–231.

Kramer, K. 1933. Die natuurlijke verjonging in het Goenoeng Gedeh complex. Tectona 26:156–185.

Kramer, P. J., and J. P. Decker. 1944. Relation between light intensity and rate of photosynthesis of loblolly pine and certain hardwoods. Plant Physiol. 19:350–358.

Kramer, P. J., and T. T. Kozlowski. 1960. Physiology of Trees. McGraw-Hill, New York. 642 pp.

Küchler, A. W. 1966. Potential natural vegetation of the United States, including Alaska and Hawaii. pp. 89–92. In U.S. Department of Interior, Geological Survey. National Atlas of the United States. [Map, in color, republished (1978) by USDA For. Serv., Washington, D.C.]

Kukla, G. J., R. K. Matthews, and J. M. Mitchell, Jr. 1972. The end of the present interglacial. Quat. Res. 2:261–269.

Kutzbach, J. E. 1970. Large scale features of monthly mean northern hemisphere anomaly maps of sea-level pressure. Monthly Weather Rev. 104:1255–1265.

Ladd, J. N., and M. Amato. 1980. Studies of nitrogen immobilization and mineralization in calcareous soils. IV. Changes in the organic nitrogen of light and heavy subfractions of silt and fine clay-size particles during nitrogen turnover. Soil Biol. Biochem. 12:185–189.

Lamb, D. 1975. Patterns of nitrogen mineralization in the forest floor of stands of *Pinus radiata* on different soils. J. Ecol. 63:615–625.

Lamb, H. H. 1972. Climate: Present, Past and Future. Methuen, London. 613 pp.

Lamb, H. H. 1977. Climate: Present, Past and Future, Vol. 2, Climatic History and the Future. Methuen, London. 835 pp.

Lamb, I. M. 1970. Antarctic terrestrial plants and their ecology. pp. 733–751. In M. W. Holdgate (ed.), Antarctic Ecology, Vol. 2. Scientific Committee on Antarctic Research. Academic Press, New York.

Lane, P. A., G. H. Lauff, and R. Levins. 1974. The feasibility of using a holistic approach in ecosystem analysis. pp. 111–128. In S. A. Levin (ed.), Ecosystem Analysis and Prediction. Proc. SIAM-SIMS Conf. (1974), Alta, Utah.

Langford, A. N., and M. F. Buell. 1966. Integration, identity and stability in the plant association. Adv. Ecol. Res. 6:83–135.

Lanner, R. M., and S. B. Vander Wall. 1980. Dispersal of limber pine seed by Clark's Nutcracker. J. For. 78:637–639.

Lawrence, D. B. 1951. Recent glacier history of Glacier Bay, Alaska, and development of vegetation on deglaciated terrain with special reference to the importance of alder in the succession. Am. Phil. Soc. Yearbook 1950:175–176.

Lawrenz, R. W. 1975. The development of Green Lake, Antrim County, Michigan. M.S. Thesis. Central Michigan University, Mount Pleasant, Michigan. 78 pp.

Laws, R. M. 1970. Elephants as agents of habitat and landscape change in East Africa. Oikos 21:1–15.

Laws, R. M., I. S. C. Parker, and R. C. B. Johnstone. 1975. Elephants and Their Habitats: The Ecology of Elephants in North Bunyoro, Uganda. Oxford University Press, London. 376 pp.

Leak, W. B. 1964. An expression of diameter distribution for unbalanced, uneven-aged stands and forests. For. Sci. 10:39–50.

Leak, W. B., and C. W. Martin. 1975. Relationship of stand age to streamwater nitrate in New Hampshire. USDA For. Serv. Res. Note NE-211. Upper Darby, Pennsylvania. 4 pp.

LeBarron, R. K. 1939. The role of forest fires in the reproduction of black spruce. Minn. Acad. Sci. Proc. 7:10–14.

LeBarron, R. K. 1944. Influence of controllable environmental conditions on regeneration of jack pine and black spruce. J. Agr. Res. 68:97–119.

Lebron, M. L. 1979. An autoecological study of *Palicourea riparia* in Puerto Rico. Oecologia 42:31–46.

Lebrum, J., and G. Gilbert. 1954. Una classification ecologique des forests du Congo. Publ. Inst. Nat. Etude Agron. Congo. Belge. Ser. Sci. 63. 89 pp.

Lee, J. J., and D. L. Inman. 1975. The ecological role of consumers—An aggregated systems view. Ecology 56:1455–1458.

Lee, K. E., and T. G. Wood. 1971. Termites and Soils. Academic Press, New York. 251 pp.

Lejeune, R. R. 1955. Population ecology of the larch sawfly. Can. Entomol. 87:111–117.

Levin, S. A. 1970. Community equilibria and stability, and an extension of the competitive exclusion principle. Am. Nat. 104:413–423.

Levin, S. A. 1976. Population dynamic models in heterogeneous environments. Ann. Rev. Ecol. Syst. 7:287–310.

Levin, S. A., and R. T. Paine 1974. Disturbance, patch formation, and community structure. Proc. Acad. Nat. Sci. U.S.A. 71:2744–2747.

Levins, R. 1974. The qualitiative analysis of partially specified systems. N. Y. Acad. Sci. Ann. 231:123–138.

Levitt, J. 1972. Responses of Plants to Environmental Stresses. Academic Press, New York.

Lewontin, R. C. 1978. Adaptation. Sci. Am. 239:212–230.

Lidicker, W. Z., Jr. 1979. A clarification of interactions in ecological systems. BioScience 29:475–477.

Lieth, H. 1974. Primary productivity of successional stages. pp. 185–193. In R. Knapp (ed.), Vegetation Dynamics. Handbook of Vegetation. Science, Vol. 8. W. Junk, The Hague.

Liew, T. C. 1973. Occurrence of seeds in virgin forests top soil with particular reference to secondary species in Sabah. Malay. Forester 36:185–193.

Lindeman, R. L. 1942. The trophic-dynamic aspect of ecology. Ecology 23:399–418.

Little, E. L., Jr. 1949. Important forest trees of the United States. pp. 763–814. In Trees. USDA Yearbook of Agriculture, U.S. Government Printing Office, Washington, D.C. 944 pp.

Little, E. L., Jr. 1953. Checklist of Native and Naturalized Trees of the United States Including Alaska. USDA Agriculture Handbook No.41, USDA For. Serv., Washington.

Little, E. L., Jr., and F. H. Wadsworth. 1964. Common Trees of Puerto Rico and the Virgin Islands. USDA For. Serv. Agriculture Handbook 249. 548 pp.

Little, I. P., T. M. Armitage, and R. J. Gilkes. 1978. Weathering of quartz in dune sands under subtropical conditions in eastern Australia. Geoderma 20:225–37.

Livingston, R. B. 1972. Influence of birds, stones and soil on the establishment of pasture juniper, *Juniperus communis*, and red cedar, *J. virginiana* in New England. Ecology 53:1141–1147.

Lofty, J. R. 1974. Oligochaetes. pp. 467–488. In C. H. Dickinson and G. J. E. Pugh (eds.), Biology of Plant Litter Decomposition, Vol. 2. Academic Press, New York.

Lohm, U., and T. Persson. 1977. Soil organisms as components of ecosystems. Ecol. Bull. 25 (Stockholm). 614 pp.

Long, J. N., and J. Turner. 1975. Aboveground biomass of understory and overstory in an age sequence of four Douglas-fir stands. J. Appl. Ecol. 12:179–188.

Longman, K. A., and J. Jenik. 1974. Tropical Forest and Its Environment. Tropical Ecology Series. Longman, London. 196 pp.

Loomis, R. S., W. A. Williams, and W. G. Duncan. 1967. Community architecture and the productivity of terrestrial plant communities. pp. 291–308. In A. San Pietro, F. A. Greer, and T. J. Army (eds.), Harvesting the Sun. Academic Press, New York.

Loope, L. L., and G. E. Gruell. 1973. The ecological role of fire in the Jackson Hole Area, northwestern Wyoming. Quat. Res. 3:425–443.

Lopez-Quiles, M., and C. Vazquez-Yanes. 1976. Estudio sobre germinacion de semillas en condiciones naturales controladas. pp.250–262. In A. Gomez-Pompa, C. Vazquez-Yanes, S. del Amo, and A. Butanda (eds.). Regeneracion de Selvas. CECSA, Mexico, D.F.

Lotan, J. E. 1976. Cone serotiny-fire relationships in lodgepole pine. Proc. Tall Timbers Fire Ecol. Conf. 14:267–277.

Loucks, O. 1970. Evolution of diversity, efficiency, and community stability. Am. Zool. 10:17–25.

Loucks, O. 1976. The Study of Species Transients, Their Characteristics and Significance for Natural Resource Systems. Institute of Ecology, Indianapolis, Indiana. 62 pp.

Lugo, A. 1970. Photosynthetic studies on four species of rain forest seedlings. pp. I-81–102. In H. T. Odum and R. F. Pigeon (eds.), A Tropical Rain Forest: A Study of Irradiation and Ecology at El Verde, Puerto Rico. U.S. Atomic Energy Commission, Oak Ridge, Tennessee.

Luke, R. H., and A. G. McArthur. 1978. Bushfires in Australia. Australian Government Publishing Service, Canberra.

Lussenhop, J., R. Kumar, D. T. Wicklow, and J. E. Lloyd. 1980. Insect effects on bacteria and fungi in cattle dung. Oikos 34:54–58.

Lutz, H. J. 1956. Ecological effects of forest fires in the interior of Alaska. USDA For. Serv., Alaska Forest Research Center Tech. Bull. 1133. 121 pp.

Lyon, L. J., and P. F. Stickney. 1976. Early vegetal succession following large northern Rocky Mountain wildfires. Proc. Tall Timbers Fire Ecol. Conf. No. 14:355–375.

Macaloney, H. J. 1966. The impact of insects in the northern hardwoods type. USDA For. Ser. Res. Note NC-10. North Central Forest Experiment Station,

St. Paul, Minnesota. 3 pp.

MacArthur, R. H. 1972. Geographical Ecology. Harper and Row, New York. 269 pp.

MacBrayer, J. F. 1973. Energy flow and nutrient cycling in a cryptozoan food-web. Ph.D. Thesis. University of Tennessee, Knoxville.

MacFadyen, A. 1975. Some thoughts on the behavior of ecologists. J. Animal Ecol. 44:351–363.

Mackey, H. E., Jr., and N. Sivec. 1973. The present composition of a former oak-chestnut forest in the Allegheny Mountains of western Pennsylvania. Ecology 54:915–919.

MacLean, D. A., and R. W. Wein. 1976. Biomass of jack pine and mixed hard-wood stands in northeastern New Brunswick. Can. J. For. Res. 6:441–447.

MacMahon, J. A. 1976. Species and guild similarity of North American desert mammal faunas: A functional analysis of communities. pp.133–148. In D. W. Goodall (ed.), Evolution of Desert Biota. University of Texas Press, Austin.

MacMahon, J. A. 1979. North American deserts: Their floral and faunal components. pp. 21–82. In R. A. Perry and D. W. Goodall (eds.), Arid-land ecosystems: Structure, Functioning and Management. Vol. 1. International Biological Programme 16. Cambridge University Press, Cambridge, England.

MacMahon, J. A. 1980a. Ecosystems over time: Succession and other types of change. pp. 27–58. In R. Waring (ed.), Forests: Fresh Perspectives from Ecosystem Analyses. Oregon State University Press, Corvallis, Oregon.

MacMahon, J. A. 1980b. Response of desert ecosystems to the environment: The role of timing of events. pp. 261–269. In R.Perry and D. W. Goodall (eds.), Arid-Land Ecosystems: Structure, Functioning and Management, Vol. 2. Cambridge University Press, Cambridge, England.

MacMahon, J. A. 1981. Succession of ecosystems: A preliminary comparative analysis. In M. A. Hemstrom and J. F. Franklin (eds.), Proceedings, Second United States Union of Soviet Socialist Republic, Symposium in Biosphere Reserves. USDA Forest Service. Gen. Tech. Rep. (in press).

MacMahon, J. A., D. L. Phillips, J. V. Robinson, and D. J. Schimpf. 1978. Levels of biological organization: An organism-centered approach. BioScience 28:700–704.

MacMahon, J. A., and D. J. Schimpf. 1981. Water as a factor in the biology of North American desert plants. pp. 114–171. In D. Evans and J. Thames (eds.), Water in Desert Ecosystems. US/IBP Synthesis Series No. 12. Dowden, Hutchinson and Ross, Stroudsburg, Pennsylvania.

MacMahon, J. A., and F. H. Wagner. 1981. The Mojave, Sonoran and Chihuahuan deserts of North America. In M. Evenari and I. Noy-Meir (eds.), Warm Desert Ecosystems, Vol. 12, Ecosystems of the World. American Elsevier, New York. (in press).

MacMahon, J. A., and T. F. Wieboldt. 1978. Applying biogeographic principles to resource management: A case study evaluating Holdridge's life zone model. pp. 245–257. In K. Harper (ed.), Intermountain Biogeography, Symposium. Great Basin Naturalist Memoirs, No. 2, Brigham Young University Press, Provo, Utah.

Madge, D. S. 1965. Leaf fall and litter disappearance in a tropical forest. Pedobiologia 5:273–288.

Madgwick, H. A. I. 1976. Mensuration of forest biomass. pp. 13–27. In H. E.

Young (ed.), IUFRO Forest Biomass Studies. University of Maine Press, Orono.

Maikawa, E., and K. A. Kershaw. 1976. Studies on lichen-dominated systems. XIX. The postfire recovery sequence of black spruce-lichen woodland in the Abitau Lake region, N. W. T. Can. J. Bot. 54:2679–2687.

Maissurow, D. K. 1935. Fire as a necessary factor in the perpetuation of white pine. J. For. 33:373–378.

Maissurow, D. K. 1941. The role of fire in the perpetuation of virgin forests of northern Wisconsin. J. For. 39:201–207.

Malloch, D. W., K. A. Pirozynski, and P. H. Raven. 1980. Ecological and evolutionary significance of mycorrhizal symbioses in vascular plants (a review). Proc. Nat. Acad. Sci. (U.S.A.) 77:2113–2118.

Mann, L. K. Personal communication. Oak Ridge National Laboratory, Oak-Ridge, Tennessee. April 25, 1980.

Mares, M. A., and M. L. Rosenzweig. 1978. Granivory in North and South American deserts: Rodents, birds and ants. Ecology 59:235–241.

Margalef, R. 1963a. On certain unifying principles in ecology. Am. Nat. 97:357–374.

Margalef, R. 1963b. Succession in marine populations. Adv. Frontiers Plant Sci. 2:137–188.

Margalef, R. 1968. Perspectives in Ecological Theory. University of Chicago Press, Chicago. 111 pp.

Marie-Victorin, F. 1929. Le dynamisme dans la flore du Quebec. Contrib. Inst. Bot. Univ. Montreal No. 13. 77 pp.

Marks, P. L. 1974. The role of pin cherry (*Prunus pensylvanica* L.) in the maintenance of stability in northern hardwood ecosystems. Ecol. Monogr. 44:73–88.

Marks, P. L., and F. H. Bormann. 1972. Revegetation following forest cutting: Mechanisms for return to steady-state nutrient cycling. Science 176:914–915.

Marschner, F. J. 1930. The original vegetation of Minnesota. Map, full color, approx. 48–60 in. (text on reverse side by M. L. Heinselman). USDA For. Serv., North Central Forest Experiment Station, St. Paul, Minnesota, 1974.

Marshall, J. F., and B. G. Thom. 1976. The sea level in the last interglacial. Nature 243:120–121.

Marshall, J. K. 1978. The belowground ecosystem: A synthesis of plant-associated processes. Range Sci. Dept. Ser. 26. Colorado State University, Fort Collins. 351 pp.

Martin, J. L. 1966. The insect ecology of red pine plantations in central Ontario. IV. The crown fauna. Can. Entomol. 98:10–27.

Martin, P. S. 1958a. Pleistocene ecology and biogeography of North America. pp. 375–420. In C. Hubbs (ed.), Zoogeography. Am. Assoc. Adv. Sci., Publ. 15.

Martin, P. S. 1958b. Taiga-tundra and the full-glacial period in Chester County, Pennsylvania. Am. J. Sci. 256:470–502.

Martin, P. S. 1969. Pollen analysis and the scanning electron microscope. pp. 89–102. Proc., 2nd Annual Electron Microscope Symposium. IIT Research Institute, Chicago, Illinois.

Maruyama, K. 1971. Effect of altitude on dry matter production of primeval Japanese beech forest communities in Naeba Mountains. Mem. Fac. Agr. Niigata Univ. 9:87–171.

Maser, C., R. G. Anderson, K. Cromack, Jr., J. T. Williams, and R.E.Martin.

1979. Dead and down woody material. pp. 78–95. In J. W. Thomas (ed.), Wildlife Habitats in Managed Forests. USDA For. Serv. Agr. Handbook No. 553. Washington, D.C.

Maser, C., J. M. Trappe, and R. A. Nussbaum. 1978a. Fungal-small mammal interrelationships with emphasis on Oregon coniferous forests. Ecology 59:799–809.

Maser, C., J. M. Trappe, and D. C. Ure. 1978b. Implications of small mammal mycophagy to the management of western coniferous forests. Trans. 43rd N. A. M. Wildlife Nat. Res. Conf., Washington, D.C.

Mattson, W. J., and N. D. Addy. 1975. Phytophagous insects as regulators of forest primary production. Science 190:515–522.

Maxwell, J. A., and M. B. Davis. 1972. Pollen evidence of Pleistocene and Holocene vegetation on the Allegheny Plateau, Maryland. Quat. Res. 2:506–530.

May, R. M. 1976. Models for single populations. pp. 4–25. In R. M. May (ed.), Theoretical Ecology: Principles and Applications. Saunders, Philadelphia.

Mayr, E. 1961. Cause and effect in biology. Science 134:1501–1506.

McAndrews, J. H. 1967. Pollen analysis and vegetational history of the Itasca region, Minnesota. pp. 219–236. In E. J. Cushing and H. E. Wright (eds.), Quaternary Paleoecology. Yale University Press, New Haven, Connecticut.

McAndrews, J. H. 1970. Fossil pollen and our changing landscape and climate. Rotunda 3:30–37.

McAndrews, J. H. 1976. Fossil history of man's impact on the Canadian flora: An example from southern Ontario. Can. Bot. Assoc. Bull. Suppl. 9:1–6.

McAndrews, J. H. 1981. Late Quaternary climate of Ontario: Temperature trends from the fossil pollen record. In W. C. Mahaney (ed.), Quaternary Paleoclimate. Geoabstracts Ltd. (in press).

McArdle, R. E. 1949. The yield of Douglas-fir in the Pacific Northwest. USDA Tech. Bull. 201 (revised). Superintendent of Documents, Washington, D.C. 74 pp.

McArthur, A. G. 1967. Fire behaviour in eucalypt forests. Comm. Aust. For. Timber Bur. Leaflet 107.

McCauley, K. J., and S. A. Cook. 1980. *Phellinus weirii* infestation of two mountain hemlock forests in the Oregon Cascades. For. Sci. 26:23–29.

McColloch, J. W., and W. P. Hayes. 1922. The reciprocal relation of soil and insects. Entomol. Lab., Kansas State Univ. Contrib. 77, 3:288–301.

McCormick, I. 1968. Succession. Via 1:1–16.

McIlveen, W. D., and H. Cole, Jr. 1976. Spore dispersal of Endogonaceae by worms, ants, wasps, and birds. Can. J. Bot. 54:1486–1489.

McIntosh, R. P. 1963. Ecosystems, evolution and relational patterns of living organisms. Am. Sci. 246–267.

McIntosh, R. P. 1976. Ecology since 1900. pp. 353–372. In B.I.Taylor and T. J. White (eds.), Issues and Ideas in America. University of Oklahoma Press, Norman.

McIntosh, R. P. 1980a. The relationship between succession and the recovery process in ecosystems. pp. 11–62. In J. Cairns (ed.), The Recovery Process in Damaged Ecosystems. Ann Arbor Science Publ., Ann Arbor, Michigan.

McIntosh, R. P. 1980b. The background and some current problems of theoretical ecology. Synthese 43:195–255.

McKee, A., G. LaRoi, and J. F. Franklin. Structure, composition, and reproductive behavior of terrace forests, South Fork Hoh River, Olympic National Park. In Proceedings., Second Conference on Research in the National Parks, 1980. National Park Service, Washington, D.C. (in press).

McKey, D. 1975. The ecology of coevolved seed dispersal systems. pp. 159–191. In L. E. Gilbert and P. H. Raven (eds.), Coevolution of Animals and Plants. University of Texas Press, Austin, Texas.

McLean, S. F., G. K. Douce, E. A. Morgan, and M. A. Skeel. 1977. Community organization in the soil invertebrates of Alaskan arctic tundra. Ecol. Bull. (Stockholm) 25:90–101.

M'Closkey, R. T. 1975. Habitat succession and rodent distribution. J. Mammalogy 56:950–955.

McLaughlin, S. B., D. C. West, H. H. Shugart, and D. S. Shriner. 1978. Air pollution effects on forest growth and succession: Application of a mathematical model. pp. 1–16. In H.B.H.Cooper, Jr. (ed.), Air Pollution Control Association, Vol. 71, Houston, Texas.

McNabb, D. H., and J. M. Geist. 1979. Acetylene reduction assay of symbiotic N_2 fixation under field conditions. Ecology 60:1070–1072.

McNabb, D. H., C. T. Youngberg, and J. M. Geist. 1976. N_2 fixation by plants in forest and range ecosystems of northwest Oregon. Completion Rep. Coop. Aid Agreement, Oregon State University Press (Suppl. No. 86), Corvallis. 110 p.

McNaughton, S. J. 1979a. Grazing as an optimization process: Grass-ungulate relationships in the Serengeti. Am. Nat. 113:691–703.

McNaughton, S. J. 1979b. Grassland-herbivore dynamics. pp. 46–81. In A. R. E. Sinclair and M. Norton-Griffiths (eds.), Serengeti, Dynamics of an Ecosystem. University of Chicago Press, Chicago.

Means, J. E. Personal communication. USDA For. Serv., Corvallis, Oregon. May 25, 1980.

Meentmeyer, V. 1978. Macroclimate and lignin control of litter decomposition rates. Ecology 59:465–472.

Meeuwig, R. O. 1979. Growth characteristics of pinyon-juniper stands in the western Great Basin. USDA For. Serv. Res. Pap. INT-238. Intermountain Forest and Range Experiment Station, Ogden, Utah. 31 pp.

Meikeljohn, J. 1968. Numbers of nitrifying bacteria in some Rhodesian soils under natural grass and improved pastures. J. Appl. Ecol. 5:291–300.

Methven, I. R., C. E. Wagner, and B. J. Stocks. 1975. The vegetation on four burned areas in northwestern Ontario. Can. For. Serv., Petawawa For. Exp. Sta. Inform. Rep. PS-X-60.

Meyer, H. A. 1952. Structure, growth, and drain in balanced uneven aged forests. J. For. 50:85–92.

Meyer, H. A., and D. S. Stevenson. 1943. The structure and growth of virgin beech-birch-maple-hemlock forests in northern Pennsylvania. J. Agr. Res. 67:465–484.

Meyer, W. H. 1937. Yield of even-aged stands of Sitka spruce and western hemlock. USDA Agr. Tech. Bull. 544. 68 pp.

Mielke, D. L., H. H. Shugart, and D. C. West. 1977. User's manual for FORAR, a stand model for upland forests of southern Arkansas. ORNL/TM-5767. Oak Ridge National Laboratory, Oak Ridge, Tennessee.

Mielke, D. L., H. H. Shugart, and D. C. West. 1978. A stand model for upland

forests of southern Arkansas. ORNL/TM-6225. Oak Ridge National Laboratory, Oak Ridge, Tennessee.
Mielke, H. W. 1977. Mound building by pocket gophers (Geomyidae): Their impact on soils and vegetation in North America. J. Biogeogr. 4:171–180.
Miles, I. 1979. Vegetation Dynamics. Chapman and Hall, London. 80 pp.
Miller, P. C. 1981. Resource use by chaparral and matorral. Springer-Verlag, New York. 352 pp.
Miller, R. E., and M. D. Murray. 1978. The effects of red alder on growth of Douglas-fir. pp. 356–373. In D. G. Briggs, D. S. DeBell, and W. A. Atkinson (eds.). Utilization and management of alder. USDA For. Serv. Gen. Tech. Rep. PNW-70. Pacific Northwest Forest and Range Experiment Station, Portland, Oregon.
Miller, R. S., and D. B. Botkin. 1974. Endangered species: models and predictions. Am. Sci. 62:172–181.
Mills, P. F. L. 1966. Effects of nitrogen on the yields and quality of hay from three types of veld at Matopos. Rhod. Agr. J. 63:9–12, 18, 21.
Minore, D. 1972. Germination and early growth of coastal tree species on organic seedbeds. USDA For. Serv. Res. Pap. PNW-135. Pacific Northwest Forest and Range Experiment Station, Portland, Oregon. 18 pp.
Mitchell, H. L., and R. F. Chandler. 1939. The nitrogen nutrition and growth of certain deciduous trees of northeastern United States. Black Rock For. Bull. 11.
Moldenke, A. R. 1976. California pollination ecology and vegetation types. Phytologia 34:305–361.
Molina, R. 1979. Pure culture synthesis and host specificity of red alder mycorrhizae. Can. J. Bot. 57:1223–1228.
Muller, C. H. 1947. The effect of thinning, age and site on foliage, increment and loss of dry matter. J. For. 45:393–404.
Muler, C. H. 1954. The influence of thinning on volume increment. pp. 5–44. In Thinning Problems and Practices in Denmark. State University College of Forestry, Syracuse, New York.
Monk, C. D., G. I. Child, and S. A. Nicholson. 1970. Biomass, litter, and leaf surface area estimates of an oak-hickory forest. Oikos 21:138–141.
Monsi, M., and Y. Oshima. 1955. A theoretical analysis of the succession process of plant community, based upon the production of matter. Jpn. J. Bot. 15:60–82.
Mooney, H. A., and S. L. Gulmon. 1979. Environmental and evolutionary constraints on the photosynthetic characteristics of higher plants. pp. 316–337 In O. T. Solbrig, S. Jain, G. B. Johnson, and P. H. Raven (eds.), Topics in Plant Population Biology. Columbia University Press, New York.
Moore, J. M., and R. W. Wein. 1977. Viable seed populations by soil depth and potential site recolonization after disturbance. Can. J. Bot. 55:2408–2412.
Moore, R. M. 1960. The management of native vegetation in arid and semi-arid regions. pp. 173–190. In Plant-Water Relationships in Arid and Semi-Arid Conditions: Reviews of Research. UNESCO, Paris.
Moreno-Casasola, P. 1976. Viabilidad de semillas de arboles tropicales y templados: Una revision bibliografica. pp. 471–566. In A.Gomez-Pompa, C. Vazquez-Yanes, S. del Amo, and A. Butanda (eds.), Regeneracion de Selvas. CECSA, Mexico, D.F.

Morris, R. F. 1963. The dynamics of epidemic spruce budworm populations. Mem. Entomol. Soc. Can. 31:1–332.

Mott, R. J. 1975. Palynological studies of lake sediment profiles from southwestern New Brunswick. Can. J. Earth Sci. 12:273–288.

Muller, C. H. 1940. Plant succession in the Larrea-Flourensia climax. Ecology 21:206–212.

Muller, C. H. 1952. Plant succession in arctic health and tundra in northern Scandinavia. Bull. Torrey Bot. Club 79:296–309.

Munger, T. T. 1930. Ecological aspects of the transition from old forests to new. Science 72:327–332.

Munger, T. T. 1940. The cycle from Douglas-fir to hemlock. Ecology 59:799–809.

Munro, P. E. 1966. Inhibition of nitrite-oxidisers by roots of grass. J. Appl. Ecol. 3:227–229.

Murdoch, W. W., F. C. Evans, and C. H. Peterson. 1972. Diversity and pattern in plants and insects. Ecology 53:819–829.

Mutch, R. W. 1970. Wildland fires and ecosystems—a hypothesis. Ecology 51:1046–1051.

Nagel, E. 1952. Wholes, sums and organic entities. Philosophical Studies 3:17–32.

Nehmeth, J. C. 1968. The hardwood vegetation and soils of Hill Demonstration Forest, Durham County, North Carolina. J. Elisha Mitchell Sci. Soc. 84:482–491.

Newbould, P. J. 1967. Methods for estimating the Primary Production of Forests. IBP Handbook No. 2. Blackwell, Oxford. 62 pp.

Newton, M., B. A. El Hassan, and J. Zavitkovski. 1968. Role of red alder in western Oregon forest succession. pp. 73–84. In J.M.Trappe, J. F. Franklin, R. F. Tarrant, and G. H. Hansen (eds.), Biology of Alder. USDA For. Serv. Pacific Northwest Forest and Range Experiment Station, Portland, Oregon.

Nichols, G. E. 1923. A working basis for the ecological classification of plant communities. Ecology 4:11–23, 154–180.

Niering, W. A., and F. E. Egler. 1955. A shrub community of *Viburnum lentago*, stable for twenty-five years. Ecology 36:356–360.

Niering, W. A., and R. H. Goodwin. 1974. Creation of relatively stable shrublands with herbicides: Arresting "succession" on rights-of-way and pastureland. Ecology 55:784–795.

Niering, W. A., R. H. Whittaker, and C. H. Lowe. 1963. The Saguaro: A population in relation to environment. Science 142:15–23.

Noble, I. R. Personal communication. Australian National University, Canberra, ACT. August 25, 1980.

Noble, I. R., G. A. V. Bary, and A. M. Gill. 1980a. McArthur's fire danger meters expressed as equations. Aust. J. Ecol. 2:201–203.

Noble, I. R., H. H. Shugart, and J. S. Schauer. 1980b. A description of BRIND, a computer model of succession and fire response of the high altitude *Eucalyptus* forests of the Brindabella Range, Australian Capital Territory. ORNL/TM-7041. Oak Ridge National Laboratory, Oak Ridge, Tennessee. 96 pp.

Noble, I. R., and R. O. Slatyer. 1977. Post fire succession of plants in Mediterranean ecosystems. pp. 27–36. In H. A. Mooney and C.E. Conrad (eds.), Proc., Symposium on the Environmental Consequences of Fire and Fuel

Management in Mediterranean Ecosystems. USDA For. Serv. Gen. Tech. Rept. WO-3.
Noble, I. R., and R. O. Slatyer. 1980. The effects of disturbance on plant succession. Proc. Ecol. Soc. Aust. 10. (in press).
Norrish, K. Personal communication. Division of Soils, CSIRO Adeliade, South Australia, Australia, August 4, 1979.
Noy-Meir, I. 1973. Desert ecosystems: Environment and producers. Ann. Rev. Ecol. Syst. 4:25–51.
Noy-Meir, I. 1982. Stability of plant-herbivore models and possible applications to savanna. In B. J. Huntley and B. H. Walker (eds.), Ecology of Tropical Savannas. Springer-Verlag, Heidelburg. (in press).
Nye, P. H., and P. B. Tinker. 1977. Solute Movement in the Soil-Root System. University of California Press, Berkeley. 342 pp.
Odum, E. P. 1960. Organic production and turnover in old field succession. Ecology 41:34–49.
Odum, E. P. 1962. Relationships between structure and function in the ecosystem. Jpn. J. Ecol. 12:108–118.
Odum, E. P. 1964. The new ecology. BioScience 14:14–16.
Odum, E. P. 1968. Energy flow in ecosystems. A historical review. Am. Zool. 8:11–18.
Odum, E. P. 1969. The strategy of ecosystem development. Science 164:262–270.
Odum, E. P. 1971. Fundamentals of Ecology, 3rd ed. Saunders, Philadelphia, Pennsylvania. 574 pp.
Odum, E. P. 1977. The emergence of ecology as a new integrative discipline. Science 195:1289–1293.
Odum, H. T. 1962. Man and the ecosystem. Bull. Connecticut Agr. Exp. Sta., New Haven, Connecticut. 652:57–75.
Odum, H. T., B. J. Copeland, and R. Z. Brown. 1963. Direct and optical assay of leaf mass of the lower montane rain forest of Puerto Rico. Proc. Nat. Acad. Sci. 49:429–434.
Odum, H. T., and R. C. Pinkerton. 1955. Time's speed regulator, the optimum efficiency for maximum output in physical and biological systems. Am. Sci. 43:331–343.
Ogden, J. G., III. 1966. Forest history of Ohio: I. Radiocarbon dates and pollen stratigraphy of Silver Lake, Logan County, Ohio. Ohio J. Sci. 66:387–400.
Ogden, J. G., III. 1967. Radiocarbon and pollen evidence for a sudden change in climate in the Great Lakes region approximately 10,000 years ago. pp. 117–127. In E. J. Cushing and H. E. Wright, Jr. (eds.), Quaternary Paleoecology. Yale University Press, NewHaven, Connecticut.
Ohmann, L. F., and D. F. Grigal. 1979. Early revegetation and nutrient dynamics following the 1971 Little Sioux forest fire in northeastern Minnesota. For. Sci. Monogr. 21. 80 pp.
Oliver, C. D., and E. P. Stephens. 1977. Reconstruction of a mixed species forest in central New England. Ecology 58:562–572.
Olson, J. S. 1958. Rates of succession and soil changes on Southern Lake Michigan sand dunes. Bot. Gaz. 119:125–170.
O'Neill, R. 1976. Paradigms of ecosystem analysis. pp. 16–19. In S. A. Levin (ed.). Ecological Theory and Ecosystem Models. Institute of Ecology, Indianapolis, Indiana.
Oosting, H. J. 1942. An ecological analysis of the plant communities of Pied-

mont, North Carolina. Am. Midland Nat. 28:1–126.
Oosting, H. J., and P. J. Kramer. 1946. Water and light in relation to pine reproduction. Ecology 27:47–53.
Opler, P. A., H. G. Baker, and G. W. Frankie. 1977. Recovery of tropical lowland forest ecosystems. pp. 379–419. In J. Cairns, K. L. Dickson, and E. E. Herricks (eds.). Recovery and Restoration of Damaged Ecosystems. University Press of Virginia. Charlottesville.
Orians, G. H. 1962. Natural selection and ecological theory. Am. Nat. 96:257–263.
Orians, G. H. 1975. Diversity, stability, and maturity in natural ecosystems. pp. 139–150. In W. H. van Dobben and R. H. Lowe-McConnell (eds.), Unifying Concepts in Ecology. W. Junk, The Hague.
Orians, G. H. 1980. Micro and macro in ecological theory. BioScience 30:79.
Otto, J. H. 1938. Forest succession in the southern limits of early Wisconsin glaciation as indicated by a pollen spectrum from Bacon Swamp, Marion County, Indiana. Butler Univ. Bot. Stud. 4:93–116.
Overton, W. S. 1975. Ecosystem modeling approach in the coniferous biome. pp. 117–138. In B. C. Patten (ed.), Systems Analysis and Simulation in Ecology, Vol. 3. Academic Press, New York.
Ovington, J. D. 1957. Dry matter production by *Pinus sylvestris L.* Ann. Bot., Lond. n.s. 21:287–314.
Ovington, J. D., and H. A. I. Madgwick. 1959. The growth and composition of natural stands of birch. 1. Dry matter production. Plant Soil 10:271–283.
Ovington, J. D., and J. S. Olson. 1970. Biomass and chemical content of El Verde lower montane rain forest plants. pp. H-53–78. In H. T. Odum and R. F. Pigeon (eds.). A Tropical Rain Forest: A Study of Irradiation and Ecology at El Verde, Puerto Rico. U.S. Atomic Energy Commission, Oak Ridge, Tennessee.
Owen, D. D. 1860. Second Report of a Geological Reconnaissance of the Middle and Southern Counties of Arkansas. Sherman, Philadelphia, Pennsylvania. 207 pp.
Owen, D. F., and R. G. Wiegert. 1976. Do consumers maximize plant fitness? Oikos 27:488–492.
Parascandola, J. 1971. Organismic and holistic concepts in the thoughts of L. H. Henderson. J. Hist. Biol. 4:63–113.
Parkinson, D., T. R. G. Gray, and S. T. Williams. 1971. Methods for Studying the Ecology of Soil Micro-Organisms. IBP Handbook 19. Blackwell, Oxford. 116 pp.
Parkinson, D., S. Vissier, and I. B. Hawker. 1979. Effects of collumbolan grazing on fungal colonization of leaf litter. Soil Biol. Biochem. 11:529–535.
Parr, J. F., D. Parkinson, and A. G. Norman. 1967. Growth and activity of soil microorganisms in glass micro-beads. II. Oxygen uptake and direct observations. Soil Sci. 103:303–310.
Patten, B. C. 1975a. Ecosystem linearization: An evolutionary design problem. Am. Nat. 109:529–539.
Patten, B. C. 1975b. Ecosystem as a coevolutionary unit: A theme for teaching systems ecology. pp. 1–8. In G. S. Innis (ed.), New Directions in the Analysis of Ecological Systems. Sim. Counc. Proc. Ser. Vol. 5, No. 1. Simulation Councils, La Jolla, California.
Patten, B. C. 1975c. Discussion. pp. 234. In G. S. Innis (ed.), New Directions in

the Analysis of Ecological Systems. Sim. Counc. Proc. Ser. Vol. 5, No. 2. Simulation Councils, La Jolla, California.

Patten, B. C., D. A. Egloff, and T. H. Richardson. 1975. Total ecosystem model for a cove in Lake Texoma. pp. 205–421. In B.C.Patten (ed.), Systems Analysis and Simulation in Ecology, Vol. 3. Academic Press, New York.

Paul, E. A., and J. A. van Veen. 1978. The use of tracers to determine the dynamic nature of organic matter. pp. 61–102. In Symposium of 11th International Congress of Soil Sci., Alberta, Canada, Vol. 3.

Peek, J. M. 1974. Initial response of moose to a forest fire in northeastern Minnesota. Am. Midland Nat. 91:435–438.

Peet, R. K. 1981. Forest vegetation of the northern Colorado Front Range: Community composition and dynamics. Vegetatio. (in press).

Peet, R. K. 1981. Forest vegetation of the northern Colorado Front Range: Community composition and dynamics. Vegetatio. (inpress).

Peet, R. K., and N. L. Christensen. 1980a. Succession: A population process. Vegetatio 43:131–140.

Peet, R. K., and N. L. Christensen. 1980b. Hardwood forest vegetation of the North Carolina Piedmont. Verffentl. Geobot. Inst. ETH Stiftung Rubel, Zrich 69:14–39.

Peet, R. K., and O. P. Council. 1981. Rates of biomass accumulation in North Carolina piedmont forests. North Carolina Energy Institute, Triangle Research Park, North Carolina. (in press).

Penfound, W. T. 1964. The relation of grazing to plant succession in the tall grass prairie. J. Range Management 17:256–260.

Perry, T. O., H. E. Sellers, and C. O. Blanchard. 1969. Estimation of photosynthetically active radiation under a forest canopy with chlorophyll extracts and from basal area measurements. Ecology 50:39–44.

Petal, J. 1978. The role of ants in ecosystems. pp. 293–325. In M. V. Brian (ed.), Production Ecology of Ants and Termites. International Biological Programme 13. Cambridge University Press, Cambridge, England.

Peters, A., and T. Webb, III. 1979. A radiocarbon-dated pollen diagram from west-central Wisconsin. Ecol. Soc. Am. Bull. 60:102.

Peterson, D. L., and F. A. Bazzaz. 1978. Life cycle characteristics of *Aster pilosus* in early successional habitats. Ecology 59:1005–1013.

Phillips, D. C. 1970. Organisicism in the late nineteenth and early twentieth centuries. J. Hist. Ideas 31:413–432.

Phillips, J. F. V. 1930. Fire: Its influence on biotic communities and physical factors in South and East Africa. S. Afr. J. Sci. 27:352–367.

Phillips, J. F. V. 1974. Effects of fire in forest and savanna ecosystems of sub-Saharan Africa. pp. 435–481. In T. T. Kozlowski, and C. E. Ahlgren (eds.). Fire and Ecosystems. Academic Press, New York.

Phillipson, J. (ed.). 1971. Methods of Italy in Quantitative Soil Ecology: Population Production and Energy Flow. IBP Handbook 18. Blackwell, Oxford.

Pianka, E. R. 1978. Evolutionary Ecology. Harper and Row, New York, 397 pp.

Pianka, E. R. 1980. Units of natural selection. Science 207:1339–1340.

Pickett, S. T. A. 1976. Succession: An evolutionary interpretation. Am. Nat. 110:107–119.

Pickett, S. T. A. 1980. Non-equilibrium coexistence of plants. Bull. Torrey Bot. Club. 107:238–248.

Pinchot, G. 1906. The lumber cut of the United States: 1906. USDA For. Serv. Circ. 122. 42 pp.
Pinchot, G. 1907. Consumption of tanbark and tanning extract in 1906. USDA For. Serv. Circ. 119. 9 pp.
Pinchot, G., and W. W. Ashe. 1897. Timber trees and forests of North Carolina. N.C. Geol. Surv. Bull. No. 6. 224 pp.
Place, I. C. M. 1955. The influence of seed-bed conditions in the regeneration of spruce and balsam fir. Canada Dept. North Affairs Natl. Resources, Forestry Branch Bull. 117. 87 pp.
Platt, W. J. 1975. The colonization and formation of equilibrium plant species associations on badger disturbances in a tall-grass prairie. Ecol. Monogr. 45:285–305.
Plochmann, R. 1956. Bestockungsaufbau und Baumartenwandel nordischer Urwlder dargestellt an Beispielen aus Nordwestalberta/Kanada. Forstwiss. Forsch. 6:1–96.
Plumb, T. 1961. Sprouting of chaparral by December after a wildfire in July. USDA For. Ser. Tech. Pap. PSW-57, Pacific Southwest Forest and Range Experiment Station, Berkeley, California. 12 pp.
Polach, H. A. Personal communication. Radiocarbon Laboratory, The Australian National University, Canberra, ACT, Australia. October 27, 1978.
Polunin, N. 1937. The birch forests of Greenland. Nature 140:939–940.
Postgate, J. R. 1971. The acetylene test for nitrogenases. pp.311–316. In J. R. Postgate (ed.), The Chemistry and Biochemistry of Nitrogen Fixation. Academic Press, London.
Poulson, J. Personal communication. Department of Ecology and Systematics, Cornell University, Ithaca, New York. February 3, 1980.
Powell, P. E., G. R. Cline, C. P. P. Reid, and P. J. Szaniszlo. 1980. Occurrence of hydroxamate siderophore iron chelators in soils. Nature 287:833–834.
Pryor, L. D., and R. M. Moore. 1954. Plant communities. pp. 162–177. In Canberra, A Nation's Capital. Angus and Robertson, Sydney.
Purchase, B. S. 1974a. Evaluation of the claim that grass root exudates inhibit nitrification. Plant Soil 41:527–539.
Purchase, B. S. 1974b. The influence of phosphate deficiency on nitrification. Plant Soil 41:541–547.
Putman, R. J. 1979. Review of R. M. May. "Theoretical Ecology" (Blackwell, Oxford, 1976). Bull. Brit. Ecol. Soc. 10:88–89.
Radvanyi, A. 1970. Small mammals and regeneration of white spruce forests in western Alberta. Ecology 51:1102–1105.
Raison, R. J. 1980. Possible forest site deterioration associated with slash burning. Search 11:68–72.
Raup, H. M. 1957. Vegetational adjustment to the instability of site. pp. 36–48. In Proc. 6th Technical Meeting (1956), International Union for Conserving Nature and Natural Resources, Edinburgh.
Reeve, R., and I. F. Fergus. 1980. Characterization of waters entering and leaving podzolised sand dunes in south-eastern Queensland. Aust. Soc. Soil Sci. Inc. Water. Conf. Pap. 115, Sydney.
Reeve, R., I. F. Fergus, and C. H. Thompson. 1980. Studies in landscape dynamics in the Cooloola-Noosa River area. 5. Chemistry of the Waters. CSIRO (Australia), Div. Soils, Div. Rep.

Reichle, D. E., and S. I. Auerbach. 1972. Analysis of ecosystems. pp. 260–280. In J. A. Behnke (ed.), Challenging Biological Problems, Directions Towards Their Solution. Am. Inst. Biol. Sci. Arlington, Virginia.

Reichman, O. J. 1979. Desert granivore foraging and its impact on seed densities and distributions. Ecology 60:1085–1092.

Reiners, W. A. 1981. Nitrogen cycling in relation to ecosystem succession. In F. E. Clark and T. Rosswall (eds.). Terrestrial Nitrogen Cycles. Processes, Ecosystem Strategies, and Management Impacts. Ecol. Bull. (Stockholm) 33:(in press).

Reiners, W. A., and G. E. Lang. 1979. Vegetational patterns and processes in the balsam fir zone, White Mountains, New Hampshire. Ecology 60:403–417.

Richard, P. 1977. Histoire Post-Wisconsinienne de la Vegetation du Quebec Meridional par l'Analyse Pollinique, Tome 1, Tome 2. Gouvernement du Quebec, Ministere des Terres et Forets, Service de la Recherche.

Richards, B. N. 1974. Introduction to the Soil Ecosystem. Longmans, Green, New York. 266 pp.

Richards, P. W. 1952. The Tropical Rain Forest. Cambridge University Press, Cambridge, England. 450 pp.

Richardson, J. L. 1977. Dimensions of Ecology. Williams & Wilkins, Baltimore. 412 pp.

Richardson, J. L. 1980. The organismic community: Resilience of an embattled ecological concept. BioScience 30:465–471.

Ricklefs, R. E. 1977. Environmental heterogeneity and plant species diversity. Am. Nat. 111:376–381.

Ridley, H. N. 1930. The dispersal of plants throughout the world. William Clones, London. 432 pp.

Rieger, S., J. A. Dement, and D. Saunders. 1963. Soil survey of the Fairbanks area, Alaska. USDA Soil Conservation Service, No. 25, 41 pp.

Rieger, S., D. B. Schoephorster, and C. E. Furbush. 1979. Exploratory soil survey of Alaska. USDA Soil Conservation Service. 213 pp.

Rigler, F. H. 1975. The concept of energy flow and nutrient flow between trophic levels. pp. 15–26. In W. H. Van Dobben and R. H. Lowe-McConnell (eds.), Unifying Concepts in Ecology. W. Junk, The Hague.

Ritchie, J. C. 1962. A geobotanical survey of northern Manitoba. Arctic Institute of North America, Tech. Pap. 9. 24 pp.

Robinson, J. V. 1981. The effect of architectural variation in habitat on a spider community: An experimental field study. Ecology 62:73–80.

Rodin, L. E., and N. I. Bazilevich. 1967. Production and mineral cycling in terrestrial vegetation. Oliver & Boyd, Edinburgh. 288 pp.

Rogers, R. S. 1978. Forests dominated by hemlock (*Tsuga canadensis*): Distribution as related to site and postsettlement history. Can. J. Bot. 56:843–854.

Root, R. B. 1973. Organization of a plant-arthropod association in simple and diverse habitats: The fauna of collards (*Brassica oleracea*). Ecol. Monogr. 43:95–124.

Rose, S. L. 1980. Root symbionts and soil microorganisms associated with active mycorrhizal plants. Ph.D. Thesis. Oregon State University, Corvallis, Oregon.

Rose, S. L., and C. T. Youngberg. 1981. Tripartite associations in snowbrush: Effect of VA mycorrhizae on growth, nodulation and N fixation. Can. J. Bot. 59:34–39.

Ross, B. A., J. R. Gray, and W. H. Marshall. 1970. Effects of long-term deer exclusion on a *Pinus resinosa* forest in North-central Minnesota. Ecology 51:1088–1093.

Ross, D. J. 1977. Dry sieve analyses of soil samples from the Cooloola sandmass, Queensland. CSIRO (Aust.) Div. Soils Technical Manual-9/77. 25 pp.

Ross, H. H. 1962. A synthesis of evolutionary theory. Prentice Hall, Englewood Cliffs, New Jersey. 387 pp.

Ross, P. J. Personal communication. Division of Soils, CSIRO, Brisbane, Queensland, Australia. April 25, 1980.

Ross, R. 1954. Ecological studies of the rain forest of southern Nigeria. III. Secondary succession in the Sasha Forest Reserve. J. Ecol. 42:259–282.

Roux, E. 1969. Grass—A story of frankenwald. Oxford University Press, Cape Town. 212 pp.

Rowe, J. S. 1961. The level of integration concept and ecology. Ecology 42:420–427.

Rowe, J. S. 1970. Spruce and fire in northwest Canada and Alaska. Proc., 10th Annu. Tall Timbers Fire Ecol. Conf. 10:245–254.

Rowe, J. S. 1972. Forest regions of Canada. Can. Dept. Environ., Can. For. Serv. Publ. 1300. 172 pp.

Rowe, J. S. 1979. Large fires in the large landscapes of the north. pp. 8–32. In Proceedings, Symposium of Fire Management in the Northern Environment. U.S. Dept. of Interior, BLM, Alaska State Office.

Rowe, J. S. 1981. Concepts of fire effects on plant individuals and species. In R. W. Wein and D. A. MacLean (eds.), Symposium Proceedings, Fire in Northern Circumpolar Ecosystems. Wiley, New York. (in press).

Rowe, J. S., J. L. Bergsteinsson, G. A. Padbury, and R. Hermesh. 1974. Fire studies in the Mackenzie Valley. Can. Dept. Indian Affairs Northern Develop., INA Publ. No. QS-1567-000-EE-Al. ALUR 73-74-61. 123 pp.

Rowe, J. S., and G. W. Scotter. 1973. Fire in the boreal forest. Quat. Res. 3:444–464.

Rowe, J. S., D. Spittlehouse, E. Johnson, and M. Jasieniuk. 1975. Fire studies in the Upper Mackenzie Valley and adjacent Precambrian Uplands. Can. Dept. Indian Affairs Northern Develop., ALUR 74-75-61. 90 pp.

Ruehle, H. L., and D. H. Marx. 1979. Fiber, food, fuel and fungal symbionts. Science 206:419–422.

Runkle, J. R. 1979. Gap phase dynamics in climax mesic forests. Ph.D. Thesis, Cornell University, Ithaca, New York.

Rushworth, J. E. 1975. The floristic, physiognomic and biomass structure of Kalahari sand shrub vegetation in relation to fire and frost in Wankie National Park, Rhodesia. M.Sc. Thesis, University of Rhodesia, Salisbury.

Russell, E. J. 1921. Soil conditions and plant growth. Rothamsted Monographs on Agricultural Sciences. 90 pp.

Russell, E. W. 1973. Soil Conditions and Plant Growth (10th ed.). Longmans, Green, London. 849 pp.

Salomonson, M. G. 1978. Adaptations for animal dispersal of one-seed juniper seeds. Oecologia 32:333–339.

Salt, G. W. 1979. A comment on the use of the term emergent properties. Am. Nat. 113:145–148.

Santos, P. F., E. DePree, and W. G. Whitford. 1978. Spatial distribution of litter and microarthropods in a Chihuahuan desert ecosystem. J. Arid Environ. 1:41–48.

Sargent, C. S. 1884. Report on the Forest of North America. Misc. Doc. 42, Part 9. U.S. Government Printing Office, Washington, D.C. 612 pp.

Satchell, J. E. 1967. Lumbricidae. pp. 259–322. In A. Burgess and F. Raw (eds.), Soil Biology. Academic Press, London.

Schaeffer, R., and R. Moreau. 1958. L'alternance des essences. Soc. For.. France Compte Bull. 29:1–12, 76–84, 277–298.

Schimpf, D. J., J. A. Henderson, and J. A. MacMahon. 1980. Some aspects of succession in the spruce-fir forest zone of northern Utah. Great Basin Nat. 40:1–26.

Schlesinger, W. H., and D. S. Gill. 1978. Demographic studies of the chaparral shrub, *Ceanothus megacarpus*, in the Santa Ynez Mountains, California. Ecology 59:1256–1263.

Schmelz, D. V., and A. A. Lindsey. 1965. Size-class structure of old-growth forests in Indiana. For. Sci. 11:731–743.

Schoenike, R. E. 1976. Geographic variations in jack pine (*Pinus banksiana*). Univ. Minnesota Agr. Exp. Sta. Tech. Bull. 304. 14 pp.

Schopmeyr, C. 1974. Seeds of Woody Plants in the United States. USDA For. Serv. Agriculture Handbook No. 450. Washington, D.C.

Schroeder, M. J., and C. C. Buck. 1970. Fire weather—a guide for application of meteorological information to forest fire control operations. USDA For. Serv. Agriculture Handbook 360. 229 pp.

Schulz, J. P. 1960. Ecological studies on rain forest in northern Suriname. The vegetation of Suriname. Verhandel. Koninkl. Ned. Akad. Wetensch. Afdel. Natuurk. Sect. 2, 53:1–267.

Scotter, G. W. 1964. Effects of forest fires on the winter range of barren ground caribou in northern Saskatchewan. Can. Wildlife Serv. Wildlife Management Bull. Ser. 1, No. 18. 111 pp.

Shane, L. C. K. 1975. Palynology and radiocarbon chronology of Battaglia Bog, Portage County, Ohio. Ohio J. Sci. 75:96.

Shane, L. C. K. 1980. Detection of a late-glacial climatic shift in central midwestern pollen diagrams. pp. 171–172. AMQUA Abstr. Program, 6th Biennial Meetings. Orono, Maine.

Sharpe, D. M., K. Cromack, Jr., W. C. Johnson, and B. S. Ausmus. 1980. A regional approach to litter dynamics in southern Appalachian forests. Can. J. For. Res. 10:395–404.

Shelford, V. E. 1963. The Ecology of North America. University of Illinois Press, Urbana. 603 pp.

Shelford, V. E., and S. Olson. 1935. Sere, climax and influent animals with special reference to the transcontinental coniferous forest of North America. Ecology 16:375–402.

Shreve, F. 1951. Vegetation of the Sonoran Desert. Vegetation and Flora of the Sonoran Desert (F. Shreve and I.L.Wiggins), Vol.1. Carnegie Institute Publ., Washington, D.C. 591 pp.

Shugart, H. H., Jr., T. R. Crow, and J. M. Hett. 1973. Forest succession models: A rationale and methodology for modeling forest succession over large regions. For. Sci. 19:203–212.

Shugart, H. H., W. R. Emanuel, D. C. West, and D. L. DeAngelis. 1980b. Environmental gradients in a simulation model of a beech-yellow poplar stand. Math. Biosci. 50:163–170.

Shugart, H. H., M. S. Hopkins, I. P. Burgess, and A. T. Mortlock. 1981. The development of a succession model for subtropical rainforest and its application to assess the effects of timber harvest at Wiangaree State Forest, New South Wales. J. Environ. Management. 11:243–265.

Shugart, H. H., A. T. Mortlock, M. S. Hopkins, and I. P. Burgess. 1980a. A computer simulation model of ecological succession in Australian subtropical rainforest. ORNL/TM-7029. Oak Ridge National Laboratory, Oak Ridge, Tennessee. 48 pp.

Shugart, H. H., and I. R. Noble. 1981. A computer model of succession and fire response of the high altitude *Eucalyptus* forests of the Brindabella Range, Australia Capital Territory. Aust. J. Ecol. 6:(in press).

Shugart, H. H., and R. V. O'Neill. 1979. Systems Ecology. Dowden, Hutchinson and Ross, Stroudsburg, Pennslyvania. 368 pp.

Shugart, H. H., and D. C. West. 1977. Development of an Appalachian deciduous forest succession model and its application to assessment of the impact of the chestnut blight. J. Environ. Management. 5:161–179.

Shugart, H. H., and D. C. West. 1980. Forest succession models. BioScience 30:308–313.

Siegel, S. 1956. Non Parametric statistics for the Behavioral Sciences. McGraw-Hill, New York. 312 pp.

Silvester, W. B. 1980. Personal communication. Biological Sciences Department, University of Waikato, Hamilton, New Zealand.

Silvester, W. B. 1978. Nitrogen fixation and mineralization in Kauri (*Agathis australis*) forests in New Zealand. pp. 138–143. In M.W.Loutit and J. A. R. Miles (eds.), Microbial Ecology: Proceedings in Life Sciences. Springer-Verlag, Berlin.

Simard, A. J. 1975. Wildland fire occurrence in Canada (map). Can. Dept. Environ., Can. For. Serv., Montreal.

Simberloff, D. S. 1980. A succession of paradigms in ecology, essentialism to materialism and probabilism. Synthese 42:3–39.

Sinclair, A. R. E. 1975. The resource limitation of trophic levels in tropical grassland ecosystems. J. Animal Ecol. 44:497–520.

Sinclair, A. R. E., and M. Norton-Griffiths. 1979. Serengeti: Dynamics of an Ecosystem. University of Chicago Press, Chicago. 309 pp.

Singh, G. 1966. Ectotophic mycorrhizae in equatorial rain forests. Malay. For. 29:13–16.

Siren, G. 1955. The development of spruce forest on raw humus sites, in northern Finland and its ecology. Acta For. Fennica 62:1–363.

Sirkin, L. A., C. S. Denny, and M. Rubin. 1977. Late Pleistocene environment of the central Delmarva Peninsula, Delaware-Maryland. Geol. Soc. Am. Bull. 88:139–142.

Sjörs, H. 1959. Bogs and fens in the Hudson Bay lowlands. Arctic 12:2–19.

Sjörs, H. 1961. Surface patterns in boreal peatland. Endeavour 20:217–224.

Slagsvold, T. 1980. Habitat selection in birds: On the presence of other bird species with special regard to *Turdus pilaris*. J. Animal Ecol. 49:523–536.

Small, E. 1972. Photosynthetic rates in relation to nitrogen recycling as an adaptation to nutrient deficiency in peat bog plants. Can. J. Bot. 50:2227–2233.

Smith, A. J. 1974. Invasion and ecesis of bird-disseminated woody plants in a temperate forest sere. Ecology 56:19–34.

Smith, C. C. 1970. The coevolution of pine squirrels (*Tamiasciurus*) and conifers. Ecol. Monogr. 40:349–371.

Smith, C. C. 1975. The coevolution of plants and seed predators. pp. 53–77. In L. E. Gilbert and P. H. Raven (eds.), Coevolution of Animals and Plants. University of Texas Press, Austin, Texas.

Smith, C. C., and D. Follmer. 1972. Food preferences of squirrels. Ecology 53:82–91.

Smith, F. E. 1970. Analysis of ecosystems. pp. 7–18. In D. E. Reichle (ed.), Temperate Forest Ecosystems. Springer-Verlag, Berlin.

Smith, F. E. 1975a. Comments revised—or, what I wish I had said. pp. 231–235. In G. S. Innis (ed.), New Directions in the Analysis of Ecological Systems. Sim. Counc. Proc. Ser. Vol. 5, No. 2. Simulation Councils, La Jolla, California.

Smith, F. E. 1975b. Ecosystems and evolution. Bull. Ecol. Soc. Am. 56:2–6.

Smith, F. E. 1980. Personal communication. Department of Ecology and Systematics, Cornell University, Ithaca, New York.

Smith, K. G., and J. A. MacMahon. 1981. Bird communities along a montane sere: Community structure and energetics. Auk 98:8–28.

Smith, T. M., H. H. Shugart, and D. C. West. 1981. FORHAB. A forest simulation model to predict habitat structure for non-game bird species. In D. E. Capen (ed.), Symposium on the Use of Multivariate Statistics in Studies of Wildlife Habitat. University of Vermont Press, Burlington. (in press).

Sneath, P. H. A., and R. R. Sokal. 1973. Numerical taxonomy: The principles and practice of numerical classification. Freeman, SanFrancisco. 573 pp.

Snedaker, S. C. 1970. Ecological studies on tropical moist forest succession in eastern lowland Guatemala. Ph.D. Thesis. University of Florida, Gainesville.

Snow, D. W. 1971. Evolutionary aspects of fruit-eating by birds. Ibis 113:194–202.

Soderstrom, B. E. 1979. Seasonal fluctuations of active fungal biomass in horizons of a podzolized pine-forest soil in central Sweden. Soil Biol. Biochem. 11:149–154.

Soil Survey Staff. 1975. Soil taxonomy: A basic system of soil classification for making and interpreting soil surveys. USDA Agriculture Handbook 436. U.S. Govt. Printing Office, Washington, D.C. 754 pp.

Sokal, R. R., and F. J. Rohlf. 1969. Biometry. Freeman, San Francisco. 776 pp.

Sollins, P., and R. M. Anderson. 1971. Dry-weight and other data for trees and woody shrubs of the southeastern United States. ORNL-IBP-71-6, Oak Ridge National Laboratory, Oak Ridge, Tennessee. 80 pp.

Sollins, P., C. C. Grier, F. M. McCorison, K. Cromack, Jr., R. Fogel, and R. L. Fredriksen. 1980. The internal element cycles of an old-growth Douglas-fir ecosystem in western Oregon. Ecol. Monogr. 50:261–285.

Solomon, A. M. 1970. Effects of suburbanization upon airborne pollen in northeastern New Jersey. Ph.D. Thesis. Rutgers University, New Brunswick, New Jersey.

Solomon, A. M., H. R. Delcourt, D. C. West, and T. J. Blasing. 1980. Testing a simulation model for reconstruction of prehistoric forest-stand dynamics. Quat. Res. 14:275–293.

Solomon, A. M., and D. F. Kroener. 1971. Suburban replacement of rural land uses reflected in the pollen rain of northeastern NewJersey. New Jersey Acad. Sci. Bull. 16:30–44.

Solomon, A. M., and J. B. Harrington, Jr. 1979. Palynology models. pp. 338–361. In R. L. Edmonds (ed.), Aerobiology: The Ecological Systems Approach. Dowden, Hutchinson and Ross, Shroudsburg, Pennsylvania.

Sousa, M. 1964. Estudio de la vegetacion secundaria en la Region de Tuxtepec, Oax. Publ. Esp. Inst. Nal. Invest. Forest. 3:91–105.

Southwood, T. R. E., V. K. Brown, and P. M. Reader. 1979. The relationships of plant and insect diversities in succession. J. Lin. Soc. Biol. 4:327–345.

Spear, R. W., and N. G. Miller. 1976. A radiocarbon-dated pollen diagram from the Allegheny Plateau of New York State. J. Arnold Arbor. 57:369–403.

Sprugel, D. G. 1976. Dynamic structure of wave-regenerated *Abies balsamea* forests in the north-eastern United States. J. Ecol. 64:889–912.

Spurr, S. H. 1954. The forests of Itasca in the ninteenth century as related to fire. Ecology 35:21–25.

Spurr, S. H. 1960. Photogrammetry and Photo-Interpretation, 2nd ed. Ronald Press, New York. 235 pp.

Spurr, S. H., and J. H. Zumberge. 1956. Late Pleistocene features of Cheboygan and Emmet Counties, Michigan. Am. J. Sci. 254:96–109.

Spycher, G., and J. L. Young. 1977. Density fractionation of water-dispersible soil organic-mineral particles. Commun. Soil Sci. Plant Anal. 8:37–48.

Spycher, G., and J. L. Young. 1979. Water-dispersible soil organic-mineral particles. II. Inorganic amorphous and crystalline phases in density fractions of clay-size particles. Soil Sci. Soc. Am. J. 43:328–332.

Stace, H. C. T., G. D. Hubble, R. Brewer, K. H. Northcote, J.R.Sleeman, and M. J. Mulcahy. 1968. A Handbook of Australian Soils. Rellum Tech. Publ. Glenside, S. Australia. 435 pp.

Stark, H. 1978. Man, tropical forests, and the biological life of a soil. Biotropica 10:1–10.

Stephens, E. P. 1956. The uprooting of trees: A forest process. Soil Sci. Soc. Am. Proc. 20:113–116.

Stephens, G. R., and P. E. Waggoner. 1970. The forests anticipated from 40 years of natural transition in mixed hardwoods. Bull. Connecticut Agr. Exp. Sta. New Haven 707:1–58.

Stevens, P. R., and T. W. Walker. 1967. The chronosequence concept and soil formation. Quart. Rev. Biol. 45:333–350.

Stocks, B. J., and G. R. Hartley. 1979. Forest fire occurrence in Ontario. Great Lakes Forest Research Centre, Canadian Forestry Serv. (map with text).

Stone, E. L., W. T. Swank, and J. W. Hornbeck. 1978. Impacts of timber harvest and regeneration systems on stream flow and soils in the eastern deciduous forest region. pp. 516–535. In C.T.Youngberg (ed.), Forest Soils and Land Use. Colorado State University Press, Fort Collins, Colorado.

Strang, R. M. 1973. Succession in unburned subarctic woodlands. Can. J. For. Res. 3:140–143.

Strang, R. M. 1974. Some man-made changes in successional trends on the Rhodesian highveld. J. Appl. Ecol. 11:249–263.

Stratton, G. E., E. W. Uetz, and D. G. Dillery. 1979. A comparison of the spiders of three coniferous tree species. J. Arachnol. 6:219–226.

Strong, D. R., Jr. 1977. Epiphyte loads, tree-falls, and perennial forest disruption: A mechanism for maintaining higher tree species richness in the tropics without animals. J. Biogeogr. 4:215–218.

Strong, D. R., Jr. 1979. Biogeographic dynamics of insect-host plant communities. Ann. Rev. Entomol. 24:89–119.

Strong, D. R., Jr., and D. A. Levin. 1979. Species richness of plant parasites and growth form of their hosts. Am. Nat. 114:1–22.

Stubblefield, G., C. D. Oliver, and C. Dearing. 1978. Silvicultural implications of the reconstruction of mixed alder/conifer stands. pp. 307–320. In D. G. Briggs, D. S. DeBell, and W.A.Atkinson (eds.). Utilization and Management of Alder. USDA For. Serv. Gen. Tech. Rep. PNW-70. Pacific Northwest Forest and Range Experiment Station, Portland, Oregon.

Swain, A. M. 1973. A history of fire and vegetation in northeastern Minnesota as recorded in lake sediment. Quat. Res. 3:383–396.

Swain, A. M. 1978. Environmental changes during the past 2000 years in North Central Wisconsin: Analysis of pollen, charcoal, and seeds from varied lake sediments. Quat. Res. 10:55–68.

Swaine, J. M. 1933. The relation of insect activities to forest development as exemplified in the forests of eastern North America. Sci. Agr. 14:8–31.

Swank, W. T., and H. T. Schreuder. 1974. Comparison of three methods of estimating surface area and biomass for a forest of young eastern white pine. For. Sci. 20:91–99.

Swank, W. T., and J. B. Waide. 1980. Interpretation of nutrient cycling research in a management context—Evaluating potential effects of alternative management strategies on site productivity. pp. 137–158. In R. Waring (ed.), Forests Fresh Perspectives from Ecosystem Analysis. Proc. 40th Biol. Colq. Oregon State University Press, Corvallis, Oregon.

Swift, M. J., O. W. Heal, and J. M. Anderson. 1979. Decomposition in terrestrial ecosystems. Studies in Ecology, Vol. 5. University of California Press, Berkeley. 372 pp.

Switzer, G. I., I. E. Nelson, and W. H. Smith. 1966. The characterization of dry matter and nitrogen accumulation by loblolly pine (*Pinus taeda L.*). Proc. Am. Soc. Soil Sci. 30:114–119.

Symington, C. F. 1933. The study of secondary growth on rain forest sites in Malaya. Malay. For. 2:107–117.

Tamm, C. O. 1964. Growth of *Hylocomium splendens* in relation to tree canopy. Bryologist 67:423–426.

Tande, G. F. 1979. Fire history and vegetation pattern of coniferous forests in Jasper National Park, Alberta. Can. J. Bot. 1912–1931.

Tansley, A. G. 1920. The classification of vegetation and the concept of development. J. Ecol. 8:118–149.

Tansley, A. G. 1935. The use and abuse of vegetational concepts and terms. Ecology 16:284–307.

Tarrant, R. F., and R. E. Miller. 1963. Accumulation of organic matter and soil nitrogen beneath a plantation of red alder and Douglas-fir. Soil Sci. Soc. Am. Proc. 27:231–234.

Tauber, H. 1965. Differential pollen dispersion and the interpretation of pollen diagrams. Dan. Geol. Unders. Ser. 2, 89:1–69.

Tauber, H. 1977. Investigations of aerial pollen transport in a forested area. Dan.

Bot. Archive 32(1):1–121.
Taylor, R. D., and B. H. Walker. 1978. A comparison of vegetation use and animal biomass on a Rhodesian game and cattle ranch. J. Appl. Ecol. 15:565–581.
Terasmae, J., and T. W. Anderson. 1970. Hypsithermal range extension of white pine (*Pinus strobus L.*) in Quebec, Canada. Can. J. Earth Sci. 7:406–413.
Tergas, L. E., and H. L. Popenoe. 1971. Young secondary vegetation and soil interactions in Izabal, Guatemala. Plant Soil 34:675–690.
Tevis, L., Jr. 1956. Pocket gophers and seedlings of red fir. Ecology 37:379–381.
Thom, B. G., and J. Chappell. 1975. Holocene sea levels relative to Australia. Search 6:90–93. land use conflicts. Proc. R. Soc. Queensland 86:109–120.
Thompson, C. H. 1980. Podzols on coastal sand dunes at Cooloola, Queensland. Aust. Soc. Soil Sci. Inc. Nat. Conf. Pap. 117, Sydney. 24 pp.
Thompson, C. H., and G. D. Hubble. 1979. Sub-tropical podzols (spodosols and related soils) of coastal eastern Australia: Profile form, classification and use. pp. 181–197. In Proc. Com. IV and V, Int. Soil Sci. Soc. CLAMATROPS Conf. August 1977, Kuala Lumpur, Malaysia.
Thompson, C. H., and A. W. Moore. 1980. Studies in landscape dynamics in the Cooloola-Noosa River area. 1. Introduction and general description. CSIRO Australia, Div. Soils, Div. Rep. No. 70. 47 pp.
Thompson, C. H., and W. T. Ward. 1975. Soil landscapes of North Stradbroke Island. Proc. R. Soc. Queensland 86:9–14.
Thompson, J. N., and M. F. Willson. 1978. Disturbance and the dispersal of fleshy fruits. Science 200:1161–1163.
Thompson, P. J. 1975. The role of elephants, fire and other agents in the decline of *Brachystegia boehmii* woodland. J. S. Afr. Wildl. Management Assoc. 5:11–18.
Thornburgh, D. A. 1969. Dynamics of the true fir-hemlock forests on the west slope of the Washington Cascade Range. Ph.D. Thesis, University of Washington, Seattle, Washington.
Thrower, N. J. W., and D. E. Bradbury. 1977. Chile-California Mediterranean Scrub Atlas. Dowden, Hutchinson, and Ross, Stroudsburg, Pennslyvania. 237 pp.
Tinley, K. L. 1977. Framework of the Gorongosa ecosystem. D.Sc. Thesis, University of Pretoria, S. Africa.
Tomlinson, P. B., and M. H. Zimmerman. 1978. Tropical trees as living systems. Proc. 4th Cabot Symp. Harvard Forest (1976). Cambridge University Press, Cambridge, England. 675 pp.
Trapnell, C. G. 1959. Ecological results of woodland burning experiments in Northern Rhodesia. J. Ecol. 47:129–168.
Trappe, J. M. 1979. Mycorrhiza-nodule-host interrelationships in symbiotic nitrogen fixation: A quest in need of questers. pp.276–286. In J. C. Gordon, C. T. Wheeler, and D. A. Perry (eds.), Symbiotic Nitrogen Fixation in the Management of Temperate Forests. Oregon State University Press, Corvallis.
Trejo, L. 1976. Diseminacion de semillas por aves en "Los Tuxtlas" Veracruz. pp. 447–470. In A. Gomez-Pompa, C. Vazquez-Yanes, S. del Amo, and A. Butanda (eds.), Regeneracion de Selvas. CECSA, Mexico, D.F.
Triska, F. J., and K. Cromack, Jr. 1980. The role of wood debris in forests and streams. pp. 171–190. In R. H. Waring (ed.), Forests: Fresh Perspectives from

Ecosystems Analysis. Proc. 90th Biol. Colloq. Oregon State University Press, Corvallis, Oregon.

Trollope, W. S. W. 1974. The role of fire in preventing bush encroachment in the Eastern Cape. Proc. Grasslands Soc. S. Afr. 9:67–71.

Tubbs, C. H. 1973. Allelopathic relationship between yellow birch and sugar maple seedlings. For. Sci. 19:139–145.

Tubbs, C. H. 1976. Effect of sugar maple root exudate on seedlings of northern conifer species. USDA For. Serv., Res. Note NC-213. North Central Forest Experiment Station, St. Paul, Minnesota. 8pp.

Turcek, F. J., and L. Kelso. 1968. Ecological aspects of food transportation and storage in the Corvidae. Commun. Behav. Biol. A 1:277–297.

Turchenek, L. W., and J. M. Oades. 1979. Fractionation of organo-mineral complexes by sedimentation and density techniques. Geoderma 21:311–343.

Turner, J. 1975. Nutrient cycling in a Douglas-fir ecosystem with respect to age and nutrient status. Ph.D. Thesis. University of Washington, Seattle, Washington.

Turner, J. 1977. Effect of nitrogen availability on nitrogen cycling in a Douglas-fir stand. For. Sci. 23:307–316.

Ugolini, F. C. 1968. Soil development and alder invasion in a recently deglaciated area of Glacier Bay, Alaska. pp. 115–140. In J. M. Trappe, J. F. Franklin, R. F. Tarrant, and G. M. Hansen (eds.), Biology of Alder. USDA For. Serv. Pacific Northwest Forest and Range Experiment Station, Portland, Oregon.

Ugolini, F. C., and D. Mann. 1979. Biopedological origin of peatlands in southeast Alaska. Nature 281:366–368.

Ulanowicz, R. E. 1979. Prediction, chaos and ecological perspective. pp. 107–117. In E. A. Halfon (ed.), Theoretical Systems Ecology. Academic Press, New York.

Ulanowicz, R. E., and W. M. Kemp. 1979. Toward canonical trophic aggregations. Am. Nat. 114:871–883.

United States Geological Survey. 1965. Monthly average temperatures for January and July. The National Atlas of the United States of America (1970). U.S. Govt. Printing Office, Washington, D.C. 417pp.

Usher, M. B., and T. W. Parr. 1977. Are there successional changes in arthropod decomposer communities? J. Environ. Management 5:151–160.

Valentine, L. W. 1968. The evolution of ecological units above the population level. J. Paleontol. 42:253–267.

Valiela, I. 1971. Food specificity and community succession: Preliminary ornithological evidence for a general framework. Gen. Syst. 16:77–84.

Valiela, I. 1974. Composition, food webs and population limitation in dung arthropod communities during invasion and succession. Am. Midland Nat. 92:370–385.

Van Cleve, K., T. Dyrness, and L. A. Viereck. 1980. Nutrient cycling in interior Alaska floodplains and its relationship to regeneration and subsequent forest development. pp. 11–18. In M.Murray and R.M.VanVeldhuizan (eds.), Forest Regeneration at High Latitudes. Gen. Tech. Rep. PNW-107. 52 pp.

Van Cleve, K., and L. Noonan. 1975. Litterfall and nutrient cycling in the forest floor of birch and aspen stands in interior Alaska. Can. J. For. Res. 5:626–639.

Van Cleve, K., and L. A. Viereck. 1972. Distribution of selected chemical elements in even-aged alder (*Alnus*) ecosystems near Fairbanks, Alaska. Arctic Alpine Res. 4:239–255.

Van Cleve, K., L. A. Viereck, and R. L. Schlentner. 1971. Accumulation of nitrogen in alder (*Alnus*) ecosystems near Fairbanks, Alaska. Arctic Alpine Res. 3:101–114.

Van der Drift, J., and M. Witkamp. 1959. The significance of the breakdown of oak litter by *Enoicyla pusilla* Burn. Arch. Neerl. Zool. 13:486–492.

van der Hammen, T., T. A. Wijmstra, and W. H. Zagwijn. 1971. The floral record of the late Cenozoic of Europe. pp. 391–424. In K.K.Turekian (ed.), The Late Cenozoic Glacial Ages. Yale University Press, New Haven, Connecticut.

van der Pijl, L. 1972. Principles of Dispersal in Higher Plants, 2nd ed. Springer-Verlag, Berlin. 162 pp.

Van Dyne, G. M. 1966. Ecosystems, systems ecology, and systems ecologists. ORNL-3957. Oak Ridge National Laboratory, Oak Ridge, Tennessee. 29 pp.

Van Dyne, G. M. 1972. Organization and management of an integrated ecological research program—with special emphasis on systems analysis, universities, and scientific cooperation. pp. 111–172. In J. N. R. Jeffers (ed.), Mathematical Models in Ecology. Blackwell, Oxford.

Van Loon, H., and J. Williams. 1976. The connection between trends of mean temperature and circulation at the surface: Part 1, Winter. Monthly Weather Rev. 104:365–380.

Van Soest, P. J. 1963. Use of detergents in the analysis of fibrous feeds. II. A rapid method for the determination of fiber and lignin. J. Assoc. Off. Agric. Chem. 46:829–835.

van Steenis, C. G. G. J. 1958. Rejuvenation as a factor for judging the status of vegetation types: The biological nomad theory. pp.212–215. In Study of Tropical Vegetation: Proceedings of the Kandy Symposium. UNESCO, Paris.

Van Voris, P., R. V. O'Neill, H. H. Shugart, and W. R. Emanuel. 1978. Functional complexity and ecosystem stability: An experimental approach. ORNL/TM-6199. Oak Ridge National Laboratory, OakRidge, Tennessee. 126 pp.

Van Wagner, C. E. 1978. Age class distribution and the forest fire cycle. Can. J. For. Res. 8:220–227.

Van Zant, K. 1979. Late-glacial and post-glacial pollen and plant macrofossils from Lake West Okoboji, northwestern Iowa. Quat. Res. 12:358–379.

Vasek, F. C. 1980. Creosote bush: Long-lived clones in the Mojave Desert. Am. J. Bot. 67:246–255.

Vasek, F. C., A. B. Johnson, and G. D. Brum. 1975. Effects of power transmission lines on vegetation of the Mojave Desert. Madrono 23:114–130.

Vazquez-Yanes, C. 1974. Studies on the germination of seeds of *Ochroma lagopus* Sw. Turrialba 24:176–179.

Vazquez-Yanes, C. 1976a. Estudios sobre la ecofisiologia de la germinacion en una zona calido-humeda de Mexico. pp. 229–287. In A. Gomez-Pompa, C. Vazquez-Yanes, S. del Amo, and A. Butanda (eds.), Regeneracion de Selvas. CECSA, Mexico, D.F.

Vazquez-Yanes, C. 1976b. Seed dormancy and germination in secondary vegetation tropical plants: The role of light. Comp. Physiol. Ecol. 57:30–34.

Vazquez-Yanes, C. 1978. Latencia y supervivencia de las semillas. Naturaleza 9:56–60.

Vazquez-Yanes, C. 1979. Notas sobre la ecofisiologia de la germinacion de las semillas de *Cecropia obtusifolia* Bertol. Turrialba 29:147–149.

Vazquez-Yanes, C. 1980. Light quality and seed germination in *Cecropia obtusifolia* and *Piper auritum* from a tropical rain forest in Mexico. Phyton 38:33–35.

Vesey-Fitzgerald, D. F. 1960. Grazing succession amongst East Africa game animals. J. Mammal. 41:161–172.

Viereck, L. A. 1970a. Forest succession and soil development adjacent to the Chena River in interior Alaska. Arct. Alp. Res. 2:1–26.

Viereck, L. A. 1970b. Soil temperatures in river bottom stands in interior Alaska. pp. 223–233. In Ecology of the Subarctic Regions, Proceedings of the Helsinki Symposium. UNESCO, Paris.

Viereck, L. A. 1973. Wildfire in the taiga of Alaska. Quat. Res. 3:465–495.

Viereck, L. A. 1975. Forest ecology of the Alaska taiga. pp. I-1 to I-22. In Proceedings of the Circumpolar Conference on Northern Ecology. National Research Council of Canada and Scientific Committee on Problems of the Environment, Ottawa.

Viereck, L. A. 1979. Characteristics of treeline plant communities in Alaska. Holarctic Ecol. 2:228–238.

Viereck, L. A. 1981. Effects of fire in the spruce dominated ecosystem. In R. W. Wein (ed.), Symposium Proceedings, Fire in Northern Circumpolar Ecosystems. John Wiley and Son, Chichester, England. (in press).

Viereck, L. A., and C. T. Dyrness. 1979. Ecological effects of the Wickersham Dome fire near Fairbanks, Alaska. USDA For. Serv. Gen. Tech. Rep. PNW-90. Pacific Northwest Forest and Range Experiment Station, Portland, Oregon. 71 pp.

Viereck, L. A., and M. J. Foote. 1979. Vegetation analysis. pp. 25–34. In L. A. Viereck and C. T. Dyrness (eds.), Ecological Effects of the Wickersham Dome fire Near Fairbanks, Alaska. USDA For. Serv. Gen. Tech. Rep. PNW-90. Pacific Northwest Forest and Range Experiment Station, Portland, Oregon. 71 pp.

Viereck, L. A., and E. L. Little. 1972. Alaska trees and shrubs. USDA For. Serv., Agriculture Handbook 410, Washington, D.C. 265pp.

Viers, S. Personal communication. National Park Service, Arcata, California. November 26, 1979.

Vitousek, P. M., J. R. Gosz, C. C. Grier, J. M. Melillo, W. A. Reiners, and R. L. Todd. 1979. Nitrate losses from disturbed ecosystems. Science 204:469–474.

Vitousek, P. M., and J. M. Melillo. 1979. Nitrate losses from disturbed forests: Patterns and mechanisms. For. Sci. 25:605–619.

Vitousek, P. M., and W. A. Reiners. 1975. Ecosystem succession and nutrient retention: A hypothesis. BioScience 25:376–381.

Vogl, R. J. 1968. Fire adaptations of some southern California plants. Proc. Calif. Tall Timbers Fire Ecol. Cont. No. 7, 111–125.

Vogl, R. J. 1980. The ecological factors that produce perturbation-dependent ecosystems. pp. 63–94. In J. Cairns, Jr. (ed.), The Recovery Process in Damaged Ecosystems. Ann Arbor Science, Ann Arbor, Michigan.

Vogl, R. J., W. P. Armstrong, K. L. White, and K. L. Cole. 1977. The closed-cone pines and cypresses. pp. 295–358. In M. Barbourogan, Utah.

Waddington, J. C. B. 1969. A stratigraphic record of the pollen influx to a lake in the Big Woods of Minnesota. Geol. Soc. Am., Special Paper 123, pp. 263–282.

Wadsworth, F. H. 1951. Forest management in the Luquillo Mountains. I. The setting. Caribbean. For. 12:93–114.

Wadsworth, F. H. 1957. Tropical rain forest: The *Dacryodes-Slonanea* association of the West Indies. Trop. Silviculture 2:13–aogan, Utah.

Waddington, J. C. B. 1969. A stratigraphic record of the pollen influx to a lake in the Big Woods of Minnesota. Geol. Soc. Am., Special Paper 123, pp. 263–282.

Wadsworth, F. H. 1951. Forest management in the Luquillo Mountains. I. The setting. Caribbean. For. 12:93–114.

Wadsworth, F. H. 1957. Tropical rain forest: The *Dacryodes-Slonanea* association of the West Indies. Trop. Silviculture 2:13–23.

Wadsworth, F. H. 1970. Review of past research in the Luquillo Mountains of Puerto Rico. pp. 33–46. In H. T. Odum and R.F.Pigeon (eds.), A Tropical Rain Forest: A Study of Irradiation and Ecology at El Verde, Puerto Rico. U.S. Atomic Energy Commission, Oak Ridge, Tennessee.

Wadsworth, F. H., and G. H. Englerth. 1959. Effect of the 1956 hurricane on forests in Puerto Rico. Caribbean For. 20:38–51.

Waid, J. S. 1957. Distribution of fungi within decomposing tissue of ryegrass roots. Trans. Brit. Mycol. Soc. 40:391–406.

Waksman, G., M. Menard, and J. Belanger. 1975. Analyse factorielle des relations entre le milieu et la production: etude des tremblaies de la section Laurentienne. Can. J. For. Res. 5:662–680.

Waksman, S. A. 1932. Principles of Soil Microbiology. Williams & Wilkins, Baltimore. 894 pp.

Walker, B. H., and I. Noy-Meir. 1982. Aspects of the stability and resilience of savanna ecosystems. In B. J. Huntley and B.H.Walker (eds.), Ecology of Tropical Savannas. Springer-Verlag, Heidelburg. (in press).

Walker, B. H., D. Ludwig, C. S. Holling, and R. M. Peterman. 1981. Stability of semi-arid savanna grazing systems. J. Ecol. (in press)

Walker, D. 1970. Direction and rate in some British post-glacial hydroseres. pp. 117–139. In D. Walker and R. G. West (eds.), Studies on the vegetation history of the British Isles. Cambridge University Press, Cambridge, England.

Walker, D. 1971. Quantification in historical plant ecology. Proc. Ecol. Soc. Aust. 6:91–104.

Walker, J., and B. R. Tunstall. 1978. Crown separation and structural classification of vegetation. pp. 26–38. CSIRO Div. Land Use Res. Workshop on Classification of Australian Vegetation, Pap. No. 6.

Walker, N. E. 1975. Soil Microbiology. Butterworths, London. 262 pp.

Walker, P. C., and R. T. Hartman. 1960. The forest sequence of the Hartstown Bog area in western Pennsylvania. Ecology 41:461–474.

Walker, T. W., and J. K. Syers. 1976. The fate of phosphorus during pedogenesis. Geoderma 15:1–19.

Wallwork, J. A. 1976. The Distribution and Diversity of Soil Fauna. Academic Press, New York. 437 pp.

Walter, H. 1971. Ecology of tropical and sub-tropical vegetation [J. H. Burnett (ed.), transl. by D. Mueller-Dombois]. Oliver & Boyd, Edinburgh. 539 pp.

Ward, R. T. 1961. Some aspects of the regeneration habits of American beech. Ecology 42:828–832.

Ward, W. T., I. P. Little, and C. H. Thompson. 1979. Stratigraphy of two sand rocks at Rainbow Beach, Queensland, Australia, and a note on humate composition. Paleogeogr. Paleoclimatol. Paleoecol. 26:305–316.

Waring, R. H., and J. F. Franklin. 1979. Evergreen coniferous forests of the Pacific Northwest. Science 204:1380–1386.

Watt, A. S. 1947. Pattern and process in the plant community. J. Ecol. 35:1–22.

Watt, K. E. F. 1975. Critique and comparison of biome ecosystem modeling. pp. 139–152. In B. C. Patten (ed.), Systems Analysis and Simulation in Ecology. Vol. III. Academic Press, New York.

Watts, W. A. 1967. Late glacial plant macrofossils from Minnesota. pp. 89–98. In E. J. Cushing and H. E. Wright (eds.), Quaternary Paleoecology. Yale University Press, New Haven, Connecticut.

Watts, W. A. 1969. A pollen diagram from Mud Lake, Marion County, north-central Florida. Geol. Soc. Am. Bull. 80:631–642.

Watts, W. A. 1970. The full-glacial vegetation of northwestern Georgia. Ecology 51:19–33.

Watts, W. A. 1973. Rates of change and stability in vegetation in the perspective of long periods of time. pp. 195–206. In H.J.B.Birks and R. G. West (eds.), Quaternary Plant Ecology. Blackwell, London.

Watts, W. A. 1979. Late-Quaternary vegetation of central Appalachia and the New Jersey coastal plain. Ecol. Monogr. 49:427–469.

Watts, W. A. 1980. Late-Quaternary vegetation history at White Pond on the inner coastal plain of South Carolina. Quat. Res. 13:187–199.

Watts, W. A., and M. Stuiver. 1980. Late-Wisconsin climate of northern Florida and the origin of species-rich deciduous forest. Science 210:325–327.

Weaver, J. E. 1954. North american prairie. Johnson, Lincoln, Nebraska. 515 pp.

Weaver, J. E., and F. E. Clements. 1938. Plant Ecology. McGraw-Hill, New York. 601 pp.

Webb, D. P. 1977. Regulation of deciduous forest litter decomposition by soil arthropod feces. pp. 57–69. IN W. J. Mattson (ed.), The Role of Arthropods in Forest Ecosystems. Springer-Verlag, NewYork.

Webb, L. J. 1958. Cyclones as an ecological factor in tropical lowland rain forest, north Queensland. Aust. J. Bot. 6:220–228.

Webb, L. J. 1979. Review of F. H. Bormann and G. E. Likens. "Pattern and Process of a Forested Ecosystem" (Springer-Verlag, New York). Science 205:1369–1370.

Webb, L. J., J. G. Tracey, and W. T. Williams. 1972. Regeneration and pattern in the subtropical rain forest. J. Ecol. 60:675–695.

Webb, T., III. 1974a. A vegetational history from northern Wisconsin: Evidence from modern and fossil pollen. Am. Midland Nat. 92:12–34.

Webb, T., III. 1974b. Corresponding patterns of pollen and vegetation in lower Michigan: A comparison of quantitative data. Ecology 55:17–28.

Webb, T., III, and R. A. Bryson. 1972. Late and post-glacial climatic change in the northern midwest, USA: Quatitative estimates derived from fossil pollen spectra by multivariate statistical analysis. Quat. Res 2:70–115.

Webb, T., III, and J. H. McAndrews. 1976. Corresponding patterns of contemporary pollen and vegetation in central North America. Geol. Soc. Am. Mem. 145:267–299.

Webb, T., III., R. A. Laseski, and J. C. Bernabo. 1978. Sensing vegetational patterns with pollen data: Choosing the data. Ecology 59:1151–1163.

Webster, J. R. 1979. Hierarchical organization of ecosystems. pp. 119–131. In E. Halfon (ed.), Theoretical Systems Ecology. Academic Press, New York.

Webster, J. R., J. B. Waide, and B. C. Patten. 1974. Nutrient recycling and the stability of ecosystems. pp. 1–27. In F.G.Howell, J. B. Gentry, and M. H. Smith (eds.), Mineral cycling in southeastern ecosystems. CONF-740513. National Technical Information Service, Springfield, Virginia.

Wein, R. S., and J. M. Moore. 1977. Fire history and rotations in the New Brunswick Acadian forest. Can. J. For. Res. 7:285–294.

Weir, A. D. 1972. Program DIVINFRE: A divisive classification on binary data. CSIRO Div. Land Use Res. Tech. Memo 72/4. 43 pp.

Wellner, C. A. 1970. Fire history in the northern Rocky Mountains. pp. 42–64. In Proc. Symp. Role of Fire in the Intermountain West. Intermountain Fire Research Council, and University of Montana.

Wells, B. W. 1928. Plant communities of the coastal plain of North Carolina and their successional relations. Ecology 9:320–342.

Wells, B. W. 1932. Natural Gardens of North Carolina. University of North Carolina Press, Chapel Hill. 458 pp.

Wells, P. V. 1969. The relation between mode of reproduction and extent of speciation in woody genera of the California Chaparral. Evolution 23:264–267.

Wells, P. V. 1970. Postglacial vegetation history of the Great Plains. Science 167:1574–1582.

Went, R. G., and F. C. Evans. 1968. The biological and mechanical role of soil fungi. Proc. Natl. Acad. Sci. 60:497–507.

Werner, P. A., and W. J. Platt. 1976. Ecological relationships of co-occurring goldenrods (*Solidago*: Compositae). Am. Nat. 110:959–971.

West, D. C., S. B. McLaughlin, and H. H. Shugart. 1980. Simulated forest response to chronic air pollution stress. J. Environ. Qual. 9:43–49.

West, N. E. 1968. Rodent-influenced establishment of ponderosa pine and bitterbrush seedlings in central Oregon. Ecology 49:1009–1011.

West, O. 1965. Fire in vegetation and its use in pasture management, with special reference to tropical and subtropical Africa. Mimeo. Publ. No. 1/1965. Commonwealth Bureau of Pastures, Field Crops, Hurley, Berks. England. 34 pp.

West, O. 1971. Fire, man and wildlife as interacting factors limiting the development of climax vegetation in Rhodesia. Proc. Tall Timbers Fire Ecol. Conf. pp. 121–145.

West, R. G. 1961. Late and post-glacial vegetational history in Wisconsin, particularly changes associated with the Valders readvance. Am. J. Sci. 259:766–783.

West, R. G. 1970. Pleistocene history of the British flora. pp. 1–11. In D. Walker and R. G. West (eds.). Studies in the Vegetational History of the British Isles. Cambridge University Press, London.

Westman, W. E. 1978. Inputs and cycling of mineral nutrients in a coastal subtropical Eucalypt forest. J. Ecol. 66:513–531.

Westman, W. E., and R. H. Whittaker. 1975. The pygmy forest region of northern California: Studies on biomass and primary productivity. J. Ecol. 63:493–520.

Whelan, R. J., and A. R. Main. 1979. Insect grazing and post-fire plant succession in southwest Australian woodland. Aust. J. Ecol. 4:387–398.

White, P. S. 1979. Pattern, process, and natural disturbance in vegetation. Bot.

Rev. 45:229–299.

Whitehead, D. R. 1972. Developmental and environmental history of the Dismal Swamp. Ecol. Monogr. 42:301–315.

Whitehead, D. R. 1973. Late-Wisconsin vegetational changes in unglaciated eastern North America. Quat. Res. 3:621–631.

Whitehead, D. R. 1979. Late-glacial and post-glacial vegetational history of the Berkshires, western Massachusetts. Quat. Res. 12:333–355.

Whitmore, T. C. 1974. Change with time and the role of cyclones in tropical rain forest on Kolombangara, Solomon Islands. Pap.No.46, Commonwealth Forestry Institute. University of Oxford. 77 pp.

Whitmore, T. C. 1975. Tropical rain forests of the far east. Oxford University Press, London. 282 pp.

Whitmore, T. C. 1978. Gaps in the forest canopy. pp. 639–655. In P. B. Tomlinson and M. H. Zimmermann (eds.), Tropical Trees as Living Systems. Proc. 4th Cabot Symp. Harvard Forest (1976). Cambridge University Press, Cambridge, England.

Whittaker, R. H. 1951. A criticism of the plant association and climatic climax concepts. Northwest Sci. 25:17–31.

Whittaker, R. H. 1953. A consideration of climax theory: The climax as a population and pattern. Ecol. Monogr. 23:41–78.

Whittaker, R. H. 1956. Vegetation of the Great Smoky Mountains. Ecol. Monogr. 26:1–80.

Whittaker, R. H. 1957. Recent evolution of ecological concepts in relation to the eastern forests of North America. Am. J. Bot. 44:197–206.

Whittaker, R. H. 1962. Classification of natural communities. Bot. Rev. 28:1–239.

Whittaker, R. H. 1966. Forest dimensions and production in the Great Smoky Mountains. Ecology 47:103–121.

Whittaker, R. H. 1972. Evolution and measurement of species diversity. Taxon 21:213–251.

Whittaker, R. H. 1974. Climax concepts and recognition. pp. 137–154. In R. Knapp (ed.), Vegetation Dynamics. W. Junk, The Hague.

Whittaker, R. H. 1975a. Functional aspects of succession in deciduous forests. pp. 377–405. In Sukzession Forschung, Ber. Int. Sym. Int. Ver. Vegetationskunde, Rinteln (1973). J. Cramer, Vaduz.

Whittaker, R. H. 1975b. Communities and Ecosystems, 2nd ed. Macmillan, New York.

Whittaker, R. H. 1975c. The design and stability of plant communities. pp. 169–181. In W. H. van Dobben and R.H.Lowe-McConnell (eds.), Unifying Concepts in Ecology. W.Junk, The Hague.

Whittaker, R. H. 1977a. Animal effects on plant species diversity. pp. 409–425. In R. Tuxen (ed.), Vegetation and Fauna. Ber. Int. Sym. Int. Ver. Vereinigung, Rinteln (1976). J. Cramer, Vaduz.

Whittaker, R. H. 1977b. The role of mosaic phenomena in natural communities. Theor. Pop. Biol. 12:117–139.

Whittaker, R. H., and S. A. Levin. 1977. The role of mosaic phenomena in natural communities. Theor. Pop. Biol. 12:117–139.

Whittaker, R. H., F. H. Bormann, G. E. Likens, and T. G. Siccama. 1974. The Hubbard Brook Ecosystem Study: Forest Biomass and Production. Ecol. Monogr. 44:233–254.

Whittaker, R. H., and G. E. Likens. 1973. Carbon in the biota. pp. 281–302. In G. M. Woodwell and E. V. Pecan (eds.), Carbon and the Biosphere. US-AEC CONF-720510. National Technical Information Service, Springfield, Virginia.

Whittaker, R. H., and P. L. Marks. 1975. Methods of assessing terrestrial productivity. pp. 55–118. In H. Lieth and R.H.Whittaker (eds.), Primary Productivity of the Biosphere. Ecological Studies 14. Springer-Verlag, New York.

Whittaker, R. H., and W. A. Niering. 1975. Vegetation of the Santa Catalina Mountains, Arizona. V. Biomass, production, and diversity along the elevation gradient. Ecology 56:771–790.

Whittaker, R. H., and G. M. Woodwell. 1968. Dimension and production relations of trees and shrubs in the Brookhaven Forest, New York. J. Ecol. 56:1–25.

Whittaker, R. H., and G. M. Woodwell. 1972. Evolution of natural communities. pp. 137–159. In J. A. Wiens (ed.), Ecosystem Structure and Evolution. Oregon State University Press, Corvallis, Oregon.

Wiegert, R. G., and D. F. Owen. 1971. Trophic structure, available resources and population density in terrestrial vs. aquatic ecosystems. J. Theor. Biol. 30:69–81.

Wielgolaski, F. E. 1975. Comparison of plant structure on grazed and ungrazed tundra meadows. pp. 86–93. In F. E. Wielgolaski (ed.), Fennoscandian Tundra Ecosystems, Part 1. Plants and Microorganisms. Ecological Studies 16. Springer-Verlag, Berlin.

Wiens, J. A. 1974. Habitat heterogeneity and avian community structure in North American grasslands. Am. Midland Nat. 91:195–213.

Wierman, C. A., C. D. Oliver, and C. Dearing. 1979. Crown stratification by species in even-aged mixed stands of Douglas-fir-western hemlock. Can. J. For. Res. 9:1–9.

Wilde, S. A., and H. H. Krause. 1960. Soil-forest types of the Yukon and Tanana Valleys in subarctic Alaska. J. Soil Sci. 11:266–274.

Williams, A. S. 1974. Late-glacial and post-glacial vegetational history of the Pretty Lake region, northeastern Indiana. Geol. Surv. Prof. Pap. 686-B. 23 pp.

Wilson, D. S. 1976. Evolution on the level of communities. Science 192:1358–1360.

Wilson, D. S. 1980. The Natural Selection of Populations and Communities. Benjamin Cummings, Menlo Park, California. 186 pp.

Wilson, E. O. 1978. Introduction: What is sociobiology? pp. 1–12. In M. S. Gregory, A. Silvers, and D. Sutch (eds.), Sociobiology and Human Nature. Jossey-Bass, San Francisco.

Wilton, W. C. 1963. Black spruce seedfall immediately following fire. For. Chronol. 39:477–478.

Wimsatt, W. C. 1980. Reductionistic research strategies and their biases in the units of selection controversy. pp. 1–55. In T.Nickles (ed.), Scientific Discovery Vol. II.: Case Studies. Proceedings of the 1st Leonard Conference on Scientific Discovery, Dordrecht, Reidel.

Wing, L. D., and I. O. Buss. 1970. Elephants and forests. Wildlife Monogr. 19:1–92.

Woods, K. D. 1979. Reciprocal replacement and the maintenance of codominance in a beech-maple forest. Oikos 33:31–39.

Wright, H. E., Jr. 1968a. History of the prairie peninsula. pp. 78–88. In R. E. Bergstrom (ed.), The Quaternary of Illinois. Univ. Ill. Coll. of Agr. Spec.

Publ. No. 14, Urbana, Illinois.
Wright, H. E., Jr. 1968b. The roles of pine and spruce in the forest history of Minnesota and adjacent areas. Ecology 49:937–955.
Wright, H. E., Jr. 1970. Vegetational history of the Central Plains. pp. 157–172. In W. Dort, Jr. and J. K. Jones, Jr. (eds.), Pleistocene and Recent Environments of the Central Great Plains. Dept. Geol., Univ. Kansas Spec. Publ. 3, University of Kansas Press, Lawrence, Kansas.
Wright, H. E., Jr. 1971. Late Quaternary vegetation history of North America. pp. 425–464. In K. Turekian (ed.), The Late Cenozoic Glacial Ages. Yale University Press, New Haven, Connecticut.
Wright, H. E., Jr. 1972. Interglacial and post-glacial climates: The pollen record. Quat. Res. 2:274–282.
Wright, H. E., Jr. 1974. Landscape development, forest fires, and wilderness management. Science 186:487–495.
Wright, H. E., Jr., and W. A. Watts. 1969. Glacial and vegetational history of northeastern Minnesota. Minnesota Geol. Surv. Spec. Publ. SP-11. 59 pp.
Wright, H. E., Jr., T. C. Winter, and H. L. Patten. 1963. Two pollen diagrams from southeastern Minnesota: Problems in the regional late-glacial and post-glacial vegetational history. Geol. Soc. Am. Bull. 74:1371–1396.
Wyatt-Smith, J. 1955. Changes in composition in natural plant succession. Malay. For. 18:44–99.
Yaalon, D. H. 1971. Paleopedology: Origin, nature and dating of paleosols. pp. 29–40. In D. H. Yaalon (ed.). Age of Parent Materials and Soils. Int. Soc. Soil Sci. (1970). Amsterdam.
Yeaton, R. I. 1978. A cyclical relationship between *Larrea tridentata and Opuntia leptocaulis* in the northern Chihuahuan desert. J. Ecol. 66:651–656.
Yoda, K., T. Kira, H. Ogawa, and H. Hozumi. 1963. Self thinning in overcrowded pure stands under cultivation and natural conditions. J. Biol. Osaka City Univ. 14:197–129.
Young, J. L., and G. Spycher. 1979. Water-dispersible soil organic-mineral particles: I. Carbon and nitrogen distribution. Soil Sci. Soc. Am. J. 43:324–328.
Youngberg, C. T., and A. G. Wollum, II. 1976. Nitrogen accretion in developing *Ceanothus velutinus* stands. Soil Sci. Soc. Am. J. 40:109–112.
Zahner, R., and N. A. Crawford. 1965. The clonal concept in aspen site relations. pp. 229–243. In Forest Soil Relationships in North America, Proceedings, Second North American Forest Soils Conf. Oregon State University Press, Corvallis, Oregon.
Zasada, J. C. 1971. Natural regeneration of interior Alaska forests—seedbed and vegetative reproduction considerations. pp. 231–246. In Proceedings, Fire in the Northern Environment—a Symposium, Fairbanks, Alaska. USDA Forest Serv. Pacific Northwest Forest and Range Experimental Station, Portland, Oregon.
Zasada, J. C., L. A. Viereck, and M. J. Foote. 1979. Black spruce seedfall and seedling establishment. pp. 42–50. In L. A. Viereck and C. T. Dyrness (eds.), Ecological Effects of the Wickersham Dome fire near Fairbanks, Alaska. USDA For. Serv. Gen. Tech. Rep. PNW-90. Pacific Northwest Forest and Range Experiment Station, Portland, Oregon. 71 pp.
Zavitkovski, J. 1971. Dry weight and leaf area of aspen trees in northern Wisconsin. pp. 191–205. In H. E. Young (ed.), IUFRO Forest Biomass Studies.

University of Maine Press, Orono.
Zavitkovski, J., and M. Newton. 1968a. Ecological importance of snowbrush (Ceanothus velutinus) in the Oregon Cascades. Ecology 49:1134–1145.
Zavitkovski, J., and M. Newton. 1968b. Effect of organic matter and combined nitrogen on nodulation and nitrogen fixation by red alder. pp. 209–223. In J. M. Trappe, J. F. Franklin, R.F.Tarrant, and G. M. Hansen (eds.), Biology of Alder. USDA For. Serv., Pacific Northwest Forest and Range Experiment Station, Portland, Oregon.
Zavitkovski, J., and R. D. Stevens. 1972. Primary production of red alder ecosystems. Ecology 53:235–242.
Zedler, P. H. 1977. Life history attributes of plants and the fire cycle: A case study in chaparral dominated by *Cupressus forbesii.* pp. 451–458. In H. A. Mooney and C. E. Conrad (eds.), Proceedings, Symposium on the Environmental Consequences of Fire and Fuel Management in Mediterranean Ecosystems. Forest Service, USDA Gen. Tech. Rep. WO-3. Washington, D.C. 498 pp.
Zedler, P. H., and F. G. Goff. 1973. Size association analysis of forest successional trends in Wisconsin. Ecol. Monogr. 43:79–94.
Zimmerman, I. 1978. The feeding ecology of Afrikander steers (*Bos indicus*) on mixed busveld at Nylsvley Nature Reserve, Transvaal. M.Sc. Thesis. University of Pretoria, South Africa.
Zobel, D. B., A. McKee, G. M. Hawk, and C. T. Dyrness. 1976. Relationships of environment to composition, structure, and diversity of forest communities of the Central Western Cascades of Oregon. Ecol. Monogr. 46:135–156.
Zoltai, S. C. 1971. Southern limit of permafrost features in peat landforms, Manitoba and Saskatchewan. Geol. Assoc. Can., Paper 9, 305–310.
Zasada, J. C., and R. A. Gregory. 1969. Regeneration of white spruce with reference to interior Alaska; a literature review. Res. Pap. PNW-79. USDA For. Serv., Pacific Northwest Forest and Range Experiment Station. Portland, Oregon.

Index